How To Tune & Modify
ENGINE MANAGEMENT SYSTEMS

How To Tune & Modify

ENGINE
MANAGEMENT
SYSTEMS

Jeff Hartman

MOTORBOOKS
INTERNATIONAL

First published in 2003 by MBI Publishing Company,
380 Jackson Street, Suite 200, St. Paul, MN 55101-3885 USA

MBI Publishing Company books are also available at discounts in bulk quantity for industrial or sales-promotional use. For details write to Special Sales Manager at Motorbooks International Wholesalers & Distributors,
380 Jackson Street, Suite 200, St. Paul, MN 55101-3885 USA

On the Title Page: This Toyota 3S-GTE engine has been stroked to 2.1 liters. Boost pressure moves directly from the turbocharger's compressor into the top-mounted air-water intercooler and directly out the other side into the big-bore RC throttle body.

On the back cover: This is the installation of a Powerhaus chip into the Motronic ECU of a Porsche 993. This is a twin-turbocharged Ferrari F50 on the dyno, controlled by a Motec EMS.

ISBN 0-7603-1582-5

Edited by Peter Bodensteiner
Designed by Chris Fayers

Printed in the United States

Dedication

For Gwen Sullivan

Acknowledgments

Thanks in particular to Corky Bell, Mark Dobeck, Tadashi Nagata, Bob Norwood, and Brice Yingling, whose help and advice was vital in producing this book. Thanks also to the many good companies and organizations that provided information, photographs, and illustrations about their products and services for possible use in this book.

CONTENTS

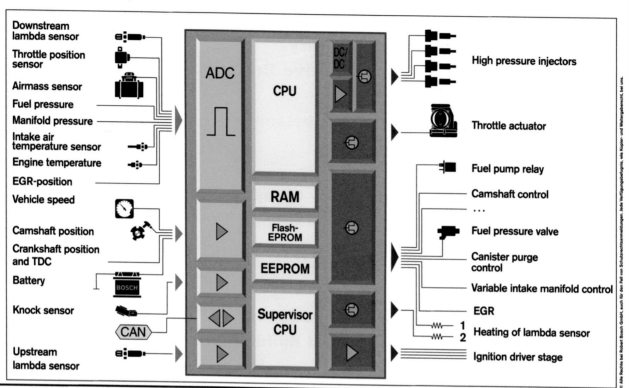

DI-Motronic

Typical state-of-the-art Bosch DI-Motronic injection crunches data from multiple lambda (O_2) sensors, mass airflow (MAF) and manifold absolute pressure (MAP) sensors, the throttle position sensor, and a standard complement of OBD-II Motronic sensors to control high-pressure injectors spraying directly into the cylinders. The system also controls a fly-by-wire throttle actuator, variable cam timing actuator, electronic fuel pressure regulator, and ignition driver stage. The system can provide homogenous charge mixtures for maximum power at wide-open throttle or a stratified-charge for maximum fuel economy in which a richer mixture in the vicinity of the spark plug lights off a leaner mixture elsewhere. Bosch

This how-to book is designed to communicate the theory and practice of designing and redesigning performance engine management systems that work. In the last 20 years electronic engine and vehicle management (General Motors now calls this discipline "eMotion") has been the most interesting, dynamic, and influential field in automotive engineering. This makes it a moving target for analysis and discussion. Electronic control systems have evolved at light speed compared to everything else on road-going vehicles. This has paved the way for unprecedented levels of reliable, specific power, efficiency, comfort, and safety that would not otherwise be possible. The naturally aspirated four-cylinder S2000 from Honda achieves 240 horsepower at 9,000 rpm—with just 2 liters of displacement and without power-adders. On the aftermarket end of the spectrum, 400-horsepower 5.0-liter Ferrari Testarossas are routinely modified by Dallas supertuner Bob Norwood to deliver *1,200* reliable horsepower to the street using turbochargers and aftermarket electronic engine management systems! Without sophisticated electronic engine management systems, neither would be possible.

Simply reconfiguring the internal configuration tables of an electronic engine management system can give the engine an entirely new personality. Changing a few numbers in the memory of an original equipment onboard computer can sometimes unleash 50 or 100 horsepower and open up all sorts of possibili-

ties for power increases with VE-improving speed parts and power-adders. But you have to do it right, and that can be difficult. Think brain transplant: something interesting is bound to happen. The question is, do you unleash a thing of power and beauty or create a monster? Today, hacking engine management systems offers extensive possibilities for both good and evil, for glorious successes and for ruinous disasters.

WHAT'S CHANGED SINCE *FUEL INJECTION: INSTALLATION, PERFORMANCE TUNING, MODIFICATIONS*

Motorbooks International published my original book, *Fuel Injection: Installation, Performance Tuning, Modifications*, more than a decade ago. Since then, the world of automotive performance has changed in ways that have greatly affected—and been affected by—increasingly sophisticated performance electronic engine management.
• The fleet of roadworthy carbureted performance vehicles has been decimated by age and hard living. Concurrently, a massive inventory of used zero-, first-, and second-generation factory EFI parts has accumulated in wrecking yards. A new carbureted automobile or truck has not been sold in the United States since 1987. In the new millennium, many motorcycles and boats have moved to electronic fuel injection and engine management as well.

• A new generation of young, computer-savvy hot rodders arrived on the scene in the 1990s, helping to create and fuel the sport-compact craze for small front-wheel-drive (mostly) four- and six-cylinder vehicles with high-tech power-adder-based performance enhancements. These "kids" are not intimidated by the idea of computer swapping or recalibration (though they may be by the price).

• The performance-racing industry has had time to develop an increasingly sophisticated arsenal of aftermarket hardware and software products to hack, modify, or replace factory engine controls to unleash increased performance.

• The performance industry has not kept pace, however, with the sophistication of factory engine management systems, which continue to outstrip the aftermarket. A gap has developed between automakers and hot rodders in terms of expertise, diagnostic equipment, test facilities, sophisticated modeling and simulation software, and other costly resources to hack and calibrate engine management systems to the highest standards of performance and efficiency. Calibrating standalone engine management systems well is not for the faint of heart.

• Electronic injection and ignition systems are no longer separated as they were on zero- and first-generation EFI systems. On the newest vehicles, they are typically integrated into a sophisticated, multi-processor network of electronic control systems that manage everything on the vehicle from a displacement-on-demand powertrain to antilock brakes and active directional-stability control systems, from onboard comfort systems to GPS-equipped robot cellular dialers that alert emergency crews of your location if an airbag deploys.

• Onboard Diagnostics Level II (OBD-II) engine management controls—equipped with sophisticated standardized diagnostic capabilities and uploadable flash EEPROM recalibration—give authorized users powerful self-diagnostics plus easy software upgrades, bug-fixes, or the ability to substantially reprogram the behavior and personality of an engine. OBD-II has been a requirement on all new cars sold in the United Stated since 1996. Government-mandated antitampering countermeasures provide substantial but not insurmountable barriers to entry for OBD-II reprogramming.

• Technology and the latest horsepower wars of the 1990s have created many high-output factory vehicles from automakers—some with *extremely* high specific power. Virtually *all* powerplants are now optimized and balanced so there is little or no "free" power available from swapping peripheral components like exhaust systems.

• Power-adders are the big hot rodding trick, and both the power-adders themselves and their support systems are highly developed and sophisticated. Factory superchargers are reasonably common, and turbos have made a comeback, especially on sport compact cars. Nitrous injection (often with electronic controls) still offers some of the cheapest horsepower around and is fully compatible with electronic engine management.

• At the other end of the scale, the first- generation muscle engines from the 1960s and 1970s with low-tech emissions control have aged to the point that many are now entirely exempt from emissions standards and testing. They can be legally upgraded to electronic controls for improved performance, reliability, and streetability.

• Diesel engines are now equipped with electronic engine management systems, and Mad-Max types with wildly hot-rodded turbo-diesel engines are now common.

• The Internet has had a large and increasing impact on the automotive performance aftermarket, providing everything from automotive information databases to overnight equipment loca-

Fuel injection, 1950s-style, meant Chevrolet constant-flow venturi fuel injection. The package was expensive and finicky, and many were converted to carburetion in the days before the cars were collector items. Bendix had already successfully demonstrated pulsed electronic port injection (the Electrojector system), but the pretransistor vacuum tubes had to warm up like an old radio before the car could start, and the whole system could wig out if you drove under high-power lines. A practical system required the solid-state electronics of the 1960s and beyond.

tion, ordering, and delivery, from peer-to-peer information sharing to downloadable engine management calibrations and software and used hot rod parts on eBay.

• The first hybrid gasoline/diesel electric cars have hit the streets. Just beyond the horizon are fuel-cell vehicles that the world's automakers' R&D labs are working intensely to develop.

Of course, engine management did not begin in the 1990s. For nearly a hundred years, engine management was accomplished by carburetors, points, ignitions, and human intervention (engineers were originally people who managed *engines*). Some people can still remember when you cranked an engine with a *crank* inserted into the front of the crankshaft and muscled it by hand after saying a prayer and adjusting spark advance and cold-start fuel enrichment by *hand* with a spark control and a choke lever. By the time the last vehicles equipped with points and ignitions rolled off American assembly lines in the 1970s (and carburetors in the 1980s), a century of intensive engineering had achieved remarkable performance and reliability from these primitive (but *inexpensive!*) electromechanical devices. Old-timers surely remember the dismal sound of a battery running down in winter weather from excessive and futile cranking. In those days, cold engines could flood from over-choking if you guessed wrong, and the ignition system was unable to ignite excessively wet mixtures that puddled in the cold intake manifold and made things worse.

Emotional and marketing justifications aside, there are solid technical reasons why automakers stampeded toward fuel injection and electronic engine management in the 1970s and 1980s and why, more recently, many automotive enthusiasts and hot rodders have been willing to pay good money to aftermarket suppliers of add-on electronic engine management systems to convert carbureted engines to fuel injection.

THE AUTOMAKERS AND ELECTRONIC FUEL INJECTION AND ENGINE MANAGEMENT

In the case of the car companies, electronic fuel injection was a tool that allowed engineers to improve drivability and reliability and to fight the horsepower wars of the 1980s. It also helped them comply with federal legislation that mandated increasingly stiff standards for fuel economy and exhaust emissions. The government forced automakers to warrant for 120,000 miles everything on the engine that could affect exhaust emissions, which was everything related to combustion. In other words, nearly everything. Intelligently and reliably controlling engine air/fuel mixtures within extremely tight tolerances over many miles and adapting as engines wore out became a potent tool that enabled car companies to strike a precarious balance between EPA regulations, the gas-guzzler tax, and performance-conscious consumers, who still fondly remembered the acceleration capabilities of 1960s- and 1970s-vintage muscle cars.

In the 1950s, engine designers concentrated on one thing—getting the maximum power and drivability from an engine within specific cost constraints. This was the era of the first 1-horsepower-per-cubic-inch motors. By the early 1960s, air pollution in southern California was getting out of control, and engine designers had to start worrying about making clean power. The Clean Air Acts of 1966 and 1971 set increasingly strict state and federal standards for exhaust and evaporative emissions. Engine designers gave it their best shot, which mainly involved add-on emissions-control devices like positive crankcase ventilation (PCV), exhaust gas recirculation (EGR), air pumps, inlet air heaters, vacuum retard distributor canisters, and carburetor modifications.

The resulting cars of the 1970s ran cleaner, but horsepower was down and drivability sometimes suffered. Fuel economy worsened just in time for the oil crises of 1973 and 1979. The government responded to the energy crises by passing laws mandating better fuel economy. By the late 1970s car companies had major new challenges, and they sought some new "magic" that would solve their problems.

The magic—electronic fuel injection—was actually nothing new. The first electronic fuel injection (EFI) was invented by Bendix in 1950s America. The Bendix Electrojector system formed the basis of nearly all modern electronic fuel injection systems. The Bendix system used modern solenoid-type electronic injectors with an electronic control unit based on vacuum-tube technology (which was kind of like inventing the jet engine and strapping it on a 1915-vintage Sopwith Camel biplane).

The Electrojector system took 40 seconds to warm up before you could start the engine. Sometimes it malfunctioned if you drove under high-tension power lines. In addition to the liabilities of vacuum-tube technology, Bendix didn't have access to modern engine sensors. Solid-state circuitry was in its infancy, and although automotive engineers recognized the potential of electronic fuel injection to do amazing things based on its extreme precision of fuel delivery, the electronics technology to make EFI practical just didn't exist yet. Bendix eventually gave up on the Electrojector, secured worldwide patents, and licensed the technology to Bosch.

Mechanical fuel injection had been around in various forms since before 1900. Mechanical injection had always been a "toy" used on race cars, foreign cars like the Mercedes, and a tiny handful of high-performance cars in America, like the Corvette. Mechanical fuel injection avoided certain performance disadvantages of the carburetor, but it was expensive and finicky and not particularly accurate.

In the 1960s, America entered the transistor age. Suddenly electronic devices came alive instantly with no warm up. Solid-state circuitry was fast and consumed minuscule amounts of power compared to the vacuum tube. By the end of the 1960s, engineers had invented the microprocessor, which combined

dozens, hundreds, then thousands of transistors on a piece of silicon smaller than a fingernail (each transistor was similar in functionality to a vacuum tube that could be as big as a fist).

Volkswagen introduced the first Bosch electronic fuel injection systems on its cars in 1968. A trickle of other cars used electronic fuel injection by the mid-1970s. By the 1980s, that trickle became a torrent.

In the late 1970s, the turbocharger was reborn as a powerful tool for automotive engineers attempting to steer a delicate course between performance, economy, and emissions. Turbochargers could potentially make small engines feel like big engines just in time to teach the guy with a V-8 in the next lane a good lesson about humility both at the gas pump and at the stoplight drags. Unfortunately, the carburetor met its Waterloo when it came up against the turbo. Having been tweaked and modified for nearly a century and a half to reach its modern state of "perfection," the carb was implicated in an impressive series of failures when teamed up with the turbocharger. Carbureted turbo engines that were manufactured circa 1980—the early Mustang 2.3 turbo, the early Buick 3.8 turbo, the early Maserati Biturbo, the Turbo Trans Am V-8—are infamous. If you wanted a turbocharged hot rod to run efficiently and cleanly—and, more importantly, to behave and stay alive—car companies found out the hard way that electronic fuel injection was the solution.

For automakers, the cost disadvantages of fuel injection were outweighed by the potential penalties resulting from non-compliance with emissions and Corporate Average Fuel Economy (CAFE) standards, and the increased sales when offering superior or at least competitive horsepower and drivability.

HOT RODDERS AND FUEL INJECTION
In the 1950s, the performance-racing enthusiast's choices for a fuel system were carburetion or constant mechanical fuel injection. Carbs were inexpensive out of the box, but getting air and fuel distribution and jetting exactly right with one or two carbs mounted on a wet manifold took a wizard—a wizard with a lot of time. By the time you developed a great-performing carb-manifold setup, it might involve multiple carbs and cost as much or more than mechanical injection (which achieved equal air and fuel distribution with identical individual stack-type runners to every cylinder and identical fuel nozzles in every runner). Assuming the nozzles matched, fuel distribution was guaranteed to be good with constant mechanical injection.

Mechanical fuel injection has been around in various forms since about 1900, and it has always been expensive. Mechanical injection could squirt a lot of fuel into an engine without restricting airflow, and it was not affected by lateral G-forces or the up-and-down pounding of, say, a high-performance boat engine in really rough waters, when fuel is bouncing all over the place in the float chamber of a carb. Racers used Hilborn mechanical injection on virtually every post-war Indy car until 1970.

The trouble was, mechanical injection relied on crude mechanical means for mixture correction across the range of engine speeds, loading, and temperatures. Early mechanical injection was also not accurate enough to provide the precise mixtures required for a really high-output engine that must also be streetable. GM tried Constant Flow mechanical injection in the 1950s and early 1960s in a few Corvettes and Chevrolets, but it turned out to be expensive and finicky. Bosch finally refined a good, streetable, constant mechanical injection (Bosch K-Jetronic) in the 1970s, but as emissions requirements toughened, it quickly evolved into a hybrid system using add-on electronic controls to fine-tune the air/fuel mixture at idle.

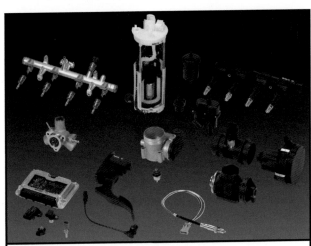

Major DI Motronic components. A powerful digital computer with large non-volatile memory space runs multiple OBD-II monitor software agents with the ability to detect problems such as combustion misfires from minor changes in crankshaft rate of acceleration. New Motronic systems are powerful and complex, but they are table-driven and extremely flexible, which makes modifications a simple programming change—if you've got access. So far, aftermarket hackers have always found a way. Bosch

By the late 1970s, carburetors had been engineered to a high state of refinement. However, there were inescapable problems intrinsic to the concept of a self-regulating mechanical fuel-air mixing system, which could only be managed by adding a microprocessor or analog computer to target stoichiometric air/fuel mixtures via pulse modulated jetting and closed-loop exhaust gas oxygen feedback.

Besides intrinsic accuracy problems and distribution issues inherent to cost-effective single-carb wet-manifold systems, carbs require restrictive venturis to create low-pressure zones that suck fuel into the charge air. By definition, this forces tradeoffs between top-end and low-speed performance. The carburetor's inability to automatically correct for changes in altitude and ambient temperature is not a problem if the goal is simply decent power at sea level. But distribution and accuracy problems can be fatal if you care about emissions, economy, or good, clean power at any altitude. Or if you want to run power-adders like turbos, blowers, and nitrous.

Throughout the 1970s, hot rodders and tuners had begun applying turbochargers to engines to achieve large horsepower gains and high levels of specific power for racing. Mainly, of course, rodders had to work with carburetors for fueling. They discovered that carbureted fuel systems are, unfortunately, problematic when applied to forced induction. It was possible to make a lot of power, but there were virtually always sacrifices of drivability, reliability, cold-running, and so forth. Nonetheless, though carbs were a problem, they were a well-understood problem, and besides, what else could you do if you couldn't afford a mechanical injection system more expensive than the engine itself?

Around this same time, car manufacturers began switching to electronic fuel injection. In 1975, GM marketed its first U.S. electronic injection as an option for the 500-ci Cadillac V-8 used in the DeVille and El Dorado. In 1982, Cross-Fire dual-throttle-body electronic injection arrived on the Corvette. The new EFI would give tuners who wanted to modify late-model cars a whole new set of headaches.

The problem was that users had no easy means to recalibrate or tune the proprietary electronic controllers that managed car manufacturer's EFI systems. Determining how electronic engine controls would tolerate various performance modifications without recalibration could not be predicted in advance.

Early EFI logic was *hardwired* into the unchangeable circuitry of an analog controller, while later digital fuel-injection controllers were directed by software logic and tables of calibration data parameters locked away in a programmable read-only memory (PROM) device that was, in many cases, hard-soldered to the main circuit board. In all of the analog electronic control units and in a fair number of the digital ECUs, changing the tuning data effectively required replacing the ECU. And even when the calibration data was located on a removable PROM chip plugged into a socket on the motherboard, the documentation, equipment, and technical expertise needed to create or "blow" new PROMs was not accessible to most hot rodders. And while an enthusiast was eventually able, in some cases, to buy a quality replacement PROM calibrated by a professional tuner with alternate tuning parameters for high-octane fuel operation or recalibrated to handle specific performance modifications, it was rarely practical for an enthusiast to tune the fuel injection himself. What's more, if you made additional volumetric or power-adder modifications to the engine, the new performance PROM was likely to be out of tune—again.

It was only in the late 1980s, as the final factory-carbureted performance vehicles aged, as the first aftermarket user-programmable EFI systems became available, and the first generation of performance EFI vehicles aged out of warranty and depreciated to the point that it was practical for more people to consider acquiring or modifying them, that hot rodders and racers began to take a hard look at the possibilities of EFI for performance and racing vehicles.

Many hot rodders and enthusiasts objected to electronic fuel injection for various reasons:

- Too expensive
- Difficult or impossible to modify
- Illegal in some racing classes
- Too high-tech (that is, complex, finicky, inaccessible, incomprehensible, mysterious, difficult to install and debug)
- Typically requires expensive auxiliary electronic equipment for diagnosis, troubleshooting, and tuning
- Regarding the carburetor: "It ain't broke, why fix it?"

Eventually, all these considerations would become much less of a factor, but for a time they put the brakes on the hot rodding of newer vehicles. Ultimately they split the sport into two evolutionary branches centered around familiar older low-tech specialty vehicles with pushrod V-8 engines—often equipped with carbureted fuel systems—and more efficient newer vehicles with high-tech computer-controlled fuel injection—often powered by smaller engines with multivalve, overhead-cam cylinder heads, some with turbochargers. The advantages of EFI created the critical mass for the 1990s sport-compact performance craze that revitalized hot rodding, but 10 to 20 years before there were actually a fair number of engines converted from early EFI backward to carburetion.

Knowledgeable racers and hot rodders soon discovered that well-tuned modern programmable EFI systems almost always produce significantly higher horsepower and torque than the same powerplant with carbureted fuel management, especially when the engine is supercharged or turbocharged. This increased performance is in addition to improved drivability, cleaner exhaust emissions, and lower fuel consumption. Early-adopter hot rodders discovered there were solid technical reasons behind the superiority of fuel injection and electronic engine management. In fact, there were—and still are—plenty of carb-to-port EFI conversions of vintage vehicles for the following highly valid reasons.

ADVANTAGES OF INDIVIDUAL-PORT ELECTRONIC FUEL INJECTION

• Greater flexibility of dry intake manifold design achieves higher inlet airflow rates and consistent cylinder-to-cylinder air/fuel distribution, resulting in more power and torque, and better drivability.

• More efficient higher engine compression ratios are possible without detonation.

• Extreme accuracy of fuel delivery by electronic injection at any rpm and load enables the engine to receive air/fuel mixtures at every cylinder that fall within the tiny window of accuracy required to produce superior horsepower.

• Computer-controlled air/fuel mixture accuracy enables all-out engines to safely operate much closer to the hairy edge.

• EFI can easily be recalibrated or adapted to future engine modifications as a performance or racing vehicle evolves. When adjustments and changes are required to match new performance upgrades made to an engine, it's often as simple as hitting a few keys on a PC to change some numbers in the memory of the onboard ECU.

• Electronic engine management with port fuel injection is fully compatible with forced induction, resisting detonation with programmable fuel enrichment and spark-timing retard and enabling huge power increases by providing the precisely correct air/fuel mixture at every cylinder.

• EFI powerplants have no susceptibility to failure or performance degradation in situations of sudden and shifting gravitational and acceleration forces that might disturb the normal behavior of fuel in a carburetor fuel system.

• Electronic injection automatically corrects for changes in altitude and ambient temperature for increased power and efficiency and reduced exhaust emissions.

• Solid-state electronics are not susceptible to the mechanical failures possible with carburetors. Tuning parameters stay as you set them, forever, with no need for readjustment to compensate for mechanical wear.

In the 1990s, many new EFI vehicles still consisted of better-performing self-contained factory EFI conversions of formerly carbureted engines. These had separate or quasi-separate distributor-based ignition systems and separate low-tech nonintegral instrumentation and chassis electrics. Virtually all onboard computer systems, with the exception of idle and light-cruise fuel-air mixture trim and idle speed stabilization algorithms, had no means of detecting if engine management strategies were successful. If the computer ordered the opening of a solenoid valve, it dumbly assumed the valve had opened. Many 1990s-vintage modified EFI engine-tuning strategies worked by stimulating the factory computer into providing more-or-less correct fuel enrichment and ignition timing on engines with upgraded volumetric efficiency during high-output operation by using clever mechanical or electrical tricks to substitute false engine sensor data (such as low engine coolant temperature) or by dynamically altering injection fuel pressure.

A few enterprising companies offered performance PROMs that could easily be swapped into factory engine-management systems such as GM's Tuned Port Injection. This provided alternate tables of rpm and load-based values for fuel injection pulse width and spark timing that improved power with premium fuel

EFI Technology's Race system rivals the sophistication of OEM electronics and algorithms—with the addition of mil-spec enclosures and connectors, FM data telemetry, fully programmable dash displays and indicators, sophisticated individual-cylinder optimization tools, and extremely sophisticated power-adder boost-control techniques. EFI Technology

calibrations or provided modified internal fuel and spark tables calibrated specifically for certain packages of hotter cams and other hot rod engine parts. Several standalone programmable aftermarket engine management systems were also available, the most successful of which was the Haltech F3. The F3 was an EFI-only system with an installed base of maybe 2,000 systems that could not manage ignition tasks at all. This was all about to change.

ELECTRONIC ENGINE MANAGEMENT IN THE 1990S AND BEYOND: THE NEW WORLD ORDER

The Clean Air Act of 1990 finally forced automakers to get serious about plans they were developing since the first serious California air pollution problems in the late 1950s. Really good, standardized onboard vehicle/engine diagnostic capabilities that would keep engines operating in a clean, efficient, peak state of tune arrived. Now that digital computers had conquered the original-equipment automotive world, there existed at last both the possibility of and the necessity for sophisticated electronic self-testing and diagnostic capabilities. In the 1980s, independent repair facilities lobbied hard for regulations to force automakers to provide open onboard computer diagnostic interfaces and to document and standardize interface protocols so independent shops servicing multiple brands did not need expensive and esoteric special scan tools for every make and model of vehicle.

The result was OBD-II (Onboard Diagnostics, Second Generation), a blessing for the typical car buyer, potentially a blessing and a curse for the hot rodder or tuner. OBD-II, which was required to be in place on new vehicles no later than 1997, implements a number of interesting capabilities. It defines standards for hardware bus connectivity to onboard computers for scan tools and laptop computers. It defines a handshake for communication between the ECU and diagnostic equipment. It defines an extensive set of standardized malfunctions that the engine management system must be able to self-detect, and it defines a standardized set of alphanumeric trouble codes. These codes must be stored by the ECU semi-permanently in nonvolatile memory that will retain its integrity even if the onboard computer loses battery power. OBD-II defines protocols for resetting such codes once a problem has been fixed.

The consequent use of large-scale electrically erasable PROMs to store calibration information and trouble codes was revolutionary because it was now feasible to flash the EEPROM with new calibration or configuration data. OBD-II thus defined a system that could be used to update the entire parameter-driven engine calibration should a bug be discovered that affected emissions or safety. It also required automakers to implement countermeasures that made it reasonably difficult to tamper with the calibration without having access to a special password known as the security seed. At the same time, OBD-II was defining a powerful mechanism that could be used to retune engines without changing any hardware or even so much as a PROM chip. Or even more: it potentially enabled recalibration of an OBD-II computer to handle an alternate engine or important additional engine systems, or even to stop doing OBD-II.

The logic of modern digital engine management systems is highly parameter driven, meaning the software design is complex, modular, universal, and all-encompassing in design. It is also highly conditional in behavior based on the status of a set of internal settings (parameters) stored in tables in memory. Changing such parameters can drastically transform functionality, giving the computer a whole new personality—for example, enabling it to manage an entirely different engine with fewer or more cylinders or one with additional power-adders and so forth.

Before OBD-II, such parameters (where they existed) were stored in read-only memory or programmable read-only memory and could be changed only by physically opening the computer and installing a new ROM or PROM, which sometimes involved soldering and desoldering and jumpering or cutting the motherboard. It would not be long before clever aftermarketers would reverse engineer the security seed and sell specialized power-programmer devices designed to connect to the diagnostic port to change parameters like top-speed-limiter or rev-limiter, or even to hack the air/fuel or ignition timing tables on GM OBD-II computers (for off-road use only, of course).

OBD-II effectively required implementation of a range of new engine and vehicle sensors to provide additional feedback to the computer so it would know when there was a problem or might be a problem. For example, where the computer might have previously commanded a valve to open to purge the charcoal canister of fuel vapors (and *assumed* it had, in fact, opened), OBD-II might have required a sensor to measure if, in fact, the valve actually *had* opened. OBD-II mandated many new diagnostic capabilities, such as the ability to detect misfires. This required precise and highly accurate crankshaft position sensors and new, more powerful, high-speed microprocessors with the computing ability to measure micro-changes in the rate of change in crankshaft speed that indicated healthy combustion or misfire events. Misfire-detection required the computational ability to correlate a transient misfire to a particular cylinder. Basically, OBD-II required the development of completely new engine management systems.

Since they were developing entirely new engine management hardware and software with a clean-sheet-of-paper approach, automakers and their OEM suppliers such as Bosch took the opportunity to develop powerful new onboard computer hardware and operating software. They made plans to implement an impressive range of new vehicle-management capabilities, including fly-by-wire throttle, traction controls, active lateral-instability countermeasures, electronic shifting, and dozens of other highly specialized and esoteric eMotion control functions.

One example is a scary-sounding thing GM calls "Adaptive Garage Fill Pressure Pulse Time and Garage Shift Pressure Control." And more: Some of the most advanced Motronic engine management systems are equipped with algorithms that deliver directly measured torque-based engine management. Some of the newest BMWs can fire up the engine without a starter motor by identifying the exact engine position, injecting fuel in the appropriately positioned cylinder, and then firing the plug.

The amount of time invested in carefully developing flawless engine calibrations for such vehicles is phenomenal, and can take years, even with complex modeling and simulation tools. The simplest millennium engine management systems used to manage economy subcompacts incorporate months or years of test-and-tune efforts on dynos, test tracks, and highways under all climatic conditions.

Meanwhile, standalone aftermarket programmable engine management systems also increased in power and complexity, with the newest systems from Motec, Electromotive, DFI, and others offering extremely sophisticated software engine modeling and highly flexible and configurable hardware with the capability to control a wide range of complex engines with a wide variety of sensors and actuators.

The most powerful aftermarket systems target pro racers and professionals building tunercars for people for whom money is not a problem. Such systems are not cheap. All involve substantial configuration, calibration, and installation efforts to approximate anything close to the observable functionality of a millennium-vintage factory vehicle (much less sophisticated little tricks like Adaptive Garage Fill Pressure Pulse Time and Garage Shift Pressure Control). Of course, conversion from OEM to programmable aftermarket engine management is not strictly legal for highway use unless the manufacturer or tuner tackles expensive CARB or Federal Test Procedure testing in order to prove that their engine management system does not degrade exhaust emissions on one or more specific vehicles in a simulated drive cycle involving a cold start and 20 minutes or so of rolling road exercise during which exhaust is captured in a big plastic bag for subsequent analysis. (Full-throttle emissions testing has *not* typically been required.)

By the time OBD-II was required on all vehicles sold in the United States, it had been 10 years since a carbureted engine had been available on a car or truck in America, and the digital microcomputer was definitely king, both in the onboard ECU (now referred to as the powertrain control module or PCM) managing the engine and in the scan tool or laptop in the hands of the diagnostician or tuner. Powerful laptop computers with graphical Windows or Mac OS interfaces were ubiquitous by 1997, and almost everyone who had been in school since 1982 had at least some degree of training and familiarity with the personal computer. This new generation arrived on the automotive-performance scene entirely comfortable with installing and manipulating user-interface and tuning software on a laptop to recalibrate engine management systems.

Unfortunately, this was only one piece of the puzzle. Ignorant jacking with calibration numbers in the computer of a vehicle with significant engine performance modifications will probably just make a bad situation worse. Tuning an engine well with good diagnostic equipment requires patience, experience, and methodical troubleshooting techniques. Tuning an engine in a car on the street without diagnostic equipment is a risky proposition that is difficult or impossible to do well without a lot of careful trial-and-error testing.

Many casual or inexperienced performance enthusiasts are simply incapable of achieving a good, safe, efficient, drivable tuning calibration from scratch, particularly on high-output engines with power-adders. If such a well-meaning but inexperienced person is lucky, they'll probably end up with an engine that fails to realize the potential of its improved volumetric efficiency. If they're unlucky, they'll end up with a polluting, gas-guzzling slug with marginal drivability or possibly a damaged engine. Even really experienced professionals—genuine *wizards*—may not have the time to achieve optimal calibration on a tuner car. What aftermarket supertuner has the days, weeks, months, even *years* to devote to a calibration task that automakers—forced by government fuel economy and emissions standards and by competitive performance pressures to put in the hard time to get perfect calibrations—can then amortize the effort over tens or hundreds of thousands of vehicles?

One response to the difficulty and complexity of recalibrating factory engine controllers or accurately installing and calibrating standalone aftermarket systems from scratch for excellent performance, drivability, and reliability is the plug-and-play strategy, as typified by AEM. The term *plug and play* is borrowed from the Windows PC environment, where users can plug in peripheral devices such as printers and Windows will immediately recognize the device, install the appropriate device driver software, and automatically configure itself and the device to properly operate or play—in many cases with little or no user intervention.

But just getting all the engine sensors and actuators *wired* to an aftermarket programmable computer can be a formidable task. It's more akin to buying a motherboard, power supply, disk drive, case, bios, and operating software from Fry's Electronics and building your own Windows system. But it's even worse, because in the case of a programmable engine management system, you'll need to locate subsystems, sensors, and actuators; route wiring; and terminate and crimp wires to connectors that need to work reliably in an environment rife with heat, cold, water, oil, vibration, and various G-forces.

AEM plug-and-play systems provide adapter wiring that allows a user to remove the stock computer and plug the stock engine wiring harness into an adapter on the programmable computer. Plug-and-play system vendors like AEM and Hondata usually provide a starter calibration that will run the engine and enable the vehicle to drive without any user intervention to calibrate it. But installation instructions warn users that virtually *any* significant modifications to the factory engine's volumetric efficiency—that is, performance modifications—require recalibration to prevent possible engine damage.

Another response to the increased complexity of 1990s-vintage OBD-II original-equipment engine management systems and the difficulty of reproducing the quality of their factory calibrations is aftermarket tuners increasingly turning to auxiliary computers or mechanical devices. Tuners are using variable-rate-

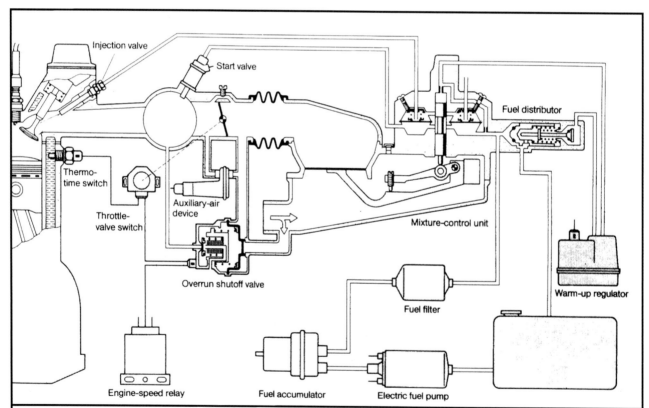

The original Electrojector fuel injection was invented by the American company Bendix in the 1950s. Bosch licensed from Bendix the concept of a constant-pressure, electronically controlled, solenoid-actuated, individual-port, periodic-timed fuel injection system and put it into production in some 1960s-vintage VWs. Bosch evolved the original concept, resulting in the newest Motronic engine management systems. However, many Americans' first exposure to Bosch fuel injection on 1970s- and 1980s-vintage VW, Porsche, Ferrari, Mercedes, and other European engines was this K-Jetronic constant-injection system (CIS), which varies fuel pressure based on a mechanical velocity air meter measuring air entering the engine. Although later K-Jetronic systems had add-on electronic trim, the system is not a true electronic engine management system, it is not easy to modify for hot rodding, and more than a few such performance vehicles still on the road have been converted to programmable EMS. Jan Norbye

only been seen in the wild turbo era of Formula One racing.

In addition to the 1,200-horsepower Ferrari 5.0-liter V-12 mentioned earlier, try 900 turbocharged rear-wheel horsepower from a methanol-powered 2.2-liter Honda VTEC four in a unibody Civic—four to five times stock power. Or try 471 rear-wheel horsepower in a 2.1-liter Toyota MR2 street car—triple the stock power. Or 500 to 700 rear-wheel horsepower from a twin-charged Toyota 1MZ-FE V-6 with both supercharging and a turbo. This is all done with electronic engine management, which is fraught with possibilities for good and for evil.

THE PURPOSE OF THIS BOOK

So this really is a how-to book, designed to communicate the theory and practice of designing and redesigning performance engine management systems that work. This book is designed to remove the mystery from electronic engine management and fuel injection. It's designed to give you the information you'll need to tune, modify, hack, or install engine management systems and components to unleash "free" power on many stock and modified engines.

Norwood Performance built this gorgeous custom twin-turbo system for a radically hot rodded Gen-1 Toyota Supra. Under Motec EMS control, with 900 horsepower on tap from 2,954 radically-boosted cubic centimeters, this streetable machine is equally at home as a dragger or a Silver State racer.

of-gain fuel pressure regulators (some even computer-controlled!) to intercept and modify or augment the actions of input-output signals from the factory onboard computer in order to change the behavior of the engine under the relatively limited operating circumstances when power-adders are in action (at wide-open throttle and higher-load operating conditions).

Hondata, Techlusion, and other vendors have perfected relatively sophisticated programmable auxiliary interceptor microprocessor boxes designed to modify or trim specific designated sectors of the factory air/fuel or spark timing curves—via laptop or, in some cases, dial pots on the processor box—without molesting the factory tune during many operating conditions. A related alternative to the interceptor or auxiliary computer is the programmable sensor or actuator that enables tuners to influence the behavior of the main engine management computer by selectively *lying* to it or by being creatively disobedient.

The complex interactions and constraints of PCM logic, PCM calibration data, PCM anti-tampering or self-protective countermeasures, fuel pump capacity and fuel pressure, injector capacity, duty cycle, electrical limitations, pressure, ignition components, and other such have made modifying engines for increased performance challenging, yet potentially rewarding. In the 1990s and beyond, the tuning of electronic engine management systems evolved to the point that supertuners like Bob Norwood have managed to achieve stupefying levels of streetable specific power that had previously

This book will provide information on how to modify engines in ways that will work with existing engine management systems and to modify and optimize engine management systems for compatibility with highly modified engines. It's designed to tell you how to design roll-your-own electronic engine management systems and convert engines to the advantages of electronic fuel injection. This book will communicate from the ground up everything you'll need to do the job yourself or to knowledgeably subcontract the work—from theory to practical installation details. It will also supply the detailed information you'll need to know about specific fuel injection and engine management control units and related products to tune factory and aftermarket fuel injection systems properly.

This book is designed to reveal secrets of factory onboard ECUs and PCMs, aftermarket programmable engine management computers, wiring and re-wiring, fuel injectors and pumps, laptop computers, PROMs, engine sensors, electronic boost controllers, chassis dyno tuning and calibration, and much more. This book provides information about what it costs to fuel inject a hot rod using various technologies, including original equipment injection systems such as GM's Tuned Port Injection. This book will help you decide whether your hot rodding plans violate government regulations. Finally, this book is designed to document how to make EFI engines more powerful and to make understanding and hacking performance engine management and fuel injection systems *fun*.

CHAPTER 1
UNDERSTANDING FUEL DELIVERY:

For many performance enthusiasts older than about 35 (and motorcyclists of any age!) the benchmark for automotive fuel delivery systems is still the carburetor. And it's true: No electronic fuel system can touch a carburetor in delivering so much performance for so little cost. Many performance carburetors are still sold.

Although the emissions carbs installed on new U.S. vehicles in the final years before the demise of automotive carburetion in the mid-1980s were horribly complex contraptions equipped with extensively convoluted vacuum systems and electromechanical add-ons that enabled carbs to trim themselves for stoichiometric (14.7:1) air/fuel ratios via feedback from an O_2 sensor, the basic carburetor was still what it always had been: a self-regulating mechanical device that uses mechanical forces to suck fuel into the air stream entering an engine.

Many people have had an easier time understanding carburetors than electronic fuel injection. Perhaps this is because you could take a carb apart with a screwdriver and actually see the air and fuel passages and the mechanical equipment that manages gas and liquid flow to reliably deliver accurate, appropriate, and repeatable air/fuel mixtures in response to changing engine conditions. If you wanted more fuel with a carburetor, you could remove the main jet and drill it out with a special bit, and you'd get more fuel.

By contrast, for most people electronic fuel injection is a black box that can only be fully understood by reading complex technical documentation (assuming it's not proprietary). It can only be modified or tuned by changing the internal software or data structures in the onboard engine management computer or with auxiliary add-on mechanical or electronic tricks that typically deceive or even work at cross purposes to the main engine management and fuel systems.

But in reality, most carburetors are actually *not* simple to understand fully or tune really well, while EFI systems are simpler than it might at first appear. Carbs are intricate and complex devices that have been engineered over the course of many decades to a high state of perfection. Though they are much less expensive than electronic injection systems, carbs are *not* simple to tune in a major way in detail. Of course, despite the several-times added cost of electronic injection, carburetors were abandoned by automotive engineers in the mid-1980s precisely because their fuel

The Edelbrock 3500 kit is the ultimate bolt-on injection conversion package, with everything you'll need to provide air, fuel, ignition, and engine management for a small-block Chevy crate motor: intake manifold/throttle body/injector rail/sensors module, computer and harness, fuel pump and filter, heated O_2 sensor, calibration module, coil driver, and more. Pretty much all you need is a V-8 long block with a distributor and fuel supply-return lines. Edelbrock

management capabilities were, in the end, inadequate to meet required new standards for low emissions with performance and economy.

The basic mechanical principle enabling fuel delivery via carburetion is the Bernoulli effect, which states that air pressure decreases when moving air speeds up to flow around a gentle curve such as an aerodynamic constriction in a tube or curvature of a wing. By designing a venturi into the inlet of an engine and introducing an atmospheric-pressure fuel bleed into the area of the venturi, the reduced fuel pressure in the venturi will automatically suck fuel into the air.

If an engine always operated at one speed, temperature, and altitude, the job of a carb would be relatively simple, and a simple venturi-based carb would do the job. Indeed, boats, aircraft, and tractors operating over a very limited dynamic range can get away with simpler carburetors. But in order to handle cold starting, transient enrichment, idling, correction for the differential flow characteristics of air and fuel, full throttle enrichment, and emissions concerns in an automotive environment on engines that must accelerate very rapidly and perform well across an extremely wide speed range, street carbs contain a myriad of add-on systems. These systems include choke valves, fast idle cams, accelerator pumps, two-stage power valves, air corrector jets, emulsion tubes, idle jets, booster venturis, and air bleed screws, not to mention the pulse width–modified solenoids mentioned above that pulse open and closed under computer control to regulate fuel flow through the jet(s) of an electronic feedback carb.

Carburetion is an ancient technology, refined by more than 150 years of engineering and tinkering. The engineering context of gasoline carburetion operates within the constraint that liquid gasoline and gaseous air moving though a carburetor or intake system do not necessarily behave the same with increases or decreases in overall flow and pressure, so the tricks required to maintain the proper air/fuel mixture tend to change as rpm and loading change.

Gasoline carburetion is an esoteric, delicate world of countless moving parts and fluids, overlapping fuel/air metering systems (each of which only covers *part* of the speed and loading range), tiny tubes that introduce air bubbles into fuel bleeds to dilute fuel delivery under certain conditions, and complex interactions among all of the above. A given problem usually has a long list of possible causes that includes everything from the weather to a speck of dirt in a tiny passage, to an incorrectly sized part, to a design flaw that prevents the carb's crude mechanically derived air/fuel curve from approximating ideal air/fuel ratios under certain circumstances in the way trying to approximate a gentle curve with a straight line would always be wrong—somewhere.

To truly master carburetion requires a thorough knowledge of fluid mechanics, engine design, and air/fuel theory as well as a great deal of experience and test equipment. Carbs have a *lot* of parts and systems, and they all do something. Many are interdependent and modify the functioning and actions of each other in complex ways. Fortunately, most people do not try to build a custom carb from parts; they buy something off the shelf targeted loosely for an engine close to what they've got. Off-the-shelf carbs are easy to get running, if only in a suboptimal fashion.

In comparison, a modern electronic fuel injection system is a thing of marvelous conceptual simplicity that is quite easy to understand. In the end, it only does one basic thing: It turns on and off fuel injectors for a precisely regulated length of time at least once every power stroke. Need more fuel this engine cycle? Open the injectors a little longer. Less fuel? Pulse the injectors a shorter burst. Pure and simple. Virtually everything else going on in elec-

Constant mechanical injection predated electronic fuel injection, and it survives today. Hilborn still provides a large variety of mechanical injection components, through they now also provide stack-throttle body systems like this Ford flathead system that look mechanical but are actually electronic, with all the precision control and enhanced performance of electronic systems. Hilborn

tronic fuel injection has to do with reading the condition of the engine at a given moment, and then doing a little math to compute injection pulse width—before firing those injectors.

So let's start by understanding the easy stuff. Let's design a virtual electronic fuel injection system.

DESIGNING A VIRTUAL FUEL INJECTION SYSTEM

Unlike the carburetor, in EFI, air and fuel are metered entirely separately under computer control, combining only at the last moment in or near the combustion chamber. EFI can define essentially any air/fuel relationship from moment to moment, depending on the status of the engine—or anything else in the world. The relationship between air and fuel in EFI is defined electronically in a computer—and is, therefore, completely flexible. If we wanted, we could change the way our EFI system injects fuel based on the phases of the moon!

So, since anything is possible with EFI, once we've had a chance to discover what an EFI system looks like, we'll take a look at some alternate design choices before getting down to the business of relating specific fuel injection components to fuel, ignition, and engine theory. The goal is to understand what kind of fuel delivery is ideal under various circumstances and how injection systems go about providing it.

Assume we're starting with a pre-EFI vehicle that was originally equipped with standalone distributor ignition and carburetor but has had the carburetor and carb manifold removed. If we're going to design a fuel injection system, where shall we start? How about at the beginning, at the fuel tank?

STEP 1. GET FUEL TO THE ENGINE COMPARTMENT

The fuel is resting in a tank, ready to make power, often at the end of the vehicle opposite from the engine. Unless we want to put the fuel tank on the roof, we need a pump to move fuel to the engine. Let's put the fuel pump in the fuel tank (which will also help keep the pump cool). We'll drive our roller-vane centrifugal pump with an electric motor so it's not necessarily tied to the speed of the engine. Therefore we'll have available plenty of fuel pressure, any time—say at least 40 psi.

Our pump should be sized to continuously move enough fuel to satisfy the engine at its highest consumption—maximum horsepower—under full load. We'll run a steel fuel line from the pump to the engine and route a second fuel line from the engine area back to the fuel tank. When the engine isn't using all the fuel we're pumping, excess fuel can be routed back into the fuel tank. That way we don't ever have to turn the pump off when the motor is running—or accelerate pump wear by forcing it to pump against a deadhead of pressure during low fuel consumption—say at idle.

STEP 2. BUILD AN INTAKE MANIFOLD AND THROTTLE

OK, let's move to the engine. Currently, the intake ports are naked holes with no intake manifold installed. Time for some fabrication. We begin by attaching a short pipe (runner) to each intake port that is sized with the same internal diameter as the port itself. We collect these runners in an air chamber (plenum) that has a single circular intake throat bolted to the plenum. We regulate airflow into the plenum chamber (and subsequently into the engine) using a throttle plate (flat, circular plate, completely blocking the circular air-intake throat). This spring-loaded throttle plate is designed to pivot on a shaft and allow more or less air into the motor. A bleed screw allows an adjustable amount of air to bypass the throttle even when the throttle is fully closed so the engine will idle. We decide to name the chamber containing the throttle plate a throttle body. Now we run a cable from the throttle plate shaft to the accelerator pedal—push down, open the throttle. This will control the speed of the engine and the power output by regulating how much air enters each cylinder in the intake stroke.

Great. We can now regulate air into the engine, and we can get fuel to the engine, but we can't yet meter the fuel.

STEP 3. FIGURE OUT A WAY TO METER FUEL

We drill a hole in each intake runner near the intake port, install a miniature fuel valve in each hole aimed straight at the intake valve to inject fuel directly into the port where air is moving fast and swirling as the engine sucks in air. Each *injector* actually contains a little electromagnetic coil. When energized with direct electrical current from the battery of the car, the energized electromagnet will overcome the force of a tiny spring in the injector that holds closed a miniature check valve. The valve snaps open and allows fuel to spray through a tiny orifice at high pressure in a fine mist until the current is switched off and the valve clicks shut, stopping the fuel flow.

At high engine speeds, each injector must be able to open and close like a frantically buzzing bee—up to 400 times a second! When an injector is open, fuel always flows from it at the same rate per unit of time as long as the fuel supply pressure to the injector remains constant. Therefore, we must meter the amount of fuel going into the engine based on how long we leave the injector open per engine cycle.

Clearly we must carefully select our injectors so that the flow rate matches our engine requirements, keeping in mind the following: Any injector's high flow threshold occurs when it is constantly open, providing an absolute maximum fuel flow. Providing more fuel than this maximum would require either increased fuel pres-

sure, injectors with a larger orifice, or additional injectors. In practice, injectors need a finite amount of time to close and open, and injectors heat up when held open constantly, so an 80 percent duty cycle is a good practical design limit.

On the other hand, when the engine is idling, the same injectors must be able to open and then close quickly enough to inject the tiny amount of fuel required for a good quality idle. In fact, most injectors cannot provide accurate flow below a threshold of 1.3 milliseconds open time because the injector is turning off before it has entirely opened, which tends to produce injection pulses that are not entirely consistent. To get less fuel requires changing to smaller injectors or decreasing the fuel pressure (and, yes, it is common practice to lower fuel pressure at idle and other light engine-loading conditions). In any case, the correct injectors are ones sized to supply both idle and wide-open throttle requirements (thinking ahead, you may realize that small engines boosted to very high power levels can run into trouble finding injectors that can meet both requirements).

Pushing ahead, we plumb the fuel supply line to each injector, being careful to size and design our fuel manifold, or fuel rail, so that each injector receives steady and equal fuel-supply pressure under all circumstances. We then plumb in the fuel return line through an inline pressure regulator designed to maintain an exact fuel pressure in the fuel rail (loop) in relation to manifold pressure. This kind of regulator, usually referenced to intake manifold pressure by an air hose, is designed to pinch off fuel going back to the tank until fuel pressure builds up at the injectors to a predetermined pressure.

At this point the regulator begins to open and permit sufficient fuel to return to the fuel tank, maintaining a predetermined fuel pressure. Manifold vacuum referenced to the regulator fights the force of the regulator's spring against the fuel diaphragm at idle, lowering fuel pressure anytime manifold pressure is less than atmospheric (and raising it during boosted conditions when manifold pressure increases *above* atmospheric). As manifold pressure rises with increasing engine loads, vacuum at the regulator drops, and the full force of the regulator spring comes to bear on the diaphragm, raising fuel pressure. The injectors, therefore, always inject against manifold pressure that is fixed in relation to fuel pressure.

At this point, our experimental system is ready to handle the mechanics of air and fuel delivery into the engine: a throttle body and intake manifold to deliver and measure air, and a fuel pump supplying high-pressure fuel to a set of injectors ready to precisely spray fuel into the inlet ports in exactly the correct amount as a function of how long each is opened.

Everything from here on is related to reading and evaluating certain engine vital signs and translating this information into actual engine fuel requirements that can then be used to compute injector open times (pulse width) and pulse the injectors.

STEP 4. TURNING THE INJECTORS ON AND OFF

At 10,000 rpm, a port fuel injector must turn on and off at least 167 times a second (and possibly 333 times a second). The on time must be computed accurately to at least a tenth of a thousandth of a second to get the air/fuel mixture just right.

This is a job for a microcomputer. A modern digital microcomputer can perform millions of instructions per second. This speed is so much faster than what is happening in an engine mechanically that the microcomputer can treat the engine, at any given moment, as if it were standing still.

A microcomputer can continuously read the current status of an engine via electromechanical sensors. Based on this informa-

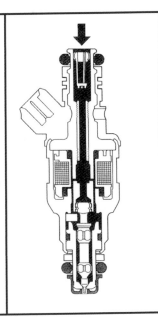

Bosch pintle-type electronic fuel injector. Pressurized fuel enters through the top and flows through the injector until it is stopped by a solenoid-actuated valve about one-third of the way from the bottom. When the injector is energized through electrical connections on the upper left, the centrally located coils magnetically lift the lower armature off its seat. This allows fuel to flow past the valve seat, around the armature body, and spray out the bottom of the injector in the area of the pintle, a tiny pin protruding through the bottom of the injector in the nozzle area used to improve the spray pattern. Bosch

tion, it can then quickly retrieve from memory reconfigured tables of numbers that provide information about the appropriate injection pulse width at all possible combinations of engine speed and loading. The microcomputer, which maintains a highly precise internal clock, combines sensor and table look-up data with arithmetic computations to schedule injector start or stop times—all before the engine has virtually moved.

Certain vital engine events, such as the crankshaft arriving at a particular position or the internal clock making another tick, can be made to automatically interrupt the microcomputer for action. As engine conditions change, the computer is able to reschedule injector start and stop times on the fly and activate electrical drivers that energize or de-energize the injector electromagnets.

OK, now we have a computer. What sensors do we need to install so the computer can understand what is happening in the engine?

STEP 5. TELLING THE COMPUTER WHAT'S GOING ON IN THE ENGINE

What does the computer need to know about the engine?

Let's start with engine position, speed, loading, and so on.

First, we'll install an optical triggering device near the camshaft or distributor designed to count teeth on a gear that is missing one or two teeth as a reference point. The cam sensor and gear are set up to send a special electrical signal to the computer each time the crankshaft passes through top dead center compression stroke on the number one cylinder. In between, the cam sensor detects the proximity of gear teeth and passes electrical signals to the microcomputer as teeth fly past. This enables the computer to know the exact position of the engine combustion cycle for each cylinder. The computer uses the toothless reference point (and offsets from it in numbers of teeth) to sequentially schedule timed fuel injection events for each cylinder coordinated with the time at which each intake valve opens.

Beyond knowing camshaft and crankshaft position and speed, in order to time a sequence of fuel delivery opening and closing events for each of the injectors, our microcomputer must know the number of injectors and cylinders and the displacement and injector flow rate. But most important of all, the microcomputer must know how much air is entering the engine at a given

moment so it can compute the basic fuel requirement in weight (1/14.7 of the air mass). The microcomputer then determines special modifications to the base fuel requirement (enrichment or enleanment) at the current engine speed and loading and then begins scheduling injector opening and closing times needed to achieve the correct pulse width, providing the correct air/fuel mixture.

To measure air entering the engine, our system might utilize a tiny heated wire located in the inlet air stream just upstream of the throttle body. This wire would measure the cooling effect of incoming air by measuring the electrical current required to achieve a precisely targeted temperature—exactly proportional to the number of air molecules entering the engine at a given moment. This is called air mass. This mass airflow (MAF) sensor's output voltage is constantly converted electronically to digital values available to the microcomputer to use in its fuel computations.

STEP 6. STARTUP

How might such a system run? Very likely the engine sputters and coughs a little, but will not start. Eventually we resort to a long shot of starting fluid straight in the throttle body while cranking. The engine struggles to life, coughing and backfiring through the intake manifold and running badly. After a few minutes of warm up it idles great.

After considering the problem, we realize another sensor is required—a coolant temperature sensor designed to give the microcomputer information about engine temperature so it can tell when there's a cold-start situation. We know that gasoline doesn't mix well with cold air, that only the lightest fractions of gasoline will vaporize well in the cold. We know that a normal but cold air/fuel mixture will not burn well in a cold engine since much of the injected gasoline has not vaporized or atomized well. The gasoline exists as liquid droplets or threads of fuel suspended in the air and coating the cold cylinder or manifold walls as dew. In fact, plain liquid fuel doesn't burn; motor fuel must be well mixed with air to burn properly. Once the engine is running, much of the fuel actually vaporizes in the hot cylinders, but during cold cranking, the cylinder walls are stone cold too, and necessary cold-start air/fuel ratio may be in the 1.5:1 to 4:1 range in cold weather!

With a coolant-temperature sensor installed, when the engine is cold the microcomputer knows to inject plenty of extra fuel so that at least some will vaporize well enough to run the engine. The microcomputer gradually diminishes this enrichment as the engine warms up.

The next time the engine is cold, we turn the key and it starts quickly and runs well while warming up.

So we back out our virtual driveway. . . .

After a few tanks of gas we noticed the car bogged badly on sudden acceleration at lower rpm, got worse than expected fuel economy, and had less than expected power under wide-open throttle. So we're back in the virtual garage installing yet another sensor—a throttle position sensor (TPS)—and reprogramming the computer one more time.

On rapid throttle opening, manifold air pressure drops, but airflow measurement systems can have difficulty keeping up with the rapid changes. With its new TPS, the computer detects when we are opening the throttle. Now the microcomputer adds transient enrichment to ordinary injection pulse width on throttle opening. The stumble goes away. We also program the computer to create a richer mixture than normal when the throttle is more wide open and to cut the pulse width a little to lean the mixture when we're running with the throttle only partly open.

The fuel economy improves, and the acceleration snaps our necks back. Cool.

STEP 7. NAMING THE SYSTEM AND MAKING IT SMART

We name our virtual EFI system sequential port injection of fuel (SPIF), and give it one more interesting capability: We give it the ability to *learn*, and we add some features that make it easier to tune and program.

Locating an oxygen sensor in the exhaust system enables SPIF to deduce the air/fuel ratio of pre-combustion charge mixtures in the cylinders based on the amount of residual exhaust gas oxygen present following combustion. And we now add code that enables the system to learn from its mistakes and make adjustments to injection times to target proper air/fuel mixtures. This mode of operation, in which the microcomputer is able to evaluate its own performance and make corrections on the fly, is called a closed-loop system, because the results of previous computations (injection pulse width) are fed back as input to future pulse width calculations along with information about the results achieved.

Just for fun, we design the operating firmware on our SPIF microcomputer so we can adjust the behavior of SPIF (reprogram it) something like a video game. We can use a mouse or a few keystrokes to change the height of graphs on the video screen of a Windows laptop computer connected by data cable to our SPIF microcomputer.

STEP 8. HOW DID WE DO?

We still haven't tried to make our virtual EFI system pass emissions, but so far, so good. We've built an electronic fuel injection system that gets the job done. And our SPIF system is, in fact, very similar to some injection systems you'll find on new cars from Detroit, Europe, and Japan.

But consider this: SPIF delivers fuel to a set of injectors aimed at the intake ports of an engine. OK, there is a certain amount of programming logic operating behind the scenes to compute pulse width under various circumstances and to make corrections according to exhaust gas feedback. But as we noted before, SPIF really does only one thing to fuel the engine: It opens and closes injectors, sometimes faster, sometimes slower, sometimes longer or shorter, but that's it. At idle—or at part throttle, full throttle during sudden acceleration, even during warm-up—the micro-computer makes its calculations, and the injectors buzz along, opening and closing. That's it.

If we were planning to get into business selling SPIF systems in the real world, we would want to make some modifications to our design for certain marketing and technical reasons, depending on the type of use we anticipated. We'd certainly need to add the ability to control ignition timing, and we might want to add the capability to control idle speed, boost pressure, and other nonfuel engine actuators. Our current version of SPIF is not quite ready for prime time. But it's very close.

STEP 9. REVISIONIST DESIGNS

OK. Our SPIF virtual EFI engine is now running pretty nicely. But before we try to take it to the races or down to the emissions lab, before we install a supercharger or nitrous-oxide injection or other wild stuff like that (which is perfectly compatible with EFI, as we'll see), it might be valuable to re-examine the engineering decisions we made in designing SPIF, and to consider other approaches we could have taken (and that have been taken in the real world) in our fuel injection system design.

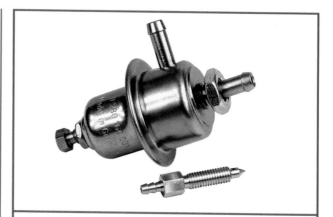

The MSD fuel pressure regulator is designed to maintain a precise fuel pressure at the injector feeds. Fuel flow enters through the side hose barb, building pressure until it overcomes a spring in the regulator housing, at which point additional fuel bleeds off through the end hose barb for return to the fuel tank. The bolt-type needle fitting on the opposite end can be adjusted to alter the spring preload and raise or lower fuel pressure. Alternately, the optional adjuster-fitting with small hose barb for a manifold pressure reference line raises and lowers fuel pressure as an offset from the basic spring pressure as a function of engine loading. This maintains a fixed relationship between fuel pressure and manifold pressure so fuel is always spraying into the same relative air pressure. MSD

ALTERNATE INJECTION LOCATIONS

SPIF injects fuel into each intake port, directly at the inlet valve, in time with the intake stroke. Alternatively, we could have injected fuel directly into each cylinder like a diesel engine—also referred to as direct-injection gasoline. Or, taking a different approach, we could have injected the fuel for all cylinders farther upstream in the intake manifold at a single point near the throttle body, referred to as Throttle body injection (TBI).

Injecting fuel directly into a cylinder to achieve some of the advantages of both a diesel and a spark-ignition powerplant requires a fuel pump with extremely high operating pressure capabilities. Under some circumstances the fuel pump will be injecting against compression cylinder pressures and must get the fuel into the cylinder extremely quickly. Such a pump tends to be noisy and robs power. Injection timing is critical, since fuel must be injected during the compression stroke and given time to atomize. This is a small window of opportunity on high speed engines operating near redline—a few thousandths of a second! Nevertheless, such a system was used for many years on gasoline-engine Mercedes cars until the 1970s.

As the new millennium approached, automakers developed several direct-injection gasoline engines that offered dramatically improved fuel economy or power compared to the same engines with port injection. New engine families such as GM's four-cylinder Ecotec had direct-injection capabilities protected in the basic design for possible later implementation.

With port fuel injection, timing is not necessarily critical. Old Bosch L-Jetronic systems injected half the required fuel from all injectors simultaneously twice per complete engine cycle. There are some fairly minor fuel consumption and emissions advantages to timed sequential port injection systems like SPIF, but many EFI systems batch it and perform very well. Batch injection would have enabled us to dispense with the crank sensor, since the microcomputer could easily determine engine speed by counting sparks at the ignition coil (and timing is irrelevant to batch injection).

Throttle body injection need not be timed either (although it must use an injection strategy that pulses often enough to provide for equal fuel distribution to the various cylinders). With any such single-point injection there is plenty of time on the way to the cylinder for the fuel to atomize. However, as in a single carbureted fueling system, all single-point injection systems must deal with the opposite problem. Fuel can separate from air on the long (and usually unequal length) path to the cylinders, wetting the surface of some sections of the intake manifold, which results in uneven and unpredictable mixtures in various cylinders at any given moment.

Single-point systems such as this can suffer from the difficulties of single-carburetor engines in achieving equal air/fuel mixtures in individual cylinders, particularly since getting equal amounts of air to all the cylinders does not always equate to getting equal amounts of air and *fuel* to each one (fluid mechanics of air and fuel are not the same). Heavier fuel is more likely to gravitate to the outside of a manifold bend, for example, compared to air.

ALTERNATE AIR MEASUREMENT

Another alternate injection design would have us eliminate the hot-wire MAF sensor measuring the amount of air entering the engine. Instead, the microcomputer could have done a good job setting pulse width using data from an airflow meter measuring the *velocity* of air entering the engine, based on air pressure against a spring-loaded flapper valve suspended like a door in the air stream and linked to a linear potentiometer (something like a volume control). Air velocity is not necessarily the same as air mass, but it is when corrected for the temperature of the intake air and atmospheric pressure (two additional easy-to-install electronic sensors). Bosch L-Jetronic systems uses velocity air metering.

In fact, an alternate version of SPIF could throw out the entire concept of actually metering air, and indirectly *deduce* how much air is entering the engine based on less-expensive sensors measuring manifold pressure and engine speed, with corrections for air temperature—which, for a given displacement and engine volumetric efficiency, equates to air mass flow. Although engines always displace the same volume of air in the cylinders, at various throttle positions, speeds, and loading, and with various camshafts, manifolds, and so on, an engine varies in efficiency at pulling in total mass of air.

An engine operating at high vacuum (with the throttle nearly closed) is unable to pull in as much air as the same engine operating at wide-open throttle with higher manifold pressure. The engine is also less efficient on a hot day or at the top of a mountain where the air is thinner. Speed-density injection systems sense engine speed and manifold pressure plus ambient air temperature to deduce the air mass entering the engine (or skip worrying about VE and just look up injection pulse width in a table for the current combination of engine speed and manifold pressure). This is the same figure supplied by SPIF's MAF sensor but is arrived at using a different method.

One advantage of speed-density systems is that they do not restrict inlet airflow like velocity air meters (with the pressure drop that exists at the air metering door). Speed-density systems also compensate automatically for any air losses or leaks in the engine's inlet system, which is definitely not true of MAF sensors. Speed-density injection systems, without MAF sensors, may allow greater flexibility of inlet air plumbing to the throttle body, particularly on turbocharged engines. On the other hand, speed-density injection systems can have problems at idle if engine vacuum fluctuates and wanders (as it might on engines with high lift, high overlap cams).

Cadillac offered one of the first speed-density EFI systems as an option on its 1976 500-ci motors. Earlier 1980s Ford 5.0 V-8s, all GM TBI (throttle-body injection) V-8s, as well as some TPI (tuned port injection) aftermarket systems like the E6K from Haltech, the Electromotive TEC[3] injection system, DFI's Gen 7 system, and, in fact, almost all other aftermarket systems allow the use of manifold absolute pressure-sensing speed-density methods to deduce mass airflow into the engine.

A few of the more sophisticated aftermarket systems from companies such as Motec are capable of using MAF (or even direct-torque load sensing) in the base fuel calculation. After a schizophrenic period some years back when car companies such as Ford and GM skipped back and forth between MAF and MAP load sensing and couldn't seem to decide which they liked better, many of the newest factory engines are equipped with *both* MAP and MAF sensors.

VIRTUAL ALTERNATE INJECTORS

It may already have occurred to you that adjusting the on-time of injectors is only one way to regulate how much fuel ends up in the cylinders of an engine. Instead of using injectors that always flow the same amount of fuel (like SPIF)—and regulating how *long* the microcomputer turns them on (pulse width)—we could use a set of injectors that spray continuously while varying the pressure to determine how much fuel is forced through the injectors in order to meet engine requirements. Such injectors could be purely mechanical devices, using a spring-loaded check valve to stop fuel from dripping or spraying out the injector until a threshold pressure forced the injector open.

Bosch KE-Jetronic fuel injection used a microcomputer and a set of sensors—including an electromechanical air meter—to vary fuel flow through a set of injectors that flow continuously on CIS systems. Based on measured airflow, the computer controls a valve regulating fuel pressure exerted against the slits of a metering valve on the way to continuous port fuel injectors. More fuel through the slits means more fuel out the injectors—on a continual basis.

MECHANICAL INJECTION

If we were using a continuous injection scheme instead of SPIF, we could actually have eliminated the computer entirely—as Bosch did in the original K-Jetronic that pre-dated the Bosch KE-Jetronic. Fuel was regulated to the injectors using entirely mechanical means. A carefully designed velocity air metering flapper valve suspended in the inlet air stream mechanically actuated a fuel control plunger contained in a vertical fuel barrel (something like a piston in a cylinder) such that vertical slit openings in the sides of the fuel barrel were progressively uncovered as the control plunger moved upward, allowing more fuel to flow through the greater unblocked area of the slits and on to the injectors. For a given amount of air entering the engine at any given moment, the fuel metering plunger in K-Jetronics reaches equilibrium at a point where the force of air entering the engine—as transmitted mechanically to the metering plunger—exactly balances fuel pressure tending to force the plunger the opposite direction.

SINGLE-POINT INJECTION

But as discussed earlier, why inject fuel at the ports at all? A single large-capacity injector, pointing straight down at the throttle body, can provide fuel for all the cylinders. This injector could operate

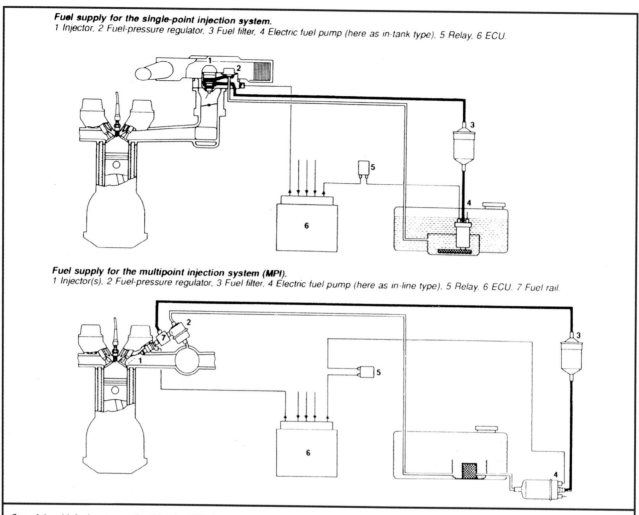

Fuel supply for the single-point injection system.
1 Injector, 2 Fuel-pressure regulator, 3 Fuel filter, 4 Electric fuel pump (here as in-tank type), 5 Relay, 6 ECU.

Fuel supply for the multipoint injection system (MPI).
1 Injector(s), 2 Fuel-pressure regulator, 3 Fuel filter, 4 Electric fuel pump (here as in-line type), 5 Relay, 6 ECU, 7 Fuel rail.

One of the chief advantages of multi-point EFI is the freedom to design dry intake manifolds that only need to handle air rather than air and fuel, which negotiate curves at varying engine speeds differently and therefore require extensive manifold modifications to keep fuel from puddling on runner surfaces and re-entering the air stream at random intervals. Jan Norbye

continuously and at reasonably low pressures since there would be no need to force fuel through a multi-injector fuel rail at high pressure to ensure accurate fuel delivery and pressure at multiple injector locations.

In fact, we might even decide to operate our single injector at *atmospheric pressure*. To do this, we could pump fuel into a reservoir near the throttle body, using a float-operated check valve (a miniature version of the valve in a toilet tank) to keep the reservoir full but prevent the fuel pump from overfilling the reservoir. The fuel in the reservoir would then exist at ambient atmospheric pressure. Cleverly using Bernoulli's principle, we could then constrict the inlet air passage slightly above the throttle plate, just above the maximum height of the fuel reservoir, which would create a minor restriction in airflow. Air headed into the engine would speed up as it passes through the restriction, or venturi, and slow down again on the other side.

As we discussed earlier, Bernoulli's principle states that when air speeds up as it flows past a solid surface, it exerts decreased pressure. Presto! We've created a small area that always has lower pressure than atmospheric when the engine is running, with the pressure reduced in proportion to the amount of air flowing through (the more air entering the engine, the faster the air will

rush through the restriction and the greater the pressure drop will be in the restriction compared to atmospheric).

We would then install a tiny injector inside the inlet constriction and run a fuel line to the reservoir. No check valve is needed in the injector anymore as is required on Bosch K-Jet systems. Since the injector is located just above the fuel level in the reservoir, fuel flow will cease flowing the instant pressure rises to atmospheric in the venturi when the engine stops. We could calibrate the flow rate of the passage connecting the injector and reservoir in order to tune the correct air/fuel mixture.

Oh no. We just invented the carburetor. But does it suck or does it blow?

It is, in fact, the higher pressure of the earth's atmosphere that forces fuel out of the float chamber and into the low-pressure area of the venturi. We can conclusively say the carb blows.

FOUR CYCLES OF AN ENGINE

In the *first cycle* of a four stroke engine—intake—the piston is moving down, drawing in air/fuel through the open intake valve. Somewhere near the bottom of the stroke, the intake valve closes. In most cases, the intake valve actually opens a little in advance of the initial downward movement of the piston and doesn't close until a bit after the piston has passed bottom dead center. This allows the incoming air/fuel mixture to continue filling the cylinder due to the inertia of the intake mixture at high inlet velocities as the crankshaft continues its rotation, but the piston has barely begun to accelerate away from bottom dead center.

Particularly in engines designed to operate at high speeds, the valve stays open a long time in order to make more power as explained above (long-duration cam profile). However, at low speeds, this late closing can allow some of the intake mixture to be pumped out of the combustion chamber back into the intake manifold, particularly at small throttle openings where high vacuum exists in the intake manifold. This pump-back phenomenon means that engines with longer cams have higher manifold pressure (lower vacuum) at lower rpm ranges, although airflow through the engine is no greater.

A computer sensing engine speed and manifold pressure has no way of distinguishing between a short-cammed engine at a heavier load and a longer-cammed engine operating at a lighter load. Yet very different fueling is required for best power and economy in these two situations.

In the *second cycle*—compression—the piston moves upward, compressing the mixture. The spark plug fires as the piston nears the top of the stroke, timed in such a way that the maximum cylinder pressure produced by combustion peaks at about 15 degrees after top center in order to produce best torque. It is necessary to fire the plug while the piston is moving up because it takes *time* for the fuel-air mixture to burn, and the flame that starts when the plug fires initially exists as a tiny kernel that is expanding relatively slowly.

Combustion is a chain reaction of sorts. Once a few percent of the compressed mixture is burning, the rest of combustion proceeds quickly. Depending on various considerations, spark advance can be anywhere from 5 to more than 50 degrees of crankshaft rotation before top dead center. Normally, engine designers strive to correlate spark advance so that, for all combinations of speed and loading, the engine produces mean best torque (MBT). If the engine begins to knock before MBT is achieved, it is considered knock limited.

In the *third cycle*—power—the piston is pushed down by pressure produced in the cylinder by combustion, typically over 600 psi.

In the *final stroke*—exhaust—the exhaust valve opens near bottom dead center, and as the piston rises, exhaust gases are pushed out past the exhaust valve. As mentioned earlier, on high-speed engines, there is usually a certain amount of valve *overlap*, in which the exhaust valve stays open after top center and the intake opens before top center to make use of inertial forces acting on the moving gases. This produces greater volumetric efficiencies at high rpm. At low rpm, particularly with high manifold vacuum (low pressure), there is some reverse flow comprised of mainly exhaust gases flowing back from the higher-pressure exhaust into the low-pressure intake.

At lower rpm, a cam with more duration and overlap results in higher manifold pressures (less vacuum) but lower airflow through the engine.

In fact, cam specifications have a large effect on spark advance and air/fuel mixture requirements. Other factors affecting mixture and timing are compression ratio, fuel octane rating, and, of course, engine operating conditions (speed, temperature, and loading). Spark advance requirements tend to increase with speed up to a point. Advance requirements decrease with loading, since dense mixtures produced by the high volumetric efficiencies of wide-open throttle or boosting power-adders like turbocharging

burn more quickly. Higher compression also increases the density of the charge mixture and the flame speed. With a big cam, the advance at full throttle can be aggressive and quick. Low VE at low rpm results in slow combustion. Exhaust dilution lowers combustion temperatures and the tendency to knock.

Part throttle advance on big-cam engines can also be aggressive due to these same flame speed reductions resulting from exhaust dilution of the inlet charge due to valve overlap.

Before computer-controlled spark advance, mechanical rotating weights advanced spark with engine speed by exerting centrifugal force against a cam-actuated advance mechanism constrained at lower speeds by the force of mechanical springs. At the same time, light engine loading would independently increase vacuum forces acting against a spring-loaded diaphragm and advance timing under conditions of high vacuum.

Limitations on the complexity of this type of engine-management-by-mechanical-means introduced tradeoffs that meant, practically speaking, on some engines in order to prevent spark knock under some conditions the spark advance curve might have to be suboptimal at other times. If the engine had a big cam, low idle manifold pressure might mean that there was no good way to achieve required advance at low speeds without over-advancing timing at mid- and high-rpm ranges.

The chemically idle air/fuel mixture by weight in which all air and gasoline are consumed occurs with 14.64 parts air and 1 part fuel. Chemists refer to such a perfect ratio of reactants as *stoichiometric*. Mixtures with a greater percentage of air are called lean mixtures and occur as higher numbers (14.7 and up). Richer mixtures, in which there is an excess of fuel, are represented by smaller numbers. The rich burn limit for a gasoline engine at normal operating temperature is 6.0, the lean burn limit for an EFI engine is above 22. The following table, courtesy of Edelbrock, indicates characteristics of various mixtures. In practice, stoichiometric mixtures are rarely optimal for fueling an engine.

Air/fuel Mixture and Characteristics

AFR	Comment
6.0	Rich burn limit (fully warm engine)
9.0	Black smoke/low power
11.5	Approximate rich best torque at wide-open throttle
12.2	Safe best power at wide-open throttle
13.3	Approximate lean best torque
14.6	Stoichiometric AFR (chemically ideal)
15.5	Lean cruise
16.5	Usual best economy
18.0	Carbureted lean burn limit
22+	EEC/EFI Lean Burn Limit

Note: These figures do not indicate anything about the effect of various mixtures on exhaust emissions.

The actual optimal air/fuel mixture requirements of an engine vary as a function of temperature, rpm, and loading. Cold engines require enrichment to counteract the fact that only the lightest fractions of fuel will vaporize at colder temperatures, while the rest exists as globules or drops of fuel that are not mixed well with air. Most of the fuel will be wasted. At temperatures below zero, the air/fuel mixture may be as low as 4 to 1, and at cranking as rich as 1.5 to 1!

Big cams result in gross dilution of air/fuel mixture at idle, which burns slowly and requires a lot of advance and a mixture as rich as 11.5 to 1 to counteract the lumpy uneven idle resulting from partial burning in some cycles. Short-cam engines may run at stoichiometric mixtures at idle for cleanest exhaust emissions.

Coming off idle, a big-cam engine may require mixtures nearly as rich as at idle to eliminate surging, starting at 12.5–13.0 and leaning out with speed or loading. Mild cams will permit 14–15 mixtures in off idle and slow cruise.

With medium speeds and loading, the bad effects of big cams diminish, resulting in less charge dilution, which will allow the engine to happily burn air/fuel mixtures of 14 to 15 and higher. At the leaner end, additional spark advance is required to counteract slow burning of lean mixtures.

At higher loading with partial throttle, richer mixtures give better power by making sure that all air molecules have fuel present to burn. Typical mixtures giving best drivability are in the range of 13.0 to 14.5, depending on speed and loading.

At wide-open throttle, where the objective is maximum power, all four-cycle gasoline engines require mixtures that fall between lean and rich best torque, in the 11.5 to 13.3 range. Since this best torque mixture spread narrows at higher speeds, a good goal for naturally aspirated engines is 12 to 12.5, perhaps richer if fuel is being used for cooling in a turbo/supercharged engine.

The main difference between computer-controlled engines and earlier modes of control is that the computer's internal tables of speed and loading can have virtually *any* desired degree of granularity and can generate spark advance and fueling that are essentially independent of nearby breakpoints, something that is unrealistic with mechanical control systems. (Note: Most engine management control algorithms actually mandate some smoothing of fuel and timing matrices by forcing a certain degree of averaging between adjacent breakpoints.)

Bottom line, computer-controlled engines eliminate compromises of mechanical fuel delivery and spark controls. Multi-port injection eliminates problems with handling wet mixtures in the intake manifold that are associated with carbs, resulting in improved cold running, improved throttle response under all conditions, and improved fuel economy without drivability problems.

THEORY OF ELECTRONIC FUEL AND ENGINE MANAGEMENT

All computer-controlled engines basically function using this four-step process:

1. Accumulate data in the ECU from engine sensors.
2. Derive engine status from sensors that provide data on engine temperature, speed, loading, intake air temperature, and other important factors.
3. Determine and schedule the next event(s) to control spark timing and fuel delivery.
4. Translate computer output signals into electrical signals that directly control actuators (injectors, coil driver, idle-air bypass control, fuel pump, and so on). Go back to 1.

The microprocessor handling this process executes a program that can be divided into the logical strategy that understands sensor input and conditionally computes what to do next based on sensor data and internal calibration tables, and the calibration data acted upon by the conditional logic, which drives the conditional logic to execute in particular ways to achieve precise results in a specific engine application.

EFI control systems can be divided into several types, all of which break down fuel and spark event scheduling based on engine speed, but which use different methods to estimate engine loading:

1. Throttle-position or Alpha-N systems use throttle angle and engine speed to estimate engine loading.
2. Mass airflow systems estimate loading by sensing engine speed and throttle angle and directly measuring air entering the intake manifold.
3. Speed-density systems estimate loading based on rpm and manifold pressure.
4. Some of the newest engine management systems can directly measure engine loading with torque-sensitive strain gauges or even in-cylinder combustion-pressure transducers. Some aftermarket systems can switch between the type of load measurement system, depending on conditions.

Naturally aspirated race cars use throttle position and speed to estimate loading. Alpha-N systems are not bothered by a lack of (or fluctuating!) manifold pressure on engines with wild cams. There is no airflow restriction on the inlet. However—particularly on engines with very large throttle area (such as one butterfly per cylinder)—tiny changes in throttle angle can result in huge changes in engine loading, meaning that the Alpha-N may have resolution problems at small throttle angles.

MAF systems are used by the car companies and are optional on a few sophisticated aftermarket engine management systems like Motec. These systems measure air mass entering the engine based, for example, on the cooling effect of the air on a hot wire held electronically at a fixed temperature. These systems *are* sometimes restrictive to inlet air. Designers may find it challenging to locate the meter so it will be unaffected by reversion pulses in tuned inlet systems.

MAF meters tend to be slow to respond to rapid changes in load (sudden wide-open throttle) and therefore require supplemental means of control to manage engines under rapidly changing conditions. Advantages of MAF control include the ability to compensate for reductions in VE as an engine wears over time and the ability to compensate for modest user modifications that affect VE, such as camshaft changes. MAF systems automatically correct for altitude and air density changes.

Velocity airflow metering uses a spring-actuated door that is forced open by the action of air rushing into the intake through the air meter. A linear potentiometer measures the opening of the door, measuring air velocity into the engine. This must be converted to air mass by considering charge air temperature. This system is similar to MAF metering, with the additional disadvantages of additional inlet flow restriction from the air door and the fact that the meter has no sensitivity to increased airflow beyond the point at which the air door is fully open.

Speed-density systems have also been used by car companies for engine control. These systems may attempt to deduce mass airflow (volumetric efficiency) into the engine with a computation involving engine speed and manifold pressure with added correction for air temperature. Speed-density systems use these same parameters as an index into a look-up table of fueling values at the various operating points, performing calculations to interpolate fueling and spark for operating points that occur between points on the look-up table grid. Look-up speed-density systems, particularly, are not able to compensate for modifications to the base engine configuration or changes in VE caused by wear over time. Speed-density control systems respond quickly to changes in load, offer good resolution at low loading, and do not restrict inlet airflow like air metering systems.

Torque and cylinder-pressure sensing systems are one level closer to the actual functioning of an engine than all of the above systems, in that they *know* when engine management strategies make more combustion pressure, as opposed to a reliance on more indirect phenomena like exhaust gas oxygen levels, engine speed, or even mass airflow. Rather than focusing on optimizing fuel delivery to chase a particular air/fuel ratio (itself typically a deduction based on exhaust gas oxygen content), such systems have the potential to *learn* to make more power, on the fly.

CHAPTER 2
UNDERSTANDING AUTOMOTIVE COMPUTERS AND PROMs

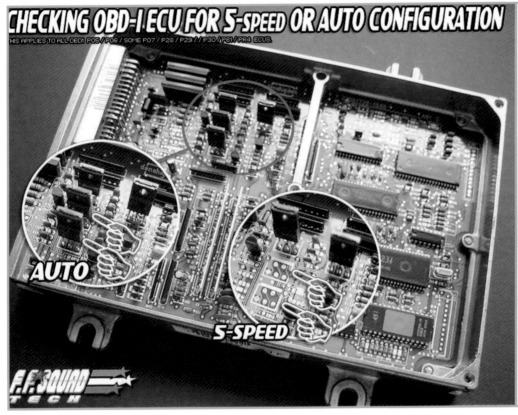

CHECKING OBD-I ECU FOR 5-SPEED OR AUTO CONFIGURATION

THIS APPLIES TO ALL OBD1 P05 / P06 / SOME P07 / P28 / P29 / P30 / P61 / PR4 ECUS.

AUTO

5-SPEED

F.F. SQUAD TECH

This ECU photo is doctored by ff-squad to illustrate how additional components or jumpers can be added to a circuit board to customize an application. Note that the red circled area is stuffed with two additional components in the automatic transmission version of the board versus the five-speed. The circuitry and logic are designed to function differently depending on the presence or absence of these components. ff-squad

It's time to take a harder look at the demands of real engines and how the logic of engine management systems work to satisfy them. Let's see how a sample engine-management computer operates.

A digital microprocessor is essentially an array of tiny switches and circuits that have the ability to do a few simple things, like switch on one wire if voltage is present in another wire. From this type of extremely basic decision-making ability comes all functionality of modern computers: from the simplest arithmetic to complex algorithms that actually simulate human thought. In the same sense that electrons, protons, and neutrons can be combined to make atoms—and atoms combined to make molecules—the basic electronic switching abilities of a digital computer can be combined to perform more complex functions. Arithmetic and logical operations can be combined into a series of steps to solve a problem, like setting the pulse width of an intermittent fuel injection event.

Microcomputers use a quartz crystal that resonates at a particular frequency to coordinate internal operations. Some microcomputers with very fast crystals and circuitry can perform millions or even billions of instructions per second. Using amplification circuitry, a digital microcomputer has the capability to activate or deactivate external circuits. For example, to send a strong voltage to a fuel injector, or stop such a voltage, as well as to read the status of an external circuit, microcomputers store this status internal memory and use this value in additional mathematical calculations.

Random-access memory (RAM) is vital to the operation of a digital computer. Memory once consisted of tiny magnetic cores consisting of tiny ferrous rings surrounding the intersection of three wires. These cores could be written to or read from with electrical currents in the same way that a tape recorder writes and reads electrical signals representing music on a magnet tape.

Today memory is contained in semiconductors, silicon chips containing hundreds of thousands, millions, or even billions of locations that are set either on or off, electrically speaking. RAM, which can be written to and read from, is extremely fast. Data in RAM is lost when electric power is turned off (unless there is a battery backup), so other means are required to store data while the computer is turned off.

Computers use RAM as a scratchpad to store data used by or generated by a computer program. RAM also stores the instructions that tell the arithmetic-logic unit of the computer how to manipulate data (which, collectively, are referred to as a program or software). A simple program might fetch a few data values (numbers), add them, output a signal of a certain duration, start over again, and repeat this cycle again and again. This is the basic function of a fuel injection computer. Different results can be obtained from the same computer circuitry by using other programs or different data.

Programmable read-only memory (PROM) retains its data when power is turned off, but it can only be read from, not written to. PROM is ideal for storing programs and data that essentially never change, such as engine management logic instructions, configuration information such as the number of cylinders, and the basic fuel and timing curves for an engine. PROMs are initially written to, or blown, in a special machine that uses ultraviolet light as the writing medium. PROM may contain numbers that represent commands or instructions to the microcomputer and numbers that are simply numbers (data). The microcomputer transfers these instructions and data into RAM at power-up time.

Electrically Erasable PROM (EEPROM) is slower in access speed and in the past was not practical for storing huge amounts of data, but it could be both read from and written to. Now, the most modern engine management computers store all calibration data in flash EEPROM. Double-E PROM is ideal for storing data values that must occasionally be updated or changed by a computer user or program. For example, configuration data or volumetric efficiency specifications about an engine on an aftermarket programmable engine management computer would change if a user made a cam change.

The arithmetic-logic unit (ALU) of a computer, as you'd expect, performs arithmetic and logical operations based on instructions from a program and generates the condition codes (status bits) that let the program know whether the operation was successful. The ALU performs its operations using its own special high-speed memory called registers. Typically, the ALU loads an instruction that tells it which data to store in its registers, what simple operation to perform on the registers (such as arithmetic shift left, or logical OR), and where to store the result of the operation—plus a data bit to indicate the status (success) of the operation. The ALU is a subset of the central processing unit (CPU). The CPU controls the operation of the computer, and today is usually located in the microprocessor on a single silicon chip.

The I/O (input/output) system controls the movement of data between the computer and the outside devices with which it communicates—in order to have an effect on the external world. A fuel injection computer receives data from sensors via the I/O system that give it information about the world. The computer also controls actuators that enable it to make things happen in the world—such as fuel injectors—via the I/O system. The EFI computer must read data from engine sensors, perform internal processing on this data, and then send data via the I/O system to an injector driver that will activate a solenoid-type injector (actuator) for a precise amount of time.

Analog-to-digital circuitry onboard the microprocessor converts analog voltages to digital numbers that are meaningful to the computer and useful in computations. Some inputs are likely to be directly digital. Digital output from the computer is similarly converted back to analog voltages and amplified to become powerful enough to actuate a physical device such as an injector or an idle air control motor. Some automotive computers have direct-digital outputs for controlling digital peripherals like some boost controllers. The input and output devices, RAM, ROM, A-D circuitry, and the CPU are connected together using a data/address bus that enforces a strict protocol defining when a device on the bus may move data to another device so that the bus can be shared by all.

A typical EFI computer contains programs and data that enable it to compute raw fueling parameters in line with engine volumetric efficiencies. The computer can modify these with additional enrichments or enleanments for cold or hot cranking, after-start operation, warmup operation, and acceleration. Software routines additionally regulate closed-loop idle and cruise air/fuel calibration. ECUs designed to control ignition events also compute speed and loading-based spark advance and knock sensor-controlled ignition retard. An ECU may also contain programming for wastegate control, control of emissions devices such as EGR, and activation of nitrous oxide injection and fuel enrichment.

COMPUTERS AND FUEL INJECTION

In can be useful to compare modern high-performance aftermarket EFI systems to the Hilborn injection system used for decades on all postwar winning Indy cars until about 1970. Hilborn systems are still sometimes used in certain classes of racing because they're cheap (roughly $3,000) and easy to install. A critical thing to understand about fueling an engine is the fact that peak torque occurs at the rpm at which an engine is achieving its best volumetric efficiency (cylinder filling with air). Engines usually make more horsepower at engine speeds exceeding this, but only because they are making more power pulses per unit of time; the cylinders are actually not filling as well with air. Therefore, fuel flow as a function of rpm no longer increases as steeply (the elbow in the curve occurs at peak VE).

The problem with such constant-flow mechanical injection systems is that the mechanical fuel pump, driven by the engine, increases in speed with the engine. Assuming a linear increase in fuel pumped with pump speed (fuel pumped doubles as pump speed doubles), the engine will begin to run rich as rpm increases beyond peak torque and fuel flow continues to increase at a constant rate while airflow into the engine increases more slowly as VE decreases.

Hilborn systems use multiple fuel bypass systems to alter the shape of the fuel curve by returning excess fuel to the fuel tank. The main bypass contains a restrictor (a jet, sometimes referred to as a pill). A secondary bypass returns fuel based on throttle

The Motec M800 is a typical single-board computer box. It consists of a metal enclosure that protects the circuitry and functions as a heat sink to dissipate heat from the power supply and output drivers. It has a big multi-pin connector on one end. Inside is the processor, memory, and A-to-D input-output circuitry. It has no hard drive or other moving parts. Most computer boxes are not designed to survive the harsh heat, vibration, and moisture of an engine compartment. Motec

position—through a barrel valve, which serves to send varying amounts of fuel to the injectors based on slits or orifices progressively uncovered in the barrel valve as throttle position changes. A high-speed bypass uses an adjustable-diaphragm regulator to prevent excessive richness at higher rpm when fuel requirements begin to level off.

Mechanical fuel injection of this sort is great at delivering perfect distribution to all cylinders with great atomization. It is easy to understand how changing jets and adjusting regulators will mechanically alter fuel delivered by a mechanical injection system. It's straightforward to optimize peak power using this system, but it's difficult or impossible to optimize the air/fuel mixture at all speeds and engine loading. But who cares? Race cars are always running with 100 percent power or 100 percent braking.

Calibrating EFI requires a more sophisticated strategy. Optimizing the look-up tables that affect the functionality of computer software in a programmable aftermarket EFI system that controls the operation of electronic injectors requires, at the very least, the ability to understand the basic principles of fueling and the ability to understand and operate the graphical interface on a personal computer or interface module connected to the ECU.

To reprogram the operation of many OEM ECUs, which are specifically designed to discourage tampering and to protect proprietary logic from easily falling into the hands of competitors, requires engineering-type skills. One needs to be able to use complex electronic debugging tools like microprocessor emulators that connect to the ECU circuitry, or be able to deduce how undocumented, proprietary data structures might be organized. One also needs to have the ability to experiment and deduce how these data structures affect the operation of the computer in controlling fueling and everything else in a late-model vehicle—ignition advance, emissions-control devices, turbo wastegates, cooling fans, torque converter lockups, as well as other factors that affect engine power, economy, and exhaust emissions.

This sort of scientific-quality reverse engineering is complex and typically requires bright, experienced engineers or micro-programmers to achieve success. Once the existing OEM ECU operation is well understood, the reverse engineer has to determine the impact of making changes to the internal data tables based on good knowledge and theory and, ultimately, trial and error. This is the sort of thing aftermarket performance chip designers have to go through. Of course, the process is much easier if you can hire an ex-employee of the OEM engine manufacturer (or have a *good* friend who works for one) who can obtain, through whatever means, access to the OEM ECU documentation (this is sometimes called industrial espionage).

The builder of an aftermarket ECU has a task that is at once simpler and more complex than reverse engineering and reprogramming an OEM ECU. An ECU builder doesn't have the hassle of reverse engineering an existing ECU, but must design a complex system from scratch (probably after purchase and close study of existing competitive products).

For a given rail pressure, the amount of fuel squirting out of an actuator (electronic injector) is determined by the injector open time, which is determined by the length of time (referred to as injection pulse width) electricity is supplied to the actuator's solenoid electromagnet under control of the computer. The computer must vary pulse width on the fly in response to engine status—speed, temperature, and loading; throttle position; air density and temperature; and exhaust-gas oxygen content—which it determines in a fraction of a second via sensors.

Fuel actually delivered by the injector per given length pulse is greatly affected by the fuel pressure supplied to the injector. Fuel

Sometimes the computer box itself functions as the user interface. This Carabine ECU from Hilborn is designed for tuners who are more comfortable tuning with a screwdriver than with a laptop. LEDs are used to display information, and the position of the various potentiometers provides information to the internal EMS logic the same way entering numbers from a keyboard defines the operation of the EMS on other systems. This type of user interface is much less flexible and powerful than a laptop but is simpler to tune and can work very well for some applications. Hilborn

pressure is held constant on some systems *in relation to the air pressure of the manifold* into which the fuel is being injected. Other systems, such as the Lucas-Bosch design used on early injected V-12 Jaguars, have used constant-pressure EFI that is *not* referenced to manifold pressure. Some of the newest OEM systems use returnless deadhead fuel supply systems that employ algorithms that vary voltage to the fuel pump to loosely target required fuel mass. Such systems are very tolerant of varying fuel pressure, because they use an overarching architecture that monitors actual rail pressure and provides reliable fuel delivery via ongoing pulse width compensation for pressure changes. None of these architectures are trivial to implement, and the stakes are high because design flaws could burn down an engine.

Many modern sensors use mechanical force or energy to change the resistance of some medium to a flow of electricity. Therefore, the electrical output or flow through the sensor varies continuously in response to some physical phenomena, such as air pressure. Analog-to-digital circuitry (today usually embedded in tiny microprocessors or even located on the main processor chip) converts this voltage to discrete binary numbers that a computer can use in calculations. The computer is programmed to understand the particular characteristics of a sensor. Some sensors output a signal that varies in *frequency* as the parameter being measured varies, a common example being knock sensors and some throttle position sensors.

Fuel injection computers running engines execute logical routines over and over in an endless loop when an engine is operating. The computers sample engine sensors (input and status calculation), compute injector pulse width and other actuator events (output calculation), and direct the operation of the actuators (output to actuators). Beginning again, they evaluate what may have changed from new sensor data, recalculating pulse width, and so on. Endlessly.

Many complex and interrelated variables affect the operation of an engine. The ECU control logic cannot anticipate and process every actual running condition with infinite granularity, and, in fact, ECUs frequently fail in their goal to achieve perfect engine operation. When this happens, there may be many possible causes. Perhaps a sensor is malfunctioning. Perhaps something very subtle is going on in the engine that cannot be detected by existing sensors. Perhaps the control logic is not complex enough (smart enough) to anticipate every possible contingency. Perhaps there is a bug in the system hardware or microcode. Perhaps the computer's actuators or sensors cannot react quickly enough, or perhaps the computer itself is slow enough that it views the engine status as snapshots rather than viewing it constantly in real time. Perhaps the computer, sensors, and actuators are too slow to keep up with changing conditions in the engine so by the time the system can react, take action, and produce a fuel injection/engine management state, conditions have changed so much that now the system can only overreact in a different direction.

We've all seen an engine hunting at idle, unable to achieve a stable operating environment. Future ECUs that have the speed required to consider the state of all engine sensors in near real time—ECUs coupled with fast sensors and fast engine actuators—would eliminate hunting problems by allowing the system to treat the engine as if it were standing still. It would reevaluate conditions and make corrections much faster than conditions can change.

EMS builders design their systems around a microprocessor. Electronic circuits, and especially the microscopic circuits in the microprocessor, tend to be sensitive to heat, which can cause them to fail. The faster the circuitry, the more sensitive it is to heat, which must be dissipated in order to keep the circuitry running without failure. Emitter-coupled-logic and other fast micro-circuitry build up a lot of heat. The problem is made worse when microprocessor logic is so fast that a few extra inches of distance, even at the speed of light will slow things down too much. In this case, circuitry must be packed more densely. Most large, high-speed computers have been designed to operate at about 68 degrees Fahrenheit, plus or minus a few degrees! Many large computers in the past have been water cooled or even Freon cooled, unlike the Windows PC used to word process this book, which was designed to operate at room temperature.

Consider the harsh environment of an automobile engine. The engine compartment may easily heat up to 200 or 300 degrees on a hot day when the engine is working hard. Even the passenger compartment can easily heat up to temperatures approaching 150 degrees or more with the windows shut on a hot sunny day with a black interior. In the winter in cold climates, the temperature might be well below zero when the engine first starts. Vibration can be extreme, and the system is subject to high humidity and possibly dust. Offshore boats, aircraft, military combat vehicles, and racing vehicles may operate in still harsher environments. Bottom line, the microprocessor controlling any engine must be

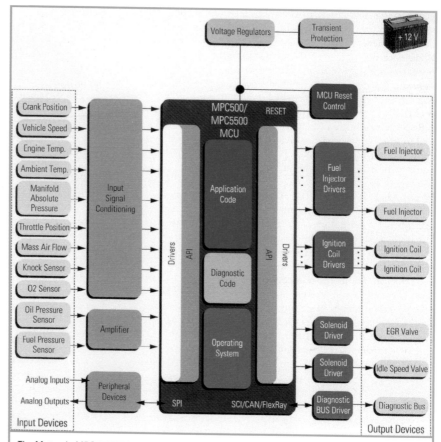

The Motorola MPC500/5500 system diagram illustrates how signals and data from various sensors are conditioned and amplified and then processed by application and operating system logic via application program interface (API) and low-level driver routines. Events are analyzed and evaluated using state-transition logic, then translated into actions via output APIs and actuator driver routines that manage injectors, ignition coil drivers, and various solenoid-driver hardware that in turn manages idle speed, EGR, diagnostics, and other actuators. Motorola

very rugged. This is why the mass-production Intel Pentium-type chips used in many Windows PCs aren't practical for engine management computers. They're not sturdy enough.

ECUs tend to use rugged microprocessors that operate at slow clock speeds to minimize logic errors due to heat-failure processor errors brought on by computation speeds that vary with temperature. Although fast microprocessors can perform millions of instructions per second (MIPS), automotive-type microprocessors generally work at slower speeds. Consequently, some ECUs approach their limit in handling sequential injection within a granularity of a few ten-thousandths of a second, while monitoring and controlling 10 or 15 other engine actuators and sensors on an engine like a motorcycle or Indy car capable of revving to 10,000-plus rpm.

Ford's 1980s-vintage EEC-IV, using a proprietary special-purpose microprocessor, was designed specifically for automotive use. It was considered by many experts to be one of the best ECUs of its time (Ford refers to it as an electronic control *module*). The EEC-IV was built for Ford by Intel, the same company that makes most of the chips for DOS-type PCs. The EEC-IV microprocessor operates at a frequency of 15 to 18 megahertz, and is capable of operating at pretty close to real time speed, evaluating the engine in a feedback loop that takes 15 to 40 milliseconds. This ECU was eventually used in all Ford cars, running various sub-

Manifold
Absolute
Pressure

Input
Signal
Conditioning

Application
Code

Fuel Inject

sets of engine management software and 15 to 56K of memory, depending on the complexity of the engine.

GM's newest 2004 Gen III LS1/LS6 Powertrain Control Module for the Corvette 5.7-liter powerplant is based on a 24-megahertz variant of the Motorola 6833x microprocessor. This microcomputer is a GM design purpose-built for automotive usage. The CPU has 19.5K RAM for working-storage, with access to 1MB read-only memory. The PCM maintains twin 16K blocks of RAM that simulate EEPROM with battery backup, which are used to store the learned adaptives and static values. System logic consists of 112,000 lines of GM Modula source code, which requires approximately 500K of storage in the microcomputer. The fastest engine feedback loops take 6.25 milliseconds. Internal bus speed is limited by the 70-nanosecond speed of the flash device, and the required 1-wait state per memory read, which gives the system an effective internal bus speed of 12 megahertz. The PCM connects to the engine environment with twin 80-pin connectors, an architecture supporting a large number of input (sensor) and output (actuator) devices.

In order for a microprocessor to manage engine control devices, the ECU contains circuits called *latches* (something like relays), which respond to a control signal from the microprocessor by turning on, for example, a 12-volt signal with a square wave format to activate injector drivers. Other circuits convert the continuously varying analog voltages signal from sensors and convert them to digital format for the microprocessor. Today, these A/D converters are usually integrated circuits and may in fact be built into the microprocessor itself (the Ford EEC-IV ECU, for example, uses a 10-bit A-D converter on the microprocessor chip itself). The engine management microprocessor contains multiple I/O *channels* that enable digital data—perhaps from a A/D converter—to be stored directly in the microprocessor's memory.

Injector drivers like the Motorola MC3484 are capable of producing 4 amps of *peak* current in less than a millisecond (a thousandth of a second) to bang open an injector quickly. In 200 microseconds the driver then would drop the current back to 1 amp to hold the injector open. Peak and hold drivers take advantage of the fact that it takes much less current to hold open an injector once the solenoid is initially energized. Such drivers consume much less power than saturated circuit drivers that hold continuously at full voltage, yet are able to open the injectors rapidly. The MC3484 uses five leads for power in, power out, ground, trigger, and reference voltage. Any time voltage in the trigger exceeds the reference, the driver will immediately route voltage from power in to power out, which can be used to fire an injector.

An operating ECU is endlessly executing its way through a loop of software code to sample sensors, evaluate their status, make output calculations, and activate control devices. If-then-else branching allows the microprocessor to execute various sections code in the loop, bypassing others, depending on conditions. Data affecting the operation of the microprocessor comes from internal look-up tables (fuel maps, VE tables, and so on) and sensors. There are various sensing strategies used to evaluate engine airflow; this logic acts upon the internal ECU data structures.

Most aftermarket ECUs are speed-density systems that deduce the number of air molecules entering the engine (hence, fuel requirements) from engine speed and manifold pressure—or by using speed and pressure and temperature as indexes into look-up raw fueling tables. Air sensors, the first aftermarket ECU, directly measure air mass via an airflow meter and can calculate fuel requirements based on this by consulting VE tables. A few aftermarket ECUs like the old Haltech F7 and the analog Holley Pro-Jection deduce airflow entering the engine based on engine speed and throttle position. Many aftermarket systems can be configured to do this.

Virtually all OEM ECUs and some aftermarket ECUs have the ability to operate in closed-loop mode, which means they can factor in the actual air/fuel mixture achieved in the last injection event, as measured by an exhaust gas oxygen sensor to influence pulse width in future injection events (which is necessary to meet emissions requirements). Some ECUs effectively modify their base fuel data tables as they learn what works in a particular engine (or with a particular driver) or as injectors and other parts age or wear out (disconnecting the battery normally erases this learned behavior).

VOLUMETRIC EFFICIENCY ENRICHMENT

An engine management computer's PROM is loaded during manufacture or initial setup with a table of numbers that correspond to the volumetric efficiency (VE) of the engine across the entire operating range. VE represents the percentage of cylinder filling that occurs compared to the displacement of the engine.

Each VE value will be interpreted or translated by the EMS microcomputer as a number representing the amount of fuel required at a certain combination of rpm and engine loading that—relative to fuel requirements as 100 percent VE—represents enrichment or enleanment requirements according to engine speed and loading. Given that engine speed and engine loading or airflow are both infinitely variable across the possible range of operation, there is, of course, a theoretically infinite number of such values. In reality, a VE table usually contains from several dozen to several hundred values logically arranged in rows and columns corresponding to rpm and engine loading.

In the case of a programmable system, the VE table numbers may have been originally entered one-by-one into the EMS microcomputer's EEPROM on the same or a similar vehicle by someone typing them in on a keyboard or using graphical means to get the data into a personal computer connected to the engine management ECU by data link, with the data then optimized during R&D testing operations on a chassis or engine dynamometer. On the other hand, Windows software running on a PC may have automated the original VE data entry, allowing a tuner to enter a few values and then copy or interpolate the data to other values in between using a Fill or Copy feature, as a human might use a ruler to draw a continuous line through several points on a graph.

On new cars and trucks, engineers or inventors build one set of PROM or EEPROM data for a particular engine-vehicle combination and simply duplicate the data over and over for other vehicles of the same type. This data is then referred to as the engine map or calibration. If you have a common variety of hot rod engine that is stock or equipped with common modifications, you'll probably find pre-configured maps for your engine available from the engine management system manufacturer. For the R&D engineer or performance enthusiast developing a unique engine configuration, it still almost always makes sense to start with fuel and timing maps from as similar as possible an engine, perform any obvious configuration changes to scale for widely divergent injector sizes and so forth, and then modify the engine management maps by laptop PC while the engine is running. This enables the tuner to see immediate changes in the way the engine performs in response to the map changes.

PROM-based factory fuel-injection systems (designed to be economical rather than easily user-programmable) typically require a device called an emulator for making dynamic calibration changes while the engine is running. An emulator plugs into the PROM socket of the OEM injection microcomputer in place of the stock PROM and contains a socket to accept the PROM, which is uploaded into emulator RAM for modification, test, and eventually offloaded into a new replacement PROM.

Emulators enable an operator to modify PROM data interactively, instantaneously testing the effect of changes on a running powerplant. Virtually all newer aftermarket EMS and newer factory systems store calibration maps in EEPROM, which can be overwritten from the OBD-II diagnostic port if you know the security seed. If the EEPROM is socketed and can be unplugged, it can be hacked with an emulator exactly like a PROM.

Some VE tables are actually filled with injection pulse width data, which must be translated to VE by dividing each number into the derived pulse width for 100 percent VE. Some engine management systems such as the Electromotive TEC[3] enable a tuner to toggle the user-presentation of the table between injection pulse width numbers, VE-type percentage numbers, and plus or minus offset from 100 percent VE (achievable on performance engine boosted with power-adders or intake and exhaust tuned to take advantage of high-rpm resonation effects).

Let's take a slightly harder look at what is going on behind the scenes in the VE table section of an engine management microcomputer's calibration.

The stoichiometric air/fuel ratio of 14.7 pounds of air to 1 pound of gasoline is theoretically ideal for combustion. Under the right conditions, this mixture will burn to release all chemical energy stored in the fuel, leaving no unused oxygen and no unburned hydrocarbons. Stoichiometric combustion produces energy, carbon dioxide, and water vapor—nonpolluting and harmless. Theoretically. Under ideal conditions that naturally assume factors such as enough time for combustion to finish completely, perfectly mixed air and fuel, and so on.

But what if the goal is to produce maximum fuel efficiency, especially at part throttle? In the real world of fires burning at high speeds among the swirling mass of air and fuel atoms in the combustion chamber of an engine, it is easy to miss a few atoms of fuel or oxygen here and there—particularly when throttle position and engine speed are changing rapidly and the air and fuel will *not* be perfectly mixed.

Since oxygen is virtually always the scarce component that limits power production, if the goal is maximum power output from an engine, you want to be sure there is enough fuel present in the air/fuel mixture so that every molecule of oxygen that has managed to get drawn or pushed into a cylinder manages to be used in combustion. Therefore, for peak power you should richen the mixture with extra fuel to shorten the odds you don't miss using any of the oxygen. (It's a little like inviting extra girls to a dance to improve the odds that every guy will find a partner!) In the same way, if you're trying for maximum efficiency, you definitely don't want to waste any unburned fuel by blowing it out the exhaust. So to make sure every molecule of fuel gets burned, you'd like to lean out the mixture (increase the ratio of oxygen to fuel) to improve the likelihood of making use of every last fuel molecule (like inviting extra guys to the dance).

There is a logical correspondence between volumetric efficiency and enrichment. The engine operates at a higher volumetric efficiency (cylinders fill more completely) when working harder (wide-open throttle), which is exactly when you want best power. At part throttle and lower speeds, the engine operates at a lower volumetric efficiency—just when you want economy.

It is a relatively simple matter to construct a microcomputer-based table of volumetric efficiencies at various engine rpm and loading values. The microcomputer then reads rpm and engine loading from its sensors and uses these to index into the VE table whose values correspond exactly to enrichments or enleanments required at a certain rpm and engine load. When the engine is operating at rpm/load points that fall between set points in the VE table, the computer is able to locate the closest VE table values and use geometric formulas (typically a running-average linear interpolation) to estimate exact volumetric efficiency and, therefore, mixture adjustment required.

A carburetor approximates this sort of VE-based air/fuel adjustment by sizing the main jet for primary regulation of fuel delivery to the venturi, trimming this fuel flow by selecting air corrector jets and emulsion tubes of precise size to progressively dilute the main jet fuel stream with air in order to limit fuel flow (which otherwise increases faster than airflow as engine speed and loading increase). A power valve in the carburetor opens under high manifold pressure (high VE) to allow additional fuel to bypass the main jet and flow to the venturi. As you can imagine, these physical correction systems in a carb are a blunt instrument compared to the precision of an electronic engine management system.

CHOKING ENRICHMENT

The microcomputer's internal tables also contain auxiliary tables of choking values that correspond to special enrichments required when a driver starts a cold engine under conditions in which fuel would not vaporize completely (any time the engine is not at full operating temperature). These enrichment offsets are independent from each other and consist of cold-cranking enrichment, after-start enrichment, and warm-up enrichment. The computer adds this enrichment to normal injection pulse width.

Carbs typically provide cold-start enrichment utilizing a choke plate located above the venturi like a second throttle, which is designed to physically restrict airflow through the carb. Cold-air activates this choking action based on its effect on a bimetallic spring (thermostat) connected to the choke plate. The closed choke plate increases vacuum in the area of the venturi to create additional pressure differential there compared to atmospheric—thereby forcing additional fuel out of the float chamber and into the inlet air stream at the venturi (producing a vastly enriched air/fuel mixture).

ACCELERATION ENRICHMENT

Additional tables in engine management computer PROM contain, in the first dimension, values corresponding to enrichment required on acceleration, usually as a percentage of pulse width, which can usually be varied at a function of rpm and loading on the latest, most sophisticated engine management systems.

A second dimension of enrichment values in this table provides for specification of rate of decay in acceleration enrichment once the throttle plate is no longer turning. Acceleration enrichment prevents a flat spot on sudden acceleration. Many automobile carburetors provide transient enrichment with a diaphragm pump activated by the throttle shaft that squirts raw fuel directly into the throat(s) of the carb while the throttle is opening.

ENRICHMENT SUMMARY

Enrichment values stored in tables in PROM, therefore, include choke enrichments, acceleration enrichments, and volumetric efficiency enrichments (and enleanments). These values together represent a fueling map for a certain engine. Density enrichments or enleanments for air temperature and altitude (not needed in MAF sensor systems like SPIF), and exhaust gas oxygen sensor enrichment and enleanment values originate in the electronic sensors, not in tables (although the ECU may contain translation tables for these). The microcomputer sums all enrichment values

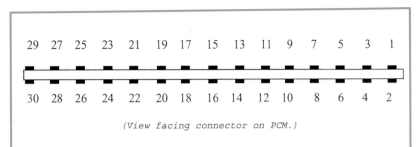

(View facing connector on PCM.)

The Motorola 8061 and other EEC processor components support extensive diagnostics and control capabilities through special hardware pins that Ford was kind enough to incorporate in a diagnostic connector accessible externally by plugging into the J3 test port in the EEC-IV enclosure. The performance aftermarket has used the J3 port extensively to hack into the EEC system and modify calibration and configuration data that affects engine tuning and EMS operation.

Ford EEC-IV ECM TEST PORT (J3)

PIN	SIGNAL / FUNCTION	MCU PIN	CPU 8061	RAM 81C61	EPROM 8763	notes
29	PWR GND	40,60				
27	VPWR	37,57				battery +
25	DI		57	22	22	
23	IT		58	21	21	
21	STROBE\		53	20	20	
19	D7		68	10	10	
17	D6		67	11	11	
15	D5		66	12	12	
13	D4		65	13	13	
11	D3		64	14	14	
9	D2		63	15	15	
7	DI		62	16	16	
5	D0		61	17	17	
3	KA5V				7	
1	VREP (+5)	26				
30	PWR GND					
28	VPWR	37,57				battery +
26	BSO					
24	BS3					
22	TSTSTB\				1	1K TO +5 only
20	PROGRAM					
18	RAMDISABLE\					
16	EPROMDISABLE\				2	10K TO +5 only
14	ERASE					
12	MRESET\				9	
10	RESET\		3	9		
8	PAUSE\		60			1K TO +5 only
6	EXTINT					
4	(high for access)					IC4-74001 pin 13
2	VCC	25				

ECM Test Port (J3) Pinout

to arrive at a total enrichment for the current fuel injection event that is represented by G (for the Greek letter *gamma*). The total enrichment (or enleanment) is an offset from the theoretical (stoichiometric) air/fuel ratio based directly on air mass.

EQUATIONS

The representative basic equation (based on Electromotive system documentation) that the microcomputer solves to compute injection pulse width is:

Pulse Width = [(MAP voltage ÷ 5) x UAP] + POT

in which pulse width is obtained by modifying the load percentage of the MAP sensor voltage (0 to 5 volts, correlating to air mass) by the user-adjustable pulse width (UAP), and modifying the result by pulse width offset time (POT). UAP is defined as the pulse width required at peak torque (full load) for a given engine displacement with a given injector size, and is the longest pulse width the engine will ever require. Pulse width offset time is a modification to correct pulse width at idle; it is an additive fudge factor that must be non-zero if the required pulse width at idle varies in any way from linearity between the UAP at full loading and the theoretical zero pulse width at zero loading (even idle requires more than zero loading, since turning the engine requires overcoming friction and other losses).

The above equation assumes no rpm-dependency for the raw fuel curve derivation, since pulse width for an engine with a perfectly flat torque curve will not vary with speed. Therefore, an engine working at 100 percent load would require 100 percent of the pulse width defined by UAP. An engine operating at 50 percent load would require 50 percent UAP, and so on. In practice, any engine in which the torque curve can be approximated reasonably well with a straight line can be run reasonably well with this simplistic assumption.

What's more, the linearity assumption of pulse width with rpm breaks down on engines that have torque curves that are very nonlinear with engine speed. Turbocharged engines that do not start to make boost at lower rpm ranges and super-high-output racing engines with poor low-end torque are two examples of such engines. A VE table allows the EMS microprocessor to make corrections to the basic pulse width calculation at any ranges of engine speed and loading where the volumetric efficiency of the engine strays badly from linearity using the following equation:

Pulse Width = [(MAP Voltage ÷ 5) x UAP x (VE Absolute % ÷ 100)] + POT

In reality, any EMS microprocessor must actually correct MAP values to mass airflow by correcting for temperature (on MAF-based systems this is unnecessary, since the MAF value is already mass airflow, and requires no correction). The computer must always adjust pulse width to include injector turn-on time (ITO) and changes in ITO due to changes in the system battery voltage (BTO) by directly summing ITO and BTO into pulse width.

The EMS computer must also factor in transient enrichment offsets from throttle position changes (TPS), engine coolant temperature (CTS), intake air temperature (IAT), air/fuel corrections based on exhaust gas oxygen (EGO), and starting enrichment requirements (SE), all of which add percentage changes to VE absolute percentage except battery voltage changes, which add a specified amount to injection pulse width in response to battery voltage changes. After all enrichments, Electromotive's pulse width derivation is as follows:

Pulse Width =[(MAP Voltage ÷ 5) x UAP x (VE Absolute% ÷100) x TPS% x CTS% x IAT% x EGO% x SE%] + POT + BTO

PROMS

Programmable read-only memory (PROM) and electrically erasable programmable read-only memory (EEPROM) are used to store data and instructions for use by a computer's central processing unit. All microcomputers without a disk drive must have PROM or EEPROM to store the instructions and data they need to begin doing work on power-up, which includes all ECUs running engine management systems.

Typically, an ECU is configured with the code (machine instructions that are executed over and over again when the engine is running) and some data tables or configuration parameters already stored in PROM. The EMS logic makes use of such data in calculations. Clever software routines can be designed to be very data-dependent in their operation. Sections of code are executed only conditionally, on the basis of separate data parameters.

In the case of aftermarket EFI computers, there will be some method that can be used to get additional data into the EEPROM look-up tables, often from a Windows laptop, or perhaps a special keypad and LCD display that is bundled with the other EMS hardware. This facility lets the user customize the system's data to the individual engine. Software routines will use certain data for EFI calibration computations, and treat other data as switches. Based on the value of a certain byte or bit of data, code might (or might not) be executed to handle, for example, computation of additional fuel enrichment when nitrous injection is armed. Modern EMS ECUs are very data dependent. They will behave in many different ways depending on the data in their PROM/EEPROM.

When GM designed its original 1980s-vintage performance EFI, called Tuned Port Injection (TPI), they were kind enough to design the PROM so it could be changed easily. GM TPI PROMs plug into a socket that is accessible without even violating the ECU's enclosure. By contrast, Ford's EEC-IV uses a PROM that is soldered in place, and therefore not easily replaceable. Fortunately, an EEC-IV ECU includes a multipin diagnostic plug that can be used to load code and data that could be used in place of the code and data in the onboard PROM. Aftermarket entrepreneurs have designed small modules equipped with replacement PROM chips that plug into the EEC-IV diagnostic port to supersede the onboard EEC-IV PROM.

Older OEM engine management systems were often set up from the factory with PROM calibrations that caused them to behave very conservatively. They were designed to be foolproof and to keep negligent or ignorant operators from hurting their engines with poor driving methods, poor quality gasoline, and poor preventive maintenance. However, a knowledgeable operator who is willing to take more care, willing to operate a little closer to the edge, to use good gas, to avoid lugging the engine, perhaps to suffer degraded gas mileage or emissions would discover it is often possible to make more power available via a higher state of tune than stock, which could be arranged by simply substituting a PROM with revised data.

The newest vehicles are equipped with engine management systems that are highly optimized, and it is uncommon to find an engine vehicle combination where calibration changes alone can create significant additional power. And all OBD-II (Onboard Diagnostic Level II) engines sold in the United States since 1996 are much better than older systems at detecting performance modifications that might adversely affect exhaust emissions.

The good news is that digital EMS logic has always been designed in such a way that the code takes into account many different possible (and configurable) characteristics of a vehicle and engine package in managing the engine. The bad news is that if the operator has made major internal changes to the engine, it may be mandatory to change the PROM data, the ECU itself, or add auxiliary electronics or mechanical changes to fuel-delivery and ignition components so the engine will operate properly.

In fact, it is essential to understand the operating envelope of the ECU logic and the specific calibration data. Many factory EFI cars are designed with logic and tuning that provides some built-in flexibility for the actual volumetric efficiency of an individual engine. Minor changes to the VE, perhaps a low-restriction air filter, will often work well with an MAF-based EFI system, since the computer has built-in ability to compensate for slight VE changes. But after installing an aftermarket PROM, the EMS might be operating so close to knock limits, suboptimal air/fuel ratios, or other constraints that a change in VE or other engine characteristics could cause it to operate poorly. Without access to system documentation, the tuner has no idea what the envelope looks like and where the system's flexibility ends.

Therefore, it is essential to get good consulting from the people who know, such as factory performance experts at places like the Ford Motorsport Tech Hotline at (586) 468-1356, or aftermarket wizards, who have reverse engineered or otherwise acquired internal system knowledge and have tested modified parts and PROMs on dynos and on the street. The whole package must work together or you may end up worse off than you started if you tamper in any way with a factory EFI vehicle, either in PROM changes or in engine changes.

Many aftermarket entrepreneurs offer packages of replacement PROMs and parts that are designed to work well together on a specific car. An increasing number are approved for legal use on the street, which is nice, considering that all parties concerned—from manufacturer to jobber to user—are potentially liable for any vehicular tampering that might affect emissions.

HOT ROD PROM PACKAGES FOR EFI SMALL-BLOCK CHEVYS

"Our first level package is the easiest to get started with," wrote TPIS in the early 1990s, "because it involves inexpensive parts and simple techniques. But if you're looking for more muscle, check out Levels II, III, or IV. And when you're ready to blow the doors off everybody in town, look into Level V, which is completely customized to your individual car." In the new millennium, TPIS is still in business and still focused on selling hot rod and replacement parts for performance fuel-injected Chevy V-8s.

The name TPIS is an acronym derived from the company's original name, Tuned Port Injection Specialties, and its original business is hot rod computer and specialty performance parts for the tuned port injection EFI small-block Chevy V-8 used in F-Bodies, Corvettes, and a few other vehicles from 1985 to 1993. The follow-up 1993 to 1997 LT1 Chevy small-block V-8 engine had a different (and somewhat compatible) induction system but was still, in most respects, a traditional small-block Chevy V-8 (which is *not* true of the newest generation LS-1 powerplants that are entirely new in virtually every way). All TPI V-8s built from 1985 on and some early LT1 powerplants had factory EMS computers with replaceable PROMs.

Arizona Speed and Marine focused on much the same market of fuel-injected Chevy-type V-8 performance. There were so many L98-type engines and vehicles built that there are still many driving around on the streets of America, and many hot rodders are still working with them as raw material. And ASM can still supply custom PROMs for these engines.

What was and is possible with PROM replacement performance packages on performance-oriented Chevy and Ford small-block V-8s?

Aftermarket PROM vendors have typically advertised a higher state of tune for targeted stock vehicles and recalibrated fuel and ignition curves for modified vehicles. For people with ongoing performance projects, the upgrade-PROM vendors have sometimes protected the PROM investment by enabling customers to upgrade

Manifold Absolute Pressure → Input Signal Conditioning → Application Code → Drivers → Fuel Injec

The new 2004 Corvette LS1 PCM incorporates 24-megahertz Motorola 6833x processor power and a large 1 MB of ROM. Drivers sink heat directly into a finned aluminum casing capable of surviving in the harsh environment of an engine compartment.

to another new PROM if they make additional modifications to an engine—for just a small upgrade fee. Most of the companies selling PROMs offer phone consulting to users in order to help them get the combination of parts and PROM right.

Typically, a first stage TPI or LT1 PROM might be designed to work well with a low-restriction air filter, a low-temperature thermostat, a throttle-body air foil, MAF screen removal, and modified air filter housing. TPIS claims a 20-horsepower gain with the above package on a 5.7-liter TPI V-8. This translates into a 0.6-second faster quarter mile with 2-miles per hour faster top end. It is very important to use the right PROM with lower-temperature thermostats (since the ECU could otherwise assume that the car is still warming up) and keep warm-up enrichment activated, which wastes fuel, cuts power, and wears out the cylinders fast from wetting down the cylinder walls.

Stock engine management tuning in the 1980s and early 1990s tended to be very conservative (safe) to protect automakers from warranty claims. Engineers would design an overly rich wide-open-throttle fuel map in order to protect the engine from knocking and thermal stress, and to guard against aging fuel pumps and injectors eventually producing dangerously lean conditions. Typically, stage one PROMs had calibrations affecting wide-open throttle, advancing the timing for premium fuel and improving the mixture (which, in cars like the 5.0-liter Mustang, involves *leaning* the mixture that is too rich with factory fuel curves). In addition to retuning fuel and ignition curves, a replacement PROM chip might also change the circumstances under which torque converters lock up, eliminating jerkiness in shifting into or out of high gear.

With a few additional minor pieces and a stage two PROM, companies like TPIS and ASM advertised power gains of up to 45 horsepower, cutting a second off the quarter mile.

A third level typically involved low-restriction exhaust (headers, free-flow mufflers, and possibly removal of the catalyst for off-road use), additional MAF modifications, and yet a different PROM.

The next level was designed to significantly improve engine breathing with intake porting, larger (and sometimes shorter)

runners, higher lift rockers, and a new PROM, yielding as much as 100 horsepower and a 2-second faster quarter. Drivability could still be excellent, with across-the-board performance improvements at all rpm levels.

A stage five PROM package might add ported heads, a hotter cam, and an appropriate PROM, resulting in 130 horsepower over the stock 305 or 350 Chevy TPI motor, with quarter times in the low 12s and 112.5 miles per hour in a 'Vette. These sorts of modifications are designed to work very well with big-displacement 383 or 406 Chevy small-block motors, which *need* the improved breathing and fuel. TPIS used to advertise that their demonstrator 'Vette achieved almost 23 miles per gallon at speeds as high as 80 to 100 miles per hour, yet turned a 12.1 at 112.5 in the quarter with a stock displacement 350 motor.

With a performance PROM calibration installed, it is essential to treat such a vehicle as the more highly tuned beast that it now is, filling up with the best gasoline around, listening for knock, and giving the vehicle excellent maintenance. It is essential to check with the manufacturer before assuming that any replacement chip would work with a different engine/vehicle combination (for example, installing a high-output Camaro chip in an LG4 or standard 305 V-8 engine).

PROM changes have always required you to know what engine you have. This sometimes requires checking the vehicle identification number (VIN), which contains a code that indicates the type of original engine installed at the factory. If engine changes have been made (or if you think the PROM calibration is nonstandard), check with the manufacturer, and have all internal and external engine documentation available. Most PROM vendors that are still in business (and most probably are) should maintain documentation and serial numbers for all PROMs they've sold. If there is any doubt, they can probably still read your PROM and tell you what you've got years down the line.

PROM INSTALLATION

GM TPI PROMs can typically be installed in less than 30 minutes using a small flat head screwdriver, a Phillips screwdriver, a 1/4-inch socket, and sometimes 7- and 10-mm sockets. Locate the ECU (which GM calls an electronic control module or ECM), remove the mounting screws, and pull it out for easier access. GM computers have a small cover that allows access to the PROM. Remove it and gently unplug the stock PROM from its socket (it is *very* easy to bend or break the multiple prongs, especially without a PROM-puller tool). Now *carefully* plug in the new PROM and reinstall the cover plate.

In the case of Ford EEC-IV vehicles, you'll be adding a power module available from companies like Hypertech, which plugs into the diagnostic port on the EEC-IV computer, superseding the stock chip that stays in place. This procedure requires the same tools as a GM PROM change, plus maybe a 9- or 10-mm Torx bit. Locate and dismount the ECM. Remove the service port cover and side screw from the computer (this is on the end of the computer with the warranty label). The power module plugs into the end of the ECM and includes longer screws that hold it in place using the side screw holes. Reinstall the computer in its place and go have fun.

Some Japanese ECMs require soldering and additional electronics installed by an expert technician before the PROM calibration can be changed. But once these changes are made, a PROM change involves unplugging and removing the ECU, opening it, and swapping in a new PROM chip (see the Toyota MR2 photography in this book).

CHAPTER 3
SENSORS

Sensors are the eyes and ears of an engine management ECU. An engine management system makes decisions about fuel injection and ignition events based on data. Such data sometimes exists simply as tables of stored numbers in memory, for example a chart of volumetric efficiencies for a specific engine. This type of data doesn't change unless someone recalibrates or reprograms the ECU. But data may also exist in the form of numbers or values in memory that are more or less constantly being refreshed according to the status of various engine sensors.

Sensors assign an electric voltage to the status of external systems or events such as a change in the temperature of the cylinder head. A sensor voltage is often converted from an analog physical magnitude to a discrete digital number by A-to-D circuitry (usually now onboard the microprocessor) and made available in random-access memory for processing, exactly like the table data discussed above. Some sensors directly generate digital data ready to be input to the processor. Sensor data heavily affects both which software instructions are executed by the microprocessor as well as the results of the logical routines that are executed.

Let's take a look at how sensors operate and their effect on engine management. A variety of types of sensors may be used for engine management:

- Positional sensors
  - Cam angle
  - Crank angle
  - Throttle position
- Exhaust gas composition sensors
  - Narrow-band O_2
  - Wideband O_2
  - Other gases
- Temperature sensors
  - Coolant
  - Air
  - Oil
  - Exhaust gas
- Torque sensors
  - Strain gauges
- Pressure sensors
  - Manifold absolute pressure
  - Barometric pressure (BARO)
  - Fuel pressure
  - Cylinder combustion pressure (pressure transducers)
- Air-metering sensors
  - MAF
  - Vane airflow
- Knock sensors
- Other
  - Vehicle speed sensor
  - Wheel speed
  - Engine oil
  - Accelerometers
  - Actuator-status sensors (for OBD-II; did the actuator work as it should have?)

POSITION SENSORS
Cam Sync/Crank Angle

Hall Effect sensors, along with optical and magnetic sensors, are used by engine management systems to determine the position of the engine. They may be located in the distributor, on a camshaft, or on the crankshaft. They have been used in the primary circuit of a distributor ignition to fire the coil, to monitor engine rpm, or to time sequential injection or fire individual coils in relation to firing order on a direct coil-on-plug or waste-spark ignition.

Monitoring engine position from the crankshaft is the most accurate method because slop in the timing chain or timing belt wear and stretch will allow the camshaft and distributor to wander and scatter in relation to crankshaft position. On older engines with adjustable distributors, the distributor could be clocked, or rotated, to adjust ignition (or injection) timing, meaning distributor position is only as accurate as the skills and care of the tuner clocking the distributor. However, exactly determining crankshaft position is ambiguous in terms of the firing order, because it says nothing about whether the engine is on the compression or exhaust stroke.

Modern engine management systems use a sync-pulse engine position sensor on the camshaft/distributor to alert the EMS to where the number one cylinder is in the firing order. EMSs then use a precision position sensor on the crankshaft to precisely time all other engine spark and injection events.

Unlike conventional magnetic sensors (which respond to changes in the magnetic flux as a flying magnet passes the sensor, which produces an alternating current output that varies in voltage with speed), the Hall Effect principle senses the actual magnetic field strength. Hall Effect sensors produce a high-quality signal of essentially binary voltage that changes state abruptly from constant maximum voltage to nearly zero and vice versa regardless of rpm. Because of this, Hall Effect sensors are sometimes referred to as switches rather than sensors because the on-off square-wave voltage signal they produce is directly usable by digital circuitry without requiring conditioning or interpretation. As such, the sensor determines magnet position regardless of rotational speed and is less sensitive to variations in gap between the sensor and wheel. Timing scatter is greatly reduced, and timing variations with rpm are eliminated.

Hall Effect sensors normally have three wires for ground, output signal, and reference voltage (from the vehicle's ECU, 5

Temperature sensors change resistance with temperature. This sensor is optimized to respond quickly to changes in air temperature. Coolant or oil temperature sensors have a solid metal bulb that must be immersed in the fluid being measured. Bosch

or 12 volts). Like flying magnet pickups, Hall Effect sensors make use of the interaction between a magnet and ferrous metal coming into close proximity. But Hall Effect sensors are based on a phenomenon discovered in 1879 by the American scientist Edwin H. Hall. Hall discovered that when an electric current was applied to metal inserted between two magnets, it created a secondary voltage in the metal at a right angle to the applied voltage. In a modern Hall Effect sensor, the voltage change occurs in a silicon chip placed at a right angle to a magnetic field.

When a magnet moves into close proximity of the Hall Effect sensor (or when a moving metal shutter blade uncovers a magnet located near the sensor), the sensor's output voltage suddenly jumps to full scale. When the magnetic field is removed, the voltage abruptly drops to zero. Additional circuitry conditions the input power that is supplied to the sensor and amplifies the output signal. Auxiliary circuitry can also produce a sensor that outputs 0 volts in the *presence* of a magnetic field, otherwise full-scale voltage. A notch in a rotating pulley, a rotating gear tooth (or a *missing* tooth!), or even a rotating magnetic button can serve the same purpose as a shutter blade to disrupt a Hall Effect sensor's magnetic window and trigger the switch in an automotive engine management system.

Depending on how its circuitry is designed, a Hall Effect sensor may be normally *on* or *off*. GM's crank position sensors are of the normally *on* variety, producing a steady voltage output when the magnetic window is open or unobstructed. Voltage output drops to near zero when a blade enters the magnetic window and blocks the field. In Ford distributor-less ignition systems, profile ignition pickup (PIP) and cylinder identification (CID) Hall Effect sensors work in the opposite manner: When a shutter blade passes through the window, blocking the magnetic field, the Hall Effect sensor's internal circuitry switches the output signal from near zero (*off*) to maximum voltage (*on*).

Throttle Position

TPS sensors are variable resistors (potentiometers) that change resistance as a shaft is physically turned. When linked to the butterfly of a carb or throttle, a TPS indicates the position of the throttle via the specific voltage output. By evaluating the change of throttle position over time, the computer can determine the *rate* of throttle opening or closure. Some older fuel injections systems such as the Lucas/Bosch EFI on early Jaguar V-12s also incorporate throttle microswitches, or nose switches, that indicate idle (closed throttle) or wide-open throttle (WOT). The TPS may be attached directly to the throttle shaft on the outside of the throttle body or carburetor, or possibly inside the carb or throttle body.

The TPS is typically a three-wire, variable resistor that changes resistance as the throttle opens and closes. The ECU provides the TPS with a voltage reference signal (VRef), usually 5 volts, and a ground. As the position of the throttle changes, the corresponding change in the TPS's internal resistance alters an output voltage signal that returns to the ECU via the third wire. Therefore, the computer receives a variable voltage signal that changes in direct proportion to the throttle position. Most TPS sensors provide slightly under 1 volt at idle with the throttle closed, and up to 5 volts at wide-open throttle.

On virtually all modern emissions-controlled passenger cars, the TPS is used to activate transient fuel enrichment on sudden throttle opening, in order to prevent bog—like the accelerator pump on a carb or enleanment/fuel cutoff on deceleration.

On some normally aspirated aftermarket EFI systems (such as the Holley Pro-Jection) and certain racing EFI systems (Alpha-N or throttle-position engine management systems), the ECU combines throttle angle with engine speed sensing to estimate air

Magnets flying past a coil induce a voltage spike that a computer can detect and time stamp for use in determining engine position while running. Most engine position sensors use Hall Effect or magnetic sensors (though occasionally you'll find an optical engine position trigger). MSD

entering the engine, which translates to engine loading and, hence, fuel injection requirements.

A big throttle angle for a given engine speed implies higher manifold pressure/higher engine loading (and, therefore, more air entering the engine). Throttle position engine management systems tend to be very responsive and therefore good for racing. This system has no ability to automatically correct for changes in engine VE unrelated to throttle angle. This makes it entirely unsuitable for turbocharging applications where a given throttle angle and engine speed cannot distinguish between a wide range of possible engine loadings.

Exact TPS calibration is critical on throttle position (Alpha-N) load-sensing engine management systems. The system specification requires a specific voltage signal that tells the computer that the throttle is closed. Calibration procedures at closed throttle require loosening the TPS and rotation while watching the TPS resistance with an ohmmeter. At the correct voltage, the tuner would tighten down the TPS mounting screws.

Today, many modern aftermarket engine management systems allow calibrating the TPS with software by manually instructing the EMS when the throttle is closed and then when it's wide open so the system can correlate specific sensor electrical outputs with these states. On some speed-density or MAF systems the TPS does not require calibration, since the TPS is only used to detect changes in throttle position, for transient enrichment—not absolute position.

Some ECUs require a linear potentiometer TPS, which is more expensive than a nonlinear potentiometer TPS (where resistance in the TPS might, for example, change rapidly with initial increments of change in throttle angle and change less with additional throttle-angle changes the farther the switch moves toward wide open). Today, many aftermarket ECUs can understand nonlinear TPS data; a mathematical function corrects the nonlinear voltage from the TPS to accurate specific throttle angles for use in fuel injection pulse width calculations.

A faulty TPS can cause drivability problems that include hunting or erratic idle, hesitation, stalling, detonation, failure of torque converter lockup, hard starting, intermittent check engine

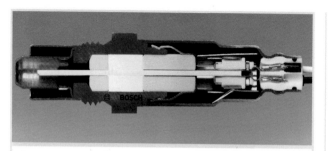

O_2 sensor cutaway. A standard O_2 sensor is essentially an "air battery" in which exhaust gases act as an electrolyte that affects the voltage output of the sensor based on oxygen content in the same way the specific gravity of the liquid in a lead-acid car battery affects voltage. Given a sensor voltage, EMS can correlate the exhaust gas oxygen content to the likely air/fuel ratio that would have resulted in this amount of residual oxygen. Bosch

SENSORS

(MIL) light during driving, poor fuel economy, and generally poor engine performance.

If the TPS attachment screws are loose, the TPS could produce an unstable signal as if the throttle were opening and closing, causing an unstable idle and intermittent hesitation. This condition probably won't set a trouble code.

On older applications with adjustable TPS, the initial adjustment is very important. If not set correctly to specifications, it can have an adverse effect on the fuel mixture, particularly on engines with Alpha-N engine management systems.

A shorted TPS produces an output signal equivalent to the high voltage of constant wide-open throttle. This will typically result in a set malfunction indicator light and cause the fuel mixture to run excessively rich.

If the TPS is open, the computer may assume the throttle is closed and stationary, and the resulting fuel mixture may be too lean and a malfunction code that corresponds to a low-voltage TPS may be set. With an inability to detect sudden throttle openings, the engine will bog and stumble during acceleration, particularly from lower rpm. Dead spots in the TPS (which are a common condition) can cause flat spots in engine performance that only occur at certain throttle positions.

EXHAUST GAS SENSORS
Oxygen Sensors

The oxygen sensor (also called a lambda or exhaust gas oxygen [EGO] sensor) is critical to modern factory engine management systems because its data enables the ECU to constantly trim or tune the engine on the fly in a feedback or closed-loop mode of operation. Because the O_2 sensor is the critical component in the feedback loop used to trim injection pulse width based on exhaust gas oxygen content, an O_2 sensor failure or a problem in the sensor's wiring circuit will prevent the system from going into closed-loop mode, causing a rich-fuel condition and increased emissions (particularly carbon monoxide) and fuel consumption. A bad coolant temperature sensor can also prevent the system from going into closed-loop.

The O_2 sensor is mounted in the exhaust manifold or header to monitor residual (unburned) oxygen in the exhaust. The O_2 sensor tells the ECU if the fuel mixture is burning rich, with a surplus of fuel in the charge mixture (indicated by less exhaust gas oxygen), or lean, with a surplus of oxygen (more exhaust gas oxygen). The ECU reads the sensor's voltage signal and alters the fuel mixture, creating a feedback loop that constantly retrims the fuel mixture.

When a cold engine is first started, the computer ignores the signal from the O_2 sensor, running open-loop. The default fuel mixture is set to run rich of the chemically correct (stoichiometric) 14.7:1 air/fuel mixture and stays that way until the system goes into closed-loop and starts reading the O_2 sensor signal to vary the fuel mixture in quest of the stoichiometric mixture. Most late-model O_2 sensors are electrically heated so they will warm up and reach operating temperature (roughly 600 degrees) sooner to reduce emissions. Heated O_2 sensors typically have three or four wires; older single-wire O_2 sensors are not heated.

In closed-loop mode, the EMS can quickly evaluate the effect of changes to engine operating parameters to make corrections on the fly, resulting in improved emissions and fuel economy. The O_2 sensor operation is so important that some technicians routinely replace the O_2 sensor before attempting to go any further with tuning since they figure an O_2 sensor is not terribly expensive and O_2-sensor-oriented problems can be so confusing and waste so much time. The computer leans the mixture when a lack of oxygen indicates a rich mixture or richens the mixture when a relative presence of oxygen indicates a lean mixture.

The O_2 sensor used in most vehicles is a voltage-generating sensor that can be thought of as a sort of "air battery." In the lead-acid battery used to start a car, an acid medium induces a current flow between two dissimilar metals, freeing energy stored in the acid. The chemical composition of the medium changes as the battery discharges, affecting the output or voltage of the battery. One could, therefore, deduce the chemical composition of the acid (state of charge) based on the output voltage of the battery. In a similar way, the tip of an oxygen sensor has a zirconium ceramic bulb coated with porous platinum on the inside and on the outside. The coated inside and outside serve as the electrodes of the air battery. The bulb contains two platinum electrodes and is vented through the sensor housing to the outside atmosphere. When the bulb is exposed to hot exhaust, the difference in oxygen levels across the bulb creates a voltage.

The zirconium-dioxide element in the tip of an O_2 sensor thus produces an electrical voltage that varies according to the difference in oxygen content between the atmosphere and the exhaust gases. The sensor can generate as much as approximately 0.9 volts when the fuel mixture is rich. When the mixture is lean, the sensor's output voltage can drop to as low as 0.1 volts. When the air/fuel mixture is chemically balanced (about 14.7:1) prior to combustion, the exhaust gas oxygen sensor will generate around 0.45 volts.

In a lean mixture, there is a surplus of oxygen (not enough fuel for all oxygen molecules to participate in combustion); the opposite is true of a rich mixture. The higher the concentration of unburned oxygen, the less differential there is in the zirconium element of the sensor and the lower the voltage output of the sensor. The sensor's output varies continuously from a lean value of 0.1 volts to a rich value of 0.9 volts, with the perfect 14.7 stoichiometric air/fuel mixture producing a sensor voltage in the middle. By design, the sensor bounces from rich to lean every few seconds, but an average of 0.5 volts indicates a correct stoichiometric mixture.

When the engine is running in closed-loop, a full-functioning O_2 sensor will produce a voltage signal that is constantly changing back and forth from rich to lean. The transition rate is configured in the EMS data tables according to the ability of the various fueling systems to execute a fuel trim action, stabilize, and get meaningful feedback. The rate is slowest on engines with feedback carburetors, typically one per second at 2,500 rpm. Engines with throttle-body injection are somewhat faster (two or three times per second at 2,500 rpm), while engines with multi-port injection are the fastest (five to seven times per second at 2,500 rpm).

Post-1994–96 late-model vehicles equipped with OBD-II are equipped with multiple oxygen sensors, with an oxygen sensor located on each exhaust manifold and an additional oxygen sensor mounted behind each catalytic converter to monitor cat operating efficiency. An LS1 Corvette equipped with dual cats thus has four oxygen sensors. A Toyota Avalon with the 1MZ-FE V-6 and single cat has three O_2 sensors. A 2003 Honda VTEC four-cylinder has two O_2 sensors. The OBD-II system compares the O_2 sensor readings before and after the cat, and may set a malfunction indicator lamp (MIL) code if there is no significant change in exhaust gas oxygen readings to indicate the converter is functional.

Limitations of O₂ Sensors

The O_2 sensor will not function until its temperature reaches 600 degrees. This means the sensor is not available immediately on startup when the rich mixtures required to start and sustain a cold engine produce the most pollution. Virtually all oxygen sensors or 1995-plus engines now contain heating elements to help the sensor reach operating temperature sooner and to keep the sensor from cooling off too much at idle.

Tetraethyl lead will rapidly kill a conventional platinum/ceramic O_2 sensor. Lucas, Bosch, and others have introduced sensors that resist lead contamination for tuning or monitoring racing engines. Special Bosch O_2 sensors for European use with leaded fuel are able to tolerate up to 0.4 grams of lead per liter, which is a lot. O_2 sensors can also become clogged with carbon or fail prematurely due to solvents from some RTV silicone sealing compounds.

An O_2 sensor normally lasts 30,000 to 50,000 miles. As the sensor ages, drivability problems can result because the sensor loses its ability to react as quickly to changes in exhaust gas oxygen. This results in improper mixtures that produce loss of power, rough idle, poor gas mileage, and high emissions.

The computer's ability to correctly deduce air/fuel ratios from exhaust-gas oxygen can be adversely affected by air leaks in the inlet or exhaust manifold, by malfunctioning injector(s), or even by a misfiring spark plug resulting in huge amounts of oxygen from unburned charge mixtures being released into the exhaust. One or more cylinders misfiring or running significantly leaner or richer than the rest can fool the O_2 sensor and computer about the overall air/fuel ratio, which can cause the computer to make incorrect closed-loop adjustments.

Oxygen sensors produce a sawtooth voltage that can be useful to the ECU in the 1 to 2 millisecond time range with clever computer algorithms. However, as the ECU makes injection pulse width corrections, the pulse width and actual air/fuel mixture tend to "hunt," varying around the stoichiometric mixture in a wave pattern.

Emissions compliance is so critical that original equipment (OE) computers tend to believe the O_2 sensor before any other sensor (though they also monitor the sensor output for implausible electrical activity indicating failure or partial failure, and will set a malfunction indicator light [check engine soon] in the case of a suspected failure). An OEM ECU is constantly making judgments about the accuracy of various sensors and actually builds a correction table for each sensor in memory in an attempt to maintain the best possible functioning. When the battery is disconnected, the correction factors are reset to 1 and reevaluated as the engine runs. This enables the system to function reasonably well in spite of aging equipment and manufacturing tolerances.

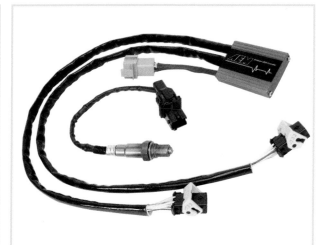

AEM's Universal Exhaust Gas Oxygen (UEGO) sensor and controller measure exhaust gas oxygen over a wide range compared to ordinary O_2 sensors, which are essentially just rich-lean indicators in the vicinity of 14.7:1, the chemically ideal air/fuel ratio for combustion. An oxygen sensor is basically a type of battery that generates a varying voltage depending on the amount of oxygen in exhaust gases (from which it is possible to deduce what the air/fuel ratio may have been). By re-normalizing such sensors to nonstoichometric mixtures across the bandwidth, manufacturers of wideband and UEGO sensors claim accurate measurement of rich mixtures approaching 10:1 (very little oxygen in the exhaust), or mixtures in the 17–20:1 range. AEM

When hot rodders or tuners attempt to tamper with sensor data in order to defeat performance limitations of an engine management system, on newer systems the computer may work against them in correcting suspect sensor data in order to bring the system back to EPA-legal calibration. It is not uncommon on the newest systems for some types of modifications to air/fuel ratio, manifold pressure, fuel pressure, timing, or acceleration to be defeated with EMS countermeasures not immediately but in a short distance or time.

Titania O₂ Sensors

Some late-model vehicles use O_2 sensors with titania elements that operate effectively at lower temperatures and do not operate relative to outside air as dozirconium sensors, because there are no reference vents to get blocked. Due to the lower operating temperatures, titania O_2 sensors warm up in as little as 15 seconds and allow the system to begin closed-loop (feedback) operation almost immediately. They don't cool below operating temperature at idle, and they can be located farther downstream from the engine or used with turbochargers, which bleed off a lot of heat from the exhaust gases.

Titania O_2 sensors do not generate a voltage like zirconium sensors; rather they change resistance dramatically as the mixture changes from rich to lean. The change is not gradual. The mixture changes from low resistance of less than 1,000 ohms when the mixture is rich, to high resistance of more than 20,000 ohms when the mixture goes lean. Therefore, the ECU supplies a 1-volt reference signal that passes through the titania sensor and is then evaluated by the ECU.

The effect of the sensor is such that at rich mixture (and low resistance) the computer reads back at nearly 1 volt. As the resistance dramatically increases with the change to lean mixture the voltage switches suddenly to 0.1 volt. This type of sensor is therefore binary: the sensor tells the ECU that the mixture is either too rich or too lean—and not how much. But a clever ECU can make educated guesses as to actual mixture based on pulse width changes required to flip rich or flip lean. However, the titania sensors is obviously not as well suited to closed-loop operation at nonstoichiometric mixtures (such as wide-open throttle in turbocharged vehicles when 14.7 is dangerously lean).

Wideband O$_2$

Bosch and others have developed extended-range O$_2$ sensors for closed-loop operation at wide-open throttle. Such wideband sensors have the problem that as the mixture moves further away from stoichiometric, changes in exhaust gas oxygen content are extremely subtle. Moving quickly in the rich direction from stoichiometric, there is little or no oxygen in the exhaust. Beyond this, additional richness simply adds more unburned fuel to the exhaust gases, hardly affecting the oxygen content.

Existing air/fuel mixture diagnostic equipment makes use of expensive EGO or UEGO sensors that offer fast operation over a wide range of mixtures. The problem for production use in cars has been the expense. Another problem for some ECUs for reading wideband O$_2$ sensors is that the wideband sensors typically return a voltage between 0 and 5 volts, whereas a narrowband sensor usually returns a 0 to 1 volt signal, so using a wideband sensor may require conditioning circuitry to connect a wideband sensor to the ECU and have it work.

TEMPERATURE SENSORS
Coolant or Cylinder Head Temperature

The coolant or engine temperature sensor is sometimes called the master sensor because the ECU uses engine temperature data along with data from the oxygen sensor to go into the closed-loop mode of controlling the fuel mixture.

The coolant sensor is used by the ECU to deduce the operating temperature of the engine and therefore the applicability of cold operation enrichment. In addition, engine temperature may also determine the following:

when the computer is willing to enter closed-loop operation (not while the engine is cold—but as soon as possible as it warms up in order to monitor and keep emissions low) the operation of emissions-control devices such as the early fuel evaporation heating grid or thermactor-type systems control of ignition timing (full advance not possible while cold) EGR flow (not allowed while cold to improve drivability) evaporative canister purge (not allowed while cold) transmission torque converter lockup fast idle (only while cold or when the air conditioning compressor is active)

Engine Temperature Sensor Functions

• **Start-up fuel enrichment on fuel-injected engines**. Injector pulse width increases to create a richer fuel mixture when the coolant sensor indicates the engine is cold.
• **Spark advance and retard**. Spark advance is often limited for emissions purposes until the engine reaches normal operating temperature.
• **EGR** flow is blocked while the engine is cold to improve cold drivability.
• **Canister purge** does not occur until the engine is warm to improve cold drivability.
• **Energizing the electric heater grid** under the carburetor on older engines to improve early fuel evaporation when the engine is cold.
• **Operation of the throttle kicker** or idle speed when the engine is cold.
• **Transmission torque converter clutch** lockup when the engine is cold.
• **Operation of the electric cooling fan** (if a separate fan thermostat isn't used) when a certain temperature is reached.

On water-cooled engines, the main temperature sensor usually protrudes into the coolant water jacket through the head or intake manifold. Variable resistance or thermistor-type sensors accurately and predictably change resistance with temperature, modifying a 5-volt reference signal (VRef) from the ECU that is then returned to the ECU and evaluated to determine engine temperature. Most of these are the negative temperature coefficient type, which means the sensor's resistance decreases as the temperature goes up. The sensor's resistance is high when cold and drops 300 ohms for every degree rise in temperature. Obviously, a CTS sensor must be located where it accurately reflects the actual engine temperature, rather than a spurious hot or cold spot in the cooling system. At the very least it must be proportional to the real engine temperature.

Some vehicles also use switch-type coolant temperature sensors that open or close at a certain temperature in a binary fashion like a thermostat—either on or off—to control the operation of something like an electric cooling fan. The switch-type sensor may be designed to remain closed within a certain temperature range (say between 55 and 235 degrees Fahrenheit) or to open only when the engine is warm (above 125 degrees Fahrenheit). Switch-type coolant sensors can be found on older GM T-car minimum function systems, Ford MCU, and Chrysler lean burn systems.

A faulty coolant temperature sensor or circuit can cause a variety of symptoms, given the sensor's global effect on so many engine functions. Symptoms might include stalling when cold from wrong mixture, retarded timing or slow idle speed, poor cold idle from wrong mixture, lack of heated inlet air or early fuel evaporation (EFE) enhancement, stumble or hesitation from lack of EFE or too early EGR, poor gas mileage due to extended cold mixture enrichment, lack of open-loop operation, and failure to activate full spark advance when warm.

Modern ECUs often incorporate "limp-home" operations strategies in which the ECU will disregard sensor data that it decides is suspect. In the 1970s, ECUs could be fooled into providing fuel enrichment during turbo or blower boost operation by grounding out temperature sensors to make the ECU think the engine requires cold-running enrichment. Today, a CTS sensor suddenly indicating cold temperatures is more likely to cause the ECU to disregard the sensor data and default to 70 degrees Celsius as the computer's trouble strategy takes effect. Modern tuners/hackers have had to develop alternate means.

Air Temperature

Some engine management systems measure the temperature of inlet air in order to correct for changes in its density. Bosch L-Jetronic and many other EFI systems have used a Bosch pressure vane air meter to measure the velocity of air entering the engine, converting this to mass airflow by correcting for altitude and air temperature: Colder air is denser than hot air; cold air contains more molecules of air per volume.

The air temperature sensor is a thermistor sensor like the coolant sensor. Placing the sensor after the turbocharger and intercooler and in a position where it will not be rendered inac-

Exhaust gas temperature (EGT) sensors are commonly used in piston-engine aircraft to manually trim the air/fuel mixture. They are also frequently used as input to engine management systems on race engines (usually to measure turbine inlet temperature). EGT sensors are typically available at each exhaust port for datalogging and readout on engine dynamometers. Healthy EGT can vary a lot, but high or rapidly rising EGT is a very good indication of looming trouble and engine damage.

curate by heat soak into the mounting surface is critical. Many aftermarket and factory speed-density EFI systems also measure inlet air temperature and use this reading for various corrections to fuel, ignition, and other engine management calculations.

Intake air temperature (IAT) sensors are similar to coolant sensors, with the exception that they are designed and made with materials intended to minimize heat soak from the surface in which they are mounted. So far as possible, IATs react quickly to measure the temperature of the air moving past the sensors on a moment-by-moment basis (for example, as a turbocharger spools and immediately heats the air, then de-spools and passes cold air again). Testing procedures are identical to thermistor-type coolant sensors.

Beyond using IAT sensors to modify engine timing and trim air/fuel mixtures, additional IAT sensors before and after an intercooler can be very useful to monitor and datalog intercooler efficiency under various conditions, with potential actions including activating intercooler fans and warning alarms.

Oil Temperature

Some OEM and aftermarket engine management systems have the ability to monitor oil temperature via sensors that are virtually identical to coolant temperature sensors. Some aftermarket engine management systems can be set up to turn on oil cooler fans and pumps to control oil temperature, and beyond that provide an alarm to the driver or initiating limp-home strategies or even engine shutdown.

It is a mystery why virtually no street engine management systems are set up to provide audible warnings or limp-home/shutdown actions. Like high coolant temperatures, high oil temperatures definitely call for backing off the throttle or stopping the engine *now* before the engine is damaged.

Acceptable temperature range for 10W-40 synthetic oil is 180 to 240 degrees Fahrenheit, according to Redline Synthetic Oil. A narrower range of 200 to 220 degrees is ideal for best performance. Maximum safe oil temperature is 300 degrees. Even the best mineral-based oils will encounter durability problems at 250 degrees or higher.

Exhaust Gas Temperature

Peak exhaust gas temperature (EGT) occurs when an engine is burning a stoichiometric air/fuel ratio (AFR), 14.68:1 on gasoline engines. However, best engine performance at wide-open throttle occurs at AFRs between 11.5 and 13.0, at which point EGTs will be considerably cooler. Retarding timing without making other changes will raise EGT, because the mixture is still burning later in the burn cycle or even while the exhaust gases are in the exhaust ports.

EGT begins to drop at AFRs above 15:1, so it is critical to know on which side of peak EGT you are operating. On piston engines, EGT readings of 1,300–1,500 are common at the maximum power mixture, but this can vary considerably, depending on the engine. Combustion temperature is *much* hotter in the combustion chamber than the EGT measures in the exhaust ports, which will already have cooled.

EGTs above 1,550 are cause for concern, though racers like Bob Norwood have successfully run temperatures measured in the 1,800–1,900 degree range during land speed record trials at Bonneville. No matter what the EGT, you should monitor AFR with a wideband O_2 sensor to see if the engine is running at a sensible AFR. If you subtly tune from a known suboptimally rich direction until you make peak power, there should be a range between rich and lean best torque where power stays the same.

On less radical engines, you can sometimes feel a lean-of-peak condition as the mixture gets harder to ignite and power begins to fall. Once the AFR gets close to 17 to 1 at WOT, the engine will typically start to misfire. The answer is to tune for peak power, keeping an eye on EGT. When power is falling on a boosted engine with a climbing EGT, *stop* immediately.

Most tuners always recommend to begin the programming or jetting process from a known very-rich initial setting and carefully leaning until torque falls off slightly, then going back richer to the point of maximum torque and note the EGT at this setting for future use. Keep in mind that this can be a tricky business. Bob Norwood points out that radical engines working hard on an engine dyno almost always make more and more power as you lean from an excessively rich direction, making the most power *right before they burn down*.

Exhaust gas temperature is useful as a benchmark for establishing which tuning parameters made maximum torque at wide-open throttle on a particular engine application. EGT gauges are widely used to set mixtures on specific aircraft engines used for steady-state high-power applications where operation has been studied and documented in advance.

Once optimal EGT is established for a particular application on a dyno, a tuner can establish a similar air/fuel ratio any time, the way airplane pilots lean an engine at altitude for peak power or peak fuel economy using EGT. Of course, it is really the AFR that is important, not the EGT. Most engines will make maximum power at an AFR of between 11.8 and 13.0 to 1.0, but the EGT may vary from 1,250 to 1,800 degrees Fahrenheit, depending on many factors.

Keep in mind that a target EGT is valid only on the *same engine configuration used on the test dyno*. Ignition timing, cam or piston changes, or headers may change the optimal EGT. For example, changing compression ratio to a higher static setting with no other changes will lower EGT at the same AFR, while retarding ignition timing will usually raise EGT at the same AFR.

Be aware that altitude, barometric pressure, and ambient air temperature may affect this optimal temperature to some degree. Wankel engines typically have higher EGTs than comparable piston engines due to their lower thermal efficiencies, with 1,800 degrees Fahrenheit a common optimal peak EGT. The trouble is, while one engine may make best power at 1,375 degrees, a similar engine might be happier at 1,525. But tuning purely on EGT without a frame of reference is just a waste of the money you spend on the EGT instrumentation. You need dyno torque and wideband AFR data because you should be tuning EGT for whatever temperature produces mean or lean best torque (MBT or LBT).

Which is better, an EGT gauge or an AFR meter? Once again, standard narrowband O_2 sensors are useless (or even dangerous) for peak-power tuning, though they will warn of a grossly lean condition when the AFR passes stoichiometric in a lean direction (an air/fuel ratio *way* too lean for peak power, in serious danger of knocking). Wideband/UEGO sensors combined with standalone lab-quality meters are expensive but accurate and useful, and they are essentially *required* somewhere in a responsible tuning regime (unless your EMS will read the wideband sensor). EGT gauges have the advantage of working with leaded fuel (which will kill oxygen sensors of any type, some faster than others), and they will warn you of dangerously hot combustion.

TORQUE SENSORS

Torque sensors measure the strain on an input shaft or axle to provide a snapshot of the force being applied to the rotating shaft. They are useful in any area that requires the monitoring or control of a rotating shaft, where indirect means are either unavailable or inadequate. An example is the Torductor, a noncontact sensor without moving parts. This type of sensor is actually part of the load-carrying shaft, so that the measured torque is the actual transmitted torque. A high-output signal provides integrity against electrical or magnetic interference from the surroundings, with the advantages of high accuracy with high overload capacity and fast response.

Increasing demands on performance and fuel efficiency stress the need for precise control of the combustion process. Torque sensors measure the actual mechanical output of the engine. This gives the EMS direct insight into the operation of the engine and a measure independent of wear and other uncertainties of engine components and fuel.

Some sensors are capable of measuring not only overall torque, but also subtle changes in torque during individual combustions. This makes a torque sensor capable of engine diagnostics such as misfire detection and control of cylinder balance. This measurement can also be used to calibrate engine output and to alert the driver to engine trouble at an early stage.

Torque sensors are sometimes used to provide control in power-steering systems designed to optimize feel and performance (by only working when they're really needed), and in gearshifting applications to monitor and control transients and oscillations due to backlash and wheel slip. A torque sensor can be an excellent tool to optimize manual gearshifts, and it can help eliminate the need for over-design of transmission components.

The surface of a shaft under torque will experience compression and tension. Displacement sensors may be optical via toothed wheels or magnetic via a variable coupling. Strain gauges or magnetostrictive coupling strips can also be used. One example involves placing two finely slotted concentric rings on either side of the angular deformation zone of a shaft that will twist when torque is applied, and measuring the overall electrical effect on a current of specified frequency as the slots in the two rings no longer precisely overlap due to torque on the shaft.

There are now some interesting new products like the Wheel Torque Sensor (Leboy Products, Troy, Michigan) designed to bolt to the brake drum or spindle of a vehicle, or in place of the wheel itself. The sensor uses a slip ring or rotary transformer to connect the torque sensor to an instrument or EMS in the vehicle. The Wheel Torque Sensor is a strain-gauge-based device that has been used by General Motors' truck division for on-road brake devel-

MAP Sensor Voltage and kPa Relationship				
Load Percentage	MAP Voltage	1-Bar kPa	2-Bar kPa	3-Bar kPa
0	0.00	10	8.8	3.6
5	0.25	15	18	17
10	0.50	20	28	33
15	0.75	24	38	48
20	1.00	29	48	64
25	1.25	34	58	80
30	1.50	39	68	96
35	1.75	43	78	111
40	2.00	48	88	127
45	2.25	53	98	143
50	2.50	58	108	159
55	2.75	62	118	174
60	3.00	67	128	190
65	3.25	72	138	206
70	3.50	77	148	222
75	3.75	81	158	237
80	4.00	86	168	253
85	4.25	91	178	269
90	4.50	96	188	285
95	4.75	100	198	300
100	5.00	105	208	315

opment testing.

PRESSURE SENSORS
Manifold Pressure

Manifold absolute pressure (MAP) sensors indicate air pressure in the intake manifold, which can be made to correspond to engine loading and enables the computer to deduce how much air is entering the engine and then make adjustments to fuel injection pulse width and ignition timing. Manifold absolute pressure is low when intake vacuum is high (as at idle). Pressure is high when vacuum is low (at wide-open throttle on a normally aspirated engine) or during turbocharger- or supercharger-boosted conditions when manifold pressure is above atmospheric pressure.

An engine management system without a mass airflow sensor is able to deduce the current operating volumetric efficiency of the engine by using rpm and manifold absolute pressure or throttle angle to index into a VE table explicitly calibrated in detail for the particular engine type it is managing. The VE table might then cause the computer to lean the mixture and advance spark timing under low loading conditions for best fuel economy. Under high loading/high manifold pressure, and high VE, the EMS might enrich the fuel mixture like a power valve to produce best power, while simultaneously retarding spark timing to prevent knocking like a vacuum/boost advance/retard unit on a conventional distributor.

Speed-density engine management systems (without airflow sensors) are highly sensitive to faulty MAP problems. In this case, if the MAP sensor is defective or out of calibration, severe drivability and performance problems will occur. Such engine management systems are especially dependent on the MAP sensor's signal because the ECU needs it (along with engine rpm, throttle position and ambient air temperature) to *calculate* airflow. Engines that estimate engine load using input from the airflow and throttle-position sensors do not need MAP sensors, though turbocharged engines often have both MAP and MAF sensors (which provides a certain degree of redundancy).

A MAP sensor is referenced to inlet manifold pressure by direct mounting or via a vacuum hose or pipe when the sensor is mounted remotely. A pressure-sensitive electronic circuit in the sensor generates an electrical signal that changes with air density. The sensor is not influenced by ambient atmospheric pressure like a DPS/VAC sensor, which reads the difference between manifold and atmospheric pressure and is influenced by altitude and weather. The MAP sensor is precalibrated in such a way that it measures absolute pressure regardless of other factors.

A pressure-sensitive ceramic or silicon element and electronic circuitry inside the MAP sensor generate a signal that changes voltage with changes in manifold pressure. Most MAP sensors have a ground terminal, a voltage reference (VRef) supply (typically a 5-volt signal provided by the ECU), and an output terminal for returning data to the ECU.

As engine manifold pressure changes, so does the MAP sensor's output. Typical MAP sensor output voltage might be 1.25 volts at idle and just under 5 volts at wide-open throttle. Voltage reads low when vacuum is high and increases as vacuum drops or moves into boosted territory on forced-induction engines. On naturally aspirated engines, output generally changes about 0.7 to 1.0 volts for every 5 inches of change in vacuum. On boosted engines with 2- or 3-bar MAP sensors, the same 1–5 volt range must represent two or three atmospheres of change in pressure, at a potential decline in accuracy and sensitivity in the normally aspirated range of operation.

Ford MAP sensors have been designed to produce a digital frequency signal rather than an analog DC voltage. Ford-type MAP sensors output a square-wave signal that increases in frequency as vacuum drops. A typical reading at idle might be 95 hertz (cycles per second) when vacuum is high, and 150 hertz at wide-open throttle when vacuum is low.

The MAP sensor is clearly *critical* to correct engine system functioning, since any accuracy problems will immediately affect both spark timing and fuel mixtures. A defective sensor, wiring problems in the circuit connecting the map sensor to the ECU, as well as air leaks in the manifold or plumbing connecting the MAP sensor to the manifold can result in bad MAP data. This can cause drivability problems of detonation from lean mixture and excessively advanced spark, or power loss and poor economy due to retarded spark and too rich mixture. Note: Knock sensors can mask problems with lean mixture and advanced spark by retarding the spark under detonation. By watching engine timing via a strobe while banging on the block with a hammer, it may be possible to determine if the knock sensor is functioning properly.

Barometric Pressure

BARO or BP (barometric pressure) sensors sense atmospheric air density. Frequently, a combination of a DPS and BP are used to determine absolute pressure in the manifold. These two functions have occasionally been combined in one sensor, as in the BMAP, once used by Ford in EEC-III and early EEC-IV systems. Some aftermarket ECUs have built-in barometric pressure sensors. Some, like the Motec M48, require an auxiliary external 1-bar MAP sensor vented to the atmosphere to provide altitude and weather correction.

Fuel Pressure Sensors

This type of sensor is very similar in concept to a 5-bar or 10-bar MAP sensor and can be used by an EMS to provide fuel pressure compensation with changes to injection pulse width on some EMSs. Diesel and direct-injection-gasoline system pressure sensors require the ability to operate at hundreds or even thousands of psi.

Combustion Pressure Sensors

By placing a small reference port in the head gasket or a spark plug, combustion pressure can be measured throughout the combustion cycle in real time, which can be used to detect detonation pressure spikes directly (rather than listening for indirect evidence of knock in the form of *sounds* coming from the engine). Combustion pressure sensors (also called in-cylinder pressure transducers) can also be used in feedback mode by an EMS to trim fuel and spark timing to directly optimize power production. Such sensors tend to be expensive, and, obviously, require fast controllers to condition the signal from a combustion pressure sensor for datalogging or analysis for feedback usage in a high-speed engine. This involves sophisticated engineering typically found at OEM car builders or advanced race teams and in the EMSs they use.

AIR METERING SENSORS
Velocity Air Meters

Velocity air meters measure the force of air rushing into an engine against a spring-loaded door in an air tunnel. A linear potentiometer rotated along with the hinge of the air door measures the angle of opening and sends an electrical signal to the ECU that changes as the door opens more or less. The ECU can deduce airflow into the engine based on this value after making certain corrections. The problem is that the force against the door is a function of two variables—weight of the air and speed of the air—and the ECU cannot distinguish, for a given door angle, whether the air is less dense but traveling faster or whether the air is more dense (heavier) but traveling a bit slower. Therefore, air velocity must be corrected for air density (temperature and barometric pressure) in order to reflect air mass (actual number of air molecules entering the engine per unit of time). Velocity air meters from Bosch have been commonly used on systems such as older Bosch L-Jetronic and as part of many factory systems, though they have mostly now been superseded by mass airflow sensors, which have many advantages and are less damaging to engine VE. Velocity air meters can be rather restrictive, particularly when engines have been modified for significantly greater airflow than stock. Velocity meters have the additional liability that once the door is fully open, there is no way for the sensor to measure further increases in airflow.

Because they can be restrictive and because they rapidly run out of measurement upside on the high end of airflow, VAF meters are virtually never used with aftermarket EFI, and they are often replaced on factory systems with speed-density or MAF conversions. Some programmable engine management systems like Motec's can be configured to access VAF or other data as engine-loading information.

With the L-Jetronic type airflow meter, adjustment is accomplished by tightening the spring tension. One notch is approximately equal to a 2 percent change in fuel mixture. For example, if you change to 20 percent larger injectors, you should tighten the spring by 10 notches. Verify fuel mixture by checking with a wideband meter to verify that AFRs are correct under wide-open throttle and adjust if necessary.

This type of airflow sensor was used mostly on European imports equipped with Bosch L-Jetronic fuel injection, Japanese imports equipped with Nippondenso multi-port electronic fuel injection, and Ford vehicles equipped with Bosch multi-port EFI (Escort/Lynx, Turbo Thunderbird, Mustang with the 2.3-liter turbo engine, and Ford Probe with the 2.2-liter engine).

A vane airflow sensor is located upstream of the throttle and monitors the volume of air entering the engine by means of a spring-loaded mechanical flapper door. The flap is pushed open by an amount proportional to the volume of air entering the engine. The flap has a wiper arm that rotates against a sealed potentiometer (variable resistor), allowing the sensor's resistance and output voltage to

A mass airflow sensor is designed to provide a high-speed frequency or voltage readout corresponding to the weight or mass of air moving through the sensor, from which deriving the appropriate raw fuel injection pulse width is relatively simple. Such mass airflow sensors typically measure the amount of current required to keep a wire or film heated to a predetermined temperature. The cooling effect of inlet air is exactly proportionate to its mass or number of molecules. Since the MAF sensor actually only samples a tiny proportion of the air passing though it, it is critical that the air be conditioned to eliminate turbulence. Bosch

change according to airflow—similar to the action of a throttle position sensor. The greater the airflow, the farther the flap is forced open (at least until it is fully open, at which point it cannot measure further increases in airflow). This changes the potentiometer's resistance and the resulting voltage return signal to the computer. Thus, a vane airflow sensor measures airflow directly, enabling the computer to calculate how much air is entering the engine independently of throttle opening or intake vacuum. The computer uses this information to adjust injection pulse width for the target fuel mixture.

The vane airflow sensor also contains a safety switch for the electric fuel pump relay, which is designed to stop the fuel pump in case of an accident if the engine is stopped. Airflow into the engine activates the pump. A sealed idle-mixture screw is also located on the airflow sensor. This controls the amount of air that bypasses the flap, and, therefore, the air/fuel mixture ratio.

Airflow sensors can't tolerate air leaks. A vacuum leak downstream of the VAF sensor allows un-metered air to enter the engine, which can lean-out the fuel mixture and causing a variety of drivability problems. The oxygen sensor can compensate for small air leaks once the engine warms up and goes into closed-loop, but large air leaks will cause large problems.

Vane airflow sensors are also vulnerable to dirt. Unfiltered air passing through a torn or poor-fitting air filter can allow dirt to build up on the flap shaft, causing the flap to bind or stick. Backfiring in the intake manifold can force the flap backward violently, often bending or breaking the flap. Some sensors have a backfire valve built into the flap to protect the flap in case of a backfire by venting the explosion, though the valve itself can become a source of trouble if it leaks. Vane airflow sensors are a sealed unit, preset at the factory with nothing that can be replaced or adjusted except the idle-mixture screw, which should be set using an exhaust gas analyzer to obtain the proper carbon monoxide readings.

Mass Airflow Sensors

A MAF sensor is located in the air duct between the air cleaner and throttle body where it can measure all air that is being drawn into the engine and react almost instantly to changes in throttle position and engine load.

There are two varieties of MAF sensors: hot-wire and hot-film. Unlike vane airflow (VAF) sensors, MAF sensors have no moving parts. They use a heated sensing element to measure airflow. In a hot-wire MAF, a platinum wire is heated 212 degrees Fahrenheit above the inlet air temperature. In a hot-film MAF, a foil grid is heated to 167 degrees Fahrenheit. As airflows past the sensing element, it has a cooling effect. This increases the current needed by control circuitry to keep the sensing element at a constant temperature. The cooling effect varies directly with the temperature, density, and humidity of the incoming air. The electrical current change is exactly proportional to the weight or mass of the air entering the powerplant.

Both voltage- and frequency-based MAFs have been built. Hot-wire Bosch MAF sensors, found on some import cars with LH-Jetronic fuel injection dating back to 1979 as well as 1985–89 GM TPI V-8 engines, generate an analog voltage signal that varies from 0 to 5 volts. Output at idle is usually 0.4 to 0.8 volts increasing to 4.5 to 5.0 volts at full throttle. Hot-film MAFs were introduced by AC Rochester in 1984 on the Buick turbo 3.8-liter V-6 and also used on Chevrolet 2.8-liter engines and GM 3.0- and 3.8-liter V-6 engines. They produce a square-wave variable-frequency output with frequency range varying from 32–150 hertz, with 32 hertz being typical for idle and 150 hertz for full throttle.

In 1990, GM switched most performance engines to speed-density fuel injection systems, except for the Buick 3.3- and 3.8-liter, which was switched to a Hitachi MAF. In 1994, the MAF sensor returned (though the speed-density MAP sensor remained) on GM LT1 performance V-8s and continued in the same configuration on the new LS1 in 1998. Bosch hot-wire units have a self-cleaning cycle where the platinum wire is heated to 1,000 degrees Celsius (1,832 Fahrenheit) for one second after the engine is shut down under EMS control to burn off contaminants that might otherwise foul the wire and interfere with the sensor's ability to read incoming air mass accurately.

Clean aerodynamics are necessary to maintain a smooth flow of air over the MAF wire or film to maintain proportionality to total flow through the air intake and avoid any aerodynamic forces or turbulence in the tunnel that would mean a disproportionate amount of air missed the wire or film. Modified engines with cold-air or high-flow intakes are subject to radiator fan blast or wind forces that can interfere with MAF functioning. Consequently, there can be a tradeoff between accurate air measurement at low speed and restriction at high speed. Aftermarket MAF sensors from Professional Flow Technologies use patented aerodynamics to build accurate high-flow MAF units that outperform factory sensors.

Pro-Flow points out that engines run with a steady airflow only below one-third of full throttle; in this realm, where 50 percent of all driving takes place, all pressure drop from atmosphere to cylinder is produced by the throttle plate, and all air meters seem to perform equally well.

"In the remaining throttle range, the air pulses wildly from actual reverse flow (caused by valve overlap, pneumatic hammer, and acoustic resonance) to a maximum of four times the swept volume of each cylinder (the intake stroke happens in one-half crank revolution and the piston is moving at two times the average piston speed), i.e., a 302-ci engine with eight cylinders (37.75 ci each [0.02185 cubic feet] at 5,000 rpm), the intake air peak value will be a maximum of 437 cubic feet per minute (5,000 rpm ? 0.02185 cubic feet per cylinder displacement ? 4 times the maximum swept volume per cylinder)."

If any part of the intake system limits this maximum flow, there is a potential for performance improvement.

Because of multiple screens and poor fluidics design, many

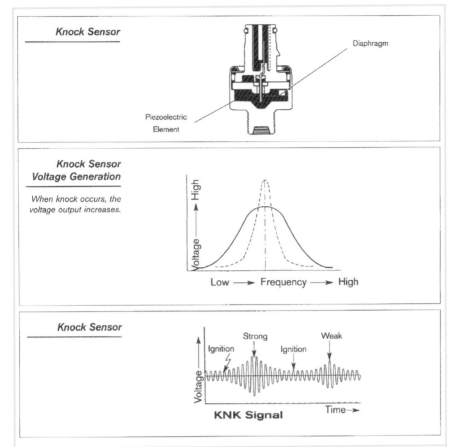

The vibrations from detonation excite a piezoelectric element in the knock sensor. The voltage output of the sensor is highest during knock, though other engine noises can also generate knock sensor output. It is up to the EMS to decide what threshold constitutes genuine knock, requiring countermeasures. Toyota

portional to airflow.

Karman-Vortex airflow sensors were used on 1987 and later Supra turbos, some Lexus engines (all except the ES 250 and 300), and some 1983 to 1990 Mitsubishis. The sensor has a five-pin connector and an integral air-temperature sensor. A light-emitting diode (LED), mirror, and photo receptor are used to count the pressure changes in these applications. The mirror mounts on the end of a weak leaf spring placed straddling a hole leading directly to the vortex generator area. When a vortex forms, the drop in pressure wiggles the spring, which causes reflected light from the LED to flicker as it is picked up by the photo receptor. This generates an on-and-off digital signal that varies in frequency in proportion to airflow. When airflow is low at idle, the signal frequency is low (30 hertz or so), but as airflow increases, the frequency of the signal increases to 160 hertz or higher.

Early Mitsubishi applications use ultrasonics to detect the pressure changes associated with changes in airflow. A small tone generator sends a fixed ultrasonic tone through the vortex to a microphone. With increasing turbulence and a higher number of vortices, the tone is increasingly disrupted before it reaches the microphone. The sensor's electronics then translate the amount of tone distortion into a frequency signal that indicates airflow. Early units contain an integral air-temperature sensor while later ones also contain an integral barometric-pressure sensor. In 1991, Mitsubishi changed to a redesigned Karman-Vortex sensor that replaces the ultrasonic generator with a sensor that measures fluctuations in air pressure directly.

mass air meters being produced have had pressure drops in the order of 12 to 24 inches of water. This reduces maximum horsepower by 5 to 7 percent due to the drop in engine VE, and it's made worse by altitude, which reduces efficiency by 0.7 percent per thousand feet.

Karman-Vortex-Type Mass Airflow Sensors

Another type of mass airflow sensor that has been used on some Japanese vehicles is the Karman-Vortex (K-V) airflow sensor. The advantage of using this type of sensor instead of a VAF sensor is that it causes less restriction. It is also simpler and more reliable than MAF sensors, where contamination of the heated wire or filament can cause problems. The K-V sensor responds more quickly to changes in airflow than other types of mass airflow sensors, which helps the EMS do a better job managing the air/fuel mixture.

A K-V sensor measures the amount of turbulence behind a small object placed in the incoming airstream to generate an airflow signal. The principle is that as airflows past a stationary object, it creates turbulence or vortexes (swirling eddies of air) behind the object similar to the wake created by a passing boat. The greater the airflow, the greater the turbulence. Turbulence can be measured electronically by sending light or sound waves through the air to detect the changes in pressure, or by measuring the frequency of the pressure changes (air turbulence). This allows the sensor to generate a signal that is pro-

KNOCK SENSORS

Knock sensors output a small electrical signal on detecting the resonant frequency vibration of engine detonation. On receiving a knock signal, the ECU (engine control unit) will temporarily retard ignition timing to prevent the condition as a countermeasure to prevent engine damage from explosive knock. The ECU will also typically advance the maximum spark timing to take advantage of higher octane fuels and higher altitudes when there is reduced tendency to knock. In some cases, the EMS will correlate knock to specific cylinders and retard only the cylinder(s) currently knocking.

The optimal ignition timing under load for a high-compression engine can be quite close to the point where engine knock occurs. However, running so close to the point of knock means that knock will almost certainly occur on one or more cylinders at certain times during the engine operating cycle. Since knock may occur at a different moment in each individual cylinder, the ECU uses a knock control processor to pinpoint the actual cylinder or cylinders that are knocking.

The knock sensor (or sensors) typically mounts on the engine block, but is sometimes located on the head(s) or intake

SENSORS

manifold. The device consists of a piezoelectric crystal measuring element that responds to engine noise oscillations. This signal is converted into a voltage signal (analogue) that is proportional to the level of knock and transmitted to the ECU for processing. The knock frequency from the sensor is usually between 6 kilohertz and 15 kilohertz. GM engines are around 8 kilohertz.

A sophisticated modern ECM analyzes the noise from each individual cylinder and sets a threshold noise level for the cylinder based upon the average noise over a predetermined period. If the noise level exceeds the reference level, the ECM initiates knock-control countermeasures.

Normal timing occurs at the optimal ignition point, but once knock is identified the knock control processor retards the ignition in steps. After knocking ceases, the timing gets re-advanced, but if or when knock returns, the ECU retards it once more. The process is one of ebb and flow, and the knock sensor and EMS monitor the engine continuously to allow for optimum performance.

For example, Bosch's Motronic KCP (knock control processor) analyzes the noise from each individual cylinder and sets a reference noise level for that cylinder based upon the average of the last 16 phases. If the noise level exceeds the reference level by a certain amount, the KCP identifies the presence of engine knock and retards the ignition timing for that cylinder or cylinders by a set number of degrees. Approximately 2 seconds after knock ceases (20 to 120 knock-free combustion cycles) the timing gets advanced in 0.75-degree increments until normal timing is achieved or knock returns. This process occurs continuously in an effort to keep all cylinders at their optimum timing.

If the system develops a fault, the ignition timing typically defaults to a setting several degrees retarded from normal as a safety precaution, which will probably lead to a reduction in engine performance. Turbocharged engines, such as the MR2 Turbo in this book, which are more susceptible to damage from spark knock without the knock sensor on line, usually implement other anti-knock, anti-performance countermeasures such as severe timing retard, low-boost turbo wastegate strategies, and disabling the variable intake system. In all cases, the EMS will turn on the check engine light.

If a fault exists in the Motronic KCP, ignition timing is then retarded 10.5 degrees by the ECU (which also provides the driver additional incentive to solve the problem, since this will severely affect performance). The Motronic ECU also limits revs to no more than 4,000 rpm.

OTHER SENSORS
Vehicle Speed Sensor

The Vehicle Speed Sensor (VSS) controls many vehicle and engine systems. In fact, the VSS is much more than a digital version of the speedometer cable. Using the VSS to generate digital speedometer readings is a relatively recent application for the VSS. Before powering a digital dashboard, the VSS was used to regulate the cruise control and control transmission shifts.
1. Vehicle speed is part of overall engine operating strategies on OBD-II systems.
2. The VSS is a factor in operation of the idle air control valve and canister-purge.
3. Automatic transmission control logic needs VSS data to provide optimal performance and mileage. In an electronically shifted transmission, VSS data—not hydraulic pressure in the valve body—is used to determine shift points.
4. Torque converter lockup clutch operation is controlled by the VSS.
5. VSS signals may be a factor used by the EMS to determine whether to run electric cooling fans.
6. Vehicle speed data is a critical factor used to operate variable steering and intelligent suspension and handling equipment.
7. VSS signals are a critical factor controlling ABS, disabling the system below a certain speed so the wheels can come to a complete stop.
8. The VSS is used to limit top speed on some vehicles via electronic throttle or fuel regulation.

The VSS is a counting, rather than a measuring sensor. The VSS typically counts revolutions of the transmission output shaft. The higher the count that the VSS sees, the higher its output signal, which is typically a square-wave pulse transmitted to the EMS. The ECM is programmed to compare the count, or signal, received from the VSS against its internal clock to determine vehicle speed on an ongoing basis. Below a threshold speed of 3 to 5 miles per hour, VSS data becomes insignificant and the EMS will ignore it.

At least two types of VSS have been used:
1. The optical VSS is located inside the speedometer. It uses a light-emitting diode (LED) and a two-blade, mirrored reflector to generate a signal. When the vehicle is moving, the speedometer cable turns the two-bladed mirror. The mirror rotates through the LED light beam, breaking the beam twice for each revolution. Each time the mirror interrupts the LED beam, light is momentarily reflected to a photocell. Whenever light hits the photocell, it generates a discrete electrical signal pulse. The faster the mirror rotates, the more electrical pulses that are generated per unit of time. The EMS transforms the number of pulses from the photocell per time to an electronic measurement of vehicle speed (based on configuration data regarding tire size, final drive ratio, and so forth).
2. A permanent-magnet VSS mounts on the transmission/transaxle case in the speedometer cable opening, in series with the speedometer cable or replacing it entirely. A permanent magnet rotates past a coil in the sensor to generate a pulsating voltage correlated to vehicle speed. The pulsating signal from a VSS creates an alternating-current (AC) voltage. At one time, VSS circuitry used an external buffer module to convert VSS output into a digital output that could input to the EMS computer. Newer systems convert the VSS signal in the ECM.

If the VSS fails, no signal from the VSS to the computer should set a trouble code in the EMS. Obvious signs that the VSS has failed include a digital speedometer that's not working, a dead cruise control, and rough and erratic automatic-transmission shifting. Other problems may include rough idling and poor fuel economy. A more difficult problem is a VSS that works but sends out an incorrect signal, which may still cause a code to be set—for example, if the VSS signal tells the computer the vehicle is traveling at a high rate of speed, but throttle position and MAP voltage tell the computer that the engine is running barely above idle. On a vehicle using the VSS as a speed-limiting safety device, a defective sensor could incorrectly indicate a too-fast condition, shutting down fuel flow at the wrong time. Symptoms could be a random or intermittent sudden loss of power and poor performance.

Typical OBD-II codes for a malfunctioning VSS include:
P0500 Vehicle Speed Sensor Malfunction
P0501 Vehicle Speed Sensor Range/Performance
P0502 Vehicle Speed Sensor Low Input
P0503 Vehicle Speed Sensor Intermittent/Erratic/High
P0716 Vehicle Speed Sensor Circuit Input Intermittent
P0718 Vehicle Speed Sensor Input Circuit Input low

Actuators are electromechanical devices that enable a computer to make things happen in the world. In the realm of EFI, these include such devices as fuel injectors, fuel pumps and pressure regulators, electronic throttles, nitrous oxide solenoids, boost controllers, EGR solenoids, idle air-bypass controllers, torque converters, engine fan controls, ignition amplifiers and coil-drivers, warning/diagnostic lights, and more as follows:

• Fuel delivery: electronic fuel injectors, fuel supply systems, pump, regulator, dual-fuel switching systems
• Air supply: throttle angle control, IAC stepper motor (idle speed control), boost control, electronic wastegate control, supercharger bypass, supercharger clutch
• Nitrous delivery: single/multi-stage solenoid, pulse-modulated solenoid
• Temperature control: radiator fans, engine compartment fans
• Ignition/park delivery: coil-driver
• Knock control: water/alcohol injection solenoids, boost timing retard
• Cam phasing/Valve lift controls
• Variable intake: plenum volume, runner length, secondary runner actuation (TVIS)
• Emissions controls (exhaust gas recirculation, air injection, etc.)
• Transmission control and auxiliary onboard computers
• Information delivery: driver displays, diagnostics

FUEL DELIVERY:
ELECTRONIC FUEL INJECTORS
Selecting Injector Capacity

Imagine a nozzle on the end of a hose on a windy day: Squeeze the handle on the nozzle, water sprays out and mixes with the air. Release the handle and the flow of water spray stops. If you imagine several nozzles on the same hose, you are visualizing the situation in a port fuel-injected engine.

Unlike most mechanical injection systems, which spray fuel constantly when the engine is running and regulate fuel delivery with *pressure* variations, electronic fuel injectors utilize a *timed* injection architecture, opening and closing extremely rapidly at least once per combustion cycle to an accuracy of a ten-thousandth of a second in order to precisely spray gasoline into the intake air rushing into a working spark ignition engine.

The goal is to deliver the charge mixture in a precise air/fuel ratio optimized to generate maximum cylinder pressures per fuel mass in order to efficiently push against the piston and cause the engine to do work. Although fuel enters the air in bursts, or pulses, in electronic injection systems, by the time the fuel and air are in the cylinder and compressed, they must be thoroughly mixed (some of this vaporization occurs inside the hot cylinders).

In order to inject fuel into the air in precisely timed pulses, electronic injector nozzles contain tiny solenoid valves operated under computer control by electromagnets that must overcome the force of a spring and fuel rail pressure to push the injector valves open. Once an injector is open, fuel pressure in the supply line leading to the injector causes fuel to spray through a tiny orifice and mist out into inlet air rushing past the injector on the way into the engine. When the solenoid is de-energized, spring

The J&S Safeguard provides both coil-driving capabilities and interceptor-type individual-cylinder knock control countermeasures using its own dedicated knock sensor. The job of a coil-driver is to condition the EMS's timing command into a powerful electrical signal optimized to trigger the coil(s) to produce a powerful spark of correct duration.

pressure immediately forces it closed. The injectors are usually aimed directly at the intake valve(s) from the intake manifold or cylinder head intake ports.

Robert Bosch GmbH, based in Germany, long ago licensed the rights to build electronic injection systems as invented by the U.S. company Bendix. Bosch has built more than 150 million electronic injectors and currently dominates the market. Several other companies, including Lucas, Rochester, and Siemens, design and manufacture large numbers of electronic injectors.

Electronic Injector Valve Architecture

Injectors consist of a housing and a valve assembly encompassing the valve itself and a solenoid armature. The housing contains the solenoid winding and electrical connections. A helical spring assisted by fuel pressure forces the valve against its seat in the valve assembly when the injector windings are not energized, blocking fuel from passing through the injector. When the ECU energizes the solenoid windings, the valve lifts by approximately 0.06 millimeter, and pressurized fuel is forced through the valve body and sprays from the injector.

According to Bosch, injectors are designed with the following goals in mind:
1. Precise fuel metering at all operating points
2. Accurate flow at narrow pulse widths, with deviation from linearity within specified tolerances
3. Broad dynamic flow range
4. Good fuel distribution and atomization
5. Valve leak tightness
6. Corrosion resistance to water, sour gas, and ethanol mixtures
7. Reliability
8. Low noise

Injectors can be divided into the following metering categories:
1. Annular orifice metering, which uses a pintle to optimize the

atomization via a conical-shaped spray pattern and which meters fuel by the size of the gap between the pintle and the valve body.

2. Single-hole metering in which fuel spray is injected directly from the drilled passage in the valve body downstream of the needle valve (atomization is not as good as with a pintle design, at worst case producing a pencil beam).

3. Multi-hole metering (C-version) that forces fuel through a stationary drilled plate located at the end of the valve body orifice, downstream from the needle valve. This design, which normally includes four precisely spaced and aligned holes, results in a good conical spray pattern.

4. Multi-hole metering for multi-valve engines is similar to 3, but aligns the metering holes so a separate spray of fuel hits each intake valve.

5. Disc metering uses a drilled disc that moves off a flat seat with the armature. Claimed benefits include resistance to deposit buildup and clogging so no cleaning is ever required, wider dynamic range for improved idle, quieter operation, lighter weight that improves response, and operation with alternative fuels like natural gas, propane, methanol, and ethanol. Atomization quality may be similar to single-hole metering.

The Bosch pintle-type injector is the earliest design but is still widely used. The valve housing in this type of injector narrows to form a seat upon which a tapered needle comes to rest when the injector is closed, choking off fuel flow. Bosch improves the spray pattern by extending the needle valve assembly into a beveled pintle assembly that protrudes out of the end of the injector through a plastic chimney. The solenoid lifts the pintle off the seat to release fuel. Bosch injector pintles and seats are always the same size. Bosch regulates flow by varying the size of the bore after the seat. Problems with the pintle design reportedly include increased chance of clogging in the small orifice area, slower response time because of the heavier armatures used to lift the pintle, and reduced service life.

In some cases, injectors will use a ball valve to seal the metering orifice. This allows the use of a lighter armature to improve response time compared to older pintle types. There is also reportedly less wear and a longer service life. The orifice can be designed with multiple openings, which allows a wider spray pattern and higher fuel delivery.

The disc-type injector eliminates the armature so the solenoid acts directly on the flat disc through the core of the injector body. The valve assembly in a disc-type injector consists of the armature attached to a disc with six tiny holes drilled around the outside. In a closed position, the disc rests on a seat, the center of the disc covering a hole in the middle of the seat that prevents any fuel from passing through the injector. When the armature is pulled open as the injector energizes, fuel passes through the six outer holes, through the hole in the seat, and sprays out of the injector. The flow of a Lucas disc injector is determined by the size of the six holes drilled in the disc. This arrangement is even lighter than the ball-type for a faster response time. This disc and seat design also results in less deposit buildup at the orifice and longer service life.

Injectors vary in the time it takes to open and close. Larger pintle-type injectors are built with heavier moving assemblies that are a little more sluggish. To compensate, they are typically designed with more powerful low-resistance windings that require peak-and-hold circuitry to power them open quickly. The Lucas Disc injectors are lighter, according to fuel system expert Russ Collins, which enables disc-type injectors to operate down to 0.8, 0.9, or 1.0 milliseconds. Most pintle injectors are limited to a minimum 1.3–2.0 milliseconds

This modern Bosch injector is designed for a fuel feed at 30–60 psi (though many will function at somewhat higher pressures). When it's time for a fuel injector to open, under onboard computer command, injector drivers in the ECU energize an electromagnetic coil in the injector and cause it to lift a spring-loaded armature off its seat and allowing fuel to spray from the injector for a precisely timed interval. Bosch

Some performance experts recommend when upgrading injectors for higher flow to switch from traditional pintle-type injectors to ball or disc-type valves.

Fuel Flow of Injectors

Injector flow is rated in pounds per hour at a given pressure, or in cubic centimeters per minute at pressure—either static (injector open continuously) or pulsed at a certain duty cycle. The amount of fuel that actually enters an engine from an injector depends on several variables.

The first is the amount of time the injector remains open for each combustion event being supplied with fuel. The longer the injector is open, the more fuel sprays out, which maxes out at a point when the injector is open all the time (since energizing of injector windings causes heat buildup, injectors may fail if operated much above an 80-percent duty cycle).

Theoretically, there is no lower limit to the length of time an injector could be opened, but practically, injectors do have a lower limit. This occurs at a point (usually somewhere between 1 and 2 thousandths of a second) when the injector is open so briefly that random factors that tend to cause injector open time and fuel flow to vary slightly each time the injector fires become significant in comparison to the planned duration of open time. The consistency of the open event begins to suffer, and you will not get a good clean repeatable spray pattern. This situation becomes even more unstable when injector open time is so short that the injector never has time to open all the way before the injector driver de-energizes the injector and the spring begins to close it.

A second factor affecting injector fuel delivery is how much fuel pressure exists in the line supplying fuel to the injector. The lower the pressure, the less fuel sprays out of an open injector—as above, effectively limited by a pressure beyond which repeatability suffers (most port injectors are never operated below 2 bar [28 psi]). On the high end, fuel delivery begins to suffer from diminishing returns as pressure increases and is absolutely constrained when the injector's internal electromagnet can no longer overcome fuel pressure to open or when the internal injector spring pressure cannot consistently overcome the pressure of fuel rushing through the injector valve under high pressure and the injectors cannot close. Many experts recommend 72.5 psi as a maximum fuel pressure.

A third important variable affecting the amount of fuel injected into an engine is the size of the valve or orifice in an injector through which fuel must flow. Most electronic injectors are rated in pounds of fuel per hour that will flow at a certain pressure (flow rate). An injector's maximum size is practically limited by the increasing weight of the injector's moving parts. Larger weights have more resistance to moving quickly and require increasing power to activate them. This follows the physical laws concerning momentum, which describe the resistance of stationary objects to movement and the resistance of moving objects to stopping.

Large-flowing injectors get very expensive. As injector size increases, each opening event has a greater effect on the hydraulics of the fuel supply system, implying larger fuel pumps, fuel rails, fuel line size, and higher pressures. And the larger an injector is, the higher the minimum flow possible—limited, as described above, to the amount of fuel that flows in 1 or 2 milliseconds at a given pressure. This is particularly important on engines with a high dynamic range, e.g., small, high-output turbocharged engines, where the amount of fuel required at maximum horsepower is high in comparison to that required at idle.

On many engines, the dynamic range of the injectors is increased by increasing the fuel pressure commensurate with load, usually by regulators referenced to intake manifold pressure. On some very large displacement engines or smaller engines with large dynamic range, it may become necessary to use multiple smaller injectors that are staged.

All of the above factors have to be considered when you are selecting injectors for an engine that has been modified in some way versus an engine for which factory engineers have already determined injector requirements.

Injector Flow Rate Calculations

Lucas offers the following formula designed to allow high-performance-oriented customers to determine which injector (flow rate) they will need for a specific engine based on horsepower output:

Flow Rate = (HP x BSFC) ÷ (#Injectors x Maxcycle)

Flow Rate is in pounds per hour of gasoline (at rated or specified pressure).

HP is projected flywheel horsepower of the engine.

BSFC is brake-specific fuel consumption in pounds per horsepower-hour. (Normally this is 0.4 to 0.6 at full throttle. Assume 0.55 for turbo-supercharged engines, 0.45 for NA engines.)

#Injectors is the number of injectors.

Maxcycle is the maximum duty cycle of the injector. (Above 80 percent, some injectors could begin to fibrillate or overheat, which could eventually cause them to lose consistency—or even fail completely!)

EXAMPLE: 5.7-liter V-8
FLOW = (240 HP x 0.65 BSFC) ÷ (8 cylinders x 0.8)
FLOW = 24.37 lb/hr injectors required

Lucas points out that their 5207011 injector, rated at 23.92 pounds per hour is close. Lucas warns, "The only problem with this equation is that to get an 'accurate' answer, you need to instrument and measure the exact engine configuration in question, which means a dynamometer with full instrumentation."

In the case of a **530-ci stroked Cadillac 500 V-8** with turbocharging that outputs 800 horsepower:

FLOW = (800 x 0.55) ÷ (8 x 0.8) = 68.75 lb/hr injectors.

Injectors this big are expensive and might not idle well on some engines. In this case, a designer might decide to use two smaller injectors per cylinder, staged so that only one injector per cylinder is operating at idle with the other gradually phasing in as engine loading and rpm increase.

Gasoline contains a certain amount of chemical energy that is released when it is burned (oxidized). If all the energy in gasoline could be converted to work, a gallon of gas could make 45 horsepower for one hour. Internal-combustion engines are notoriously inefficient, so the actual horsepower you can expect to make from a certain injector is significantly less than theoretical.

The following formula is a revision of the above equation rearranged to yield the maximum horsepower that a given injector can make at a specified fuel pressure:

HORSEPOWER = (FLOW RATE x #INJECTORS x MAXCYCLE) ÷ BSFC
213 = (20 POUNDS x 6 X 0.8) ÷ 0.55

Injector Pressure Changes

In fact, altering the system fuel pressure is a good and inexpensive way to increase the dynamic range of any fuel injector, which is why this is done on many original equipment injection systems. The normal method is to reference the fuel pressure regulator to manifold vacuum. As manifold vacuum rises and falls with decreasing and increasing engine loading, the vacuum is used to increase or decrease the effect of the spring located in the regulator pressing against the regulator's diaphragm, which dynamically changes the pressure of fuel that can force its way through the regulator. This means that fuel pressure is lowered at idle (high vacuum) and raised at wide-open throttle (WOT) when there is zero vacuum.

On an engine with high vacuum at idle, system pressure might be 39 psi at WOT, 25 psi at idle. Some regulators are not manifold vacuum-referenced but have an adjustable bolt or screw that can be turned to increase or decrease regulator spring pressure against the diaphragm to alter fuel pressure. As we'll see shortly, the static flow rate of a fuel injector is directly proportional to the change in system pressure.

In the case of a turbo/supercharged engine, one atmosphere of boost (14.7 psi) against a manifold pressure-referenced regulator would raise the system fuel pressure an equal amount above that at zero vacuum—assuming the fuel pump has the capacity to provide the fuel. Therefore, fuel pressure on a forced-induction engine would vary from 25 to roughly 55 psi at 15 psi turbo boost. At equivalent injector pulse widths, this fuel pressure change causes a tremendous difference in the amount of fuel injected into the engine. The long and short is that referenced regulators enable larger injectors to idle better, smaller injectors to produce more high-end power.

The following formula calculates the effect of changed fuel pressure on injector flow:

[SQR (SysP2 ÷ SysP1)] x Statflow1 = Statflow2

SQR means take the square root of the value in the parentheses.
SysP2 is the proposed new system pressure (psi).
SysP1 is the specified pressure for the injector's rated flow (psi).
STATFLOW1 is the rated static (nonpulsed) injector flow rate.
STATFLOW2 is the calculated new static flow rate.

For example, a Jaguar 4.2-liter Bosch injector is rated at roughly 20 pounds per hour at 39 psi.

At 25 psi:

[SQR (25 ÷ 39)] x 20 = 16 lb/hr at 25 psi!
(a 20 percent difference)

On the other hand, consider the same injector at 55 psi:

Fuel Injector Specifications

Manufacturer/ID	cc/min	lb/hr	Calculated HP per injector	kPa Test Pressure	Ohms	Vehicle Application	Engine Application
Bosch 0 280 150 208	133	13.0	26.6	300		BMW	323
Bosch 0 280 150 716	134	13.1	26.8	300			
Nippon Denso (light green)	145	14.2	29.0	255	2.4	Toyota	4KE
Nippon Denso (green)	145	14.2	29.0	255	2.4	Toyota	1GE
Bosch 0 280 150 211	146	14.3	29.2	300			
Lucas 5206003	147	14.4	29.4	300			Starlet
Lucas 5207007	147	14.4	29.4	270		Ford	1.6L
Bosch 0 280 150 715	149	14.6	29.8	300			
Nippon Denso (red/dark blue)	155	15.2	31.0	290	13.8	Toyota	
Nippon Denso (violet)	155	15.2	31.0	290	13.8	Toyota	3EE 2EE
Nippon Denso (sky-blue)	155	15.2	31.0	290	13.8	Toyota	1GFE
Nippon Denso (violet)	155	15.2	31.0	290	13.8	Toyota	4AFE
Lucas 5207003	164	16.1	32.8	300		Buick	
Lucas 5208006	164	16.1	32.8	250		Renault	
Bosch 0 280 150 704	170	16.7	34.0	300			
Bosch 0 280 150 209	176	17.3	35.2	300		Volvo	B200, B230
Nippon Denso (light green)	176	17.3	35.2	290	13.8	Toyota	4AFE
Nippon Denso (grey)	176	17.3	35.2	290	13.8	Toyota	4AFE
Bosch 0 280 150 121	178	17.5	35.6	300			
Nippon Denso (dark grey)	182	17.8	36.4	255	2	Toyota	4AGE
Nippon Denso (grey)	182	17.8	36.4	255	2.4	Toyota	4ME, 5ME, 5MGE
Bosch 0 280 150 100	185	18.1	37.0	300			
Bosch 0 280 150 114	185	18.1	37.0	300			
Bosch 0 280 150 116	185	18.1	37.0	300			
Bosch 0 280 150 203	185	18.1	37.0	300			
Bosch 0 280 50222	188	18.4	37.6	300		1985-6 GM	TPI 305 V8
AC 5235047	188	18.4	37.6	300		1985 GM	TPI 305 V8
AC 5235301	188	18.4	37.6	300		1987-8 GM	TPI 305 V8
AC 5235434	188	18.4	37.6	300		1989 GM	TPI 305 V8
AC 5235435	188	18.4	37.6	300		1989 GM	TPI 305 V8
Bosch 0 280 150 125	188	18.4	37.6	300			
Lucas 5202001	188	18.4	37.6	250		914	1.8L
Lucas 5204001	188	18.4	37.6	250		Fiat	
Lucas 5206002	188	18.4	37.6	250		Toyota	
Lucas 5207002	188	18.4	37.6	250		Chev	5.0L
Lucas 5208001	188	18.4	37.6	250		Nissan	280ZX
Lucas 5208003	188	18.4	37.6	250		Alfa	
Lucas 5208007	188	18.4	37.6	250		BMW	325E
Bosch 0 280 150 614	189	18.5	37.8	300			
Nippon Denso (dark grey)	200	19.6	40.0	290	1.7	Toyota	3SFE
Nippon Denso (beige)	200	19.6	40.0	290	1.7	Toyota	4YE
Nippon Denso (orange)	200	19.6	40.0	290	1.7	Toyota	22RE
Nippon Denso (brown)	200	19.6	40.0	290	1.7	Toyota	3VZE
Nippon Denso (pink)	200	19.6	40.0	290	2.7	Toyota	4AGE
Nippon Denso (dark blue)	200	19.6	40.0	290	13.8	Toyota	3SFE
Nippon Denso (orange/ blue)	200	19.6	40.0	290	13.8	Toyota	
Nippon Denso (brown)	200	19.6	40.0	290	13.8	Toyota	3VZFE
Nippon Denso (red)	200	19.6	40.0	290	13.8	Toyota	2VZFE
Lucas 5207013	201	19.7	40.2	270		Jeep	4.0L
Nippon Denso (blue)	210	20.6	42.0	255	2.4	Toyota	4AGE
Nippon Denso (sky blue)	213	20.9	42.6	290	13.8	Toyota	3FE
Nippon Denso (beige)	213	20.9	42.6	290	13.8	Toyota	4AGE
Nippon Denso (yellow)	213	20.9	42.6	290	13.8	Toyota	5SFE
Bosch 0 280 150 216	214	21.0	42.8			Buick	
Bosch 0 280 150 157	214	21.0	42.8	250		Jaguar	4.2L
Bosch 0 280 150 706	214	21.0	42.8	250			
Bosch 0 280 150 712	214	21.0	42.8	250		Saab	2.3l Turbo
Bosch 0 280 150 762	214	21.0	42.8	300		Volvo	B230F
AC 5235211	218	21.4	43.6	300		1986 GM	TPI 350 V8
AC 5235302	218	21.4	43.6	300		1987-8 GM	TPI 350 V8
AC 5235436	218	21.4	43.6	300		1989 GM	TPI 350 V8
AC 5235437	218	21.4	43.6	300		1989 GM	TPI 350 V8
Lucas 5207011	218	21.4	43.6	300		Chev	5.7L
Bosch 0 280 150 A9152	230	22.5	46.0	?		Alfa	Turbo
Bosch 0 280 150 201	236	23.1	47.2	300			
Lucas 5208004	237	23.1	47.2	250		Ford	
Lucas 5208005	237	23.1	47.2	250		Chrysler	BMW
Bosch 0 280 150 151	240	23.5	48.0			BMW	633

Nippon Denso(yellow/orange)	250	24.5	50.0	255	1.7	Toyota	
Nippon Denso (green)	250	24.5	50.0	290	13.8	Toyota	4AGE
Nippon Denso (violet)	250	24.5	50.0	290	13.8	Toyota	4AGE
Nippon Denso (brown)	250	24.5	50.0	255	13.8	Toyota	3SGE
Nippon Denso (violet)	251	24.5	50.2	290	13.8	Toyota	1UZFE
Bosch 0 280 150 001	265	26.0	53.0	300			
Bosch 0 280 150 002	265	26.0	53.0	300			
Bosch 0 280 150 009	265	26.0	53.0	300			
Bosch 0 280 150 218	275	27.0	55.0	300		Buick	
Nippon Denso (light green)	282	27.6	56.4	290	13.8	Toyota	2RZE
Nippon Denso (violet)	282	27.6	56.4	290	13.8	Toyota	2TZFE
Bosch 0 280 150 802	284	27.8	56.8	300		Volvo, Renault	B200Turbo
Nippon Denso (yellow)	295	28.9	59.0	255	2.7	Toyota	7MGE
Nippon Denso (pink)	295	28.9	59.0	255	1.6	Toyota	22RTE
Nippon Denso (green)	295	28.9	59.0	255	13.8	Toyota	3SGE
Bosch 0 280 150 811	298	29.2	59.6	350		Porsche	944 Turbo
Bosch 0 280 150 200	300	29.4	60.0	300		BMW	
Bosch 0 280 150 335	300	29.4	60.0	300		Volvo	B230 Turbo
Bosch 0 280 150 945	300	29.4	60.0			Ford Motorsport	
Nippon Denso (pink)	315	30.9	63.0	290	13.8	Toyota	3SGE
Nippon Denso (light green)	315	30.9	63.0	290	13.8	Toyota	7MGE
Bosch 0 280 150 804	337	33.0	67.4	300		Peugot	505 Turbo
Bosch 0 280 150 402	338	33.1	67.6	300		Ford	
Lucas 5207009	339	33.2	28.88				
Bosch 0 280 150 951	346	33.9	69.2	300		Porsche	
Bosch 0 280 155 009	346	33.9	69.2	300		Saab Turbo	
Nippon Denso (red/ orange)	346	33.9	73.0	255	2.9	Toyota	
Bosch 0 280 150 967	346	34.0					
Bosch 0 280 150 003	380	37.3	76.0	300			
Bosch 0 280 150 015	380	37.3	76.0	300			
Bosch 0 280 150 024	380	37.3	76.0	300		Volvo	B30E
Bosch 0 280 150 026	380	37.3	76.0	300			
Bosch 0 280 150 036	380	37.3	76.0	300		MB	4.51
Bosch 0 280 150 043	380	37.3	76.0	300		BMW	
Bosch 0 280 150 814	384	37.6	76.8	300			
Bosch 0 280 150 834	397	38.9	79.4	300			
Bosch 0 280 150 835	397	38.9	79.4	300		Chrysler	
Lucas 5207008	413	40.1	82.6				
Nippon Denso (black)	430	42.2	86.0	255	2.9	Toyota	7MGTE,3SGTE
Lucas 5208009	431	42.3	86.2				
Bosch R 280 410 144	434	42.5	86.8	300		Bosch R Sport	
Bosch 0 280 150 400	437	42.8	87.4	300		Ford	4.51
Bosch 0 280 150 401	437	42.8	87.4	300		Ford	
Bosch 0 280 150 041	480	47.1	96.0	300		MB	6.91
Bosch 0 280 150 403	503	49.3	100.6	300	0.5	Ford	
Lucas 5107010	530	52.0	106.1				

(SQR [(55 ÷ 39)] x 20 = 23.75 lb/hr at 55 psi
(a 19 percent difference)

On boosted engines keep in mind that some fuel pressure increase is required just to keep even with increases in manifold pressure when the boost arrives. To actually increase fuel delivery, the increase must be *above and beyond* such increases!

Therefore, simply by changing fuel pressure based on manifold pressure, the maximum flow of the injector is changed 40 percent! Most experts suggest 70 psi as a maximum system pressure, particularly if the ECU is not programmable. This is because very high pressures begin to affect the ability of the injector's internal spring to force the injector closed against fuel pressure flowing through it and may also affect the ability of the armature and windings to open the injector. As a consequence, the opening time, closing time, or repeatability of flow during the open event can change and become unpredictable. Furthermore, as mentioned previously, flow gains with increased system pressure offer diminishing returns. MSD Fuel Management discovered the following effects when varying the pressure of 72 and 96 pound-per-hour (at 43.5 psi) injectors using 85 percent duty cycle:

FUEL PRESSURE	72 PULSED FLOW (LB/HR)	96 PULSED FLOW (LB/HR)
30	56	72
40	64	84
50	71	93
60	76	101
70	76	106

The ability of the injector to realize high flow rates depends upon the ability of the fuel pump and plumbing to supply enough fuel. As the fuel pressure increases, the pump is constrained to supplying progressively lower volumes of fuel. In addition, it is worth pointing out that the fuel pump can also be limited in its ability to supply fuel by the hydraulics of supply line(s) size, filter(s) restrictions, restrictions posed by bends and turns in the supply line, and restrictions and pressure differentials in the fuel rail. The ability of the pump to supply fuel is also affected by the voltage supplied by the vehicle's electrical system (more about this in the section on fuel pumps).

Fuel Management Units

When adding turbocharging to engines with stock fuel injection systems, the ideal situation would be to maintain stock injection characteristics under low load conditions, adding some sort of fuel enrichment during the relatively rare bursts of turbo boost. This is exactly the approach taken by Bell Engineering of San Antonio and several other turbocharging specialists.

Bell Engineering supplies a special variable rate of gain (VRG) fuel pressure regulator that is inserted in the fuel return line, which raises fuel pressure roughly 7 pounds per pound of turbo boost, automatically causing the injectors to provide enrichment. This fuel management unit (FMU) is adjustable for the manifold pressure threshold at which it begins to raise fuel pressure, and it is also adjustable in terms of the rate of gain in fuel pressure for each psi increase in manifold pressure. Under naturally aspirated conditions, the regulator is dormant. Given the fact that most injectors do not function well above 70 to 100 pounds of fuel pressure, starting with nominal fuel pressure of 40 psi, an FMU enables fueling for turbo boost in the 5 to 7 psi range.

FMUs work well to increase fuel delivery in manifold pressure ranges above the range of the stock MAP sensor on a speed-density EMS, for example on a previously normally aspirated powerplant that has been converted to turbocharging. The stock MAP sensor on the NA engine is probably a 1-bar sensor that is essentially blind to pressure much above atmospheric (although it may have just enough ability to detect boost in the 0–5 bar positive range that it activates a fuel cut!). In this case, the MAP-based EMS continues to fuel for atmospheric pressure while boost pressure climbs, but the FMU increases the volume of each injector pulse to provide correct fueling under boost conditions.

FMU tuners should be aware that running an FMU on a *MAF-based* EMS can cause over-enrichment if the FMU comes into play at the same time that the MAF detects super-normal airflow and increases injection pulse width. This would be the case if the above turbo-conversion powerplant had a MAF sensor—at least until airflow exceeded the authority of the stock MAF sensor (or possibly until the ECU deduced a possible overboost condition and instituted countermeasures or went into limp-home mode and began disregarding MAF sensor output).

Carroll Supercharging sells a device marketed under the brand name Superpumper, which is essentially an electronic FMU that offers the additional sophistication of fuel pressure gain that is programmable in a nonlinear fashion with manifold pressure, enabling the Superpumper to provide fuel enrichment and to (almost) correct for nonlinear variations in engine volumetric efficiency.

Injector Electrical Characteristics

Another factor to consider when selecting injectors has to do with the electrical power requirements of fuel injectors, which must be compatible with the power supply and switching capabilities of the electronic control unit (ECU) that drives them. In addition, of course, when combined with the electrical requirements of all other onboard equipment put together, the fuel injection system power requirements must fall within the constraints of the electrical output of the vehicle's battery and charging system.

When it comes to electrical power requirements, think *plumbing*. Most car people intuitively understand fuel supply systems. They understand what pressure (pounds per square inch) means, they understand the concept of volume in pounds or cubic centimeters per minute. They understand discussions of injectors as valves, and they understand fuel line size and resistance to flow and supply restrictions and so forth. Electrical properties affecting fuel injectors and ECUs are similar to plumbing concepts, but the terms used to describe them are different. Voltage is used to describe electrical "pressure," as psi is used to describe fuel pres-

sure. Amperage is used to describe electron flow, as pounds-per-hour describes fuel flow. Resistance in ohms describes how easily current flows through something like a wire or coil, which is similar in concept to the varying resistance to the flow of water through a small pipe versus a large pipe. Ohm's law states that:

CURRENT = ELECTROMOTIVE FORCE ÷ RESISTANCE
or
AMPERES = VOLTAGE ÷ OHMS

Since injectors are activated by electromagnetic force, each injector contains a magnetic coil with a particular resistance to the flow of electricity through it. This is also referred to as its impedance, which is rated in ohms. A coil of wire rated at 16 ohms has much greater resistance to current flow than a coil of wire rated at 2 ohms. Many port fuel injectors are rated at roughly 16 ohms. A 16-ohm coil will allow much less current to flow through it for a given voltage than a 2-ohm coil.

Very large port injectors and throttle-body injectors (usually huge in comparison to port injectors) may require more current to produce enough magnetic force to operate the injector quickly and reliably. Very large injectors require large springs to force the injector armature and needles closed against the large volume of fuel flowing through them. This can be achieved by lowering the resistance of the windings of the injector coil to flow more current and by designing the ECU injector driver circuits so that they can safely provide sufficient current to drive the low impedance injectors. And there's the rub: ECUs designed to drive low-resistance injectors can easily provide the lower current needed to drive high-impedance injectors. But ECUs designed for a port injection system may encounter a problem if you want or need to change to low-impedance injectors.

For example, if you decide your modified V-6 turbo needs larger injectors because the injectors are already running close to the limit in duty cycle (that is, they're on all the time, or nearly all the time), you may discover that many of the larger flow injectors readily available are low resistance. Your ECU's power supply might be able to provide additional current, but the transistors and integrated circuits in the driver section of an ECU could blow out trying to regulate the higher electrical draw of a low resistance injector. Actually, virtually all modern aftermarket ECUs can drive low-resistance injectors, though how many is a different question; OEM powertrain control modules may *not* be capable of safely driving low resistance injectors. Since electrical circuits are often over-designed to an extent, the ECU may work fine for a time, but with a shortened life. For this reason, it is essential to match injector resistance to the capabilities of the ECU driving them.

Most high-flow, low-resistance injectors are designed for peak-and-hold drivers, which conserve power circuitry by supply a higher peak current to bang open the big injectors quickly, then reduce power to a lesser state required to simply hold the injectors open while simultaneously reducing thermal loading through the injector windings.

Injector Physical Layout

The final factor to consider in injector selection has to do with the physical layout of the injector, which must be compatible with the fuel supply or rail connections. The layout must also permit the injector to mate properly with the bosses on the intake manifold (or wherever else the injectors are located on the intake air plumbing system) or in the throttle body. Some injectors are supplied by an O-ringed side feed, others an O-ringed top feed, while others present a top-feed hose-barb fitting. Still others are designed for supply via banjo bolt. The injector's metering/spray pattern must also be compatible with the application. In terms of single-hole metering, annular orifice metering, multi-hole metering, and multi-hole metering with two sprays. Lucas has introduced a line of disc injectors in the last several years that are designed to replace Bosch pintle injectors.

FUEL SUPPLY SYSTEMS: PUMPS, FUEL LINES, REGULATORS, DUAL-FUEL, HYBRID MECHANI-CAL-ELECTRONIC

Fuel injection systems require a reliable and stable fuel supply. In the absence of a sufficient and accurate fuel supply, the most precisely calculated injection pulse width will not provide accurate fuel delivery.

The fuel pump must be capable of providing adequate fuel at the conditions of highest demand, which is wide-open throttle (WOT) at high rpm and maximum manifold pressure (peak power). Fuel pumps are rated to provide a certain fuel flow at a certain pressure. Be aware, however, that there can be pressure losses in the fuel line, through the fuel filter, and in the fuel rail. Also be aware that the fuel pump's ability to deliver a certain volume of fuel *declines* when it's forced to supply increased fuel pressure.

Where present, the fuel pressure regulator must have adequate capacity to supply the greatest engine needs as well as the capacity to precisely regulate fuel pressure by bleeding off sufficient excess fuel through a return line back to the fuel tank under the conditions of *least* fuel requirements (at idle). Regulators are equipped with a diaphragm spring that determines the default pressure that must exist upstream of the regulator in the vicinity of the injectors before the diaphragm will crack open to begin bleeding off fuel in order to hold the target pressure. Some regulators are equipped with an adjustable bolt that varies pre-load against the spring to increase (or decrease) fuel pressure. Most regulators have a vacuum-boost reference port designed to employ manifold pressure to vary fuel pressure to keep it consistent with the pressure of the intake manifold into which the injectors are spraying.

You can check the actual maximum fuel delivery capacity—with the engine off—by artificially supplying a target manifold vacuum/pressure reference to the regulator, energizing the fuel pump, and collecting the fuel that would normally return to the fuel tank in a safe container of known volume over a fixed period of time (such as 30 seconds or a minute). This will yield the maximum amount of fuel available to feed injectors without bleeding fuel rail pressure below the specification. This method takes into account any restrictions in the fuel supply plumbing that may affect the advertised fuel delivery of a known new fuel pump. It can also tell you the capacity of an unknown pump or a used pump that may have suffered inefficiencies due to wear.

Some newer vehicles are equipped with *deadhead* fuel supply systems. These have no fuel pressure regulator, no fuel return line, and a variable-speed electric fuel pump. Engine management systems controlling EFI with deadhead fuel supply systems monitor fuel rail pressure sensors and vary fuel pump voltage to increase or decrease fuel flow or pressure. Such systems typically do not attempt to maintain exact fuel pressures like a pressure-regulated system, but instead provide pulse width compensation for whatever the fuel pressure happens to be, which is similar to battery-voltage compensation.

The following fuel supply calculation can be worked backward or forward. You can start with a known engine fueling spec and calculate a minimum fuel pump capacity to provide adequate fueling. Or you can start with a pump spec and calculate the max-

An inline fuel pump operates under computer control to pressurize the injector fuel rail. Usually the EMS turns on the pump for a few seconds when the key goes on, again while cranking, and after starting only as long as the engine is running (to prevent pumping after a crash). Usually the pump is energized by a computer-controlled relay, though pump output can also be altered by managing voltage to the pump under EMS or Boost-a-Pump control. MSD

imum-horsepower engine it can supply. Suppose you have a 300-horsepower engine.

FLOW = Horsepower x BSFC

Assuming this is a turbocharged engine, most experts suggest using 0.55 as BSFC (Brake specific fuel consumption in pounds per horsepower-hour).

Therefore:

FLOW = 300 x 0.55

FLOW = 165 lb/hr

Convert pounds to gallons by dividing 165 by 6 (slightly over 6 pounds per gallon of gasoline). The answer is 27.5. Therefore, the fuel pump must supply 27.5 gallons per hour at the injector rated pressure, which is slightly under half a gallon per minute. Doing a little math, we can figure that at full throttle this engine requires a quart of fuel every 33 seconds. It is clear that any pump that could fill a container through the outlet of the regulator *at the required pressure* in less than this will do the job. The calculation looks like this:

27.5 gallons per hour divided by 3600 (seconds in an hour) equals 0.00763888 gallons per second of fuel required. Dividing 1 (gallon) by 0.0076388 tells how many of this volume it takes to make 1 gallon, or, looking at it another way, how many seconds of this flow it takes to make a gallon (130.9090909 . . .). Dividing this figure by 4 (quarts in a gallon) yields the number of seconds of this flow it takes to fill a quart container (convert this to metric on your own).

MSD Fuel Management has offered a roller vane-type electric fuel pump designed for port fuel injection that supplies 43 gallons per hour (282 pounds per hour) at 40 psi and 12 volts of power, which would be very adequate for the 300-horsepower engine discussed above. Many normally aspirated engines use fuel injection systems designed to run at 40 psi under full load, which means zero vacuum. But suppose you turbocharge this engine to 10 pounds of boost, referenced to the fuel pressure regulator, raising fuel pressure by 10 psi? It is critical to understand that this pump will deliver *less* fuel at 50 pounds of fuel rail pressure than it would at 40 psi (50 psi is what a nominally 40 psi manifold pressure-referenced regulator will produce in the fuel rail at 10 pounds of boost).

How much horsepower can the MSD pump support? At 40 psi with 12 volts of power, the MSD pump will flow 282 pounds per hour.

Horsepower = Flowrate ÷ BSFC
Assuming a BSFC of 0.55 (turbo motor),
the equation becomes:
Horsepower = 282 ÷ 0.55 = 512

The pump can support 512 horsepower at 40 psi.
But suppose the pump can only deliver 200 lb/hr at 50 psi?

Horsepower = 200 ÷ 0.55 = 364

Just when you need it most (under boost) the pump is only good for 363, *not* 512 horsepower!

There is no simple formula to compute what happens to pump fuel flow capacity with increased pressure, since this depends on various factors including on the horsepower of the pump. You have to either measure fuel flow at the target pressure or get this information from the manufacturer.

Many fuel rails are equipped with a connection allowing attachment of a fuel pressure gauge, and if there isn't one, adding one should be a simple matter for a competent fabricator. This allows a mechanic or EFI designer to observe the fuel pressure while operating the vehicle. In fact, today some ECUs can monitor fuel pressure through a sensor on the fuel rail and then display it on the engine data page (or even sound an alarm or activate a check light). Clearly, unless yours is a deadhead fuel supply system, the pressure should either remain constant under all conditions (never falling) or, if the regulator is referenced to manifold pressure, vary in exact correspondence with manifold pressure.

The performance of an electric motor increases with voltage and decreases with a voltage drop. Experiments by MSD Fuel Management show that at 40 psi and 12 volts, a particular MSD electric pump flowed 220 pounds per hour. At 13.5 volts, the flow increases to 340 pounds per hour.

The voltage delivered at the fuel pump will be a function of the capacity of the battery and charging system, possibly limited by the resistance of the wiring circuitry connecting the pump to the battery. You will get the most voltage at the pump by running a heavy-gauge low-resistance wire directly from the battery to the pump and switching this wire with a low-resistance relay controlled by the ECU. Beware: Some original-equipment fuel supply systems are equipped with a relay-switched fuel supply such that under conditions of low demand, a resistor-equipped wire supplies less fuel; under higher demand, the ECU flips the relay and the system now supplies power with an ordinary wire. If you rework the electrical supply to the fuel pump, make sure you are not inadvertently using a resistance wire that might dramatically reduce pumping capacity.

AIR/OXYGEN DELIVERY:
Electronic Throttles

Electronic throttles eliminate the need for a cable connecting the accelerator and the throttle, because the throttle is controlled by a fly-by-wire electric stepper motor under engine management control as influenced by an accelerator position sensor connected to the ECU. Electronic throttles make it easier for automakers to incorporate traction controls and they potentially enable the computer to achieve higher mileage and better emissions by smoothing out the heavy foot of a poor driver who is constantly pumping on the accelerator.

Automakers put considerable time and expense into developing electronic throttles, because it is obviously safety-critical that the electronic throttle never malfunctions (or if it does, that it is fail-safe in such a way that the throttle never sticks open).

Electronic throttles are the bane of performance tuners because they enable computers to back off the throttle if the computer thinks the car is accelerating "out of the envelope," for example, if someone has added a turbo conversion to the engine. This is great for automak-

ers that wish to avoid warranty claims if someone installs power-adders that over-boost and damage the engine, then de-installs the turbo kit and brings the vehicle in for free repair (in addition, data-logging may now be able to detect illegal modifications).

In 2003, virtually no affordable aftermarket engine management systems offered electronic throttle control. Electronic throttle is an issue for tuners working on very new modified vehicles. In some cases they have been forced into the extreme situation of installing sophisticated aftermarket engine management systems and adding mechanical throttles.

Idle Air Control

Engine idle speed affects idle quality, idle emissions, charging output, engine cooling, and the automatic transmission or transaxle—so precise idle speed is important. Before electronic idle speed stabilizations, tuners adjusted idle speed using an adjustment screw on the carburetor or throttle linkage. However, once set, an adjustment screw had no way to compensate for changes in engine load. The load from turning on the air conditioning or other high-consumption electrical accessories could lug down the engine, as could using the power steering to park, for example. Idle speed might vary as a cold engine warmed up (requiring careful adjustment of the manual choke or automatic choke pull-off linkage).

Modern engine management systems maintain idle speed within a preconfigured optimal rpm range, regardless of engine temperature or parasitic loads. This has come to be known as IAC, which originally referred to GM's idle air control system.

The IAC valve is opened and closed by an electrical stepper motor that opens and closes a small air bypass circuit that allows more or less intake air to flow around the throttle. Increasing the volume of air bypass increases idle speed—exactly like opening the throttle slightly—while reducing the bypass airflow decreases idle speed.

The IAC valve is controlled by the EMS. The ECM monitors idle speed when the throttle position sensor or throttle switch signals the computer that the throttle is closed. If idle speed is out of range of the target spec, the computer commands the IAC valve to either increase or decrease the bypass airflow. Target idle speed will typically be modified according to sensor inputs from the coolant sensor, brake switch, and vehicle speed sensor. Idle speed may also be increased when the air conditioning compressor is engaged, the alternator is charging above a certain voltage, the automatic transmission is in gear, or when power steering loads are high.

Boost Controllers

Boost controllers often use pulse width–modulation (PWM) techniques to bleed off boost pressure on its way to the reference port on the wastegate actuator diaphragm in order to—on occasion—under report boost pressure in such a way that the wastegate permits a turbocharger to build more boost pressure in the intake than it otherwise could. In effect, a boost-control solenoid valve lies to the wastegate under ECU control. The boost control solenoid contains a needle valve that can open and close very fast. By varying the digital control frequency to the solenoid, the solenoid valve can be commanded to be open a certain percentage of the time. This effectively alters the flow rate of air pressure through the valve, changing the rate at which air bleeds out of a T in the manifold pressure reference line to the wastegate. This effectively changes the air pressure as seen by the wastegate actuator diaphragm.

The wastegate control solenoid can be commanded to run at a variety of frequencies in various gears, engine speeds, or according to various other factors in a deterministic open-loop mode. Or—by monitoring manifold pressure in a feedback loop—the engine management system can monitor the efficacy of PWM changes in the boost control solenoid bleed rate at altering boost pressure in the intake manifold, increasing or decreasing the bleed rate to target a particular maximum boost.

The basic algorithm sometimes involves the EMS "learning" how fast the turbocharger can spool and how fast boost pressure increases. Armed with this knowledge, as long as boost pressure is below a predetermined allowable ceiling, the EMS will open the boost control solenoid to allow the turbocharger to create overboost beyond what the wastegate would normally allow. As overboost approaches the programmable maximum, the EMS begins to decrease the bleed rate through the control solenoid to raise boost pressure as seen at the wastegate actuator diaphragm so the wastegate opens enough to limit boost to the maximum configured level of over-boost.

Cam Phasing and Valve-Lift Control

Although a few tuners have modified the basic operating equipment of variable valve systems, this type of activity involves expertise on the level of sophisticated engine design and construction. For the vast majority of tuners, the actuator system itself is a given, and what might be considered for modification is the EMS control portion (though the main issue is likely to be duplicating the factory control strategy exactly when an aftermarket programmable EMS gets installed for other reasons).

The earliest variable valve systems from Honda, Porsche, and others were essentially binary. The valvetrain was in either high-rpm or low-rpm mode. Under certain conditions, the computer would essentially flip a switch that would open a valve to provide oil pressure that would drive forward a plunger to push against the timing chain in order to advance the intake cam relative to the exhaust cam.

In the case of Honda's patented VTEC system, the EMS would open (or close) a valve that pushes into place a set of pins that lock a complement of cam followers into place on all cylinders such that a more radical set of cam lobes comes into play with higher lift and longer duration, which improves performance at higher rpm. Honda's three-stage VTEC system is similar to the two-stage system, except that the EMS must control two binary valves, one to begin opening the second intake valve on a four-valve combustion chamber, the second to switch the intake cams to a more radical profile. So the EMS simply opens the two electrical switches at various combinations of engine rpm and loading, and the factory hydraulics do the rest.

The newest cam phasing systems from Honda, Toyota, GM, and others are more sophisticated. Honda's i-VTEC (intelligent VTEC) is exactly like earlier DOHC VTEC systems with the addition of infinitely variable (within a certain range) intake cam phasing, the main purpose of which is to vary valve overlap to optimize midrange torque and smooth over torque dips that tend to occur when the VTEC system is switching between the low-rpm and high-rpm cam lobes.

The VTC (Variable Timing Control) portion of the Honda system uses a spring-loaded intermediate hydraulic piston sliding on a helical gear between the intake camshaft and the cam sprocket.

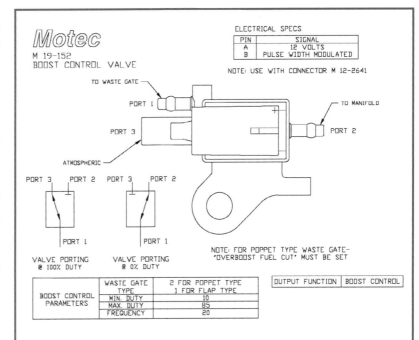

Essentially a controlled leak, under pulse width—modulated computer control, a PWM boost control valve flutters the valve open at high speed to vary the amount of air that bleeds through the valve to produce a target pressure on the other side. Such a valve is typically used to control a Deltagate-type wastegate by feeding a percentage of pressure to the Deltagate port to fight pressure tending to open the wastegate at the standard port, or to directly limit boost pressure to the standard port (i.e., to lie about manifold pressure). Motec EMSs have the capability to configure whether the boost control valve being open increases boost/manifold pressure or closed increases pressure. Motec

When the intermediate piston moves sideways along the helical splines, it is forced to rotate slightly, changing the phasing of the cam relative to the sprocket (and, of course, the exhaust cam and crankshaft). This is similar to the VarioCam system used on newer liquid-cooled Porsche powerplants. When the cam phaser is commanded toward zero degrees, the spring forces it back into position.

Phasing control is simple on a binary system (advanced or retarded only), but when the cam phasing must be infinitely variable within its authority range, a control valve directs the flow of pressurized oil to either side of the phaser piston depending on the duty cycle of the PWM input. Such systems require cam phasing measurement feedback based on interrupt pulses from the crank ref and cam sync sensors. The response of the PWM valve to varying system voltage, oil pressure, oil temperature, and oil viscosity requires highly sophisticated algorithms for repeatable accurate control.

Variable Intake Volume and Length

Systems like the ZR1 Corvette's secondary intake runner control and Toyota's MR2 Turbo TVIS (Toyota Variable Intake System) function to actuate a set of butterflies that open or close to allow intake airflow to move through one or two intake runners for super-high cylinder-filling low-speed intake velocities or for greater CFM at higher speed. This is a binary function: the secondary runners are either open or closed, and the opening or closing work is done with solenoid-actuated pneumatic pressure under computer control. Same with the variable plenum system

This sophisticated Bosch tri-mode virtual dash provides three different styles of display for standard street vehicle information. Several aftermarket EMS suppliers such as Motec market a "dash," which is a combination multigauge display, diagnostic readout, and warning alarm. It is entirely feasible to embed a complete LCD display in a car's instrument panel for constant EMS user-interface display.

via a relay that triggers a false temperature sensor signal to the ECU, triggering a special temperature enrichment (for an engine coolant temperature that would seldom or never be encountered, such as -20 degrees in southern California, or a high engine temperature of 250 degrees on an engine with a 180-degree thermostat). This would be relatively simple with a programmable ECU like the Haltech that enables you to specify temperature-based enrichment, but potentially possible on an older OEM ECU, which, unlike late-model ECUs, wouldn't detect and reject sudden temperature sensor changes as suspect.

TEMPERATURE CONTROL

Radiator fans, engine compartment fans, intercooler fans and water-spray systems, and oil-coolers and fans involve relay-controlled on-off functionality with feedback control based on temperature sensor input.

Many aftermarket engine management systems now provide fan-control subsystems that take hysterisis into account by providing differing configurable temperatures for fan on and fan off, depending on whether the system reaches a certain potentially triggering temperature with the fan currently on or off.

Water injection is either on or off. When it's on during boosted operations, it helps cool combustion and fight detonation. Water injection can be turned on during higher boost operations by a simple manifold pressure switch, or by the engine management system.

TRANSMISSION CONTROL

DFI and many OEM ECUs provide programmable control over torque converter lockup, activating lockup based on rpm, manifold pressure, and throttle position.

INFORMATION DELIVERY

All OEM EMSs provide a check engine light that turns on to indicate a problem with the engine or its EMS, and can be forced into a diagnostic mode in which it will flash out a list of specific trouble codes indicating what the EMS thinks is wrong. A few aftermarket systems have a check light, though most don't, given the powerful laptop-based user interface virtually all provide. Some engine management systems can drive a dash display usually called something like dash-logger, which is a (somewhat) general-purpose display that can act as a combination gauge/warning alarm and one-way interface to the driver.

found on Toyota's 1MZ-FE V-6. Manifold pressure acts on a solenoid actuator to open a set of butterflies in the intake plenum in order to open and increase the plenum volume for higher CFM at higher engine speeds and loading, or to close and convert the plenum into halves that effectively elongate the intake runner to extend all the way to the throttle body for high low-speed intake velocities. Porsche's Varioram is similar in concept.

From an engine management system perspective, these systems are usually open or closed, with an electro-pneumatic switching operation easily accomplished with general-purpose binary output drivers that are either on or off.

Nitrous Injection

A nitrous solenoid is something like a giant injector that stays open constantly when activated. ECUs from DFI, NOS, and Electromotive have the capability to examine a triggering signal from the arming switch of a nitrous system, and under specified conditions, activate the nitrous solenoid, simultaneously providing fuel enrichment for the nitrous through the fuel injectors (as opposed to traditional nitrous systems that provide nitrous fueling via a second solenoid through a spray bar or port nozzles).

Appropriate nitrous solenoids are provided by the suppliers of the nitrous kits (NOS and others), and ECU suppliers have documentation regarding fuel enrichment and wiring. ECUs without this specific capability can potentially provide enrichment

Making power is simple the way dieting is simple. If you want to lose weight, the secret is to eat less and exercise more. If you want to make more power, the secret is to put more air and fuel in an engine and light it off at the right time. Of course, the devil is in the details.

The name of the game in increasing power is removing bottlenecks. Bottlenecks come in various forms: air bottlenecks, fuel bottlenecks, and control system bottlenecks. Some power bottlenecks are generic to any piston engine, while others are specific to various types of electronic fuel injection and engine management systems.

In general, well-designed modern original equipment factory EFI systems will have balanced components, without unusual bottlenecks in any particular area. However, as soon as you begin modifying components of a performance engine, the bottlenecks will begin to shift around—at first subtly, then more dramatically with basic breathing improvements like head-porting and cam changes.

Essentially, truly modern *normally aspirated* piston engines are ultimately limited in the amount of torque and horsepower by the amount of *air* they can draw into the cylinders per displacement (volumetric efficiency). It is conceivable there could be *minor* performance improvements possible via engine management recalibration alone if the engine would always be using super-premium fuel. Though even this is dubious, since modern performance engines rely on knock sensor strategies—not default tuning—to protect the engine from bad gasoline. Any engine that could benefit much from more timing and better fuel is already set up to seize the day when good fuel is there.

Almost no free power is available anymore from EMS calibration changes. In fact, the EMS on a vehicle like the C5 Corvette is so well balanced that it is unlikely that *any* external bolt-on performance parts will individually make any power, even with perfectly matching calibration. The exhaust system, the air cleaner, the air inlet—the easy things—are simply not a restriction any more on a modern factory performance engine. If any bolt-on part would help much, it's already there in the dealer showroom—for the power or for the corporate average fuel economy. If you find a bolt-on part for your new Porsche that makes power or torque or both in some rpm range, it probably takes it away into other ranges in a way most people would find unacceptable (a tuned air intake with resonation box being one such example).

Bottom line: today, unless you're talking power-adders (blowers, turbos, nitrous, or high-powered fuels), the real NA power increases come from *packages* of breathing parts that almost definitely include hotter cams and maybe high-flow heads, or a *big* displacement increase to make much more power and torque. It is amazing how much some people with late-model Vipers or 'Vettes will pay today for performance exhaust systems that can't do any good.

For reasons of emissions, drivability, and warranty, some older original-equipment EFI systems may have some air, fuel, or control bottlenecks that can be removed by the performance enthusiast or racer, and some of these have external breathing parts that are suboptimal. But don't expect to set the world on fire.

On the other hand, factory turbo engines will set the world on fire—quite literally, if you don't know what you're doing when you overboost them too much. Factory turbocharged cars are a different story from a normally aspirated performance engine, and a much more interesting one, because an overboosted turbo engine can breathe so well that many internal and external factors can create a bottleneck. At stock power levels, forced-induction engines are almost never limited in performance by volumetric efficiency considerations but rather by artificial boost-control devices designed to protect engine components unable to withstand the severe mechanical or thermal loading of full overboost without damage. Careful hot rodding can typically unleash a *lot* of power and torque on a turbo-EFI engine or, to a lesser extent, on a supercharged powerplant.

Whether normally aspirated or forced induction, traditional methods of improving engine pumping efficiency with improvements to the valvetrain, cylinder head, and intake/exhaust systems or the addition of power-adders can work well with fuel-injected vehicles

Supercharged LT1 powerplant uses an intercooled Vortech blower to pressurize the manifold enough to make power in the 700–1,000 range. To live at this power setting, you need forged pistons and rods and precise engine management. We calibrated this engine with a DFI Gen 6 system on a Norwood Autocraft Superflow load-holding engine dynamometer.

Custom Porsche 968 turbo conversion from Norwood Performance represents the apex of today's radical club-race/outlaw street performance. Using sophisticated Motec engine management, high-boost turbocharging, and super-duty internal parts, the engine makes over 600 horsepower at the wheels.

with electronic engine controls if done properly. Of course, any improvements in VE require commensurate increases in fuel-delivery capacity, as well as appropriate spark timing.

EFI POWERPLANT CONSTRAINTS
Air
1. Throttle body (and fly-by-wire throttle body)
2. Airflow meter (VAF, MAF)
3. Air intake ram tube
4. Air cleaner
5. Cold manifold (versus coolant-heated or exhaust cross-over)
6. Intake manifold
7. Exhaust (headers, pipe[s], cat, muffler, tailpipe)
8. Displacement (bore and stroke)
9. Cams and VE (including variable lift and cam phasing, VTEC)
10. Head-porting
11. Engine change
12. Cold-air intakes

Fuel
1. Injector (size, duty cycle, pressure limitations)
2. Staged injectors
3. Fuel pump (flow @ pressure, voltage, and so on)
4. Fuel lines and filters
5. Fuel rails (rail pressure pulses, flow constraints)
6. Injector geometry
7. Fuel pressure
8. Fuel pressure regulators

Controls
1. Speed/density versus MAF versus Alpha-N
2. Recalibration versus auxiliary computer versus standalone aftermarket versus programmable sensor issues
3. Variable rate of gain regulators (fuel management units/Super-pumper, etc.)
4. Cam changes
5. EMS blind sports
6. Knock control
7. Water injection
8. Injection duty cycle
9. Emissions
10. Limp-home strategies
11. Boost limits
12. OBD-II and flash-PROM ECUs
13. Traction controls
14. Electronic (fly-by-wire) throttles
15. Engine mathematical model and complexity
16. Anti-tampering countermeasures

Power-Adders
1. Forced induction (turbos, blowers)
2. Boost control
3. Bypass/blow-off valves
4. High-powered fuels
5. Nitrous (proportional, fueling through primary injectors, and so on)
6. Blower-clutch control

INJECTION SYSTEM AIRFLOW BOTTLENECKS
Throttle body: A throttle body should have enough capacity to provide sufficient airflow at peak power without producing a pressure drop (bottleneck) through the throttle area. A throttle that is too small for a normally aspirated powerplant will kill horsepower. The situation is a little trickier on a boosted engine, because a certain amount of air might be forced into the engine through a larger throttle body at a lower boost pressure, or—assuming the compressor has some excess capacity—might still be able to be forced through a smaller throttle body with less CFM capacity with a higher boost pressure.

So why not just install a *huge* throttle body guaranteed to have more capacity than the engine can use? Throttle bodies that are much too big for an engine lose their authority way below full throttle, because once the engine is fully loaded, it simply cannot use more air (think of driving up a long, steep hill below peak power rpm). Throttle and full throttle may produce exactly the same power because the engine cannot develop the horsepower to use the extra air at full throttle. Conversely, at lower throttle setting, a too-big throttle body can be twitchy and over responsive, with a tiny bit of additional throttle producing a large airflow increase in a way that hurts drivability.

The way to match a throttle body with an engine is to compute the engine airflow and throttle body airflow to make sure the throttle body CFM exceeds that of the engine. The way to do this is to compute engine airflow at an assumed volumetric efficiency, or, even better, to compute current airflow based on the horsepower produced on a dynamometer. Throttle body airflow will be available from the manufacturer—in the case of a new aftermarket throttle body—or can be calculated. If you have drivability problems with a twitching throttle, progressive units have been used on some engines that accelerate or decelerate the number of degrees of throttle-shaft rotation with increasing accelerator percentage.

Some engines now have fly-by-wire electronic throttles that are driven by a computer-controlled electric stepper motor under electronic-accelerator directives rather than by a cable connected to the accelerator. Such throttles are very effective for traction-control systems. They can smooth out poor driving habits that waste fuel, and they can prevent people from hurting an engine by driving it too hard when cold.

In some cases, performance tuners have eliminated the electronic throttle in favor of an older-style cable-actuated throttle body when switching to an aftermarket EMS. If the throttle size is a constraint, there is no reason why companies like RC Engineering—with the ability to machine larger openings in a throttle for a bigger throttle plate in a traditional throttle—couldn't increase the flow potential of an electronically controlled throttle if it becomes a restriction.

Air meters: The air metering section of the injection system, located upstream of the throttle body, may itself be a bottleneck to higher horsepower by restricting airflow into the engine, thus hurting volumetric efficiency. EFI systems that are designed to measure airflow directly usually estimate mass airflow (MAF) by considering the cooling effect of intake air on a hot wire or film suspended in an orifice of fixed size through which all inlet air must flow.

In the case of older velocity airflow meters, the pressure of inlet air against a spring-loaded door suspended in the inlet air stream produces the air velocity reading. If there is a significant pressure drop through the airflow sensor, then the cylinders will

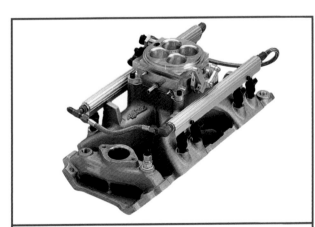

With a small-block Chevy V-8, you've got a lot of options. This turnkey EFI intake from Accel provides a high-slow single-plane intake with injector bosses cast in, injectors, fuel rails, water temp sensor, four-barrel Holley bolt-pattern throttle body, TPS, and idle air control stepper motor. This package could be used to convert a stock throttle body injection system to port injection, as the basis of an EFI conversion (with aftermarket or GM EMS), or to upgrade the airflow of a long-runner TPI setup (along with a PROM upgrade). Accel

Pro-Flow Tech Mass Air Meter Sizing	
Meter Diameter	Max HP
55mm	320
70mm	380
80mm	485
75 mm Bullet	600
87mm Bullet	650
77mm Pro M	700
80mm Pro M	800
83mm Pro M	900
92mm Pro M	1000
117mm Pro M	1500

not fill as well with air, which will lower power output. OK, why not use a gigantic airflow meter? The problem is that the larger the airflow meter, the less sensitive it is to small variations in airflow because smaller orifices magnify changes in air velocity that make it easier for the sensor to get accurate readings. Big sensors with lazy airflow through them can also have problems with turbulence that affect the readings. (It is *critical* under all circumstances that a MAF be positioned where the air cleaner, sharp bends in the intake ram pipe, and the throttle body do not interfere with smooth airflow through the MAF.)

There are several ways to eliminate MAF or VAF airflow restrictions. The first is with a high-flow performance MAF from a company like Professional Flow Tech. High-flow MAFs are available for many performance engines equipped with OEM MAF sensors, and some of them can be adjusted or programmed to optimize operation on different engines, or to compensate for larger injectors. Some engines with OEM VAF meters can be upgraded to a higher-flow mass airflow meter, a newer, superior technology that requires no correction for air temperature and is not subject to problems from manifold and intake shock waves fibrillating the mechanism like a VAF sensor.

The second way to eliminate air-meter restrictions is to eliminate the airflow sensor. HKS sells a device called the vane pressure converter (VPC), with which you can retain the stock EFI ECU while eliminating the air-metering system. The VPC is an auxiliary computer that combines data from an add-on MAP sensor with engine speed, air temperature, and a complex table of internal translation numbers to estimate mass airflow for a specific vehicle, then sends this information to the stock ECU in the proper electrical format that exactly simulates the action of an airflow meter, but without the airflow restriction of the air meter. The MR2 project in this book used a VPC to eliminate Toyota's VAF meter (until we installed an aftermarket EMS).

Yet another path to eliminating intake restriction from a MAF meter is to junk the factory EMS. Converting to an aftermarket speed-density EMS that measures manifold absolute pressure (MAP) and air temperature, calculates engine speed, and then deduces airflow or engine loading based on this data without requiring an airflow metering device enables a tuner to eliminate the metering system bottleneck. Virtually all aftermarket engine management systems permit use of speed-density engine controls as the default engine load sensing, though some alternately allow MAF or other load sensing. Several of the project vehicles in this book involve such modifications.

The main disadvantages of replacing the stock fuel injection system is that it may not be legal (for emissions reasons), and the replacement may not preserve all the capabilities of the original system.

Air cleaner and cold-air intake: It is critical that an air cleaner should clean the air well and flow enough air that the air

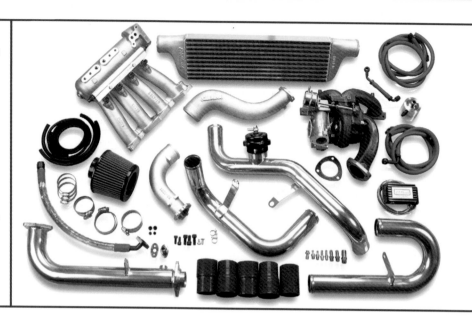

Edelbrock's intercooled Honda turbo conversion package is the shape of future hot rodding. Because there is little or no power available from tuning alone on modern four-valve engines, the choices are massive (and expensive!) "all-motor" displacement or VE changes or power-adders. Edelbrock's package converts 1.6-liter Hondas to turbo power, increasing power from 125 to more than 200 horsepower, with fuel enrichment provided via fuel pressure modifications. Boost timing retard is provided via a retard interceptor box. Edelbrock

cleaner is not a bottleneck to airflow that hurts engine VE. Dust or other foreign contaminants in charge air will *dramatically* accelerate engine wear. The only excuse for ever running any engine with open stacks or other unfiltered air intake is if it is a competition engine with severe-to-impossible space constraints.

Virtually all modern factory EFI systems use cold-air intake systems that breathe air that is 100 percent isolated from hot engine compartment air, which is very important because cold air is denser and will make noticeable power. The air cleaner should not introduce undesirable turbulence to intake air that interferes with a MAF sensor. The air cleaner system should insulate intake air from heat soaking by hot engine components.

Intake ram tube: An intake tube of the proper length can improve throttle response and actually make power on some engines. Air moving through a long tube has momentum and inertia that can help get air in the engine when you suddenly open the throttle (a very large in-line air cleaner or resonator can interfere with this). You can tune an intake ram tube on some engines to provide pressure-wave tuning benefits at some engine speeds—exactly like tuning the length of intake runners in the manifold and cylinder head (see the chapter on intake manifold construction in this book).

Intake manifold: Freedom of intake manifold design is one of the great things about multi-port electronic fuel injection. Dry manifolds can be optimized to deliver *air* into an engine without the compromises required to manage *fuel* and air. Mixture distribution is not an issue with port EFI, and there is no need to choose between a single, centrally located carburetor or the nightmare of mounting individual carbs or carb throats on each runner for perfect distribution (like a motorcycle). Dry intake manifolds do not require power-robbing coolant- or exhaust-heating chambers that help wet manifolds keep fuel vaporized in cold temperatures on carbureted and throttle body injection engines but hurt VE by heating up the intake air. Some port-EFI throttle bodies are heated by engine coolant to prevent ice from forming in cold weather, but the intake manifold should not need heating on a dry manifold. This is something to keep in mind if you are converting a wet-manifold to port EFI by installing injector bosses. Carb-type V-8 intake manifolds will not require the exhaust-heating crossover passages cast into the manifolds on many engines, and these could be blanked off to help improve engine VE.

Some four-valve intake manifolds are equipped with runner systems that close off secondary runners at low rpm with throttle butterflies in half the ports. These systems improve cylinder filling by maintaining very high port velocities in the primary runners at low engine speeds, opening the secondaries under computer control as airflow increases with high rpm and power settings.

It is common hot rodding practice on engines like the 16-valve Twin Cam MR2 Turbo to remove the staged intake system (Toyota Variable Intake System or TVIS) and siamese the two intake ports for each cylinder by band sawing the manifold open and cutting/grinding out the wall between each pair of runners. This will hurt low-end performance but contribute to power gains at high levels of boost and airflow. We built a custom-welded stroker crankshaft on the MR2 project in this book to improve low-end torque when we removed the TVIS for the big peak power. This type of radical intake surgery affects both runner tuning and flow.

Some dry intake manifolds are equipped with variable-plenum intake manifolds, which is similar to Porsche's Varioram, in which a giant butterfly can open at higher flow to convert a dual-plane intake into a single-plane intake. The MR6 project in this book used a 3.0-liter Toyota 1MZ V-6 with such a system. The Toyota system is equipped with a vacuum-operated diaphragm that actuates a rod and crank, which then rotates to close the plenum butterfly at high vacuum, opening it at higher manifold pressure or under boost. This does not require computer control to function effectively to improve low-end torque and high-end power.

Intake manifold design is so important to drivability, torque, and power that this book contains a whole chapter about it. If you are lucky you have an engine so commonly hot rodded (small-block Chevy V-8, Honda four, Ford 5.0, and so on), that there are off-the-shelf aftermarket performance intake manifolds available to change the torque and power curves on the engine. Airflow rules and tradeoffs with dry EFI intakes are identical to those of carbureted manifolds. Long narrow runners produce great low-end torque and tuning effects at lower rpm. Short, wider runners flow better and produce tuning effects at higher engine speeds.

If you fabricate a custom sheet-metal manifold or heavily modify a factory or aftermarket manifold, you should always have a cylinder head shop test it on a flow bench (before and after) to make sure there are no problems or inconsistencies that should be corrected, and to make sure you haven't done something that is performance-*negative*.

Injector placement and aim in a custom EFI manifold is important. If at all possible, fuel injectors should aim at the base of the intake valve where the stem joins the cup (or between the two intake valves on a four- or five-valve engine). TWM, manufacturers of Weber DCOE-compatible throttle bodies and intake manifolds, points out that increased distance from a port injector to valve(s) aids in fuel atomization and vaporization, and injectors have been successfully aimed straight down the throats of intake runners from across the plenum. Beware of resonation effects and "standoff" causing fuel to reverse-flow out of the intake runners back into the plenum and migrate from port injectors closest to the throttle body through the plenum in the direction of runners farther from the throttle body.

The EFI intake manifold may be an issue if you are interested in supercharging. Traditional root-type superchargers are bolted directly to a vestigial intake manifold on American performance V-8s, with one or two four-barrel carbs bolting directly on top of the blower, and the wet mixture helps to lubricate the supercharger. Today it is common practice on modern four-valve four- and six-cylinder EFI engines to develop supercharger conversions that utilize a special intake manifold casting in which the casting serves as the blower housing in which an Eaton supercharger rotor assembly in effect replaces the plenum section of the OEM manifold.

Building a custom blower conversion for an EFI engine for which none is available is a difficult engineering job that requires a thorough understanding of supercharger flow versus speed characteristics and extensive fabrication capabilities (if not the ability to build casting mold bucks) that include the ability to make a belt-drive system work with the stock crank pulley and accessories. The alternative is a centrifugal supercharger of the type that looks like a big alternator and requires dealing with blower drive-belt issues but would rarely require any changes to the EFI engine's intake manifold.

Exhaust: Traditional performance exhaust flow principles still apply to computer-controlled port EFI engines. However, many electronic engine management systems require O_2 sensors to function in closed-loop mode in order to optimize economy and drivability at idle and light cruise, and O_2 sensors don't work

CXI's LT4 'Vette became a showcase for LT performance parts under factory GM engine management controls. The main strategy was (1) a huge 420-inch stroker displacement and (2) incredible amounts of nitrous injection. This Dual-tank Nitrous Express system was designed to deliver 500 horsepower on top of the stroker powerplant's all-motor capabilities.

well until they're *hot*. OBD-II factory EMSs may have as many as four or five O_2 sensors in the exhaust headers and after the cat(s). Any modifications that steal heat from the exhaust (such as a turbo conversion) can render unheated O_2 sensors inoperative. And cats with too much flow may not stay hot enough at idle to "light off" and do their job. Really, all O_2 sensors should be heated these days. Some highly effective turbo conversions for new engines mount the turbocharger downstream of the cat. Engine management tuning strategies can use retarded timing at idle to put more heat in a cat and O_2 sensor.

Some cat-back exhaust systems (the only kind that are legal unless proven to degrade emissions in the federal test procedure) have made some power on otherwise stock engines (particularly on overboosted turbo engines), but the greatest value of such systems is as part of a balanced package of performance upgrades that push back bottlenecks all across the engine's pumping system from intake through cylinder heads and camshaft through exhaust.

Displacement: Hot rodders used to say there's no substitute for cubic inches. That's not strictly true, but it is true that displacement is essentially always a good thing in terms of street performance. How does adding bore or stroke affect the EMS on a powerplant with electronic fuel injection? In terms of percentages, without extremely radical monster-motor procedures, except in the case of V-12 engines with huge bore and stroke increases, it is difficult to add more than 10 or 20 percent more displacement (think of GM 350 V-8s bored and stroked to 383, 396, or 421—increases of 9.4, 13.1, and 20.1 percent). Without commensurate cam and head breathing upgrades, *peak power* is unlikely to increase as much as the displacement because breathing is likely to be suboptimal compared to the stock displacement.

Engine control systems with MAF sensors will probably handle 5 to 15 percent more airflow without complaining, and closed-loop O_2-sensor-based feedback will certainly trim idle and light-cruise operations with no trouble. Engine control systems with speed-density EMS will probably require fuel pressure increases as a countermeasure against lean mixtures at peak power. A few dyno runs with a wideband air/fuel-ratio meter and an adjustable fuel pressure regulator will enable a tuner to optimize the fuel pressure for best power.

Cams, Heads, and VE: Changing the camshaft and cylinder head airflow specs changes the volumetric efficiency of an engine because the purpose of such changes is to increase engine breathing—at least in some range of operation—and that's what volumetric efficiency *is*. And there is usually a cost to a cam change. If you make engine breathing better in one part of the range, you usually make airflow worse somewhere else. As long as the airflow changes are not so extreme as to move airflow out of the bounds of what is acceptable airflow to the stock calibration (and within the authority of the airflow sensor), a MAF-based engine control system will still provide good timing and fuel control. A MAP-sensor-based system will *not* be able to compensate for these cam-change or head-porting VE changes without recalibration.

Engine changes: If you can get an EMS to work with changed displacement and even changed cams and breathing, why not with a foreign engine in a complete engine swap situation? For example, is it feasible to swap in a Ford 460 big block in place of a Ford 5.0 V-8 and use the stock EMS, or, say, replace a Chevy 5.7-liter V-8 with a Chevy 7.4-liter big block? Or maybe replace a Honda 1.6 VTEC with a 2.3 VTEC? The simplest situation would be if you swapped out an engine with MAF controls to a much bigger powerplant with a similar VE curve, retaining the

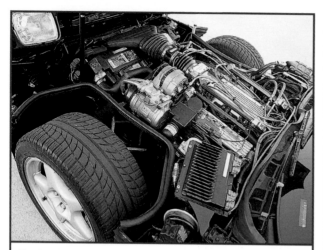

Five-hundred horses worth of "safe" nitrous and fuel at a minimum requires special port-delivery rails and solenoids and a good nitrous management strategy. This may or may not deliver nitrous fuel enrichment through the primary injectors, with secondary injectors, or in bulk via special nitrous-fuel nozzles. In many cases, nitrous systems have been an on-off situation, but EMSs with pulse width–modulation can bring in the nitrous and fuel more gradually, fluttering the solenoids to limit the shock of a massive horsepower onslaught.

MAF system, changing fuel delivery with a combination of larger injectors and fuel pressure regulation such that pulse width for the old engine size worked well for the changed engine size.

For example, changing a Ford 5.0 to a 460 (which is 152 percent larger), changing out the stock 19-pounds-per-hour injectors to 30-pounds-per-hour injectors, and reducing fuel rail pressure slightly to reduce the injector flow to 28.8-pounds-per-hour would get fuel delivery very close. Now, if you installed a programmable MAF sensor and adjusted it for across-the-board underreporting of airflow so the new engine's airflow looked to the EMS like the old engine's airflow, on a dyno with a wideband air/fuel-ratio meter, you could tune fuel pressure to optimize horsepower at peak-power rpm and then let the closed-loop system trim idle and light cruise for perfect air/fuel ratios. (Another possibility would be to build a MAF bypass tube to bypass a percentage of air equal to the increase in displacement.)

The situation with a speed-density EMS would be at once simpler and more complex. All the EMS knows about is manifold pressure—not airflow—and the radically changed airflow of a large displacement change would be invisible to the EMS. If the VE of the two engines were close, simply adjusting fuel injector size and pressure would get the replacement engine running, and then you could trim fuel pressure as above to optimize high-end power, leaving it to the closed-loop to get air/fuel ratios right at idle and light cruise.

But what about performance at power levels too high for closed-loop, but below peak power? Good luck. If drivability was acceptable, a dyno and wideband air/fuel-ratio meter would reveal the details of how good or bad the calibration was for the new engine. I would not go down this path unless there was an interceptor box or custom PROM or flash-PROM available if less heroic measures were unacceptable. Clearly, big VE changes would not work well on the above speed-density situation without interceptor or recalibration, but, of course, on some engines like the small Ford and Chevy pushrod V-8s, they built both MAF and speed-density controls, so you could also convert over if nec-

essary and use the first strategy.

INJECTION SYSTEM FUEL BOTTLENECKS

In general, getting more fuel into an engine is way less of a challenge than providing more air. This is mostly because the engine requires 15 times the air by weight, and much more by volume, and engines depend on air being pushed in—normally—by atmospheric pressure alone. However, internal-combustion engines do require a certain amount of fuel per horsepower, exactly how much varies somewhat according to the brake specific fuel consumption (BSFC) of the engine—related to the efficiency of the combustion chamber and how much fuel needs to be wasted cooling off combustion and the quality of the state of tune.

Injector capacity: Injectors need to be small enough to provide a good engine idle at minimum achievable pulse width (never less than about 0.8 millisecond, often as much as 2 milliseconds). Yet injectors need to be large enough to provide sufficient fuel for peak power at a fuel pressure that is within the limits of the injector's ability to open and close against the pressure (figure 90–110 psi; see the section on injectors in the chapter on actuators) and at a duty cycle (percentage of time open) between about 10 and 80 percent.

It is easy to find higher-flow injectors that will fit in place of the stock injectors on most factory engines. It is a different matter to recalibrate the control system to adjust the opening time and pressure so they provide the right amount of fuel. The injectors must be electrically compatible with the ECU firing them. If the dynamic range required of injectors on a small engine with few cylinders that's capable of extremely high power is beyond what's possible, even with dynamic pressure changes, the solution may require multiple staged injectors per cylinder, with one injector operating at idle and a second (possibly of higher flow) coming into play as the engine requires more fuel.

Fuel pumps: The high-pressure port-EFI electric fuel pump (and its plumbing!) must be up to the job of moving sufficient fuel to maintain the required fuel pressure (which may vary according to engine speed and loading). Fuel pumps vary in capacity and are capable of less volume as pressure increases. If your fuel pump will not pass muster, there may be a higher-flow in-tank unit available, or you'll need to supplement the stock in-tank unit with an auxiliary in-line pump.

An auxiliary pump can be activated off a factory EMS as needed (for example, to boost pressure) via a relay from the primary fuel pump power supply activated by a simple boost-pressure switch. Alternately, many aftermarket ECUs can turn on an auxiliary fuel pump at a certain manifold pressure or MAF setting in much the same way that ECUs turn on cooling fans at a certain configurable threshold temperature and then off at a second user-configurable pressure.

Devices such as the Kenne Bell Boost-a-Pump can widen the dynamic range of a fuel pump by varying power supply voltage with increasing manifold pressure, which is similar to the action of modern OEM "deadhead" (returnless) fuel supply systems that control fuel pump voltage in stages or continuously with the main ECU to target approximate required fuel pressure according to the pressure data from a fuel rail pressure sender, then trimming injection pulse width to precisely fine-tune the fuel delivery.

Fuel pressure regulators: On looped high-pressure fuel systems, pressure regulators determine how much fuel sprays out of an injector per length of pulse by pinching off fuel flow returning to the tank to the degree required to target a certain fuel pressure in the injector manifold (fuel rail). Some pressure regulators are adjustable by turning a jam bolt that varies the pre-load against the diaphragm spring in the fuel pressure regulator. And most reg-

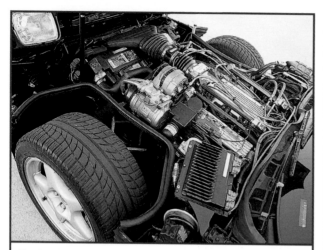

Quatro-valve 3.0–3.5-liter Ferrari V-8s make good hot rodding material, says Bob Norwood, who likes the combination of high compression and turbocharging on a small-displacement four-valve powerplant to keep the torque up until the turbo kicks in. It is a rather simple matter to remove the K-Jet mechanical injectors and open up the bosses for electronic injectors. In its day, there were people who tried to hot rod the K-Jet system, but an EFI conversion is much better equipped to handle power-adders like turbocharging.

ulators dynamically raise and power fuel pressure according to a manifold pressure reference line that supplies vacuum or boost pressure that hinders or helps the action of the diaphragm spring.

Adjusting injection pressure can be a highly effective way of tuning the final air/fuel ratio under open-loop conditions. Fuel management units (FMUs) take this principle one step further by increasing fuel pressure as a *multiple* of boost pressure (or in one big step in the case of nitrous activation). Electronic FMUs like the Carroll Superpumper go beyond this by varying pressure under programmable electronic control according to configurable

graphs in the control unit.

Be aware that new deadhead factory engine management systems may adjust injection pulse width to compensate for rail pressure, defeating efforts to tune AFRs with fuel pressure changes. With even higher authority, EMS closed-loop mixture-control strategies will similarly trim pulse width to achieve certain target levels of residual exhaust gas oxygen at idle and light cruise.

CONTROL SYSTEM BOTTLENECKS

Basic control strategies: If an engine uses an EMS with MAF sensor load sensing, then additional air entering the engine at a given speed will automatically call for more fuel, and most stock EFI systems can deliver some additional fuel for a given speed. For example, older Mustang 5.0 MAF systems and the GM TPI system can deliver a small percentage of extra fuel for heavier-breathing engines—about 10 or 15 percent. This is enough for minor modifications like small cam changes, air filter changes, and so on, but not enough to support fueling for significant turbo boost.

The situation is more complicated if the EFI system is not programmable and uses a speed-density system or Alpha-N control strategy. Changing cams may alter the manifold pressure for a given engine speed and loading, causing the ECU to be confused or wrong about estimating how much air the engine is consuming and therefore injecting the wrong amount of fuel. Alpha-N control systems that determine engine load purely from engine speed and throttle position similarly have perhaps even less ability to compensate for modifications that affect engine volumetric efficiency, including—notably—the varying effects of a turbocharger on engine VE.

Throttle body control strategies are great for engines with big cams that have difficulty idling smoothly due to fluctuating manifold pressure (which can cause MAP or MAF systems to hunt—as the ECU chases it own tail with a wandering load signal). Throttle body load-sensing strategies cannot work well with turbochargers, which can produce huge VE changes while throttle and engine speed remain steady (except in the case of sophisticated aftermarket engine management systems like the Motec that can switch over from Alpha-N to MAP sensing only when a turbo

This C3 'Vette uses twin turbos without intercooling to rack up the boost. Fuel cooling via rich mixtures is a viable trick to fight detonation.

tion. This typically requires high processing speed and highly accurate engine-position sensing of the sort used to detect misfires for OBD-II.

Some aftermarket knock control systems like the J&S Safeguard are capable of associating detonation with specific cylinders and launching knock control countermeasures (spark timing retard) for the knocking cylinders *only* while the other cylinders remain at peak efficiency—simultaneously protecting the engine but maintaining higher power levels throughout. Some aftermarket EMSs can trigger water injection at higher levels of boost as an effective anti-detonation countermeasure. Keep in mind that the volume of water required for knock control is small, which makes port water injection impractical and makes poor water distribution difficult through a dry intake manifold.

Some modern aftermarket ECUs warn tuners when the basic calibration reaches 100 percent duty cycle driving fuel injectors, which means the injectors are open constantly and the air/fuel ratio is at least temporarily out of control. However, if an EMS even *nears* 100 percent injector duty cycle, there may be no ability to add enough fuel—if a turbo makes a little more boost, for instance, or if the engine is cold and makes too much boost—without the injectors reaching 100 percent and going static, which could then cause lean detonation or some other type of meltdown.

Fuel experts like Russ Collins, owner of injector re-flow specialists RC Engineering, seem to agree that it is not a good idea to run pintle-type injectors at more than 80 percent duty cycle. This is because they may not have enough time to close all the way, and thus can fibrillate somewhere around 85 percent duty cycle (see actuator section on fuel injections), with fuel flow becoming less predictable and even diminishing in fuel flow in a dip around 85 percent until the injector goes static and fuel flow reaches maximum.

Obviously, the ECU is essentially blind as to whether an individual injector is open or not or fluttering somewhere in the middle (or even whether it is stuck completely open or stuck closed). The ECU has no way of knowing whether the installed injector can accurately carry out commands to open for very short intervals (say, shorter than the injector's combined open and closing time) to idle well on small, high-power engines. However, with this said, engines are run every day at 100 percent duty without adverse consequences. If the engine is equipped with a fuel management unit, keep in mind that *it* will continue to add fuel enrichment by pumping up the fuel pressure to the static fuel injectors held wide-open.

Lack of complexity in the engine model of an EMS or lack of processing speed can cause blind spots. An EMS with insufficient resolution in the speed or loading tables may in effect be unable to distinguish between subtly different operating conditions that really require different engine management strategies. Does the EMS have the brains to correlate knock sensor data with detonation in specific cylinders to target countermeasures at the cylinders that need them and not the others? How much time—or how many degrees of crankshaft movement—does it take before the EMS reexamines the state of engine sensors and executes new actions to affect what is happening?

Limp-home strategies: When an engine management system decides it has detected a serious problem such as a failed MAF, MAP, or TPS sensor, high levels of overboost, uncontrollable knock (or total lack of a knock sensor signal), or serious overheating (or lack of oil pressure!), the engine management strategy becomes one of keeping the engine operating in a way least likely to cause engine damage.

Engines equipped with redundant MAP and MAF sensors

begins making boost).

An engine management system's blindness to modification-induced changes in engine volumetric efficiency can cause problems as above with speed-density controls, but the opposite can also be true. While a MAF-sensed system's awareness of airflow increases can be good in the case of moderate airflow increases within the authority of the MAF and EMS to manage, this awareness can also produce incompatibilities with control modification strategies such as fuel management units if *both* attempt to increase fueling in certain operating ranges. And airflow awareness can also alert a factory EMS to levels of overboost that trigger limp-home countermeasures that disregard sensor values and attempt to protect the engine with safe-mode engine management procedures that kill performance—particularly on an EMS with OBD-II capabilities.

The solutions to the above problems include the typical array of PROM recalibrations, interceptors, and piggyback ECUs, aftermarket standalone EMSs (which might take over all or just some functions from the main ECU), programmable sensors and actuators, additional injector controllers, and so forth—some of which will not be legal for street use.

EMS blind spots: Knock sensors are essentially sensitive microphones, and many OEM sensors are matched to detect specific resonant vibrations from the particular engine they are protecting. The problem is that engine modifications—particularly valvetrain items or other components that affect other parts—can produce vibrations and resonation that can appear to be a knock to the engine control system. Some aftermarket knock sensor controllers have a sensitivity or tuning adjustment that can be used to help filter out vibrations unrelated to actual knock.

Keep in mind, true detection of actual knock requires in-cylinder pressure transducers that detect the steep pressure spikes of genuine detonation. These are expensive but obtainable and can be found integrated into special spark plugs. A second-best method of detecting genuine knock involves carefully monitoring micro-changes in crankshaft acceleration and deceleration to detect signs of explosive combus-

HOT RODDING EFI ENGINES

Bob Norwood made a pretty good living hot rodding Ferraris like this 1,800-horse modified 308. Using Motec engine management, the bow-tie big-block GM powerplant managed to push the 288-GTO-paneled 308 to a one-way record of 267 miles per hour at Bonneville.

can be run with excellent performance with one or the other having failed, and some turbo engines like VW's 20-valve 1.8T will tolerate more overboost without going into limp-home mode with one or the other disconnected (the MAF, in the case of the 1.8T). However, uncontrolled overboost, knock, or overheating is so dangerous to the health of an engine that the EMS will immediately do anything and everything it can think of to avoid engine damage.

Examples include the following:
• Turning on the check engine light (so the driver can help out).
• Defaulting questionable or failed sensors to "reasonable" *safe* values to keep sensors that have failed in exotic ways from driving insanely dangerous engine management strategies.
• Retarding timing the maximum amount to kill possible knock and limit performance.
• Attempting to cool off combustion.
• Killing knock or preignition and limiting performance by injecting tons of fuel into the engine.
• Limiting turbo boost to the maximum degree possible (for example, if there is a dual-diaphragm wastegate).
• Closing any secondary intake runners to limit intake airflow to minimal volumetric efficiency—which also limits performance.
• Executing "camel mode," (in the case of the newest GM V-8s) in which coolant loss or overheating triggers the EMS to switch off fuel and spark to four cylinders on alternate banks at intervals that optimize the ability of intake air to cool the resting bank to the maximum extent before the operative bank is damaged by excessive heat.

Naturally, you don't want any of this to happen just because the ECU is unhappy you have installed a turbo kit or pumped up the boost too much on a powerplant that is otherwise fully capable of making serious power.

Overboosting: What is overboost and how does a factory EMS work to prevent it? Most turbocharged engines have internal calibrations that provide a good calibration to the limit of the fueling capacity of the fuel injectors, even if this is higher than "design" street power. It makes sense to provide plausible engine fuel and ignition mapping to the limits of the factory fuel injectors, which would typically be sized sufficiently large to provide sufficient fuel to the airflow/boost limits of the stock turbocharger. This is because the engine can overboost if the wastegate manifold-pressure reference line is cut (melted, crushed, disconnected, and so

on), and a "safe" calibration—even in forbidden overboost territory—is one more countermeasure designed to help avoid engine damage from lean mixtures if the engine does overboost in case fuel cut strategies fail.

The first line of defense against overboost is typically the wastegate setting, which will bleed off exhaust gas pressure to limit turbo boost to safe "normal" levels, for example at 9 psi. The second line of defense will be a fuel cut when overboost reaches levels that could possibly damage the engine if it were fueled with regular unleaded gasoline, for example 12 psi. The last line of defense would be anti-knock/limp-home countermeasures described elsewhere in this book, which implement performance-killing timing retard and over-fueling (while there is still enough fuel capacity to produce excessively rich combustion-fooling mixtures), and limp-home breathing limitations (secondary ports close, and so on).

The fuel cut tactic is implemented when airflow or MAP sensor data signals reach levels defined in parameters in the EMS configuration as overboost. This implies that every time the ECU looks at the current MAF/VAF or MAP data, it checks to see if airflow or boost exceeds the overboost threshold to see if the system should supply normal fueling or cut fuel as defined in a configurable (internal) parameter that defines overboost. If the EMS detects overboost, it will set the overboost flag, which can only be cleared by turning off the engine (or, in some cases, by removing power from the ECU by temporarily removing a fuse or a battery terminal).

Outlaw tuners working with factory ECUs on overboosted turbo engines or turbo or supercharger conversions of normally aspirated powerplants typically work to disable the fuel cut using several possible methods.

1. Manual or electronic air bleeds and pressure regulators interposed between the wastegate actuator diaphragm and the manifold pressure reference source enable a tuner to "lie" to the wastegate, causing the wastegate's mechanical controls to raise boost pressure (assuming the turbo compressor has the airflow capacity)—whether or not the factory EMS has any control over wastegate operation. Alternately, electronic valve controllers like HKS's EVC can pulse modulate manifold reference pressure or a bleed orifice in the reference in order to dynamically lie to the wastegate—as can an EMS (like the Motec system used on several projects in this book) via tables that provide PWM control to a boost control solenoid like the one GM uses for factory boost control. Pulse width–modulated solenoids flutter an electromagnetic valve open and closed at varying frequencies to limit airflow and pressure through the valve to produce a desired flow.

2. Electronic voltage clamp devices like the HKS Fuel Cut Defenser used on the MR2 project in this book typically use variable-compressor (or resistor) circuits that truncate or transform the MAP or MAF signal, minimizing it to the level required to make sure the factory ECU cannot detect overboost when it happens. However, this tactic also means the factory ECU is also calculating fuel and timing as if the engine were making less boost than it actually is, which is dangerous without some other method of providing fuel enrichment during overboost. Not only is a "defensed" engine overboosting, it is being inaccurately underfueled for the actual overboost.

3. Check valves can prevent positive manifold pressure from reaching a factory engine management system's MAP sensor. This is useful for systems like the factory Honda engine control system used on two project vehicles in this book, which has the ability to measure about 0.6 bar (almost 9 psi) positive boost pressure, even though the Honda engines being controlled are all normally

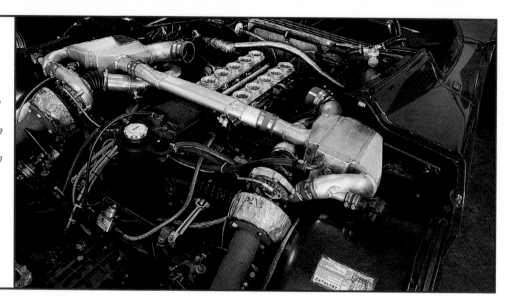

John Carmack's boosted F50 uses a 60-valve high-compression V-12 and twin-turbo conversion to make some serious horsepower at medium boost pressure. In a gutsy move aimed at achieving emissions legality and avoiding detonation problems, Bob Norwood converted the car over to E80 fuel, a blend of ethanol and gasoline that has high octane, good intercooling capabilities, and a clean burn.

aspirated. Unfortunately, Honda EMS *hate* seeing boost and will complain and limp at anything over a pound or two of positive boost pressure. A one-way check valve and bleed allows negative manifold pressure up to atmospheric to reach the sensor. Once the reference line from the intake manifold begins developing boost pressure in the other direction on a turbo or blower conversion, the valve closes so the maximum pressure ever seen on the MAP sensor is atmospheric. The bleed is there to prevent vacuum or boost from being trapped in the reference line between the check valve and MAP sensor.

4. A disconnected MAF sensor is enough to prevent VW 1.8Ts from noticing that overboost is happening.

Increasing turbo boost/fueling enrichment: With a vehicle that is factory turbocharged, it is often a simple matter to increase the power output by increasing the maximum boost—which is generally possible if there is a way to raise the pressure at which the wastegate opens to cut off further increases in turbo boost. A turbocharger increases torque about 8 percent per pound of turbo boost. The MR2 project in this book produced massively more power and torque at 16-psi boost than at the stock 7 to 9 psi. HKS sells an electronic valve controller (used on several projects in this book) to enable a tuner to overboost an engine to make more power.

An EVC or similar boost controller fools the stock turbo wastegate by underreporting manifold pressure to the wastegate under electronic control. Mechanical means of increasing boost include mini regulators that restrict manifold pressure to the stock wastegate; pressure bleeds that prevent full manifold pressure from arriving at the wastegate actuator; increasing the diaphragm spring pressure of certain stock adjustable wastegates; and replacement wastegates or larger turbochargers. Most factory turbochargers have excess airflow capacity that can be used with the above wastegate tricks.

Gasoline bottlenecks: Methanol and ethanol (alcohols) can be controlled by electronic injectors. The main engine management issues have to do with the large volume of fuel required, since stoichiometric air/fuel ratios are much richer with alcohols (9:1 and 6:1), which are already partly oxygenated. Alcohol fuels are expensive, and you burn a lot because of the low air/fuel ratios. But specific power is higher than gasoline because a given weight of air can burn more alcohol (remember, it's already partially oxygenated), and air is *always* more scarce in a piston engine.

Nitromethane, which can actually combust as a mono-pro-

pellant in the complete absence of oxygen (the molecular structure is rearranged, liberating heat), requires fuel in such copious quantities even when burned with oxygen that even gigantic 1,600 cc/min electronic injectors used for methanol injection are far too small for nitromethane. Indy cars have used mixtures of mostly methanol with nitro as a power-adder.

Nitropropane has been mixed with gasoline (nitromethane will not mix with gasoline, but will mix with methanol), and the mixture could possibly be electronically injected, but nitropropane is too rare, expensive, and unavailable to be practical as an automotive fuel except under unusual circumstances in odd competition classes.

Methanol, by the way, is a harsh compound that is corrosive to ordinary gasoline-fuel systems, which typically must be upgraded to stainless steel pipes and other components. Nevertheless, it is common practice to purge the fuel supply systems of methanol-fueled race cars after a race to protect injectors and other delicate components from being damaged. Methanol burns with a clear, invisible flame. It is a cold-blooded fuel that can be difficult to start in cold weather, requiring a gasoline mixture to fire up. Despite these rather nasty disadvantages, methanol is a great racing fuel because it makes more power than gasoline, it has an extremely high octane rating, and it burns with a cool, smooth, even combustion that is easy on piston engines.

A happy combination of electronic engine management, methanol, and high-performance turbochargers is currently enabling tiny 1.8-liter Honda fours to run at more than 900 horsepower in import drag racing classes. Interestingly, O_2 sensors and wideband O_2 sensors work fine with alternate fuels like alcohols, because even though the target air/fuel ratios may be greatly different than gasoline, the target exhaust gas oxygen is quite similar.

Propane and natural gas, by the way, are *not* high-powered fuels, because they make less power per pound of air than gasoline, alcohols, or nitro fuels, However, propane and natural gas are very clean burning (meaning they're street legal with almost no emissions devices), and they have high octane ratings, so they are capable of making a lot of power with an adequately large turbocharger.

Since they're gases, propane and natural gas mix well with air. Propane carbs (mixers, really) and intake manifolds do not have to deal with liquid-gas problems like gasoline-fuel engines, so there is not much of an advantage to using standard electron-

Powerhaus 968 turbo was designed to duplicate the 968 Turbo Porsche coulda-shoulda-woulda built but never did, so of course it had to use Motronic controls.

David Raines and Powerhaus techs sorted through the 944T, 968, and Motronic parts bin to put together a slick package of turbo conversion parts for the 968 that followed the Porsche design philosophy. They then worked with Autothority to build a calibration for the 3.0-liter powerplant good for 400–500 horsepower with excellent streetability. A 944 Turbo brain box was more easily adapted to handle the larger size and four-valve 968 breathing than the 968's normally aspirated Motronic ECM.

ic fuel injectors. What's more, the volume of propane or natural gas required for combustion is higher than what's feasible with gasoline-type injectors. The low temperature of vaporized LP or natural gas makes high-speed pulse width–modulation dicey.

Industrial engines that run on gases today typically use special mixers to introduce the fuel and electronic engine management for timing and other engine management functions. Propane is commonly used as a power-adder for turbo-diesel engines. And for this purpose, electronic solenoids similar to nitrous-oxide solenoids provide the required on-off constant-flow functionality required.

Street diesel engines are now universally managed with computers, which manage turbo boost, fuel pressure, and even electronic diesel throttles that help reduce emissions and noise.

Nitrous injection maintains its long time reputation as an easy and effective power-adder on the most modern OBD-II engines, the only real downside being the limited life of a tankful and the fact that nitrous is really only safe and effective at wide-open-throttle.

ENGINE DESIGN CONSIDERATIONS AND THE JOB OF ENGINE MANAGEMENT

Spark timing and advance have a large impact on the efficiency and exhaust emissions of an engine. As an engine turns faster, the spark plug must fire earlier in the compression stroke in *degrees before top dead center (BTDC)* in order to allow time for the mixture to ignite and achieve a high burn rate and maximum cylinder pressure by the time the piston is positioned to produce best torque at roughly 15 degrees past top dead center. The mixture does not take any more time to burn at higher rpm for a given air/fuel ratio, but the time it does take to burn *takes place over more degrees of crankshaft rotation.*

Spark advance was traditionally accomplished with a centrifugal advance mechanism in the distributor, which incrementally turned the distributor's cam mechanism by action of a set of spring-loaded rotating weights so the points opened sooner in crankshaft degrees as rpm increased.

Another factor affecting the need for spark advance as rpm increases is the need to modify ignition timing corresponding to engine loading and consequent volumetric efficiency variations. This is due to the throttle's effect position on cylinder filling and corresponding variations in air/fuel mixture requirements related to variations in *manifold pressure* and the effect of MAP on *cylinder pressure* during the ignition and burn process. A *denser* mixture burns more quickly, and a *leaner* mixture requires more time

to burn (two independent variables).

Traditionally, a spring-loaded canister is attached to a metal rod that progressively retards spark advance with increased manifold pressure by rotating the points breaker plate and points. Many carburetors have special ports for timed spark that delivers full manifold pressure reference at all speeds except idle, at which time the port source is covered by the throttle plate. This means that referencing a distributor vacuum advance canister to this special port did not produce any vacuum advance at idle (but did produce a surge of advance [and power] as the throttle opened).

Some engines are designed with suboptimal spark advance as an anti-emissions strategy. One tactic is to retard ignition timing at idle, for example, sometimes locking out vacuum advance in lower gears or during normal operating temperature but allowing more advance if the engine is cold or overheating.

Since oxides of nitrogen are formed when free nitrogen combines with oxygen at high temperature and pressure, retarded spark reduces NOx emissions by lowering peak combustion temperature and pressure. This strategy also reduces hydrocarbon emissions. However, retarded spark combustion is less efficient, causing poorer fuel economy and higher heating of the engine block as heat energy escapes through the cylinder walls into the coolant.

The cooling system is stressed as it struggles to remove the greater waste heat during retarded spark conditions. Fuel economy is hurt since some of the fuel is still burning as it blows out the exhaust valve, necessitating richer idle and main jetting to get decent off-idle performance, and because if the mixture becomes too lean, higher combustion temperatures will defeat the purpose of ignition retard—producing more NOx.

Inefficiency of combustion under such conditions also requires the throttle to be held open farther for a reasonable idle speed, which, combined with the higher operating temperatures, could lead to dieseling (not a problem on fuel-injected engines that immediately terminate injection when the key is switched off). By removing pollutants from exhaust gas, three-way catalysts tend to allow more ignition advance at idle and part throttle.

Valve timing has a large effect on the speeds at which an engine develops its best power and torque. Adding more lift and intake/exhaust valve opening overlap allows the engine to breathe more efficiently at high speeds. However, the engine may be hard to start, idle poorly, bog on off-idle acceleration, and produce poor low-speed torque.

All this occurs for several reasons. At low speeds, increased valve overlap allows some exhaust gases still in the cylinder at higher than atmospheric pressure to rush into the intake manifold, diluting the inlet charge exactly like EGR. This dilution continues to occur until rpm increases to the point where the overlap interval is so short in time that reverse pulsing becomes insignificant. The charge dilution of reversion tends to make an engine idle badly.

Valve overlap also hurts idle and low-speed performance by lowering manifold vacuum, which can cause problems on engines with carbs or throttle body injection. Since the lower atmospheric pressure of high vacuum tends to keep gasoline vaporized better, racing cams with low vacuum may have distribution problems on engines with wet intake manifolds and a wandering air/fuel mixture. This, again, may require an overall richer mixture in order to keep the motor from stalling. Changed or wandering vacuum will affect OEM speed-density fuel injection systems but would have less effect on MAF-sensed engine management unless excessive pulsing and turbulence interferes with the MAF's ability to accurately read airflow.

Temperature affects fuel injection because colder air is denser than hotter air, colder air inhibits fuel vaporization, and colder air affects combustion temperatures. Engine management systems normally have sensors to read the temperature of inlet air and adjust the pulse width of injection to compensate. Engines will make noticeably more power on a cold day because the cold, dense air increases engine volumetric efficiency, filling the cylinders with more molecules of air. This is bad news for a carb, which has no ability to self-compensate for air density changes, the only means of compensation being jet changes (one size per 40 degree temperature change).

As you'd expect, racing automotive engine designs always endeavor to keep inlet air as cold as possible, and even stock street cars now almost universally make use of cold air inlets, since each 11 degrees Fahrenheit of increase reduces air density 1 percent. On the other hand, gasoline does not vaporize well in cold air. Oil companies do change their gasoline formulation in cold weather to increase vapor pressure, which could cause a rash of vapor lock in sudden winter warm spells in the days of carbureted engines. Vapor lock, however, is rare on injected engines.

Cold-start fuel enrichment systems on EFI are designed to produce a rich enough mixture to run the vehicle even when much of the fuel exists not as burnable vapor mixed with air, but as drops of nonflammable liquid fuel suspended in the air or clinging to the intake manifold runner walls. Most air pollution is produced by cold vehicles burning cold-start and cranking mixtures as rich as 3 or 4:1 (even 1:1 during cranking). Electronic fuel injection systems sense coolant temperature in order to provide cold start enrichments (cranking, after start, and warm-up). Port EFI systems usually inject fuel straight at the intake valve and into the swirling, turbulent high-velocity air in that vicinity, which greatly improves atomization and vaporization. EFI does not normally need the exhaust gas heating that carbs require to provide acceptable cold-start operation. EFI manifolds may use coolant

Powerhaus' 968 Turbo package required some "special" parts—like the reworked Porsche manifold, throttlebody and MAF sensor.

heating to increase vaporization at idle in very cold weather.

Air density varies with temperature, altitude, and weather conditions. Hot air, with greater molecular motion, is less dense. Air at higher elevations is less dense, as is air with a higher relative humidity. Air is less dense in warm weather, but air that is heated for any reason on the way into the engine becomes less dense and will affect the injection pulse width required to achieve the correct air/fuel ratio. Intake system layout can have a great effect on the volumetric efficiency of the engine by affecting the density of the air the engine is breathing.

Air cleaners that suck in hot engine-compartment air will reduce the engine's output and should be modified to breathe fresh, cold air from outside. Intake manifolds that heat the air will deliver less dense air into the cylinders, although a properly designed heated intake manifold will quickly be cooled by intake air at high speed and can improve distribution at part throttle and idle when hot.

High compression ratios squash the inlet air/fuel mixture into a more compact, dense mass, resulting in a faster burn rate. Turbochargers and superchargers produce effective compression ratios far above the nominal compression ratio by pumping additional mixture into the cylinder under pressure. Either way, the result is a denser mass of air and fuel molecules that burns faster and produces more pressure against the piston. The peak pressures can also produce more NOx pollutants.

Lower compressions ratios (as are typical of forced-induction engines) raise the fuel requirements at idle because there is more clearance volume in the combustion chamber that dilutes the intake charge. Because fuel is still burning longer as the piston descends, lower compression ratios raise the exhaust temperature and increase stress on the cooling system.

Until 1970, high-performance cars often had compression

Dee Howard's '32 Pierce Arrow land-speed-record car runs a heavily modified P-A flathead V-12 with twin turbos and methanol fuel. The original fuel system was constant mechanical injection. For improved performance at the Bonneville Salt Flats, Bob Norwood converted the car to Motec electronic injection.

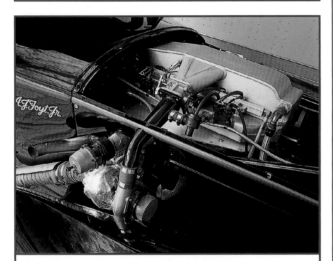

Speed-record '32 Pierce-Arrow V-12 is equipped with a custom plenum and twin throttlebodies. Gigantic 1,600-cc/min electronic injectors spray methanol into the modernized V-12 powerplant at a stoichiometric air/fuel ratio of 6.5:1, roughly twice that of gasoline. Methanol makes about 10 percent more specific power per air mass than gasoline, mainly due to the significant cooling effect of vaporizing alcohol, which greatly increases air density and can negate the need for an intercooler on efficient turbo systems running modest boost pressure. Gasoline-type injectors will work with methanol, but must be large to handle the volume.Bosch

Hot rod twin-turbo Norwood big-block powerplant uses turbochargers the size of a spare tire to achieve the prodigious levels of power required to approach 300 miles per hour in the thin, 6,300-foot elevation of the Bonneville Salt Flats. The Motec engine management system's altitude compensation greatly reduces the tuning effort required when carbureted or mechanical-injection vehicles arrive at the high desert dry lake in the summer

explode as combustion pressures and temperatures rise. Environmental concerns caused congress to outlaw gasoline lead in the early 1990s, completing a trend of many years toward unleaded gasoline requirements in cars. Precisely controlled mixtures in all cylinders via port fuel injection are a key tactic in avoiding lean cylinders with increased tendency to knock.

HOT RODDING CHEVY'S TPI

Chevrolet's L98 Tuned Port Injection engine was the last traditional small-block Chevy V-8, of which there were about 60 million built since the mid-1950s. TPI will fit on *any* traditional small-block Chevy, which gives it importance even though the last TPI engine was built in 1993. TPI will also fit many TBI truck motors built in the 1990s and a few in some big vans built beyond the 2000. It is, therefore, still important. The Jag-rolet project in this book involved swapping a TPI engine into a late-model Jaguar sedan, and hot rodding the power with a centrifugal supercharger and other significant performance modifications.

Chevy designed Tuned Port Injection for a 305-ci engine that would be driven on the street. Design goals included responsive acceleration with an automatic transmission in a heavy car under average driving conditions (which means low to medium rpm), plus good fuel economy and emissions (which implies tall gears). As it should with these goals, the TPI system makes great low-end torque that dies out rapidly above 4,000 rpm. This is a situation that leaves hot rodders and racers looking for ways to make more power at higher rpm on a TPI-injected Chevy.

It's important to keep in mind that torque is what makes a car fun to drive. Horsepower is a measure of how much work an engine can do at a certain speed, such as lifting a weight of a certain size. Torque measures the moment of a force, that is, its tendency to produce torsion and rotation around an axis. In other words, twisting force—the kind of force that causes a car to leap forward when you nail the gas because of the instantaneous rotational force of the engine's crankshaft. Torque is critical to an

ratios of up to 11 or 12 :1, safely accommodated with the vintage gasolines readily available in those days with octane as high as 100 ((R+M)/2). By 1972, engines were running compression ratios with 8–8.5:1. In the 1980s and 1990s, compression ratios in computer-controlled fuel-injected vehicles were in the 9.5–10.5:1 range based on fuel injection's ability to support higher compression ratios without detonation coupled with its precise air/fuel control and catalysts that are able to keep emissions low.

Higher compression ratios or effective compression ratios *demand* whatever octane and other anti-knock countermeasures are required to prevent detonation, which at worst case can shatter pistons and rings. Previously, high-octane gasoline contained the additive tetra-ethyl lead to slow down gasoline's tendency to

The multi-stage crank-driven mechanical pump on GM Racing's Ecotec drag motor pumps at least 800 horsepower worth of methanol into the engine and also provides dry-sump lubrication and scavenging of crankcase windage and blow-by. The great thing about a mechanical pump is that the output automatically increases with rpm.

engine's ability to accelerate.

Peak torque usually occurs at the engine's point of highest volumetric efficiency. Both race and street engines need to accelerate well from a variety of speeds; therefore, good torque over a broad range is important, not just peak torque. High horsepower achieved at high engine speeds over a narrow range (way above the peak torque at low VE where the cylinders are no longer filling as well, just more often) tends to produce cars that aren't flexible or fun to drive—you have to drive it like you're mad at it to get any performance.

Tuned-runner port injection is great at making torque, but there are trade-offs with various manifold/runner designs, just as there are with intake manifolds designed for carburetors. Port EFI manifolds do not have to deal with keeping fuel suspended properly in air as wet air carburetors or throttle body injection manifolds must do, which gives designers more flexibility with runner length, geography, and cross section—that is, the ability to tune the intake for the best torque curve for a given application.

However, it is a fact that intake runner length and cross section (as well as exhaust port/header length and cross section) do affect the torque curve of an engine. There are tradeoffs. Longer, narrower runners accelerate air to high velocities with high momentum that produces higher air pressure at the intake valve. If the engine is turning slowly, time is not a critical factor in cylinder filling and VE is higher with longer, narrower runners. At higher rpm, time is critical. There is not time for a long column of air to accelerate and enter the cylinder to fill it well. The narrow cross section of a runner implies less volume and more pressure drop at high rpm. Again, less cylinder filling.

If the motor is stock, and always will be, it is hard to beat the GM TPI system, which works well up to 4,600 rpm. Stock TPI manifolds negate the effect of better heads and cams and headers for higher-rpm power. If the motor is not stock or you need power above 4,600 rpm, the stock TPI manifold will have to be modified with larger runners and a larger throttle body or a complete new manifold (available from Arizona Speed and Marine, Accel/DFI, TPI Specialties (TPIS), and others). You'll need an aftermarket ECU or a modified PROM for the stock GM com-

puter that is matched to the new engine modifications. And if the engine changes again, you may need yet another PROM. With new intake systems, the heads will become the flow bottleneck. As always, every part of the complete system must be designed to work well together or some unexpected bottleneck can negate the effect of other expensive modified parts. A mild stock 305 engine is well matched to the stock TPI system.

But let's say you've improved the cam a little, and installed headers, perhaps even upgraded to a crate 350 HO motor with somewhat better cam and heads, designed to make roughly 350 horsepower. It's easy to install runners from ASM or TPIS and a matching larger throttle body. Typically, the TPIS runners are 0.190 inches in diameter larger than stock and 0.325 inches longer, which adds considerably higher rpm torque and power from 3,000 rpm up—in one test 75 ft-lbs over stock across the board from 3,700 rpm on up to redline!

Runners are available not only in larger sizes, but with adjacent ports siamesed so a cylinder can share two runners for better breathing at rpm (which hurts lower rpm power and torque). Serious hot rodders may decide to extrude hone the manifold base to match the flow capabilities of the new runners and higher cfm throttle body—or switch to a large manifold base like the TPIS Big Mouth, with its round 1.750 entry sizes and 1.960/1.200 exits.

A serious big-displacement Chevy small-block motor with a good 240-degree duration 0.550–0.600 cam and compression over 10:1, with some seriously ported good high-rpm aftermarket aluminum heads will require a complete new manifold to really take advantage of this configuration's ability to make high-rpm power. Accel, TPIS, and others make such manifolds, designed with considerably shorter and fatter runners and much higher-volume plenums. The Accel SuperRam runners are more than 3 inches shorter than the stock TPI (4.125 versus 7.250) incorporating D-section geometry of 1.878 x 1.970 (versus stock 1.470 round runner). The one-piece TPIS Mini-Ram is a serious high-rpm manifold with still shorter runners (3.5 inch) and huge 2.600 x 1.350 entry, tapering to 1.960 x 1.200 exits.

Properly sized carburetion can make virtually as much maximum horsepower as port EFI. Unfortunately, by sizing the venturis in such a way that they are not restrictive on a hot motor at high rpm, the carburetor would provide terrible low- and midrange power and torque. Good, big-runner TPIS manifolds, even short-ram EFI manifolds, and stock TPI will all be making 100 ft-lbs more torque at 2,500 rpm compared to a higher-rpm carb and manifold, which is power you can feel as well as measure.

HOT RODDING FORD'S MPI

The Ford 221/260/289/302 arrived on the scene in the early 1960s and, like the small-block Chevy, the Ford Windsor was built in enormous numbers. The MPI 5.0 is the system Ford used on 1985–93 Mustangs and some Explorers and trucks into the later 1990s. The MPI 5.0, like Chevy's TPI system, will retrofit to older Windsor V-8s. There is still a large number of Ford Windsor V-8s on the road, and it is still an important motor.

In order to get the 5.0-liter Ford V-8 air metering to work properly in low-range airflow, Ford restricted the high end, which created a market niche for companies like Professional Flow Tech, which builds performance MAF meters. The Pro-M MAF line uses patented aerodynamics that use resonation effects to get excellent metering at low airflow while handling 1,000 or more cfm on the high end. This translates to a bolt-on 15–20 horsepower on the 5.0-liter engine. The Pro-M 77 will work on any MAF EEC-IV engine, including the Mustang, the Supercoupe, and the SHO Taurus. Professional Flow Tech is currently design-

Norwood Autocraft propane-fueled Cadillac Seville used a 520-ci GM-Rat-type V-8 and twin turbos to make enormous power at 20-psi boost. Most engine management systems for gaseous fuels like propane use simple mechanical air-gas mixing valves to combine air and fuel, since the volume of fuel is much too large for traditional pintle-type solenoid injectors. On the other hand, propane and air do not tend to separate in the intake tract like gasoline and air, and an engine management system can control an electronic pressure regulator to control constant injection of propane. Electronic ignition is identical to gasoline systems except that propane is very high octane with its own flame speed, requiring unique spark timing for optimal performance.

ing a system for Ford speed-density EFI trucks that converts the system to Mustang high-output programming, including a high-performance MAF.

Many of the same principles that apply to GM TPI also apply to the Ford 5.0 system, which is designed to make good low-end and midrange torque. And, like the TPI systems, many parts are available to pump your 5.0. A good place to start is Ford, whose Motorsport program offers high-performance GT-40 upper- and lower-intake manifold pairs, as well as GT-40 iron and SVO aluminum heads, which will increase the breathing of a 5.0. Other common mods on Mustang 5.0s include cams; 77-mm mass air kits; Pro-M 77-mm complete conversion kits (with wiring, computer, and 77-mm sensor); SVO complete mass air conversions; 65- and 75-mm throttle bodies; 24- and 30-pound injectors; adjustable pressure regulators; the Crane Interceptor; shorty headers; cat back exhaust systems; and high-volume fuel pumps. Cartech, Vortech, Spearco, and others offer turbocharger and supercharger kits for the Mustang, some of which are even street legal in California. Most of these kits use add-on fuel pressure regulators that massively increase fuel pressure under boost to provide adequate fueling and enrichment.

The problem in modifying Ford 5.0 engines is that maximum horsepower is limited by the stock injection components: The 19 pound per hour stock injectors can make 300 horsepower at 45 psi. The stock fuel pump will supply sufficient fuel for 300 horsepower at 45 psi (the SVO pump will supply 375 horsepower), and the stock MAF meter will supply air (and voltage signal to the EEC-IV) for about 310 horsepower. The limit with stock EFI components is roughly 300 horsepower. Any strategy to increase power levels above this must address these limitations.

EEC-IV PROMs are not replaceable, but there are systems that use a performance module inserted between the EEC-IV and the wiring harness to alter the performance of the OEM computer, while other devices reprogram the EEC-IV by plugging into the diagnostic port.

CHAPTER 6
RECALIBRATING FACTORY ECUs

Virtually all digital engine management systems use table-driven software designed to be sufficiently general purpose that the basic software logic is potentially capable of running a variety of different engine applications. In fact, it is the calibration tables and configuration parameters—the data—that give the EMS a substantial portion of its personality managing a particular engine application.

This book has several projects in which we recalibrated the stock factory EMS as a tactic in a major performance increase. These include the Honda del Sol, Toyota MR2 Turbo, VW Golf 1.8T, and the BMW 325ci/M3.

Keep in mind that all engine management systems—even aftermarket programmable systems—have an overall hardware and software architecture that is specialized enough that it will limit the flexibility to manage some types of engines. This is especially true of mass-produced OEM systems, where cost is important. As an example, EMS hardware and software might have the ability to run either speed-density or MAF-based engine management, or both, or neither, in the case of some Alpha-N bike computers.

In some cases, the software architecture may only provide the ability to enrich fuel delivery above the raw, air-meter-based fuel calculation (this was true of the Toyota MR2 project in this book when it was running the factory computer). Some important basic operating considerations such as the number of cylinders, which are typically configurable on an aftermarket programmable EMS, are likely to be fixed in the system logic. But they might not be if, as is increasingly true, a particular manufacturer decides it is simpler and

cheaper to maintain a single software architecture across the entire landscape of vehicle and engine platforms.

There are many examples of software logical limitations that limit the recalibration flexibility of the EMS. And it is not practical or cost effective to modify the software architecture, that is, to change the actual machine instructions that the onboard computer executes. The only choice is switching to a different ECU.

Factory engine management systems have utilized a variety of hardware architectures that affect the difficulty of modifying the factory calibration in various ways.

THE MANY LEVELS OF PROM HACKING

The simplest kind of hacking is to buy a **performance chip or module** from one of the many suppliers that offer a variety of off-the-shelf upgrades for factory-stock vehicles or vehicles with standard packages of VE modifications (which they may also sell) and stick it in your car's computer. Such a solution should plug and play, though it may not if your application is at all nonstandard. More likely, the problem is that it won't do much all, producing reduced fuel economy or resistance to knock in favor of really *minor* power gains at or above the stock redline.

Today, except in the case of turbocharged engines, there is usually little or no "free" power available from a simple chip change. But hope springs eternal, and firms like Superchips, Hypertech, Jet, and others make a nice living selling chips and upgrade modules.

A second option is to buy a **custom-made chip or module** from an expert designed specifically for your engine and its modifications. Since it may take a few iterations to get it right, if you are remote from the expert, you may need to install the modifications and then run the vehicle on a chassis dynamometer with a coupled wideband air/fuel-ratio sensor and send the dyno/AFR chart back to the expert to be analyzed and optimized—and possibly repeat the procedure more than once. ASM, Autothority, Active Autowerke, and others sell custom chips and calibrations.

Yet another option is to acquire a **calibrator software package** that can modify or edit the calibration from the PROM or EEPROM in your factory EMS (as well as acquire the hardware you'll need to offload the calibration data to a personal computer and then reload it into the EMS when the tuning is complete). This is sophisticated software that intimately understands the calibration data structures of your engine application and provides an intuitive, user-friendly human interface.

On the hardware side, you may need a communications cable, PROM burner, emulator, OBD-II diagnostic link, or other equipment. In some cases, you (or someone) may need to extract the motherboard from the ECU and remove or install components using advanced soldering techniques. As always, the tuning process itself will be a typical iterative session of science plus trial and error, with all the standard risks and pitfalls of dyno or road tuning. Calibration software-hardware packages are available from Hondata, ZDyne, Techtom/Technosquare, Bonneville Motor Werks (Motronic Editor), Carputing (LS1edit), Shiftmaster (EEC-Tuner), and others.

A significant increase in difficulty involves using **general-purpose low-level PROM-hacking software** (probably hardware too, as

Hondata provides a package of hardware and software that converts various stock Honda EMSs to user programmability. The trick is to socket the motherboard for an auxiliary PROM, add two jumper resisters, and attach a Hondata datalogger/interface box to the Honda ECU. With the PROM data in a PROM-emulator box plugged into the PROM socket, tuning is completely interactive from a laptop PC. The alternative is an iterative process of logging acceleration air/fuel data, making changes offline to the calibration, and then burning a replacement PROM for immediate installation and retest.

Hondata's "Hondalogger" box allows a tuner to datalog the complement of stock Honda and modified sensors during test runs, including, optionally, wideband O_2 data. Note the 9-pin serial connector for laptop PC user interface and connections to Honda ECM. Datalogging is the true path to good optimized recalibration of any engine. Hondata

above) to display the hex data from selected PROM or EEPROM memory addresses, and then to modify the data to alter the behavior and tuning of the EMS. This is true hacking—though you may not be entirely on your own, since there is probably documentation available from similar ECMs that have been successfully hacked that will be at least partially applicable to your application.

For example, no one may have documented the memory locations of critical EMS tables and parameters in ROM for your application, but someone may know roughly what the tables should look like. You can probably get some help from the person who wrote the PROM editing software or other enthusiasts who love hacking car computers and are working on other related projects with the same editing package. Programs like CarPROM, LS1edit, EEC-Tuner, FreeScan, Tunercat, Winbin, and Winhex are able to edit entire ROM images (as well, in many cases, as automatically finding calibration tables on known applications for display and modification).

The next step beyond this level of hacking can only be described as reverse engineering., which is done by the guys who write the software used by the hackers. Some are car guys who got started in ECU hacking as obsessed hobbyists with serious day-job computer or electrical engineering skills. They may have begun the hacking process by dumping all ROMs on a factory ECU into a workstation, and then exhaustively working to separate code from data by processing the ROM data with a disassembler that attempts to convert machine instructions back into assembly language source code, tagging the target addresses of instructions that move data out of the code address space (presumably into calibration tables), working with logic probes to observe how memory addresses change when you diddle various sensors, and so forth.

"I'm going to simulate this using some standard hardware," wrote Jim Conforti some years back when the Motronic Editor was barely a gleam in his eye. "Run the speed signal through a PLL (with div. by 3 in the feedback) setup to provide a 3x freq . . . count this in a counter, while comparing it to some latched values (dwell and spark) in 8-bit identity comparators . . . the comps output to a flip-flop which through an igniter fires the coil . . . a third comparator will provide the reset signal to the counter . . . all the CPU will need do is look at inputs, go to a MAP or three, and output two values to the latched comparator registers . . . the hardware does the timing for me."

Gulp.

Fortunately for car guys who are not super-geeks, there are people out there who *love* to hack car computers and who post the results on the Internet for free or a reasonable fee for the goodies needed to delve deeply into the knickers of your onboard computer.

The original ECUs from the late 1960s and 1970s were **analog computers** that did not have separate hardware and software instructions like a modern digital computer but whose logic was instead purely rooted in the hardware circuitry that simply processed varying input voltage signals to generate output signals of varying magnitude and duration. Such an analog computer is permanently deterministic (unless you started ripping out and changing discrete components, which is unrealistic, given the alternative of a stand-alone digital aftermarket ECU that would cost less).

The only realistic way to modify the behavior of such a black box is to add external equipment to modify the signals to and from the sensors and actuators to selectively lie to the computer or selectively disobey it. The project Jaguar XKE in this book was originally carbureted, but was later run by an analog Bosch L-Jetronic EFI system with an additional injector controller (and eventually converted to aftermarket programmable digital engine management). The Cadillac 500-ci engine in the GMC motorhome project in this book was also initially run by a GM factory analog computer.

The earliest **digital engine management systems** typically used read-only memory (ROM or PROM) to store the logic (executable instructions), with a programmable read-only memory (PROM or EPROM) chip used to store calibration tables and parameters. This enabled carmakers to manufacture a computer box with one or two circuit boards that could be given a personality for a particular type of engine application depending on the PROM data.

In some cases—where flexibility was not important or cost was critical—manufacturers soldered the PROM onto the circuit board. In other cases, the circuit board was socketed so a PROM specific to a particular engine could later be inserted into the board without requiring soldering. So, for example, if the ECU on a particular vehicle had to be replaced, the original PROM calibration could easily be moved to the new computer box so the parts inventory system would not need to stock ECUs with every type of calibration.

GM was kind enough to locate the socketed PROM of first-generation digital ECMs in such a way that the PROM was accessible externally *without even opening the computer box* on early GM digital computers in 1980s-vintage cars like the TPI Corvette and F-body.

If a PROM is soldered to the circuit board, modifying the calibration requires desoldering the old PROM and (probably) replacing it with a socket into which a new PROM chip with the modified calibration can be plugged—or swapping the computer box to a different unit with the required calibration preinstalled.

In some cases, manufacturers have gone out of their way to make it especially difficult and expensive to modify the calibration. For example, in some cases the processor and PROM are designed so the EMS will not function without the original PROM chip in place. In this case, more expensive solutions are required to substitute a new calibration for the original factory PROM data. For example, in the case of recalibrating the Toyota computer in a Gen-II MR2, we were forced to install a special miniature daughterboard (TechTOM's ROMboard) on the factory motherboard in place of the original PROM.

The ROMboard installation procedure is to desolder the stock

Toyota PROM, socket the circuit board where the original PROM had been, install the original PROM into a socket in the small ROMboard, and then plug a second aftermarket PROM into the daughterboard with the modified calibration. A special processor on the daughterboard instructs the stock Toyota computer on a regular basis to run in diagnostic mode, causing the main Toyota processor to read modified calibration data from an alternate memory address on the auxiliary PROM.

In order to make changes to a PROM-based calibration while the EMS is running an engine, you'll need an emulator, which is a fast microprocessor device with a cable/connector made to plug into the PROM socket of an ECM or other microcomputer in place of the PROM. A PROM socket in the emulator accepts the PROM that is to be emulated, and the emulator subsequently stores the PROM data as a duplicate image of the PROM data in the emulator's RAM memory where it can be manipulated via laptop-based user-interface software.

Hondata, in fact, which markets a package of hardware and software that converts a wide variety of Honda ECUs to full programmability with datalogging capabilities (including the project del Sol turbo conversion in this book), embeds specific emulator commands in the Hondata ROMeditor graphical calibrator interface laptop software in a pull-down menu that include functions such as: REALTIME UPDATE, INCREASE CURRENT CELL, DECREASE CURRENT CELL, GOTO CURRENT CELL, DOWNLOAD CURRENT TABLE, DOWNLOAD WHOLE ROM, and REBOOT ECU.

Even if you will not be testing calibration changes in real time while the engine is running, an emulator can greatly speed up the process of testing new changes to the calibration. Without an emulator, every time you want to try out a change to the calibration, the PROM-based EMS tuner must burn a new PROM and install it in the ECU (though some EPROMs—erasable PROMs—can be reused multiple times). When the tuner arrives at the final calibration, the emulator must dump the PROM image to a PROM burner to blow a new PROM for installation in the ECU in place of the emulator. Some inexpensive PROM burners can be found at places like Radio Shack for around $100.

Not all emulators are the same. They vary in both speed and flexibility in the types of PROM devices they can emulate. Really fast emulators can be expensive, but expensive emulators are better at keeping up with the high demands of feeding data to the ECU in real time without stumbling while simultaneously interfacing with a tuner making calibration changes from a laptop.

Starting about 1994, many automakers began moving to powertrain control modules (PCM), which are engine control modules with some automatic transmission control capabilities. PCMs consist of hardware with **electrically erasable EEPROM** used to store calibration data. This architecture became standard on the OBD-II computers found on all 1996 and later U.S.-market vehicles (and

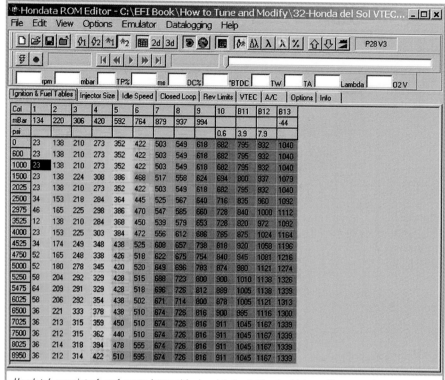

Col	1	2	3	4	5	6	7	8	9	10	B11	B12	B13
mBar	134	220	306	420	592	764	879	937	994				-44
psi										0.6	3.9	7.9	
0	23	138	210	273	352	422	503	549	618	682	795	932	1040
600	23	138	210	273	352	422	503	549	618	682	795	932	1040
1000	23	138	210	273	352	422	503	549	618	682	795	932	1040
1500	23	138	224	308	386	468	517	558	624	694	800	937	1079
2025	23	138	210	273	352	422	503	549	618	682	795	932	1040
2500	34	153	218	284	364	445	525	567	640	716	835	960	1092
2975	46	165	225	298	386	470	547	585	660	728	840	1000	1112
3525	12	138	210	284	368	450	539	579	653	728	820	972	1092
4000	23	153	225	300	384	472	556	612	688	785	875	1024	1164
4525	34	174	249	348	438	525	608	657	738	818	920	1058	1196
4750	52	165	248	338	426	518	622	675	754	840	945	1081	1216
5000	52	180	278	345	420	520	649	696	783	874	980	1121	1274
5250	58	204	292	329	428	515	688	723	800	900	1010	1138	1326
5475	64	209	291	329	428	518	696	726	812	889	1005	1138	1339
6025	58	206	292	354	438	502	671	714	800	878	1005	1121	1313
6500	36	221	333	378	438	510	674	726	816	900	995	1116	1300
7025	36	213	315	359	450	510	674	726	816	911	1045	1167	1339
7500	36	212	315	362	440	510	674	726	816	911	1045	1167	1339
8025	36	214	318	394	478	555	674	726	816	911	1045	1167	1339
8950	36	212	314	422	510	595	674	726	816	911	1045	1167	1339

Hondata's user interface formats internal fuel and timing tables for high- and low-speed cam operation on a VTEC engine. Configurable parameters for injector size, idle speed, closed-loop operation and other operations are conveniently located in tabs above the table. Alternately, a user can view data in graphical format.

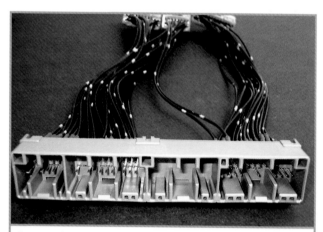

For some OBD-II Honda vehicles, the simplest way to recalibrate for competition use is to convert to a pre-OBD-II ECM. Hondata markets converter harnesses like this one, good to plug and play an OBD-I computer in an OBD-II Honda. Hondata

some earlier 1994 and later vehicles such as the stock EMS used on the Lexus/Camry 1MZ V-6 engine used in the project MR6 in this book), which are universally equipped with EEPROM. EEPROM acts like ordinary PROM or EPROM, except that the microprocessor itself can write to the EEPROM to modify its data (no special PROM burner required).

Since it can be problematic for technicians to install PROM chips without damaging the chip or the circuit board, EEPROM-

Selected Re-Calibration Applications

Model	Year	Engine	Features	Vendor	Req. Hdw
ACURA-HONDA					
Integra	1992-1995	B18	FUEL,IG,REV	Hondata	Socket +
	1996-	B18	FUEL,IG,REV,VT	Hondata	Socket +
Civic/del Sol	1993-1995	B16	FUEL,IG,REV	Hondata	Socket +
	1996-	B16	FUEL,IG,REV,VT,SPD	Hondata	Socket +
Prelude/Accord	1992-1995	H22	FUEL,IG,REV	Hondata	Socket +
	1996-	H22	TBA	Hondata	Socket +
BMW					
325	1993-1995	2.5L six	FUEL,IG,REV	Active Auto	PROM
328	1996+	2.8L six	FUEL,IG,REV	Active Auto	FLASH
BUICK					
V-6 Turbo	1984+	3.8 V6 turbo	FUEL,IG,BC,REV,SPD	Tunercat	PROM
CHEVROLET					
Corvette/F-Body	1985-1991+	L98	FUEL,IG,REV,FAN,CLOOP,ANTI	Tunercat	PROM
	1994-1995	LT1	FUEL,IG,REV,SPEED,DISP,INJ-SCALE,TRACTION,FAN,CLOOP,ANTI	Carputing	FLASH
	1997+	LS1/6	FUEL,IG,REV,SPEED,DISP,INJ-SCALE,TRACTION,FAN,CLOOP,ANTI	Carputing	FLASH
FORD					
Mustang	1986-1992	5.0 V-8	FUEL,IG,REV	EECtuner	J3 Module
	1993-1995	4.6 V-8	FUEL,IG,REV	EECtuner	J3 Module
SVO Mustang, T-Bird, Merkur	1984+	2.3 Turbo	FUEL,IG,REV	EECtuner	J3 Module
MAZDA					
RX-7 Twin Turbo	1993-1995	13B-REW	FUEL,IG,BC,REV,BL	Techtom	socket+board
RX-7 Turbo	1989-1991	13BT	FUEL,IG,BL	Techtom	?
	1987-1988	13BT	FUEL,IG,BL	Techtom	?
Miata MX-5/Miata	1990-1993	B6ZE	FUEL,IG,REV	Techtom	socket+board
	1994-1995	BPZE	FUEL,IG,REV	Techtom	?
MITSUBISHI					
3000GTVR-4 6MT	1993-1995	6G72BT	FUEL,IG,BL,SPD,REV	Techtom	socket+board
	1996-1998	6G72BT	TBA	Techtom	socket+board
3000GTVR-4 5MT	1991-1995	6G72BT	FUEL,IG,BL,SPD,REV	Techtom	socket+board
3000GT NA	1991-1995	6G72B	FUEL,IG,SPD,REV	Techtom	socket+board
	1996-1998	6G72B	FUEL,IG,SPD,REV	Techtom	socket+board
Eclipse GS-T	1989-1997	4G63BT	FUEL,IG,BL,SPD,REV	Techtom	socket+board
Eclipse GS	1989-1994	4G63B	FUEL,IG,SPD,REV	Techtom	socket+board
NISSAN/INFINITI					
300ZX Twin Turbo	1990-1995	VG30DETT	FUEL,IG,BL	Techtom	socket
300ZX NA	1990-1995	VG30DE	FUEL,IG	Techtom	socket
240SX	1989-1990	KA24E	FUEL,IG,REV,SPD	Techtom	socket
	1991-1994	KA24DE	FUEL,IG,REV,SPD	Techtom	socket
	1995-1997	KA24DE	FUEL,IG	Techtom	socket
PORSCHE					
911 Twin Turbo	1993+	993	FUEL,IG,BC,REV,SPD	Powerhaus	PROM
911 (996)	1996+	996	FUEL,IG,REV	Powerhaus	FLASH
944 Turbo	1985-1991	944T	FUEL,IG,BL,REV	Powerhaus	PROM
TOYOTA					
Supra Twin Turbo	1996	2JZ-GTE	FUEL,BL,REV,SPD (MT only)	Techtom	socket+board
	1993-1995	2JZ-GTE	FUEL,IG,BC,REV,SPD (MT only), BL	Techtom	socket+board
Supra NA	1993-1995	2JZ-GE	FUEL,IG,REV	Techtom	socket+board
Supra Turbo	1987-1992	7M-GTE	FUEL,IG,REV, BL	Techtom	socket+board
Supra NA		7M-GE	FUEL,IG,REV	Techtom	socket+board
MR-2 Turbo	1991-1996	3S-GTE	FUEL,IG,BL,REV	Techtom	socket+board
MR-2 NA		5S-FE	FUEL,IG,REV	Techtom	socket+board
MR-2 Supercharged	1988-1990	4A-GZE	FUEL,IG,BL,REV	Techtom	socket+board
MR-2 NA	1985-1990	4A-GE	FUEL ,IG,REV	Techtom	socket+board
VOLKSWAGEN/AUDI					
1.8T		1.8L Turbo	FUEL,IG,BC,REV,BL,SPD	Active Auto	PROM

Fuel Main (FUEL), Ignition Main (IGN), Rev Limit (REV), Boost Limit (BL), Boost Control (BC), Valve Timing (VT), Speed Limiter (SPEED), Set Number Cylinders (CYLS), Set Displacement (DISP), Injector Scaling (INJ_SCALE), MAF/MAP Scaling (AIR_SCALE), Traction Control On/Off (TRACTION), Fan Control (FANS), Closed Loop On/Off (CLOOP), Nitrous Fueling (NITROUS), Anti-Theft On/Off (ANTI), Valet Mode On/Off (VALET)

based ECUs have introduced a welcome reliability and ease to the process of upgrading a PCM or giving a replacement computer box a personality. They also greatly simplify the inventory of PROMs required in the replacement and warranty parts infrastructure. EEPROMs can only be written a finite number of times before they wear out.

OBD-II defines a protocol by which a standards-based diagnostic device can handshake with a **1996 and later OBD-II EMS** to move data of various kinds back and forth between the PCM and the diagnostic device. OBD-II defines a standard 16-pin data link connector, with some pins rigorously defined and others reserved for future or proprietary use.

OBD-II defines several different electronic protocols of varying speed and complexity that can be used to communicate between the EMS and a diagnostic device. A typical OBD-II communications message from the tool to the EMS consists of a header byte, a byte identifying the originating processor, a byte defining the destination processor, a command (code) byte, data bytes, and a final checksum byte.

For example, the hex message C4 10 F1 23 FF 06 8D from scan tool 10 to PCM device F1 REQUESTS DATA (hex 23 command) from memory address FF06, with a final checksum byte 8D. The PCM will then respond with a message such as C4 F1 10 63 FF 06 56 4E 41 41 92, which says the destination device is the scan tool, the originator is the PCM, Message Type is a RESPONSE TO DATA REQUEST (hex 63), that the address being dumped is FF06, that the data contained at this address is 56 4E 41 41, followed by a checksum byte of 92.

This type of communication can be used to retrieve diagnostic data, but it could also be used to offload, for example, some or *all calibration data*, or to *replace all* (or just some) of the calibration data in the EEPROM. Such messaging can also be used by the PCM to request a password or security seed before it will initiate certain actions—such as supplying, uploading, and overwriting *calibration data*. Obviously, this sort of proprietary procedure is not a standard OBD-II procedure, but it is permitted within the standard, and standard communication and messaging protocols provide the mechanism.

Given a little time, aftermarket firms are able to reverse engineer the proprietary messaging protocols and internal calibration data structures and develop software tools that handshake with the PCM and display the PCM calibration data in an organized tabular or graphical format for possible modification. With the help of a 16-pin OBD-II datalink connector and laptop-based communications card, PC-based software programs such as LS1edit are able to offload, modify, and replace the factory calibration on cars like the C5 Corvette—while the engine is running. This is exactly analogous to calibrating a PROM-based system with a laptop and fast emulator.

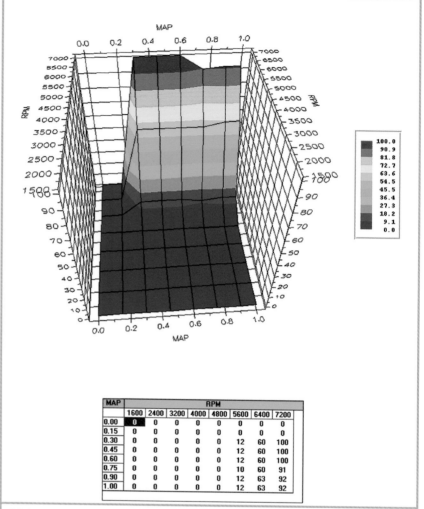

MAP	RPM							
	1600	2400	3200	4000	4800	5600	6400	7200
0.00	0	0	0	0	0	0	0	0
0.15	0	0	0	0	0	0	0	0
0.30	0	0	0	0	0	12	60	100
0.45	0	0	0	0	0	12	60	100
0.60	0	0	0	0	0	12	60	100
0.75	0	0	0	0	0	10	60	91
0.90	0	0	0	0	0	12	63	92
1.00	0	0	0	0	0	12	63	92

Brit Dave Whittaker was another enthusiast who had will and the means—i.e., extensive computer and engineering expertise—to hack the GM computer managing his Lotus Turbo Espirit. Whittaker methodically mapped out the internal data structures using extremely sophisticated reverse-engineering skills, and then developed a user interface for "Freescan," which is, indeed, free to other enthusiasts. Graphs display the various engine parameters as a function of manifold pressure and rpm for analysis and possible modification. Freescan

Some of the data (and code) used by an EEPROM-based OBD-II system may be located *in* the microprocessor itself, although there are commercially available, but possibly be hard to find, adapters that enable hackers to access and modify this data. *This is really* not for the faint of heart.

It is one thing to have the ability to *physically* change a PROM or to modify an EEPROM with a hex editor or calibrator software; it is another thing to have sufficiently detailed knowledge of the internal data structures to modify the calibration to make the car *run better*. Automakers have had the bad habit of seemingly randomly swapping the order of internal PROM data structures (calibration tables and scalars) for no apparent reason on otherwise nearly identical engine applications on identical size PROM chips (not to mention the chaos of the inexorable increase in PROM chip size as the years have gone by and EMS software engineers required more space for bells and whistles).

GM, for example, used several different sizes of PROM throughout the life of the PROM-based computers that ran 5.0-liter and 5.7-liter Tuned Port Injection and (early) LT1 small-block

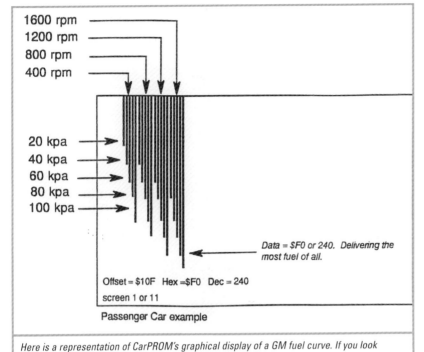

1600 rpm
1200 rpm
800 rpm
400 rpm

20 kpa
40 kpa
60 kpa
80 kpa
100 kpa

Data = $F0 or 240. Delivering the most fuel of all.

Offset = $10F   Hex =$F0   Dec = 240

screen 1 or 11

Passenger Car example

Here is a representation of CarPROM's graphical display of a GM fuel curve. If you look carefully, you'll see each rpm range, and within the rpm range each engine load is represented sequentially. A data byte specifies how much fuel will be delivered. ASM

Start of file
Offset $0000

- Prom ID data
- Checksum data

- Misc. Timing   parameters
- Main Spark advance tables

- Electronic Spark Control parameters
- Diagnostic parameters
- EGR tables
- Canister purge tables
- Transmission parameters
- Acceleration and Crank enrichment
- Main fuel delivery curves
- Blank area separating tables from executable code

- Executable code

End of file
Offset $FFFF

Using CarPROM, ASM hacked the data structures of many GM PROMs installed on GM vehicles in the 1980s and 1990s, and used the knowledge to recalibrate for traditional VE performance mods, superchargers, and turbos. This chart shows a basic memory map of a typical calibration found in a Camaro with an L98 5.7-liter engine. Unfortunately, GM had a bad habit of reordering the internal tables from vehicle to vehicle and year to year. What's more, the size and complexity of the PROM increased over the years. Consequently, CarPROM is a tool best suited for near-professional enthusiasts or pro tuners. ASM

cases the VIN) before modifying the PROM or EEPROM. Actually, some calibration software is sufficiently targeted so it will only work on one particular vehicle or engine application. In many cases, PROM hacking software (let's call it *calibration* software in this case) is tailored specifically to certain years and models of vehicles. The user interface can thus automatically find, display, and label various data structures in ways that are appropriate to the purpose of the particular structure, making it much easier for the tuner to understand and modify. Commercially available calibration programs are now typically smart enough to recognize when there is a mismatch in data locations if the tuner misidentifies the PROM engine application, or if the exact vehicle has not already been reverse engineered.

There are also hacking tools available that are not so much designed to tune an engine as to facilitate hacking or reverse engineering data structures using engineering-class scientific methods. General-purpose computer-science programs like WinHex will display the contents of any file in hexadecimal format: Each byte is displayed as one of the 256 hex numbers between 00 and FF (counting as: 1, 2, 3, 4, 5, 6, 7, 8, 9, A, B, C, D, E, F, 10, 11, 12, 13, 14, 15, 16, 17, 18, 19, 1A, et al.).

General-purpose calibration-hacking software like Winbin or ASM's CarPROM provides a visual user interface for PROM-based GM systems that displays the contents of a GM PROM in graphical format in such a way that a tuner can search out the telltale sawtooth-like fuel and timing tables, display the hex contents of any location in the map, and modify individual bytes of data by typing in replacement hex data. This sort of hacking, however, is not for the timid, it's more for the sophisticated pro tuner who can leverage the R&D investment by hacking a particular PROM calibration to produce for commercial profit.

Some calibration-hacking software is actually quite sophisticated, though it may, again, also have a powerful set of primitive, low-level capabilities that are *not* sophisticated). You tell the Windows-based LS1edit program the make and model of your GM LS1 vehicle, and it finds the tables and parameters and displays them in graphical or tabular format and provides the user-specific turnkey-type modification procedures.

Companies like Hypertech have commonly marketed turnkey retail EEPROM hacking tools that will work on one type of engine/vehicle, and can only be used to modify one vehicle at a time. With these tools, the stock calibration must be offloaded, and the programmer tool will not work on another engine until the original calibration has been returned to the PCM it came from. Such programmer modules typically have reduced capabilities in an effort to keep would-be tuners who are more in the class of shade-tree mechanics from getting into trouble. Powerful hacking tools that will modify any of a class of engine management systems—repeatedly—are marketed to pro tuners, and they are typically not cheap.

Of course, when you gain access to view and modify internal EMS tables, the trick is still to modify (*tune!*) them in ways that make sense, and this is no less trivial than on any standalone aftermarket engine management system. LS1edit, for example, is loaded with warnings that require the user to explicitly acknowledge that they understand that this or that type of hack

V-8s. What's more, even when the PROM is physically the same, even though the data structures might be conceptually similar between different engine applications, the exact locations of the fuel table, spark timing table, and other data often vary from application to application.

Therefore, it is critically important that a tuner know precisely the manufacturer, engine, vehicle, and year (and in some

The 993 twin turbo was the last air-cooled flat-six power-meister from Porsche, good for 400 horsepower. But—as with most turbo cars—there's more available with calibration changes and increased stock boost. High-boost chips for turbocharged cars running Motronic EMS are available from several sources. Powerhaus focuses intensively on Porsche performance, while Autothority and Active Autowerke can supply performance and custom chips for a wider variety of European cars with Motronic controls from BMW, Mercedes, VW, and others.

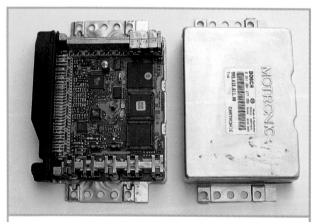

Factory recalibration is not always such a chore. Here's a stock 993 Motronic ECU at Powerhaus. Note the three large chips along the edge of the board next to the ECU enclosure.

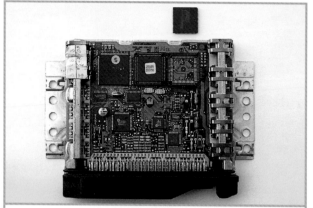

This is the Powerhaus chip, ready for installation.

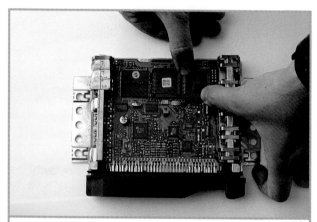

Chip installation is all thumbs, very carefully. This is a simple procedure compared with what's required on many Japanese computers, but is more involved than pre-1994 GM PROM computers, which provide access without even opening the enclosure! OBD-II systems, of course, allow reflashing the EEPROM using a communications cable and laptop PC or diagnostic box.

could actually damage the engine or transmission permanently, or that interrupting the process, say, of updating the PCM could permanently destroy it. In any case, this is the point at which hacking becomes tuning with its own art and science.

OEM ENGINE MANAGEMENT SYSTEMS

General Motors entered the digital age with certain Cadillac models in 1980, but the first true performance engine management systems were delivered shortly thereafter on Corvette and Camaro-Firebird 5.0 and 5.7-liter V-8s in the form of Tuned Port Injection on L98 Chevrolet small-block V-8s. All 1985–93 Chevrolet and Pontiac V-8s are equipped with PROM-based TPI or TBI engine management systems on which the PROMs can be changed without even opening the computer box.

The performance precursor of TPI, Crossfire Injection, a twin-throttle-body injection system similar in concept to the subsequent single-throttle TBI system found on Chevrolet trucks and low-performance car V-8s, arrived on the final C3 Corvettes and continued on in the new 1984 C4 'Vette. It is also a PROM-based system of little performance consequence beyond historical interest. The same multi-port PROM-based EMS electronics were used on the first 1992–93 LT1s, which thereafter were switched to a new EEPROM-based system (which also introduced sequential fuel injection). Throughout the life of these systems there were engineering changes that took the EMS gently away from full compatibility with the traditional small-block Chevy long block, though all pre-LT1 components can be made to work on older V-8s with minor modifications and most are not a problem at all.

The original high-energy ignition with add-on EST electronic module was eventually integrated completely into the main processor and became the entirely nonadjustable Opti-spark delivered on the first LT1s. The original MAF-based air metering system was abandoned in favor of a MAP-sensed speed-density system in 1990 and later 'Vettes and F-body cars (but later returned on the 1994 LT1 with *both* MAP and MAF).

All PROM-based GM systems are very capable of recalibration for almost any reasonable purpose, up to and including turbocharging and supercharging. If so inclined, you can downgrade 1996 and later LT1 PCMs to a pre-OBD-II 1994 EEPROM

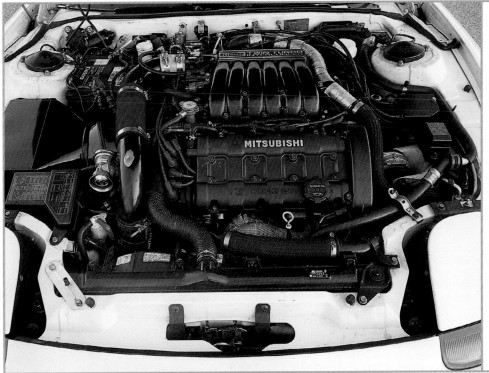

The early-1990s-vintage Mitsubishi 3000GT VR4 is an all-wheel-drive car with an SOHC twin-turbo transverse V-6 powerplant good for 320 crankshaft horsepower. The turbos are small, but there are two of them, and there is definitely more power available from the stock engine—if you can get into the calibration. I took the computer to Tadashi Nagata, founder of G-Force Engineering (now with Technosquare) to photograph Techtom programmability upgrades to the factory ECM that unlock the door to recalibration. This upgrade involves much more than simply unplugging and swapping out a chip.

ECM (for off-road use only). Aftermarket suppliers have offered wiring, MPI intake and fuel rail, and replacement PROM required to convert a TBI EMS to multi-port injection.

Like *all* closed-loop GM engine management systems, TPI and later systems may work to undo certain main fuel table calibration parameters by correcting the air/fuel ratio to 14.7:1 in closed-loop mode, and building up block-learn tables of semi-permanent (until you disconnect the battery) correction factors that the system then folds into the main fuel calculation.

GM 1994–95 LT1/LT4 EEPROM EMS must be programmed with a dedicated programmer module or the right communications cable/hardware interface combined with laptop software such as LT1edit from Carputing. LT1edit is a powerful calibration editor designed to work with all LT1 engines (16188051 or 16181333 computer, complete with name upgrade from electronic control module to *powertrain control module*) in any vehicle in the 1994 or 1995 model years (some very early cars might require an upgrade to PROM revision level C).

GM's EEPROM EMS architecture permits editor software to change the main fuel and spark tables plus an extremely powerful set of other functions, including disabling traction control, MAF-metering, crank AIR, vehicle antitheft, and charcoal canister purge. You can select an alternate decel fuel cut and modify the EMS to use just one knock sensor. You can set several idle speed and IAC parameters, define EGR functionality, and disable diagnostic tests for various DTC codes. You can turn on cylinder balance tests.

With EEPROM EMS architecture you can adjust a target air/fuel ratio table, battery voltage injection compensation, define WOT, power enrichment versus temperature and rpm, and modify a 0–7,000 rpm VE table. You can modify the calibration curve for MAF sensor voltage versus airflow, which tells the PCM the correspondence between sensor voltage and grams per second airflow. You can also define the conditions for closed-loop mode, and you can adjust the cylinder size in cubic inches

and the injector flow rate per injector. You can set a variety of spark advance table parameters, including rates of retard and advance in case of knock detection and several other knock-related parameters. There are a variety of automatic transmission parameters affecting shift points and harshness. Finally, you can adjust gear-tire scaling, rev limiting, maximum speed, and several cooling fan parameters.

GM's 1996 and later OBD-II performance engine management systems provided the new ability to detect and pinpoint individual-cylinder misfires, and were equipped with a full complement of emissions equipment including four O_2 sensors, EGR, and air pump. The LS1 engine, which appeared in 1997 'Vettes and 1998 F-bodies, arrived with direct-fire ignition, and in 1998 the familiar fuel loop was gone in favor of a supply line only to the injectors with a vestigial loop and regulator in the tank to bleed off excessive pressure, which can make fuel pressure modifications much more difficult. LT1 engines from 1996 and later can be downgraded to OBD-I electronics for off-road applications.

Editing the 1997 to 2002 LS1 is similar in concept to recalibrating the LT1, and the EMS control logic is similar to what is used on pushrod-V-8 Vortech truck powerplants. The LS1 system is more extensive and powerful than the LT1, in effect a superset of LT1 PCM's capabilities. GM allows some of the LS1 fuel parameters and tables to be modified, these include MAF calibration, injector flow rate, VE table modification, closed-loop enable temp, fuel pressure versus voltage compensation, injector offset, fuel air multiplier, stoic ratio, cylinder size, max allowed enrichment, and power enrichment.

Engineer Dave Darge of Powertrain Electronics says the following about GM powertrain electronics:

"GM OEM controllers are capable of performing on almost all engine applications. This includes Throttlebody Injection (dual and single), Port Fuel Injection (1 to 8 high-impedance injectors), with GM auxiliary injector driver 12 low-impedance

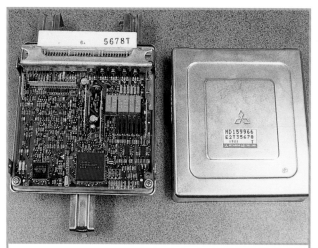

Here's the Mitsubishi ECM with cover removed. Note the large square processor/memory chip near the edge of the circuit board.

With the motherboard removed from its enclosure, Tadashi first employs a special desoldering tool to simultaneously melt solder and fasten all the multiple pins to the motherboard.

The chip is free.

injectors, batch fire or sequential, single coil or coil-on-plug and electronic transmission control. All of the GM controllers offer an OBD-I type of diagnostics with a port for a scanner. The last time I compared functions for a customer quote to a leading aftermarket controller the GM unit offered 56 additional features compared to the aftermarket unit. The customer decided the GM controller is the best for reliability and functionality. At this time the GM controller is far above the capabilities of any aftermarket controller except when triggering a nitrous system or handling functions like multiple rpm limiters.

"In fact, I cannot name any function limits that are beyond the scope of GM parameters or tables up to 9,000 rpm. On the contrary I see the GM controller as having more features and controllability than any aftermarket engine controller. Yes, there are times when the OEM-designed functions and diagnostics are hard to discern and cause problems that are difficult to determine. With several hours on a simulator the solution can usually be found.

"As you can see I am Pro-GM OEM when recommending a engine controller for any application. At times, I use two GM controllers on one engine when it is required. Most aftermarket engine controllers do not have the versatility of the GM controllers or the reliability and in most cases are not able to control the injectors precisely for dynamic situations (accel fuel and decel enleanment). An engine is a dynamic power source and the controller must be able to follow in all situations."

BOSCH ENGINE MANAGEMENT SYSTEMS (BMW, FERRARI, MERCEDES, PORSCHE, VOLKSWAGEN, AND OTHER VINTAGE EUROPEAN AND JAPANESE VEHICLES)

Bosch licensed the original Electrojector EFI system from the American company Bendix in the 1950s, and Bosch introduced commercial electronic injection in the late 1960s in the form of the D-Jetronic system. The system evolved into the much more common L-Jetronic system of the 1970s and 1980s (components of which were found in the original GM TPI EMS). L-Jet was a fuel-only EFI system, which depended on an entirely separate traditional solid-state ignition to light the fires.

Automakers as varied as Datsun (Nissan) and Jaguar used L-Jet systems on vehicles such as the 280Z and the XJ6. Bosch L-

Jetronic injection was modern in most respects except for the ECU, which used analog electronics. Bosch K-Jetronic fuel injection is an old-style constant mechanical injection system (*not* EFI), with the addition of mechanical air metering and—in later KE-Jetronic systems—O_2-sensor-based electronic fuel pressure trim to target stoichiometric AFRs at idle and light cruise.

Bosch introduced Digital Motor Electronics (DME) in the Motronic system in the early-1980s, about the same time the rest of the automotive world moved to digital electronics.

Pre-OBD-II Motronic systems managed the engines of virtually all post-K-Jet Porsche, Audi, VW, Ferrari, Mercedes, and BMW vehicles from about 1984 to 1995. These are all carmakers with a significant enthusiast interest in performance and performance improvements, and Bosch engine management systems have been hacked by firms such as Active Autowerk, Bonneville Motor Werks (DME Editor), Autothority, and others. If you have one of these vehicles, you can probably find calibration/PROM software for it, and you can definitely find someone who will make you a custom PROM or flash-calibration for it (for a price).

FORD

Ford released the powerful, modern EEC-IV EMS in 1983 on the 1.6-liter Escort, Lynx, EXP, and LN7 cars. It appeared shortly thereafter in the 1983–84 Mustang 2.3-liter GT Turbo, finding

A special tool simultaneously heats and vacuums out any remaining solder from the pinholes in the printed circuit board.

The desoldered Mitsubishi chip will have to be soldered onto this Techtom ROMboard (daughterboard), which is already equipped with a diagnostic processor that will redirect Mitsubishi memory access to twin auxiliary PROMs that will be located on the daughterboard.

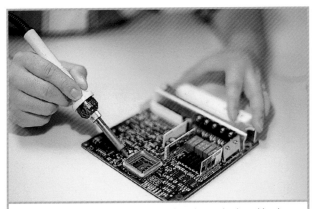

Here Tadashi solders a special socket into the motherboard in place of the chip. This type of socket enables installing a chip or daughterboard without requiring solder.

its way onto the high performance manual-trans 5.0-liter Mustang GT for 1986 (the 210-horsepower manual-trans 1985 Mustang GT was still carbureted, though the auto-trans model already had throttle body [central] injection).

The EEC-IV was a very advanced and flexible system that could be configured to handle virtually any Ford gasoline powerplant; Ford subsequently released EEC-IV applications for all four-, six-, and eight-cylinder engines, with both speed-density and mass flow controls. As was the case with GM TPI and TBI ECMs, the EEC-IV gained in complexity and power as time went on.

The J3 test port on the side of the ECU box is for developer use; this is the way aftermarket chipmakers gain access to the EEC-IV. The test connector port contains the micro-controller's multiplexed address/data bus signals, as well as a PROM disable signal. Although more expensive than the simple chip-change technology of similar-vintage GM systems, aftermarket chipmakers and tuners wishing to hack the EEC-IV need only to design a relatively simple electronic module that plugs into the J3 port, disables the EEC-IV internal PROM, and substitutes its own PROM with modified calibration.

Examples of functions controlled by the EEC processor and subject to modification via recalibration are: A:F ratio in closed-loop, acceleration and warm-up transient fueling, EGR, charcoal canister purge, thermactor operation, adaptive control system, control of OBD-I and OBD-II testing (enable/disable, change test values, and so on), main fuel, main spark, MAF calibration, VE table, injector flow, rev limits, speed limits, electronic transmission control, and many more.

The EEC-tuner package from Shiftmaster Enterprises provides EEC calibration software, cabling, and hardware to modify EEC-IV and EEC-V processors. In general, this package enables a tuner to effectively modify any hex location in PROM memory on the EEC-IV, and all locations in bank 1 and bank 8 in the EEC-V (this is similar to ASM CarPROM program for GM PROM systems). The trick is knowing where calibration tables and parameters are located in memory.

In the case of several specific common Mustang V-8 EEC processors and the Mustang 2.3 turbo, the EEC-tuner is preconfigured with the location of calibration parameter data structures in EEC ROM memory. In the case of Mustang speed-density processors and many other EEC systems, you are on your own, and this sort of reverse engineering is not for the faint of heart but rather for smart, dedicated professional engineers and tuners. As time goes on, an increasing number of EEC processors will certainly have their memories mapped out and available.

JAPANESE

The simplest way to get a handle on what's available to hack Honda, Mazda, Mitsubishi, Nissan, Subaru, Toyota, and other vehicles is to look at the application charts from Hondata, Techtom, and other firms that have focused on recalibrating Japanese engines (see appendix). In many cases, changing the PROM calibration of these vehicles requires hardware modifications to the motherboard of the electronic control module. This book details such modifications to a Mitsubishi 3000GT VR4 above, and to a Toyota MR2 Turbo project.

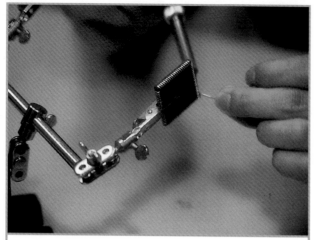

Priming the Mitsubishi chip for installation.

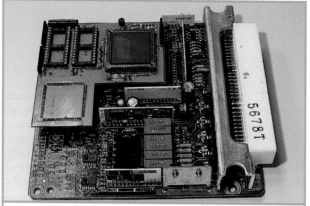

The Techtom daughterboard now plugs into the socketed Mitsubishi ECM motherboard (note two additional PROM sockets on the ROMboard).

Installing the recalibrated PROMs on the Techtom board. Recalibration is now a simple matter of a PROM change, and with an emulator plugged into the ROMboard in place of the auxiliary PROM(s), calibration changes could even be made on the fly with the engine running using a laptop PC to modify PROM image(s) stored in the emulator.

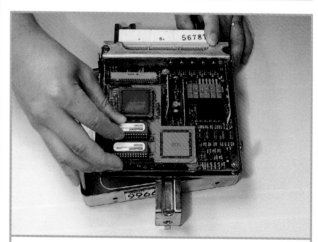

Tadashi fires up a desktop PC running MightyMAP, a PROM hex editor for producing recalibrated (re-mapped) automotive PROMs for Japanese engine management systems with Techtom modifications. The PC is connected to a PROM-burner device. After editing the PROM image with MightyMAP, Tadashi offloaded the recalibrated data to twin PROMs of a common type designed to plug into the ROMboard.

CHIP-MODULE-SOFTWARE FIRMS: SPECIAL OEM PROM AND EEPROM CALIBRATIONS/SOFTWARE

APR
Active Autowerke
Autologic
Autothority
Arizona Speed and Marine
Bonneville Motor Werks
Carputing
ChipTorque
Diablosport
Fastchip.com
Ghettodyne.com
Hondata
Hypertech
Jet Performance Products
Jim's Performance
JMS
Powertrain Electronics Co.

ProECM
Skunk2 Racing
Street and Performance
Superchips
Technomotive
Technosquare (Techtom)
TPIS
TTS
TunerCat
Wester's Garage
ZDyne
Z-Industries

Hacking Software

www.amitron.ru/opel/diagnos.htp
www.cruzers.com/~ludis
www.davewhittaker.com
www.download.com (winhex hex editor)
www.geocities.com/MotorCity/Shop/1624/95cam2.html
www.geocities.com/MotorCity/Shop/9938
www.mindspring.com/~amattei/aldl.htm
pweb.de.uu.net/pr-meyer.h/aldl.htm
www.efilive.com/downloads/index.html
www.techedge.com.au/vehicle/aldl8192/8192hw.htm
www.ttspowersystems.com
www.tunercat.com

Some people still bring a carb mentality to EFI, eager to leap into action to "re-jet" their engine management systems for the free power.

In truth, it is, and always has been, extremely difficult to perfectly optimize electronic engine management for a particular mission. Delivering great drivability under all conditions while achieving safe mean best torque numbers where it matters is not for the timid. It typically takes months of work, even *years* for automakers to build the "perfect" calibration for a particular engine in a particular application. This level of exhaustive, perfectionist work cannot be profitable on a one-off basis for aftermarket tuners, which is one reason even fully qualified professionals seldom achieve perfectly optimal results when they tamper with factory EFI. There isn't time.

On the other hand, original equipment engine and vehicle manufacturers can and *must* take the time today to tune EFI perfectly. They can afford to, because the time investment will be amortized over hundreds, thousands, or even millions of vehicles. They have the equipment and the engineering talent to do it right. What's more, with feds ready to levy hefty fines for noncompliance with emissions standards—and with the market ready to pass judgment on engines whose *performance* is not competitive—vehicle manufacturers clearly have the *incentive* to get it right. Perfection is the whole *point* of factory EFI and electronic engine management.

Factory engine management with multi-port injection is, in many respects, hard or impossible to beat, and today it's also

relatively hostile to modifications compared to the carbs and points ignitions of a generation ago. But everyone wants more power and drivability. And a vocal group of journalists, hardcore enthusiasts, and professional tuners has driven the performance aftermarket to provide highly sophisticated and expensive solutions to modifying the behavior of EFI that are great in theory, but in reality may be of such complexity that most nonprofessionals have neither the experience, tools, or diagnostic equipment to succeed in improving performance (or to even know whether they've succeeded or gone backward). The result for many would-be performance tuners is somewhere between fuel corruption and fool injection.

From a purely performance perspective, there *are* still certain tradeoffs in engine tuning, even with the most sophisticated factory electronic engine management. Some have to do with the basic chemistry of burning gasoline with air in high-speed engines. Some have to do with government regulations. There are even a few pseudo-tradeoffs that show up on inertial chassis dynamometers, but cannot be duplicated in the real world. Some have to do with cost-based design decisions. Bottom line, best-torque and lowest-emissions air/fuel ratios are still *not* one and the same, and the federal test procedure for proving emissions compliance in a standardized 15-minute suburban and city drive cycle does not give equal weight to exhaust emissions at all rpm ranges and throttle positions.

So where are the compromises in factory electronic engine management?

Definitely not at peak power on normally aspirated performance vehicles. Performance cars like the Corvette, Mustang, Porsche, S2000, NSX, Ferrari, Camaro, and virtually all motorcycles compete heavily in performance statistics: 0–60 and quarter-mile times, peak power and torque, and *every* last top-end miles per hour. Top superbike engines essentially use Formula One-type engine technology with a goal of *extremely* high levels of specific power. Stock EFI performance today is definitely optimized *exhaustively* for mean best torque air/fuel ratios at maximum power. You can safely assume peak power is already as good as it gets.

Idle and off-idle to medium cruise operating ranges are occasionally a problem, particularly on EFI motorcycles. These ranges are highly scrutinized for emissions in the federal test procedure. And there's the rub: The air/fuel mixture that optimizes best torque in this range will not produce lowest emissions. Proper tuning in this range is vital to producing responsive and fun drivability, and it must be addressed for optimal performance.

Where retuning really bears fruit is on overboosted turbo cars and engines with power-adder conversions. Such conversions combine the potential for vast power improvement with a nontrivial potential for engine damage with the wrong tuning. Stock engine management is almost guaranteed to be compromised when you add a turbo, blower, or nitrous system to an EFI car, truck, or motorcycle. Sometimes unworkably so.

THE PREHISTORY OF "FREE" POWER

Before electronic engine management, performance tuners worked on carburetors and mechanical spark-advance mechanisms in a world of tradeoffs. Tuners and engineers worked to optimize engine management by manipulating, swapping, and

Interceptors can sometimes look like a Band-Aid approach, but factory calibrations are very hard to beat for streetability. Why mess with the entire calibration if all you need to do is introduce a small delay in the timing signal to prevent knock or extend the injection pulse to add a little fuel under hard boosted conditions? ProECM

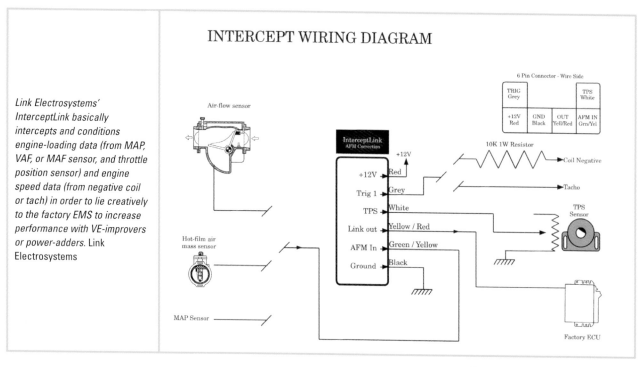

INTERCEPT WIRING DIAGRAM

Link Electrosystems' InterceptLink basically intercepts and conditions engine-loading data (from MAP, VAF, or MAF sensor, and throttle position sensor) and engine speed data (from negative coil or tach) in order to lie creatively to the factory EMS to increase performance with VE-improvers or power-adders. Link Electrosystems

adjusting a crude overlapping maze of mechanical parts and fuel delivery circuits that include main jets, needle jets, jet needles, idle jets, emulsion tubes, venturis, booster venturis, accelerator pumps, power valves, air-slides, multicarb synching, fuel pressure, float chamber depth, throttle angle, multiple points and coils, centrifugal advance springs, vacuum advance systems, and more.

Unfortunately, liquid gasoline and gaseous air simply do not behave the same as flow increases. What's more, engines require varying air/fuel ratios and spark timing for optimal power and drivability across the operating range—as well as compensation to these basic parameters for starting, warm-up, acceleration, altitude, air temperature, and barometric pressure, not to mention reasonable fuel economy and exhaust emissions. Add the impact on tuning of hot rodding modifications like headers, TT pipes, high-flow air cleaners, stroker kits, cam changes, and head porting—some of which may actually be performance-negative, particularly without radically reworked tuning—and you've got the potential for a real nightmare.

Trying to achieve optimal air/fuel ratios and ignition timing with mechanical controls was something like trying to approximate a circle with a square: You can make it perfect in a few places, but it's at least a little wrong in one direction or the other everywhere else. Adding to the nightmare was the fact that in the days of mechanical engine management, it was essentially impossible to tune while driving. And even if you could tune on the fly, in the days before affordable dynos and data acquisition systems, how would you *know* if you did get it "perfect?" Tuners focused on getting it right at idle and maximum power.

Legions of engineers and tuners labored thousands of lifetimes trying to make carbs and mechanical ignitions work better over the course of nearly a hundred years of engine development. Nevertheless, over the years, carburetors and mechanical ignitions have run so poorly in so many applications, with so many tradeoffs and compromises, that the legend now lives that "free" power lurks inside that fancy pot-metal casting any time you have the nerve to go prospecting for a higher state of tune: People still figure that "There's Free Power in Them Thar Hills," and they don't have any trouble believing

Split Second Fuel timing Calibrator

Cell Value	Mode Direct	Signal Modify
0	0.0	-2.5
1	1.25	-2.25
2	0.5	-2.0
3	0.75	-1.75
4	1.0	-1.5
5	1.25	-1.25
6	1.5	-1.0
7	1.75	-0.75
8	2.0	-0.5
9	2.25	-0.25
10	2.5	0.0
11	2.75	0.25
12	3.0	0.5
13	3.25	0.75
14	3.5	1.0
15	3.75	1.25
16	4.0	1.50
17	4.25	1.75
18	4.5	2.0
19	4.75	2.25
20	5.0	2.5

The number entered into the timing map indicates the timing retard in degrees. The value can range from zero to 20.0. The maps for both fuel and timing are overlay maps that are applied on top of the maps in the stock ECU.

that carburetors made in the last several years (on motorcycles) can often be improved or at least that the tradeoffs can be shifted around for a new mission.

AUXILIARY METHODS AVAILABLE TO TWEAK EFI AND ELECTRONIC ENGINE MANAGEMENT

There are various ways to modify the results of stock electronic injection and engine management. All work in theory, but—in the real world—come at a price in complexity of use, operating limitations, and cost.

MSD's Boost Timing Master is a "programmable" coil driver that intercepts the spark timing signal and introduces an adjustable delay under boosted conditions as an anti-knock countermeasure on turbo, blower, or nitrous conversions running factory computers. MSD offers a variety of simple and complex timing interceptor devices, many adjustable with a screwdriver or simple dials. MSD

Split Second's fuel/timing calibrator offers adjustable timing retard and fuel enrichment for power-adder conversions. Split Second

OEM computer recalibrations change the factory engine management computer's PROM data to recalibrate fuel and spark delivery. The recalibration may or may not be as good as stock—much less *better*—but whatever you get, once you drive away with new data in your PROM, it's not necessarily easy or cheap to change it again. A typical example of recalibration technology is the Tech-tom ROMboard solution that converted the project MR2 Turbo elsewhere in this book to use a removable PROM. The Hondata software and emulator package that converted the Honda del Sol project car to online programmability is yet another example.

Sensor manipulation boxes basically lie to your stock computer about engine sensor values in a way that influences the engine management system as a strategic means to change fuel delivery. For example, all factory ECUs use coolant temperature or cylinder head temperature sensors to trigger enrichment strategies that command enrichment as a percentage of the base fuel calculation. Reporting a false low temperature will produce fuel enrichment. Temperature-based enrichment schemes are obviously limited to enrichment percentages available within the range of the factory warm-up map. This scheme is typically more likely to work on older cars or trucks or less sophisticated new motorcycle EFI than on late-model OBD-II car or truck systems, which are much better at recognizing problems—like an engine that apparently never warmed up all the way and continues to

want warm-up fuel enrichment or an engine that apparently wants warm-up enrichment suddenly without the engine having shut down and cooled off. Ford SVO and others sell a piggyback microcomputer called the Extreme Performance Engine Computer (EPEC) that lets you add a boost-pressure (MAP) sensor and manipulate MAF, ECT, ACT, and TPS sensor signals to optimize timing and air/fuel ratios for turbo and blower conversions.

Mass airflow (MAF), vane airflow (VAF), and manifold absolute pressure (MAP) sensors represent another class of sensor subject to fruitful manipulation strategies. By underreporting airflow by a certain percentage, an adjustable airflow sensor can cause an ECU to command shorter injection pulse widths that compensate for fuel injectors larger by the same percentage. By selectively over or under reporting airflow in various segments of the flow range of the sensor (via trim pots or other means), an adjustable sensor lends itself to re-jetting the way you'd adjust the idle mixture, midrange, and peak-power ranges of the engine.

Examples of airflow-manipulation products include the APEXi Airflow Converter (AFC) and the HKS vane pressure converter (VPC)—used for a time on the project MR2 in this book. The VPC substitutes its own user-adjustable (via trim pot) MAP-based airflow data for the stock VAF data. The VPC can also be reprogrammed by changing the internal conversion tables that drive its overall operation. Professional Flow Tech sells the Mass Air Adapter, which converts VAF engines to (adjustable) MAF-based airflow metering, which can be configured to match the voltage requirements of the EMS and can manipulate an airflow signal to compensate for larger fuel injectors.

Sensor manipulation schemes can break down if the your engine encounters temperature and altitude or airflow conditions that require the enrichment you were depending on to enhance performance (or, possibly, drivability). For example, if you are using a false low-temperature sensor reading to achieve fuel enrichment, what happens when it *is* that low temperature? If you are over-reporting the airflow voltage to achieve fuel enrichment in some operating range, what happens if you hit the peak voltage for the sensor while the real engine airflow is still increasing?

Or suppose the entire sensor range is not normally used, but that voltages too near peak airflow voltages trigger a situation in which airflow is nearly at peak measurable flow, causing the ECU to recognize that the engine might be overboosting or through other unknown means entering a realm in which further airflow would shortly be unmeasurable, and triggering limp-home mode or other self-protective countermeasures.

Additional injector controllers or other auxiliary computers may control additional injectors or manage fuel pressure invisibly to the main EMS computer. Examples include HKS's Additional Injector Controller and Carroll Supercharging's Super-pumper-II, which is essentially an intelligent, computer-controlled fuel pressure regulator capable of managing fuel pressure with a programmable varying rate of gain over boost pressure increases (that is, X psi fuel pressure increase per pound of increase in manifold pressure). The Jaguar 4.2-liter engine featured in this book was at one time equipped with a Miller-Woods TurboGroup Fueler, an additional injector controller capable of pulsing one or two additional fuel injectors to increase fuel enrichment under boost conditions.

Auxiliary interceptor computers monitor, extend, truncate, or interrupt signals to and from an OEM computer in order to influence or modify injection or spark timing. They can be relatively expensive and may require modifications to your EFI harness. They are typically not easy to program properly and have the potential to initiate a fight with your stock computer by triggering hostile active countermeasures. Examples include the Apexi Power FC, Greddy eManage, Techlusion TFI (for motorcycles), and many others. Some interceptor boxes must be calibrated or adjusted with a laptop computer, while others, like the Techlusion, calibrate with trim pots on the processor box.

TECHLUSION'S TFI FUEL NANNY: TO INTERCEPT OR NOT TO INTERCEPT

In order to upgrade the drivability and fun factor of EFI motorcycles, Techlusion's auxiliary fuel computer needed the ability to selectively provide adjustable fuel enrichment at idle, light cruise, wide-open throttle, and during sudden increases in throttle opening. The unit had to work in the real world of real people without their own R&D lab, so it had to be easy to understand, install, set up, tune, troubleshoot, and even uninstall—and it needed to have self-diagnosing and validating characteristics. It also had to be affordable. To meet these requirements, the goal was to *keep it simple*.

The TFI system taps into the injector-driver circuitry of a stock bike's stock EFI and selectively appends extra pulse width voltage to keep injectors open the exact incremental time required for precise fuel enrichment. A bike's onboard computer and wiring harness remain intact, and no wires need be cut. The stock air/fuel map remains intact, as does its correlation to the engine's volumetric efficiency (breathing) curve.

Techlusion's fuel box is quite simple to adjust. Up to five adjustable dials (pots) and circuitry in the control unit define the breakpoint between cruise and main-jet rpm, the length of time and magnitude for auxiliary transitional enrichment, and the incremental enrichment in tenths of milliseconds for the three operating ranges. The system optionally appends an adjustable amount of additional injection pulse to various adjustable ranges of a bike's stock air/fuel map—by twisting dials that simulate changing the adjustment of idle jetting, cruise jetting, full throttle main-jetting, and accelerator-pump jetting on a carburetor. LEDs provide status information to the tuner.

The transition from midrange to main-jet operating ranges is user-selectable via dial and LED. The Techlusion piggyback computer is effectively self-diagnosing via *removal* and other simple tests. Returning to stock is simple: Turn off the unit. If adding fuel improves response, you're on the right track for a given performance range. If a dial is all the way down, and adding any fuel worsens response, the EFI was already too rich as stock (probably implying there is a problem).

By adding an adjustable fixed amount of additional pulse width to an injection event in a certain operating range, the TFI system *automatically* produces a declining *percentage* of addition to pulse width as rpm increases. Even at zero load, engines require a bigger injection as volumetric efficiency increases with rpm until you hit peak torque. Since engines become *more* sensitive to air/fuel ratio with increased rpm as the AFR-spread narrows between rich and lean best torque, a declining percentage of enrichment is exactly what's required when you're looking to fix factory compromises in air/fuel ratio. The fixed-declining addi-

Turbo Performance Center uses the Split Second FTC1 to control additional injector and retard timing on its Porsche 993 supercharger kit. Split Second

tion method of pulse width–modification provides a user the ability to fix problems in a jetting range without simultaneously corrupting other parts of the range.

HKS PFC

The HKS PFC (Programmed Fuel Computer) F-CON is an interceptor black box that plugs into OEM injection wiring harnesses to enable modification of stock fuel curves—anything from minor enrichment (or enleanment) to the radical pulse width changes needed to support really big injectors and high-boost forced induction. The idea is to retain as much as possible of the highly developed factory fueling map, modifying only those areas necessary to provide correct fueling for modifications such as turbo or supercharger conversions, increased turbo boost, larger turbos, or nitrous injection.

HKS offers multiple control maps for the PFC for use with various engines based on extensive testing. The unit is not generally sold for universal applications, although HKS will work with manufacturers of performance equipment outside the range of HKS's own specific product line to supply PFC maps for the new applications.

The PFC is designed to plug into factory EFI wiring harnesses and requires no fabrication or surgical skills to install. Once installed, PFC internal dip switches allow selection of gross richer or leaner percentage of the basic program, plus selection of nine separate condition programs. Beyond that, the graphic control computer (GCC) option gives the user a second black box with pots to adjust individual load points of 500; 2,000; 3,500; 5,000; 6,500; and 8,000 rpm, each of which can be adjusted to deliver up to 16 percent more fuel or 12 percent less fuel than the base map. The GCC averages values between the adjustable points for smooth fueling. HKS offers a further optional black box called the electronic valve controller to raise turbocharger maximum boost levels by electromechanically altering boost pressure levels as seen at the factory wastegate/controller. And to work with this,

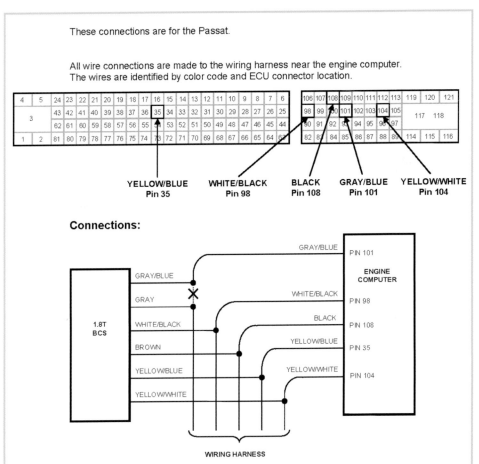

These connections are for the Passat.

All wire connections are made to the wiring harness near the engine computer. The wires are identified by color code and ECU connector location.

YELLOW/BLUE Pin 35   **WHITE/BLACK** Pin 98   **BLACK** Pin 108   **GRAY/BLUE** Pin 101   **YELLOW/WHITE** Pin 104

Connections:

This table shows the output voltage for direct and signal modify modes of operation in the fuel portion of Split Second's fuel/timing calibrator. There are actually 200 cell values, but only 20 are shown for simplicity. Numbers entered into the cells on the fuel map represent either the actual output voltage (direct mode) or the amount that the input voltage is offset (signal modify mode) to provide fuel enrichment. The number entered in the cell can range from 0.0 to 20.0. The step size in both modes is 25 mV. Split Second recommends signal modify mode for most engine management applications. Split Second

an interceptor because rather than intercepting fuel injection pulse width signals, it simply *extends* pulse width signals by taking up where the main ECU left off to provide voltage to keep the injectors open longer. In the case of timing retard, such a device simply introduces an adjustable delay in the spark timing pulse.

Somewhat less simple true interceptors require cutting wires in the injection and timing circuitry, enabling the interceptor to extend or *truncate* injection pulse width and to advance or retard timing. This type of interceptor advances timing or truncates injection pulse width by predicting the future. That is, it anticipates what the ECU is about to do based on a pattern of what it has done in the past or what it *just did* on the last revolution (likely to be very similar in timing and pulse width to what will be true one revolution in the future).

Besides intercepting and directly modifying ECU output signals on the way to ignition and/or injector circuitry, an interceptor can affect ECU output indirectly by lying to it about engine position (delaying or advancing the crank reference signal), airflow, manifold pressure, and so forth. Such strategies can cause problems in the latest OBD-II systems, which depend on micro-changes in crankshaft acceleration and deceleration to detect misfires and may be unhappy if an interceptor is playing with the crank ref signal

Advanced interceptors have the ability to alter cold-start, knock sensors, temperature compensation, air conditioning or auto-trans control and more.

INTERCEPTORS AND SELF-LEARNING ECUS

Modern ECUs use learning strategies to optimize performance, emissions, durability, and performance as engines age and wear out.

For example, strategies trim injection pulse width to target stoichiometric air/fuel ratios by using an O_2 sensor to monitor exhaust gas oxygen content, a fairly good indicator of precombustion air/fuel ratios. Over time, the ECU builds a matrix of block-learn values that direct the ECU to *anticipate* air/fuel correction required to achieve stoichiometric mixtures, minimizing the magnitude of O_2-based air/fuel correction trim—based on past correction required. By this method the ECU learns.

Closed-loop O_2 sensor strategies generally operate between a 14.1 and 15.5 air/fuel ratio (AFR), beyond which the ECU is in open-loop mode, with AFR correction beyond the authority of the O_2 sensor and fuel correction logic. Even a light stab on the throt-

they offer the fuel cut defenser (FCD) black box to defeat the factory fuel shutdown with overboost or high rpm for use with modified engines where such limits are dysfunctional.

UNICHIP AND PERFECT POWER'S SMT6

Interceptor and piggyback systems can be divided into several classes of systems. The simplest interceptors can extend fuel injection pulse width or retard timing for turbo conversions, but are unable to shorten pulse width or advance timing. The simplest of the simple systems are calibrated by hand (or screwdriver) via tuning trim pots dedicated to increasing acceleration enrichment, idle, or low-end pulse width gain, pulse width gain at peak airflow, and sometimes other adjustment ranges. The idle gain may or may not affect pulse width under all conditions, whereas the high-end gain may affect only peak airflow pulse width, with decreasing effect as airflow or rpm decreases (equivalent to changing the *slope* of the gain in pulse width with speed or loading).

Spark retard adjustments are exactly equivalent. By adjusting the combination of various tuning pots, a tuner can warp the shape and slope of various segments of the fuel and timing curves—but always in the direction of more fuel and less timing advance. Technically speaking, this type of auxiliary black box is not technically

tle can terminate closed-loop mode, calling for rich or lean mixtures and effectively bringing air/fuel ratios beyond what can effectively be monitored by a standard narrow-band O_2 sensor.

Interceptors like the Unichip allow a tuner to optimize timing at part-throttle, which Unichip says helps the ECU achieve optimal stoichiometric air/fuel ratios.

ECUs use closed-loop-type logic to work with knock sensors to implement spark-timing retard-then-advance strategies to stop detonation. Most OEM engine management systems use knock sensors specifically designed to react to the detonation frequency of the particular engine. Knock-retard strategies are critical in avoiding engine damage from detonation, particularly on turbocharged engines.

In some cases, adding performance equipment that changes the amount and kind of engine noise can cross the threshold of knock-sensor output that the ECU would suspect to be detonation. This might seem like an opportunity for an interceptor with the ability to modify knock sensor output, though it might be more fruitful and straightforward to look for a knock sensor more suited to the equipment on the engine, or to *move* the knock sensor to a location less sensitive to spurious engine noise—or to

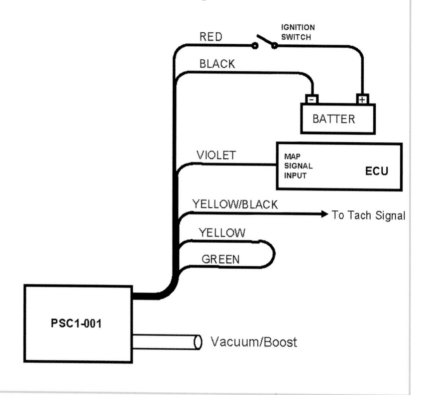

Signal Modify Configuration Using internal MAP Sensor:

Split Second recommends this configuration for most applications of its programmable signal calibrator, which is used to lie to the ECM about manifold pressure to allow injector size changes or to induce fuel enrichment by overstating engine loading to the stock calibration logic. Output to the ECU in this mode is a modified version of the internal MAP sensor signal, which is already close to the desired output. The PSC1-001 uses its internal MAP table to modify the MAP signal as necessary. The range of adjustment in this mode is +/- 2.5 volts. A cell value of 0 will subtract 2.5 volts from the signal. A cell value of 10 will not alter the signal. A cell value of 2.5 volts will add 2.5 volts to the signal. The starting map for this configuration is a map loaded with 10 in every cell. Split Second

work with an adjustable third-party knock detection and retard device like the J&S Safeguard for which there is documentation and support available for difficult pseudo-detonation problems.

Some experts flatly recommend *never* tampering with knock-sensor output with an interceptor (though *monitoring* the sensor output never hurts). If you have a sophisticated engine that is highly modified and are experiencing false knock problems, perhaps you should be working with a standalone programmable engine management system. If you are experiencing *real* knock, you should *never* interfere with the ECU's detonation counter-measures under any circumstances, but should immediately take steps to stop knock with higher octane fuel, colder plugs, piston-cooling oil squirters, upgraded combustion chamber cooling, better intercooling or colder air intake, and so on.

How does a modern ECU know if the engine is producing more power than is good for it? One leading indicator is boost pressure. If the ECU sees 5 volts from the MAP sensor, it can conclude the engine is making high power. In this case, an interceptor can truncate the MAP signal, say, to a maximum of 4 volts, preventing the ECU from adjusting boost or implementing other antipower countermeasures.

Another indicator of high power is the mass airflow signal. Even if the ECU only sees normal manifold pressure, a high MAF signal

could alert the ECU to dangerously high power levels. An interceptor can truncate the MAF signal to prevent ECU anti-power countermeasures. Of course, the ECU will probably not inject enough fuel for the true amount of air, so the tuner would have to artificially extend injection pulse width (and possibly retard timing).

The good news is that, as the interceptor company Unichip puts it, "There is no facility for the standard ECU to say that at a particular rpm point and at a certain throttle position, it should see a certain airflow. Otherwise the ECU would not cope with altitude changes, etc. In a nutshell, the standard ECU does not try to pull values back to predetermined levels. It tries to keep mixtures within limits at cruise and the ignition timing off the knock sensor circuit. Both good things to do." For tuners worried about the interceptor triggering the ECU entering limp-home mode, Unichip writes, "In normal operation this is not an issue. You would have to do something very radical to have this happen."

How can an interceptor compensate for the factory ECU's closed-loop system tuning out all the adjustments made with an interceptor? In the first place, closed-loop applies only to fuel trim, not timing. And interceptors generally do not alter air/fuel ratios (AFRs) when closed-loop is working properly but only when aggressive camshafts push AFRs out of the O_2 sensor's narrow operating band. In this case, an interceptor could be used to bring the air/fuel ratios back within range so the closed-loop could again operate correctly to achieve stoichiometric AFRs at cruise for good economy and throttle response. However, sometimes an aggressive cam requires idle mixtures too rich for stoichiometric closed-loop operation. Such an engine would require an interceptor capable of simulating a stoichiometric O_2 sensor signal even when mixtures are considerably richer.

If the airflow meter signal is being controlled by an interceptor to bring about changes to injection pulse width, might this affect ignition timing or boost control? The simple answer is, maybe. In fact, one advantage of an interceptor over a simple programmable mass airflow sensor on a super-modified engine is that a sophisticated interceptor will control fuel, advance/retard, and boost pressure—not just fuel.

Interceptors like the Unichip can also control auxiliary injectors and water injection, and injection itself can be controlled and accurately mapped with highly sophisticated air/fuel maps that rival the power of a standalone EMS. Unichip's internal software contains a large amount of powerful, table-driven Boolean logic that can be programmed via laptop user interface by setting the tables to make intelligent decisions for control of other devices.

Powerful interceptors can control boost pressure to a high degree of resolution throughout the rpm range, allowing aggressive boost pressures to be programmed where it is safe to do so (say at midrange), while limiting boost pressure at higher rpm, which must also be coordinated with the correct AFR and ignition advance. With sophisticated interceptors, you should not have to run modest boost pressure across the rpm range to keep things safe at higher engine speeds.

Some powerful interceptors can remove a high-speed cut, fuel cut at high-boost, and rev-limit. Some interceptors can datalog. Some interceptors allow multiple configurations to be stored in memory. In the case of the Unichip, the tuner is presented with 3-D ignition and fuel maps along with a boost vector. The ignition and fuel maps contain 17 rpm points, each having 12 throttle position load sites, for a total of 204 separate adjustments for ignition advance/retard and eight separate boost values spread across the rpm range.

Perfect Power points out that piggyback auxiliary engine management computers are, in the end, reliant on the existing ECU to provide fuel and ignition to the engine, and are ultimately subject to the limitations of the ECU, its architecture and logic. Such systems, says Perfect Power, typically change fueling indirectly by tampering with the load sensor or lambda sensor signal, causing the ECU to calculate an alternate injection pulse width. Ignition timing might similarly be modified by delaying or advancing the crank angle sensor's signal on the way to the ECU, which then triggers the coil driver at an alternate timing angle. Bottom line, the piggyback unit tricks the ECU into thinking it is running under different conditions, thereby allowing a different output.

Perfect Power's SMT6 is an interceptor combined with additional injector controller capabilities. SMT6 capabilities and requirements include the following:
• Designed to tune *any* engine
• Windows-based user interface for map modification
• Requires laptop
• Average of six wire connections to the OEM wiring loom

Typical Connections:

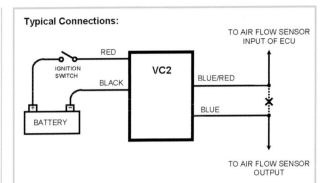

Wire Assignments:

WIRE COLOR	CONNECT TO	LABEL
Red	Switched battery positive (+12V)	BATT +
Black	Battery negative (chassis ground)	BATT -
Blue	Air flow sensor output	FLOW IN
Blue/Red	Air flow sensor input to the ECU	FLOW OUT

Clamp Function (User adjustable):

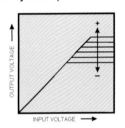

Wiring connections for Split Second's Voltage Clamp 2 (VC2). This interceptor device is designed to condition the output of voltage-based airflow meters, MAP sensors, and mass airflow sensors to avoid a fuel cut in forced-induction conversion applications. Under normal operating conditions, the VC2 outputs a signal that is identical to the signal present at its input. When the voltage of the flow signal reaches a user-adjustable clamp level, the VC2 maintains a constant output voltage at the clamp level as the input voltage rises. The VC2 differs from the VC1 by offering a hard clamp function that is user adjustable. Once in clamp mode, the VC2 maintains the clamp level with high accuracy regardless of overdrive. Split Second

• Installation of SMT6 within 30 minutes by experienced installer
• Development kit includes all software, manuals and 650-car wiring database or available unbundled
• No-tune option to hide setup and maps
• Lambda sensor signal modification via linear mode (tunes according to the voltage of the sensor) and nonlinear mode (tunes according to lambda value)
• Auxiliary output can be enabled according to rpm, airflow, temperature, and analogue deflection points
• Extra injector driver can alternately be used for proportional nitrous injection
• Input for airflow sensors and temp sensors, each having a map to manipulate fuel, extra injector, and ignition

AUXILIARY COMPUTER AND INTERCEPTOR SOURCES

034 Supplementary ECU (AIC)

APEXi Integration Air Flow Converter (AFC-VTEC), APEXi Digital Super AFC
330 W. Taft
Orange, CA 92865
(714) 685-5700   (714) 685-5701
www.apexi-usa.com

Autothority (VAF-MAF conversion + chip)
3769-B Pickett Road
Fairfax, VA 22031
(703) 323-6000 (showroom)   (703) 323 0919 (ordering)
www.autothority.com

BBR-GTi Engine Control Unit Interceptor 2000
Unit One, Oxford Road
Brackley, Northant,
NN13 7DY, England
(44) 1280 700800   (44) 1280 705339 (fax)
www.bbrgti.demon.co.uk/

Carrolls Supercharging Superpumper (intelligent FMU)
1803 Union Valley Road
West Milford, NJ 07480
(973) 728-9505   (973) 728-9506 (fax)
www.carrollsupercharging.com

ChipTorque (Xede Interceptor)
1/50 Lawrence Drive
Nerang Queensland 4211
(61) 7 5596 4204
(61) 7 5596 4228 (fax)
www.chiptorque.com.au

ECUTek (Remap hardware and software for Subaru WRX)
8 Union Buildings
Wallingford Road
Uxbridge UB8 2FR United Kingdom
(44) 1895 811200   (44) 1895 811611 (fax)
www.ecutek.com

Greddy eManage piggyback
9 Vanderbilt
Irvine, CA 92618
(949) 588-8300
(949) 588-6318 (fax)
www.greddy.com

HKS (PFC FCON, VPC, Super AFR, AIC III)
2355 Miramar Avenue
Long Beach, CA 90815
(310) 763-9600
(310) 763-9775 (fax)
www.hksusa.com

Hondata (Honda ECU programming interface platform)
2341 West 205th Street
Ste 106
Torrance, CA 90501
(310) 782-8278

Link (Interlink interceptor, UEGOLink Wideband controller)
1643 Monrovia Avenue
Costa Mesa, CA 92627
(949) 646-7461
(949) 646-7471 (fax)
www.link-electro.usa.com

Perfect Power SMT6 Piggyback
600 Texas Road
Morganville, NJ 07751
(732) 591-1245
http://www.perfectpower.com

PMS (EEC Piggyback)
Anderson Ford Motorsports
PO Box 638, Route 10 West
Clinton, IL 61727
(217) 935-2384
(217) 935-4611
www.andersonfordmotorsport.com

ProECM Powerchip PC-3
(harness-connected piggyback computer chip)
info@proecm.com
www.proecm.com

Split Second (MAF, MAP Piggyback Laptop-programmable Signal Calibrator, Pot-adjustable Air-fuel Ratio Calibrator, based on VAF, MAF, TPS or other signal input, Additional Injector Controller, Sensor Conditioners and Clamps)
1949 Deere Avenue
Santa Ana, CA 92705
(949) 863-1359   (949) 863-1363
www.splitsec.com

Techlusion TFI (Fuel Nanny Interceptor)
668 Middlegate Road
Los Vegas, NV 89105
(702) 558-5142 or (877) 764-3337
(702) 558-5396

Turbosmart (Electronic and Pneumatic Fuel Cut Defenders)
Mailing: P.O. Box 264
Croydon, NSW 2132 Australia
Home Office: 32 Milton Street North
Ashfield, NSW, Australia
(61) 2 9798 2866   (61) 2 9798 2826 (fax)
www.turbosmart.com.au

TurboXS (Subaru WRX)
8041 Queenair Drive, Unit 2
Gaithersburg, MD 20879
(301) 977-4727
(301) 977-6507 (fax)
http://www.turboxs.com

Unichip (Dastek)
The Racers Group (US Distributor)
29181 Arnold Drive
Sonoma, CA 95476
(707)935-3999   (707) 935-5889 (fax)
www.theracersgroup.com
www.dastek.co.za/

.7 -36.8 -40.7 -43.2 -41.1 -30.2 -34.4 -34.8 -22.6 -8.8 -9.3 -2.4 5.9 12.2 30.0 30.0
.7 -36.3 -41.2 -42.9 -36.3 -34.2 -38.1 -38.8 -23.7 -12.5 -13.0 -4.3 6.1 0.0 30.0 30.0
.0 -29.5 -35.7 -36.8 -26.8 -34.3 -37.6 -38.7 -24.0 -13.7 -13.6 -3.3 -4.5 -6.0 -6.0 -6.0
.0 -28.3 -28.6 -25.0 -25.5 -28.5 .4 -2 -1 -6.0 .0 -6.0 -6.0

CHAPTER 8
STANDALONE PROGRAMMABLE ENGINE MANAGEMENT SYSTEMS

Whether you're a performance enthusiast considering the acquisition of your first programmable engine management system or a veteran of multiple engine calibration efforts on many brands of standalone onboard computers, choosing the best engine management system to manage *your* performance engine can be daunting. It's unfortunate but true that the strange world in which real-time computing intersects with kick-ass performance engines is formidable in both theory and application.

Finding someone who truly understands engine management theory is tough. Finding a theorist with really solid, in-depth practical knowledge of even *one* aftermarket engine management system is impressive. And finding a wizard with superior practical knowledge of *all* the various aftermarket systems is extremely difficult because there are only a few in the world.

Dallas supertuner Bob Norwood, who was consulted for this chapter, has been calibrating and programming aftermarket engine management systems on wildly powerful exotic and standard engines ever since the first standalone systems arrived from

Air Sensors in the early 1980s. Norwood is a Motec and Accel/DFI dealer and has been the U.S. master distributor for Haltech. He has calibrated virtually every past and present engine management system on his Dallas Dynojet and has made them all work—at least so far as possible. In 2002, Norwood teamed up with racer Kenny Tran and Quaker State and used methanol fuel and a Motec engine management system to boost Tran's Honda from 600 to 900 horsepower and post a record-setting best quarter mile time of 8.56 at 172 miles per hour with the world's fastest front-wheel-drive Honda.

So how *do* you perform an apples-to-apples, real-world comparison of the virtues of multiple complex engine management systems? How do you compare the scores of features, capabilities, and options that may not be completely analogous and any one of which might be truly important or of no consequence in managing *your* engine?

A realistic comparison involves much more than simply evaluating the features. For example, in the real world, how do you compare the performance potential of a powerful but complex system—that may be difficult or impossible to tune well without expensive diagnostic equipment and extensive experience—to a less-powerful system with simpler engine-modeling algorithms that—in the real world—may achieve a higher state of tune because calibrating it well requires less time, expertise, and wizardry?

All the sexy features and complex, table-driven logic in the world simply provides the rope to hang yourself if the configuration and calibration are beyond the scope of your tuning capabilities (or those of the guy you hire). What's more, even the best professionals can seriously burn up time calibrating a complex engine management system, particularly if it's not a system they've worked on recently. Of course, an ECU with simplistic engine modeling logic may be physically incapable of managing some engines with high-quality drivability and safe power.

There are a *lot* of aftermarket engine management systems out there today. When I published my first book about custom fuel injection in 1993, there were a half-dozen or so standalone (as distinct from piggyback or interceptor) systems on the market. When researching the situation for this book there were more than 20 manufacturers building user-programmable aftermarket systems capable of managing 100 percent of the fuel and spark delivery on an engine. What's more, many of these companies build several different engine management systems each targeted to take advantage of the vast price elasticity and differing requirements in the market for performance and racing equipment. In addition, there are still many standalone programmable aftermarket systems that are no longer sold but are still out there by the hundreds or thousands faithfully running performance engines on the street or race track (or perhaps awaiting your bid on eBay).

There are also several companies that build hardware-software tools designed to hack into factory OEM engine management systems or change their behavior or in some sense modify them to

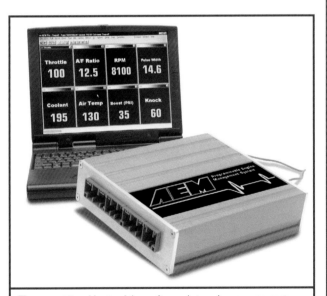

The newest trend in standalone aftermarket engine management systems is the plug-and-play concept, which is designed to allow a replacement ECU to plug into the stock ECU connector with no changes to the wiring, sensors, or actuators. AEM designed a vastly powerful ECU with great flexibility, which AEM can use to run a large variety of EFI systems by adapting the ECU pin-outs to the local requirements of a particular vehicle's EMS. Though AEM supplies generic calibrations for stock engines or common modifications, the company warns that failure to have an expert tune the calibration could damage your engine. Like most aftermarket systems, the AEM ECU is wide open to tuning with a laptop PC. AEM

.7	-36.8	-40.7	-43.2	-41.1	-30.2	-34.4	-34.8	-22.6	-8.8	-9.3	-2.4	5.9	12.2	30.0	30
.7	-36.3	-41.2	-42.9	-36.3	-34.2	-38.1	-38.8	-23.7	-12.5	-13.0	-4.3	6.1	0.0	30.0	30
.0	-29.5	-35.7	-36.8	-26.8	-34.3	-37.6	-38.7	-24.0	-13.7	-13.6	-3.3	-4.5	-6.0	-6.0	-6
.0	-28.3	-28.6	-25.0	-25.5	-28.5	-38.7	-34.9	-20.2	-17.5	-16.4	-7.1	-6.2	-6.0	-6.0	-6

R.P.M.	5106
Pressure	-0.49

be programmable. In addition, there are a number of interceptor control units on the market designed to modify the behavior of your original equipment factory engine management system by creatively lying to it about engine conditions or by hijacking the main onboard computer's output signals targeted at fuel injectors or other actuators and modifying these signals to be compatible with performance engine modifications or other revised user requirements.

There are also several electronic controllers available that piggyback on a stock computer and entirely take over *some* of the functionality of the main onboard computer (for example, fuel injection control). Still other auxiliary systems control additional injectors or other devices that remain invisible to the main computer. And then there are a number of programmable intelligent sensors or actuators that lie to or willfully disobey the main engine control unit to influence its behavior in various ways.

This chapter focuses on new standalone aftermarket engine management systems that handle all engine management tasks. It includes a summary chart to compare, contrast, and analyze the basic capabilities of standalone vendors and their systems. The chart includes one or two gray-area systems from companies like Hondata, which are not exactly standalone programmable engine management systems but essentially function to convert original-equipment systems into standalone programmable engine management systems.

This chapter is designed to provide the gist of what is out there and to get you thinking about what matters when you purchase a standalone engine management system. Like camshaft selection, an intelligent decision about the right engine management system involves much more than just reading a chart and looking for the biggest, baddest one on the page. More than just evaluating what the systems can do, an intelligent choice must also include evaluating what your engine really needs and then evaluating your own financial and technical capabilities and limitations to find a compatible marriage with the right standalone engine management system.

GENERAL CAPABILITIES

Some general capabilities of an engine management system are fairly straightforward. How many cylinders can the unit manage, and in which configurations? Can it do throttle-body injection, or sequential injection? What about staging in secondary port injectors if the fueling requirements exceed the capacity of one injector per cylinder? Which kind of ignitions can it control? What is the maximum rpm it can handle? Can it handle any shifting functions such as torque-converter lockup? Does it have soft, fuel-based rev-limiting capabilities available to keep you from hurting the engine if your supercharger conversion makes it really fun to blast hard past 8,500 rpm?

On one hand this is not rocket science. If you have a V-6, don't buy a system that only has four injector drivers. And cost is

Accel's Gen 7 EMS is the latest system from one of the old-timers in the aftermarket EMS business. Accel, which years ago acquired John Meany's DFI (Digital Fuel Injection) company, has sold many older Gen 6 systems, which were getting a little long in the tooth until the new system arrived. Meany, by the way, moved on to design the Fel-PRO EMS, which was then renamed Speed-PRO, and eventually became the FAST (Fuel Air Spark Technology) system. Accel

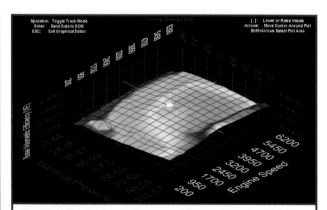

The Accel Gen 7 system has an extremely complex and powerful architecture, including the volumetric efficiency table structure, which describes how engine breathing varies from a perfectly flat torque curve with changes in rpm and manifold pressure. Accel

extremely difficult to quantify in a chart. The *cost* of an engine management kit can be really deceptive if it turns out you have to buy expensive optional components that are not included in the base price.

Here's the thing: Many ECUs these days are flexible about accepting data from a variety of types of engine sensors. Most can handle a variety of types of fuel injectors, and a variety of igniters, so your stock ones will probably work. The system may even support your stock manifold absolute pressure (MAP) sensor. But let's say your stock engine used to be naturally aspirated: The stock MAP sensor will almost definitely be a 1-bar unit

	.7	-36.8	-40.7	-43.2	-41.1	-30.2	-34.4	-34.8	-22.6	-8.8	-9.3	-2.4	5.9	12.2	30.0	30.0
R.P.M. 5106	.7	-36.3	-41.2	-42.9	-36.3	-34.2	-38.1	-38.8	-23.7	-12.5	-13.0	-4.3	6.1	0.0	30.0	30.0
	.0	-29.5	-35.7	-36.8	-26.8	-34.3	-37.6	-38.7	-24.0	-13.7	-13.6	-3.3	-4.5	-6.0	-6.0	-6.0
Pressure -0.49	.0	-28.3	-28.6	-25.0	-25.5	-28.5	-38.7	-34.9	-20.2	-17.5	-16.4	-7.1	-6.2	-6.0	-6.0	-6.0

Autronic is one of the newer players in standalone engine management, yet another company from Down Under with significant automotive engineering talent. Recognizing the difficulty of the calibration task, Autronic has a continuing development effort in "self-tuning" capabilities that involve wideband air/fuel sensing and various learning modes of operation and driving. Autronic

DTA is a British company that provides very sophisticated engine management for multi-cylinder direct-fire applications, including the likes of the Cosworth/Opel DTM V-6, with unequal-firing 75-degree/165-degree architecture, six direct coils, 8-bar injection pressure, and very low duty cycle—and 450 brake horsepower at 11,500 rpm from 2.5 liters. DTA

APEXi's Power FC is riding the wave of interest in plug-and-play ECUs that seamlessly plug into the stock ECU connector in place of the factory computer and then allow you to drive away. Hardware compatibility with stock sensors and actuators is not the only important thing. Excellent calibrations for specific engines and vehicles are critical, and APEXi has concentrated on doing this right. A dedicated calibration module enables recalibration or customization for future engine modifications. APEXi

PRODUCT/TECH SUPPORT

Product support is also difficult to quantify in a chart, but it is probably the most important feature of an engine management system—and this extends to the expert pre-sales consulting that should be available up front to make certain that the unit will really work well for you. OK, there may be a toll-free number (or maybe it's *not* a toll-free number) to call for product support.

But how easy is it to get through to a human being? Don't laugh. In some industries there have been documented cases of customer service numbers no one *ever* answers. How many systems do they have installed out in the field that they're supporting, and are they straight about answering such questions? How polite, how helpful is the person who picks up the phone? What about the fifth time you call, late Friday afternoon? Is there anyone there answering the phone in early evening on the East Coast when you call at 4:30 p.m. Pacific time? What about on weekends? Will the person on the phone give good advice that rapidly solves problems or will it be bad advice that causes you to break things and flame your engine. Or even worse, is he unaware that his system *cannot work* in your application?

How good is the manual (some *really* suck)? Is there a good Web site with a good knowledge base or FAQs, or is the site mainly a selling tool? Is the engineer who actually built the unit available for consulting as a last line of defense if all else fails or if your application actually uncovers a hardware or software bug in the product, or does he live in a different country or continent? Is there a dealer or shop in your area, preferably with a chassis dynamometer, who can help get the system working right on your vehicle if all else fails and you just can't get it right yourself or if the performance is disappointing?

Is there *another* dealer or tuning shop in reasonable proximity if the first dealer turns out to be a disappointment on your application? Is your application standard or are you the first person in the world lucky enough to be trying to get an XYZ Superfreak programmable engine controller to work on a 1912 John Deere tractor with 50-jigawatt magnetos that emit so much electrical interference that the rig jams AM radio stations in Calcutta?

Are you about to be an interesting experiment for a supplier or dealer hot for one more sale this month (like that Brand-X muffler guy in the old Midas ad with a really big hammer who announces, "We'll *make* it fit.")? Well-run, stable companies with

unable to measure boost much above atmospheric pressure. On a turbo or blower conversion you're going to need a 2-bar or 3-bar MAP sensor, and these cost money. That ECU you're lusting for may drive eight fuel injectors right out of the box. But maybe it will drive only *four* low-resistance performance-type peak-and-hold injectors, not the *eight* you're going to need on your V-8.

Is an ECU able to manage coil dwell (charging) time with *your* coil (or coils) and *your* igniter (some igniters provide fixed dwell, others variable). In general, be wary of kit prices. Unless you're buying a turnkey EFI conversion system like Edelbrock's carb-to-EFI conversion Pro-Flow Chevy V-8 system, there *will* be extras. There may well be extras *even on a turnkey system.*

.7 *36.6 -40.7 -43.2 -41.1 -30.2 -34.4 -34.8 -22.6   -8.8   -9.3   -2.4   5.9   12.2   30.0   3
.7 -36.3 -41.2 -42.9 -36.3 -34.2 -38.1 -38.8 -23.7 -12.5 -13.0   -4.3   6.1   0.0   30.0   3
.0 -29.5 -35.7 -36.8 -26.8 -34.3 -37.6 -38.7 -24.0 -13.7 -13.6   -3.3   -4.5   -6.0   -6.0   -
.0 -28.3 -28.6 -25.0 -25.5 -28.5 -38.7 -34.9 -20.2 -17.5 -16.4   -7.1   -6.2   -6.0   -6.0   -

| R.P.M. | 5106 |
| Pressure | -0.49 |

Engine Management Internal Tables and Parameters List

The following tables are documented within the architecture of Accel's Gen 7 EMS, and are examples of a fairly exhaustive definition.

Steady-State

- Individual Cylinder Fuel Coefficient table (fuel correction)
- Injector Correction v. Battery Voltage table
- Base Offset Injector Timing table (intake valve closing angle in crank degrees)
- Target Injector Timing table (Sequential-only end-of-injection event in crank degrees for speed and loading)
- Target Air-Fuel Ratio table (Fuel to Engine = Measured Air In / Target Air-to-Fuel Ratio)
- VE Correction to Target Air-Fuel Ratio table (richer mixtures increase charge mass, which is harder to pump)
- VE Correction coefficient versus Estimated Intake Port Temperature (density DECREASES with temp, but hotter air pumps more easily)
- VE Correction versus Engine Coolant Temp table (correction to base volumetric efficiency calculation)
- Engine Coolant Temperature VE correction vs. MAP/RPM (MAP-RP* correction to VE compensating value for ECT) (second-order correction?)
- Base VE table (engine pumping efficiency at all speeds and loading)
- Alpha-N v. Percent Throttle table (estimated MAP for given throttle position, treated as actual MAP for calculations)
- Engine Coolant Temperature versus Intake Air Temperature Difference table (contains coefficients used to estimate intake port temperature
- Estimated Intake Port Temp = ECT – IAT Coefficient x (ECT – IAT) – Manifold Surface Temp Coefficient x (ECT – MST)—captures the effects of the various heat-flow elements in the intake manifold; at idle, the incoming air picks up much of the heat from the cylinder head, while at high air-flow the outside air temp dominates
- MST-IAT Difference table (coefficients used to estimate intake port temperature, looked up from table based on intake mass air flow)

Transitional

- Model #1 (Speed-Density-based)
- Model #2 (MAF-based)
- Model #3 (Alpha-N/Throttle position)
- Acceleration Threshold table (to disregard very small requests for fuel)
- Acceleration Modifier table (adjustment to acceleration fueling for coolant temp)
- Accel Cutoff table (no more asynchronous fuel when MAP above this value—prevent fuel changes v. small MAP changes)

Feedback

- Fuel Control Proportional Gain (amplifies O2 correction control to drive EGO position toward desired response (if too high, excessive overshoot and oscillation))
- Fuel Control Integral Gain (reflects error history)
- Fuel Control Differential Gain (Reacts to the slope of the error)
- UEGO Maximum Fuel Feedback Coefficient table (Clamps fuel correction in closed loop mode versus engine load and RPM; a zero valve mandates Open Loop at RPM/load)
- HEGO Maximum Fuel Feedback Coefficient table
- HEGO Stoichiometric Air-to-Fuel Ratio (ratio for most complete combustion; must be 14.5 when running gasoline)
- Closed-Loop ECT Threshold table (prevents Closed-Loop correction during warmup; in degrees F.)
- Closed-Loop Delay v. Starting ECT (seconds to Closed-Loop depending on ECT at startup)
- HEGO Modifier table (provides smooth transition between Open- and Closed-Loop; absolute value in table is difference between stoich and target AFR)

Ignition-starting

- Ignition Startup Term table (base amount of ignition advance added to base IGN during startup; ECT is pointer into table
- Ignition Startup Decay Interval table (time to x-degree startup advance decay, as function of MAP and ECT)
- Ignition Startup Phase-in Interval table (time between 1-degree advance increment steps during cranking)

Ignition-steady state

- Port Air Temp IGN Compensation table (port air temp is pointer into table that adds advance as function of port air temperature)
- PATM/VE IGN Comp (air pressure is pointer into table that adds advance as function of atmospheric pressure VE correction)
- Base ECT IGN Compensation table (ECT is pointer into table adding advance to base IGN timing as function of coolant temperature)
- ECT Compensation Power Modifier table (how much the ECT IGN advance is trimmed for speed and load)
- Idle Spark Control Compensation table (how much IGN advance is added to assist in idle speed control)
- Base Advance-IGN table (IGN advance as a function of RPM and MAP—limited by crank offset value maximum)

Ignition-knock

- Knock Feedback Retard Limit table (clamps maximum IGN advance that can be removed during knock)
- Knock Feedback Retard Interval table (time period between ignition advance reduction steps during knock—based on RPM and MAP)
- Knock Feedback Advance Interval table (time period between ignition advance steps after knock is over—based on RPM and MAP)

Ignition-Dwell Control

- IGN Voltage Dwell Compensation table (Amount of time added to base ignition dwell as a function of ignition voltage)
- IGN Dwell Period table (length of time the ignition coil dwells as a function of RPM)

Idle Control

- Target Idle Speed table (Desired engine RPM when throttle closed, based on ECT)
- Maximum Percent Throttle Idle Mode (RPM below which idle speed control is active)
- Maximum Percent Throttle Idle Spark (RPM below which IGN Idle Spark Compensation mode is active and timing is used to help control engine speed)

.7 -36.8 -40.7 -43.2 -41.1 -30.2 -34.4 -34.8 -22.6 -8.8   -9.3   -2.4   5.9   12.2 30.0 30.0
.7 -36.3 -41.2 -42.9 -36.3 -34.2 -38.1 -38.8 -23.7 -12.5 -13.0 -4.3   6.1   0.0   30.0 30.0

R.P.M.   5106
Pressure   -0.49

.0 -29.5 -35.7 -36.8 -26.8 -34.3 -37.6 -38.7 -24.0 -13.7 -13.6 -3.3   -4.5 -6.0 -6.0 -6.0
.0 -28.3 -28.6 -25.0 -25.5 -28.5 -38.7 -34.9 -20.2 -17.5 -16.4 -7.1   -6.2 -6.0 -6.0 -6.0

- Idle Control Proportional Gain (specifies Idle bypass air response magnitude based on error and this value to command stepper motor)
- Idle Control Integral Gain (alters Idle bypass response gain based on past errors)
- Idle Control Differential Gain (alters Idle bypass change in response to sudden change in idle speed)
- Minimum IAC Position v. ECT table (Clamps minimum limit of throttle bypass air; prevents large speed undershoots upon rapid deceleration)
- Maximum IAC Position v. ECT table (Clamps maximum limit of throttle bypass air; prevents large delays in resuming idle speed when returning to idle from off-idle)
- IAC starting Position v. ECT table (amount of air bypassed while starting engine
- Idle Control Delay table (how long ECU waits after initial crank to start before commencing idle speed control)
- Throttle Follower table (bypass air as a function of throttle position for smooth transition to idle control when returning to idle)

Nitrous
- Nitrous System Enable
- Linear O2 Sensor Enable (Enable UEGO to enable or disable Lean Linear O2 Threshold table)
- Stage 1-Linear O2 Threshold (A:F) (leanest UEGO AFR to permit Stage 1 nitrous active)
- NOS BSFC—All Stages (BSFC of engine when Nitrous is delivered
- NOS Stage 1—Number Orifices
- Stage 1-NOS Diameter
- NOS Stage 1-Line pressure
- NOS Stage 1 Delay
- NOS Stage 1—Minimum Percent Throttle
- NOS Stage 1—Minimum RPM
- Total Fuel Enrichment (lbs/hr) Monitor (Displays total fuel used by engine, including NOS fuel, if active)
- NOS Fuel Enrichment (lb/hr) Monitor)
- Stage 1-IGN Retard –Ignition compensation NOS 1 table (amount of advance added to base advance when nitrous activated; RPM is pointer into table)
- Stage 1—Fuel Trim table (amount of positive or negative fuel trim (lbs/hr) added to main fuel flow—compensates for overall NOS fuel efficiency)
- NOS Stage 2... (same as above)
- NOS Stage 3 (same as above)
- Datalogging—data connection speed between ECU and Laptop/User interface
- Datalogging—logged to ECU memory or laptop memory?
- Datalogging—Maximum number parameters that can be logged
- Datalogging—Minimum time between logging events for parameters being logged
- Datalogging—Maximum number of logging events

Other
- Return or Return-less Fuel System (ECU calculate the pressure differential across the injector—like battery voltage correction)
- Fuel Injection Type

integrity will walk away from no-win business. And they have the technical and financial depth to recover from problematic situations if they made a mistake. And *all* companies and individuals make mistakes because they're human.

USER INTERFACE

First of all, do you need a laptop? This is less important now than it was a few years ago because there are a lot more laptops today, but most systems *do* require a Windows 95 or later laptop computer to interface with the onboard control unit for setup and tuning. Edelbrock's Pro-Flow and the ECUs from Simple Digital Systems do not, and you may or may not find this to be a good thing.

The friendliness of the graphical or tabular (table-based) user interface is of decreasing importance as your knowledge and experience with a system increase, but really great graphics can solve problems by making it easier to see anomalies in, for example, the volumetric efficiency or injection pulse width tables (and behind the gorgeous graphics, all tuning maps are matrices of numbers).

If you are *not* super familiar with an engine management system, a trim module, in which turning a dial richens or leans the air/fuel mixture or advances or retards timing or changes maximum turbo boost, might make tuning much easier (though some people loathe such things).

HARDWARE

Check out the environmental requirements for an engine management system: In general, engine management systems are designed to operate reliably in a temperate environment, which is

why virtually all onboard computers require placement in the cockpit with the driver. Electronic circuits hate heat, moisture, and vibration, but a few engine management systems are hardened and equipped with special waterproof connectors so it is *possible* to locate them in the engine compartment, if not desirable. The older TEC 1 and TEC 2 systems from Electromotive were examples. Pay close attention to the environmental and mounting requirements. If you don't, the system may fail.

ECUs with external indicator LEDs or check engine lights are desirable, because they can warn you of incipient or intermittent problems. And in case of a serious malfunction, they can immediately steer you in the right direction.

System power requirements provide information about the flexibility of the system to handle voltage changes (some have a much wider operating range than others). They also let you know how much current (amperage) the system consumes from the alternator (and it is *not* trivial). The power consumption depends very much on the number and type of injectors and the number and type of ignition coils and igniters.

The physical size and weight are self-evident, but remember that the ECU box is effectively larger when the connector plug(s) and wiring harness are installed.

RESOLUTION

Digital computers approximate fuel and ignition curves with a table of discrete numbers, and then compute a moving average for points in between. A naturally aspirated engine with a flat torque curve does not require many numbers in its fuel and ignition tables

continued on page 98

```
.7 -36.8 -40.7 -43.2 -41.1 -30.2 -34.4 -34.8 -22.6 -8.8 -9.3 -2.4 5.9 12.2 30.0 3
.7 -36.3 -41.2 -42.9 -36.3 -34.2 -38.1 -38.8 -23.7 -12.5 -13.0 -4.3 6.1 0.0 30.0 3
.0 -29.5 -35.7 -36.8 -26.8 -34.3 -37.6 -38.7 -24.0 -13.7 -13.6 -3.3 -4.5 -6.0 -6.0 -
.0 -28.3 -28.6 -25.0 -25.5 -28.5 -38.7 -34.9 -20.2 -17.5 -16.4 -7.1 -6.2 -6.0 -6.0 -
```

STANDALONE PROGRAMMABLE ENGINE MANAGEMENT SYSTEMS

EMS			Accel/DFI Gen 7
GENERAL	Kit Includes		ECU, Wiring Loom, Ignition Adapter, Relays, ECT, IAT, MST, O2 sensor, Cable
	Optional Components		IAC, Wide O2, Ignition Module, Fan, Tq Convt, 3-stage Nitrous, Knock, Check light, A/C clutch
	Cost (8 cy kit w/sensors & harness; 4-cy; ECU)		Jobber: $1475; $1575; $975
	Injection Modes		Sequential, Stag. Batch, Batch, 1&2 TBI, Staged Batch/Seq
	Ignition Control		Yes (GM HEI, Ford TFM, Ind Pickup, Hall Effect)
	Engine Configurations (cylinders)		4,6,8
	Load-Sense via MAP,TPS,MAF,Veloc,Torque		MAP, TPS
	Transmission/Powertrain Mgmt?		Yes-Torque Converter lockup
USER INTERFACE	Laptop Requirements		Win 98+, 400mhz, Serial Port, 20MB Disc space
	Graphical v Tabular; Tuning trim pots?		Both; No
HARDWARE	Physical Size And Weight		6" x 6" x 2"; 2lbs
RESOLUTION	# Fuel; IGN Map rpm Ranges; User defined?		16/16; yes (to 10 RPM res)
	# Load Points/RPM Range; Total points/Map		16 points; 256 pts
AIR-FUEL	Target Air-Fuel Ratio Map?		Yes
	Fuel Press. Detect & Inj Pulse Comp		No
	Individual Cylinder Fueling Adjustment?		Yes (as a function of speed and load)
CLOSED-LOOP	via O2,WideO2; High-load Closed Loop?		Yes, Yes; Yes
CALIBRATION	Utility to Create "Startup" Calibration?		VE predictor based on user input gens a "Running" calibr.
	Automated Mapping Features		%, +/- Additive Change, Abs. change via Keys or mouse
PLUG AND PLAY	Hardware, Wiring		Most applications only with dealer interface harness
	Self-Learning ("Tunes itself")		EGO & Wide O2 trim, AutoCal will rewrite exist. VE map
	Specific Applications in Calibration Library		Chevrolet 502, 454, 350, Ford 302, 351, more summer '03
DATALOGGING	Rate-Time; User selectable?; in Laptop of ECU?		?; ?; Yes; Laptop
I/O	ECU Inputs		8,
	ECU Outputs		7,
	Max # IGN-Only Coil Drivers; Coils		1 (no direct coils)
SENSORS	Sensors (Required; Optional; Programmable)		RPM, TPS, MAP, IAT, CTS, MST; EGO, UEGO, KNOCK
	Eng. Position Triggers; Pattern		Hall, Opt., Ind. mag reluctor; 5/12V square waveform, EDIS
	Maximum Recognizable Map Pressure		3-Bar sensor, extrapolation to 50psi manifold pressure
	Knock Sensor; Individual cyl retard?		Yes (1); yes
ACTUATORS	Injector Types Supported		Most
	Ignition Types Supported		Single distr, Waste-spark with external module
	Software-Configurable IGN Compatibility		Yes
	Extra Ignition Amplifier Required?		Yes (ignition igniter or module needed to direct-drive coil)
	Additional Supported Actuators		?
NITROUS	Nitrous Triggering/ Fuel via Primary Inj.		Yes (three-stage), Yes
WIRING	Wiring Harness (Generic/Specific)		Universal 4,6, or 8 cyl. harness

EMS			Autronic SMC/SM2
GENERAL	Kit Includes		ECU, harness, data cable, software, H2O temp, Air Temp
			Fuel pump relay, indicator LED, Check light output
	Optional Components		IAC, EGO, Wideband O2, Boost Controller, Ignition Module, Coils
	Cost (8 cy kit w/sensors & harness; 4-cy; ECU)		$1599 (smc)-$1899 (sm2); same
	Injection Modes		Sequential
	Ignition Control		?
	Engine Configurations (cylinders)		1-16 cyls (8 sequential),
	Load-Sense via MAP,TPS,MAF,Veloc,Torque		MAP, TPS, MAF, Alpha-N + MAP
	Transmission/Powertrain Mgmt?		No
USER INTERFACE	Laptop Requirements		Pentium, Serial Port, Floppy
	Graphical v Tabular; Tuning trim pots?		Both; No
HARDWARE	Physical Size And Weight		157 x 122 x 35 mm; 180 x 128 x 35 mm (overall)
RESOLUTION	# Fuel; IGN Map rpm Ranges; User defined?		16 x 32/16 x 32; yes
	# Load Points/RPM Range; Total points/Map		User selectable; 512 pts
AIR-FUEL	Target Air-Fuel Ratio Map?		Yes
	Fuel Press. Detect & Inj Pulse Comp		No
	Individual Cylinder Fueling Adjustment?		Yes
CLOSED-LOOP	via O2,WideO2; High-load Closed Loop?		Yes, Yes; Yes
CALIBRATION	Utility to Create "Startup" Calibration?		No
	Automated Mapping Features		Copy, % change, absolute change, injector change, etc.
PLUG AND PLAY	Hardware, Wiring		Plug-in motherboard and connector
	Self-Learning ("Tunes itself")		Yes
	Specific Applications in Calibration Library		Subaru WRX, Mitsubishi EVO, 2G series Mitsubishi
DATALOGGING	Rate-Time; User selectable?; in Laptop of ECU?		50/second; Yes; ECU
I/O	ECU Inputs		
	ECU Outputs		Inj Drivers (4 dedicated, 8 configurable)
	Max # IGN-Only Coil Drivers; Coils		4 direct or waste-spark coils
SENSORS	Sensors (Required; Optional; Programmable)		TPS, MAP, Engine speed, IAT, CTS, Cam position
	Eng. Position Triggers; Pattern		Hall Effects, Inductive Magnetic Reluctor/?
	Maximum Recognizable Map Pressure		Any
	Knock Sensor; Individual cyl retard?		No; no
ACTUATORS	Injector Types Supported		All
	Ignition Types Supported		Single Distr, Dual Distr, Waste-spark, Coil-on-Plug, Rotary
	Software-Configurable IGN Compatibility		Yes
	Extra Ignition Amplifier Required?		Yes
	Additional Supported Actuators		
NITROUS	Nitrous Triggering/ Fuel via Primary Inj.		Yes/No
WIRING	Wiring Harness (Generic/Specific)		Yes/Through Specific Dealers

```
.7  -36.8  -40.7  -43.2  -41.1  ...  ...  ...  ...  ...  ...  ...  ...  30.0  30.0
.7  -36.3  -41.2  -42.9  -36.3  -34.2  -38.1  -38.8  -23.7  -12.5  -13.0  -4.3   6.1   0.0  30.0  30.0
.0  -29.5  -35.7  -36.8  -26.8  -34.3  -37.6  -38.7  -24.0  -13.7  -13.6  -3.3  -4.5  -6.0  -6.0  -6.0
.0  -28.3  -28.6  -25.0  -25.5  -28.3  -38.7  -34.0  -20.2  -17.5  -16.4  -7.1  -6.2  -6.0  -6.0  -6.0
```
R.P.M.   5106
Pressure   -0.49

AEM Plug and Play EMS	A'PEX Power FC
ECU and Comms cable (uses factory wiring loom, relays and sensors)	ECU (plugs into stock harness)
Software is fully functional with NO added $$$ tuning options. Uses all factory components.	Optional parts include a boost control kit (with solenoid), and the hand held commander (laptop optional)
na; na; avg $1975, $1450 street	na; $980; $980
Sequential, Staggered Batch, Batch, TBI, Staged Batch	?
Yes	
1-6,8,10,Rotary-2,Rotary-3,Rotary-4	4, 6, or 2 rotors (Depending on application)
MAP, TPS, MAF, Air Vel. (Karman Vortex type)	?
Gear, shift point, firmness, lockup, Full man. valvebody shifting	No
Win 95+, 233Mhz, 64 MB, CD, & 20MB free disk	Laptop or hand held commander
Both; No	?
2.0" x 7.0" x 7.5"; ?	?
21/21; yes, fully definable and can be non-linear	?/?; ?
17 points; 357 pts	?; 200 pts
Yes (17*21 point map)	Yes
Yes via fuel trim channel (OR EGT Fuel correction)	?
Yes, via 17 point RPM table for each injector	No
Yes, Yes; Yes	?
Startup calibrations supplied for each veh. supported	?
Copy, %, 3D Interp, Fill, Add, Resize Inj, Chg fuel Press	?
Uses factory harness with no wiring necessary	Uses stock harness
Complete fuel auto mapping allows the ECU to create its own fuel map based off of any good O2 sensor input.	na
Honda, Mustang, Supra, WRX, MR2, Eclipse, Viper	Yes (exclusively)
Up to 250/second/channel; Yes; Both	Datalogging not yet available at this time
0-5V(8): TPS, MAP, MAF, AFR(x2), Baro, Spare(2); Knock(2); Temp(2); EGT(4); Speed(4); Switch = 6	Stock
Inj(10); Coils(5); Gen Purp. Low Side Drivers(12); G.P. High Side Drivers(4); Stepper motor Drivers(2); PWM(2)	Stock plus shift light, boost controller
0,5	
Std sensors, MAF, EGT, O2; EGT (x4) or fuel trim, Reassign inputs, 2 aux(0-5V) inp, 6 aux switched input	Stock
Hall, Optical, Mag (0-5v or VR type)	?
No maximum (any 0-5V or frequency type MAP)	?
Yes(2); yes (each cyl assigned to spec knock sensor)	?; no
All > 9 ohms, otherwise a resistor box is needed.	Specific applications
Single distr., waste-spark, coil-on-plug, Rotary lead-trail	Specific applications
Yes, except logic out/current source	?
No, factory ignition module used	No
Many	Stock plus shift light, boost controller
Yes, Standard, 4-stages, 357 point nitrous Map	No
Uses factory harness with no wiring necessary	Uses stock harness

Edelbrock Pro Flow	Electromotive TEC3
ECU, Wiring, Relay, Sensors (MAP, IAT, CTS, TPS), Injectors, Int manifold w/rail & Inj, Cal Module, fuel pump	ECU, Wiring Loom, CLS, MAT, MAP, HEGO), Communications Cable, Direct-fire Coils
Nothing (turnkey system)	Fuel/IGN/boost trim module, IAC, EGO, Wide O2
Dealer pricing ?	$2655; $2515 (incl. direct, wasted-spark ignition)
Semi-sequential (two injectors 90 degrees of crank rotation)	Sequential, Staggered Batch, Batch, TBI, Staged Batch
Single Distr	Wastefire Direct Coils
8,	1, 2, 3, 4, 6, 6(oddfire), 8, 12, R1, R2, R3, R4
MAP	MAP, TPS, MAF
No	No (trans control module in development)
Not required	Pentium-class, Win 95+, CD, Serial Port, Floppy, Std DB9 cable
Both; No (tuning via Calibration module)	Both; Fuel, Ignition, Boost trim modules
8" x 8" x 1 5/8"; 2lbs	5.5" x 6.4" x 1.67"; 1.8 lbs.
24 cells/6; No (fixed 1K-7K)	16/16; Yes
1 point; 24 pts	16 points; 256 pts
Pre-determined	Yes
No	Yes (with proper input configuration)
No	Yes
Yes, No; No	Yes, Yes; Yes
Pre-determined factory calibration EPROM	Via HP RPM, Eng. type, Induction, Cam, Inj. Size, Boost
Pre-determined	Universal "Tuning Wizard"
Naturally-aspirated Chevrolet V8s	?
No	Yes
Yes (based on .050 cam duration)	?
na	?
MAP, TPS, CTS, IAT, BARO, EGO	7 analog, 2 selectable (analog or digital), 2 eng. pos. inputs
Injector Drivers (4), Fuel Pump Relay	24 digital (8 Inj Drivers)
0/0	4?/8?
MAP, Engine Speed, TPS, IAC, CTS, BARO included in kit	MAP & Engine Speed required, user-defined,?
na	Hall Effects, Inductive Magnetic Reluctor/?
100 kPa (1-Bar GM)	4-bar (400kPa)
No; no	Yes (1); yes
Edelbrock injectors included	All
Single Distr	Wasted-spark, Coil-on-plug, Rotary Engine Toggling
No	No
?	No
Injectors, IAC included	?
No/na	Yes/Yes
No/Yes (custom harnesses supplied for Chevrolet V8 apps)	Yes/Custom and Application-Specific

```
.7  -36.8  -40.7  -43.2  -41.1  -30.2  -34.4  -34.8  -22.6   -8.8   -9.3   -2.4    5.9   12.2   30.0   3
.7  -36.3  -41.2  -42.9  -36.3  -34.2  -38.1  -38.8  -23.7  -12.5  -13.0   -4.3    6.1    0.0   30.0   3
.0  -29.5  -35.7  -36.8  -26.8  -34.3  -37.6  -38.7  -24.0  -13.7  -13.6   -3.3   -4.5   -6.0   -6.0   -
.0  -28.3  -28.6  -25.0  -25.5  -28.5  -38.7  -34.9  -20.2  -17.5  -16.4   -7.1   -6.2   -6.0   -6.0   -
```

STANDALONE PROGRAMMABLE ENGINE MANAGEMENT SYSTEMS

EMS		EFI Technology
GENERAL	Kit Includes	Made to Users' specifications in all respects
	Optional Components	Fuel/ign trim, IAC, O2/Wide O2, dual boost (speed/gear), Ind or CDI ign, coils, tract, launch, speed, Vtec, throt-by-wire, var int, LCD dash display w/shift, gear pos, alarm lights
	Cost (8 cy kit w/sensors & harness; 4-cy; ECU)	$3500+ (sequential V8 w/4 ignitions); $1500+; $1250
	Injection Modes	Sequential, Staggered Batch, Batch, TBI, Staged Batch
	Ignition Control	Yes (any)
	Engine Configurations (cylinders)	1, 2, 3, 4, 5, 6, 8, 10, 12, 16, R1, R2, R3, R4
	Load-Sense via MAP,TPS,MAF,Veloc,Torque	MAP, TPS
	Transmission/Powertrain Mgmt?	Traction, shift cut/solenoid cntrl, shift firm & convt lockup
USER INTERFACE	Laptop Requirements	Win 98+, 2MB disk space, CD-Rom, Serial Port, Std Cable
	Graphical v Tabular; Tuning trim pots?	Both; Fuel, Ignition trim modules
HARDWARE	Physical Size And Weight	Varies, but biggest is 4" x 6" x 2"
RESOLUTION	# Fuel; IGN Map rpm Ranges; User defined?	40/40; Yes
	# Load Points/RPM Range; Total points/Map	20 points; 800 pts
AIR-FUEL	Target Air-Fuel Ratio Map?	Yes
	Fuel Press. Detect & Inj Pulse Comp	Yes
	Individual Cylinder Fueling Adjustment?	Yes
CLOSED-LOOP	via O_2,WideO_2; High-load Closed Loop?	Yes, Yes; Yes
CALIBRATION	Utility to Create "Startup" Calibration?	Yes
	Automated Mapping Features	Many
PLUG AND PLAY	Hardware, Wiring	No
	Self-Learning ("Tunes itself")	Yes
	Specific Applications in Calibration Library	EFI Tech. provides custom Mapping
DATALOGGING	Rate-Time; User selectable?; in Laptop of ECU?	100Hz, up to 12 HOURS, cyclic logging standard; Yes; ECU
I/O	ECU Inputs	Up to 38 (22 analog, 4 digital, 4 switch, 2 knock, 2 thermocouple)
	ECU Outputs	Up to 32 (up to 12 inj drivers, 12 spark drivers), plus 2 RS232(64 channels at 115,200 baud each), 1 Current loop
	Max # IGN-Only Coil Drivers; Coils	Up to 12
SENSORS	Sensors (Required; Optional; Programmable)	TPS, CAM Position, CRANK Position
	Eng. Position Triggers; Pattern	Hall Effects, Inductive Magnetic Reluctor, 1-per rev/custom
	Maximum Recognizable Map Pressure	? Any
	Knock Sensor; Individual cyl retard?	Yes (2); individual bank
ACTUATORS	Injector Types Supported	All
	Ignition Types Supported	Single Distr, Dual Distr, Waste-spark, Coil-on-Plug, Rotary
	Software-Configurable IGN Compatibility	No
	Extra Ignition Amplifier Required?	No
	Additional Supported Actuators	Var int man, Boost contr, Throttle-by wire, Var. valve tim
NITROUS	Nitrous Triggering/ Fuel via Primary Inj.	Yes/Yes (up to four stages)
WIRING	Wiring Harness (Generic/Specific)	No/Full Custom

EMS		HKS FCON V PRO
GENERAL	Kit Includes	ECU, vehicle specific harness (if available)
	Optional Components	3 bar map sensor, intake temp sensor, scramble switch, twin injector driver module, mixture controller.
	Cost (8 cy kit w/sensors & harness; 4-cy; ECU)	na; MSRP $1895; MSRP $1595
	Injection Modes	Sequential, Staggered Batch, Batch, TBI, Staged Batch
	Ignition Control	Yes
	Engine Configurations (cylinders)	1-12 cyl, rotary -2
	Load-Sense via MAP,TPS,MAF,Veloc,Torque	MAP, TPS, MAF, Velocity
	Transmission/Powertrain Mgmt?	No
USER INTERFACE	Laptop Requirements	Win98+, USB and serial port needed
	Graphical v Tabular; Tuning trim pots?	Tabular; Opt. mixture controller
HARDWARE	Physical Size And Weight	5.5" x 6" x 1"; 15 oz
RESOLUTION	# Fuel; IGN Map rpm Ranges; User defined?	32/32; Yes (any RPM can be entered)
	# Load Points/RPM Range; Total points/Map	32 points; 1024 pts
AIR-FUEL	Target Air-Fuel Ratio Map?	Yes
	Fuel Press. Detect & Inj Pulse Comp	Yes (injection compensation)
	Individual Cylinder Fueling Adjustment?	Yes
CLOSED-LOOP	via O_2,WideO_2; High-load Closed Loop?	Yes; Yes; Yes
CALIBRATION	Utility to Create "Startup" Calibration?	Yes
	Automated Mapping Features	Copy, add/subtract, multiply, interpolate
PLUG AND PLAY	Hardware, Wiring	Yes (Japanese imports)
	Self-Learning ("Tunes itself")	Yes
	Specific Applications in Calibration Library	Japanese imports
DATALOGGING	Rate-Time; User selectable?; in Laptop of ECU?	?; Yes; Both
I/O	ECU Inputs	19 sensor/switch inputs (Crank pos, Cam pos,MAP,TPS,CTS, IAT, O2, Wideband meter, VSS, Aux analog
	ECU Outputs	Inj(8); fuel pump; PWM(3); IAC; Fan; Shiftlight; VTEC
	Max # IGN-Only Coil Drivers; Coils	
SENSORS	Sensors (Required; Optional; Programmable	TPS, MAP or MAF, IAT, CTS
	Eng. Position Triggers; Pattern	Opt, Mag Reluctor; 1 pulse, Nissan Opt, Subaru, Multi-tooth
	Maximum Recognizable Map Pressure	3 Bar
	Knock Sensor; Individual cyl retard?	Yes (1); no
ACTUATORS	Injector Types Supported	?
	Ignition Types Supported	Single distr., waste-spark, coil-on-plug
	Software-Configurable IGN Compatibility	Yes
	Extra Ignition Amplifier Required?	OE ignition amplifiers used
	Additional Supported Actuators	?
NITROUS	Nitrous Triggering/ Fuel via Primary Inj.	No
WIRING	Wiring Harness (Generic/Specific)	No; Yes

| .P.M. | 5106 |
| ressure | -0.49 |

```
.7 -36.3 -41.2 -42.9 -36.3 -34.2 -38.1 -38.8 -23.7 -12.5 -13.0  -4.3  6.1  0.0 30.0 30.0
.0 -29.5 -35.7 -36.8 -26.8 -34.3 -37.6 -38.7 -24.0 -13.7 -13.6  -3.3 -4.5 -6.0 -6.0 -6.0
.0 -28.3 -28.6 -25.0 -25.5 -28.5 -34.9 -20.2 -17.5 -16.4  -7.1  -6.2 -6.0 -6.0 -6.0
```

FAST Electronics	Haltech E6K
ECU, Wiring Loom, Relay, Cable	ECU, Wiring loom, fuse block, Relays, IAT, CTS, TPS, cable, software, instructions
IAC, Wide O2	MAP sensor, IAC, boost/fuel/ign trim module, IGN module, EGO sensor, Boost controller, IGN coils
$1300-3500; $1300-3500; ?	?
Sequential, Staggered batch	Sequential(4cyl), semi-seq(6-8cyl), Batch, TBI, Staged Batch
Yes	Yes
4,6,8	1-6,8,10,12, Rotary-2
MAP, TPS	MAP, TPS
No (separate system required)	Torque Convertor Only
Any Win95+ PC with serial port	DOS, 640k RAM, up to Windows 98
Both; No	Graphical/Tabular; Yes (Fuel, Timing, Boost trim)
App. 8" x 6" x 2"; ?	140mm (5 17/32") x145mm (5 5/8") x 41mm (1 5/8")
16/16; Yes (used-defined cell width values)	22 (500 rpm ranges) or 17 (1000 rpm ranges); 22/17; No
16 points; 256 pts	32 points; 704 pts
Yes	No
No	No
Yes (optional)	Yes, sequential only
Yes, Yes; Yes	Yes, No; No
No	Yes
Copy/paste, percentage, multiplicative, additive changes in user-selectable areas of data points	Yes
No	No
No; "Realtime" internet tuning capabilities	No, base maps available on request
?	7 samples / Sec; Yes; Laptop (viewed via software/spreadsheet)
30 mins at 10 events/sec; Yes; Laptop or Aux Module	MAP, Crank sensor, CTS, IAT, TPS, BARO(intern), Primary and sec. triggers, EGO, 2x Trim module, Spec-pur digital, Road speed
Crank pos, cam pos, MAP, TPS, CTS, IAT, Wide O2	Inj(8), Fuel pump relay, Dedicated PWM(4), IAC, Ignition, Special purpose digital
IAC, Thermofan, NOS enable, VTEC, Injector Drivers(8), Fuel Pump Relay	4 drivers; 4 coils
TPS, MAP, Eng Speed, IAT, CTS; Baro(Alpha-N only);	TPS, CTS, IAT, engine position,
Hall, Opt., Ind. mag reluctor; Single-pulse/cycle pattern	Hall, Optical, Mag Reluc
5-Bar	30psi (up 65psi available on special request)
Yes; no	No; no
Virtually Any	Up to 8 low impedance or up to 16 high impedance injectors
Single Distr, Dual Distr, Waste-spark, Coil-on-Plug	Single distr., Twin distwaste-spark(8), coil-on-plug(4), R1&2
No	Yes, compatible with most OEM and aftermarket ignitors/coils
Varies with Application	Yes, must use external ignitor
na	IAC, Boost control, thermo, turbo timer, NOS, shiftlight, anti-lag, VTEC, a/c, Aux fuel pump, intercooler fan, tacho, etc
Yes, Yes (wet or dry flow possible)	Yes, wet NOS system only
Yes (generic); Specific via Dealers	Yes, available as flying lead loom and terminated wiring

Holley Commander 950	Hondata
ECU, Wiring Loom, Relay, Sensors (MAP, O2, IAT, CT), Communications Cable, Laptop Software	Datalogging/interface box for stock Honda EMS
IAC, EGO, and Wideband O2 via third-party suppliers	G-sensor, datalogging
$1370; $860	na; $245-820;
Staggered batch, TBI	Sequential
Yes	Yes
4, 6, 8	4,
MAP, TPS	MAP
No	No
66 Mhz, Win 3.1+, DB9 serial port. Floppy, Cable incl.	Win95+,133mhz, 16MB RAM, serial port
Both; No	Both; No
5.5" x 5.5"; 2.5 lbs	OEM Honda (uses stock ECU hardware)
16 ranges; Yes and can be non-linear	20/24 (High-speed cam); 20/24; Yes
16 points; 256 pts	16 points; 320 or 384 pts
Yes (requires Race software)	Yes
No	"Built into fuel system"
No	No
Yes, Yes (requires Race Software; Yes (Race software)	Yes, No; No
No, Many base calibrations supplied on disk	Yes
Mathematical changes allowed (+,-,*,/,=)	Copy, percentage, absolute, injector, A/F correction
GM TPI, LT1, and 5.0L Ford	OEM Honda '89-'00
No	No
ZZ4 Plug-and-Play calibration, many calibrations	OEM Honda '89-'00 (NA and forced-induction)
5, 10, or 20 samples/sec; 2 min, 1 min, 30 sec; Yes; Both	?; No; Laptop
Crank Pos, MAP, TPS, CTS, IAT, BARO (Alpha-N only), EGO, Wide O2, VSS, Aux Analog (2)	OEM + 2
"Injector Drivers (4), Pump Relay, Digital Outputs (2), 2 (opt) config for RPM, TPS, MAP, Fan, etc	OEM + 2
zero/zero	1 (no direct coils)
TPS, MAP, IAT, CTS	TPS, MAP, Engine Speed, IAT, CTS (OEM Honda)
Hall Effects, Magnetic Reluctor/?	OEM Honda
300 kPa (GM 1, 2, or 3-Bar)	OEM - 1.6-Bar, GM = 3-Bar
GM sensor (1); no	Yes (1); no
High and Low Impedence	High Impedance, Low Impedance with resistors
Single Distr	OEM Honda
Yes	na
Yes if not using HEI or TFI	No
	OEM + 2
No/na	Yes via output; Yes; separate Nitrous timing Map
Yes	na

| R.P.M. | 5106 |
| Pressure | -0.49 |

```
.7 -36.3 -41.2 -42.9 -36.3 -34.2 -38.1 -38.8 -23.7 -12.5 -13.0 -4.3 6.1 0.0 30.0 3
.0 -29.5 -35.7 -36.8 -26.8 -34.3 -37.6 -38.7 -24.0 -13.7 -13.6 -3.3 -4.5 -6.0 -6.0 -
.0 -28.3 -28.6 -25.0 -25.5 -28.5 -38.7 -34.9 -20.2 -17.5 -16.4 -7.1 -6.2 -6.0 -6.0 -
```

STANDALONE PROGRAMMABLE ENGINE MANAGEMENT SYSTEMS

EMS			Link ElectroSystems USA Link 2
GENERAL	Kit Includes		ECU, Wiring Loom, Relay, Sensors?, Actuators?, Comminications Cable
	Optional Components		IAC, EGO, Wideband O2, Boost Controller, Ignition Module, Coils
	Cost (8 cy kit w/sensors & harness; 4-cy; ECU)		$1049 ECU w/MAP sensor and breakout wire loom
	Injection Modes		Sequential, batch
	Ignition Control		Yes
	Engine Configurations (cylinders)		2, 3, 4, 5, 6, 8, 10, 12, 16
	Load-Sense via MAP,TPS,MAF,Veloc,Torque		MAP, TPS
	Transmission/Powertrain Mgmt?		No
USER INTERFACE	Laptop Requirements		Win 95+, Serial Port, Std. Serial Cable
	Graphical v Tabular; Tuning trim pots?		Graphical; Virtual
HARDWARE	Physical Size And Weight		Length 150mm X width 160 mm X Height 40mm
RESOLUTION	# Fuel; IGN Map rpm Ranges; User defined?		10 ranges; Yes
	# Load Points/RPM Range; Total points/Map		20 points; 200 pts
AIR-FUEL	Target Air-Fuel Ratio Map?		Yes
	Fuel Press. Detect & Inj Pulse Comp		No
	Individual Cylinder Fueling Adjustment?		No
CLOSED-LOOP	via O2,WideO2; High-load Closed Loop?		No, Yes; No
CALIBRATION	Utility to Create "Startup" Calibration?		No
	Automated Mapping Features		No
PLUG AND PLAY	Hardware, Wiring		LS1, Honda, Mazda, Mitsubishi, Nissan, Subaru, Toyota
	Self-Learning ("Tunes itself")		Yes
	Specific Applications in Calibration Library		?
DATALOGGING	Rate-Time; User selectable?; in Laptop of ECU?		?; Yes; Laptop or data trap
I/O	ECU Inputs		10 inputs
	ECU Outputs		34 outputs (8 injector drivers)
	Max # IGN-Only Coil Drivers; Coils		Avail. 8/8/2003
SENSORS	Sensors (Required; Optional; Programmable		MAP, Engine Speed; TPS, IAT, CTS
	Eng. Position Triggers; Pattern		Hall Effects, Optical, Magnetic Reluctor/?
	Maximum Recognizable Map Pressure		300 kPa
	Knock Sensor; Individual cyl retard?		Two; no
ACTUATORS	Injector Types Supported		All
	Ignition Types Supported		Single Distr, Dual Distr, Waste-spark, Coil-on-Plug
	Software-Configurable IGN Compatibility		Yes
	Extra Ignition Amplifier Required?		Yes
	Additional Supported Actuators		?; IAC, Fans, Boost Control,
NITROUS	Nitrous Triggering/ Fuel via Primary Inj.		No/na
WIRING	Wiring Harness (Generic/Specific)		Yes

EMS			Perfect Power PRS
GENERAL	Kit Includes		ECU, wiring loom, sensors, , cable, CD
	Optional Components		?
	Cost (8 cy kit w/sensors & harness; 4-cy; ECU)		$1124; $1124; $596-$1115
	Injection Modes		Sequential, Staggered Batch, Batch, TBI, Staged Batch
	Ignition Control		Yes
	Engine Configurations (cylinders)		1-12 plus Rotary
	Load-Sense via MAP,TPS,MAF,Veloc,Torque		MAP, TPS, MAF
	Transmission/Powertrain Mgmt?		Yes-Shift, Shift firmness, Convertor lockup
USER INTERFACE	Laptop Requirements		Laptop required
	Graphical v Tabular; Tuning trim pots?		Graphical; No
HARDWARE	Physical Size And Weight		PRS2: 2.75" x 4", PRS4: 4" x 5", PRS8: 5" x 10"
RESOLUTION	# Fuel; IGN Map rpm Ranges; User defined?		?/?; ?
	# Load Points/RPM Range; Total points/Map		?; 765 pts
AIR-FUEL	Target Air-Fuel Ratio Map?		Yes
	Fuel Press. Detect & Inj Pulse Comp		Injector Compensation
	Individual Cylinder Fueling Adjustment?		No
CLOSED-LOOP	via O2,WideO2; High-load Closed Loop?		Yes, No; Yes
CALIBRATION	Utility to Create "Startup" Calibration?		Coming soon
	Automated Mapping Features		?
			No
PLUG AND PLAY	Hardware, Wiring		Yes
	Self-Learning ("Tunes itself")		Limited
	Specific Applications in Calibration Library		
DATALOGGING	Rate-Time; User selectable?; in Laptop of ECU?		?; Yes; Laptop
I/O	ECU Inputs		Crank pos., Cam pos., MAP, TPS, CTS, IAT, EGO
	ECU Outputs		Inj(12), Relay, PWM, IAC, Boost, Thermofan, Stall saver, NOS, Shiftlight, VTEC, A/C, I/C fan, Torq convt, Check
	Max # IGN-Only Coil Drivers; Coils		
SENSORS	Sensors (Required; Optional; Programmable		TPS, MAP, MAF, Engine speed, IAT, CTS
	Eng. Position Triggers; Pattern		Hall, Opt., Ind. mag reluctor; any even # teeth
	Maximum Recognizable Map Pressure		Any analog
	Knock Sensor; Individual cyl retard?		No; no
ACTUATORS	Injector Types Supported		Any 6-amp max
	Ignition Types Supported		Single Distr, Dual Distr, Waste-spark, Coil-on-Plug, Rotary
	Software-Configurable IGN Compatibility		Yes
	Extra Ignition Amplifier Required?		No (included)
	Additional Supported Actuators		Fuel Relay, IAC, Boost, Thermofan, Stall saver, NOS, Shiftlight, VTEC, A/C, I/C fan, Torque convt, Check light
NITROUS	Nitrous Triggering/ Fuel via Primary Inj.		Yes (multi-stage), Yes
WIRING	Wiring Harness (Generic/Specific)		Yes, No

.7	-36.3	-41.2	-42.9	-36.3	-34.2	-38.1	-38.8	-23.7	-12.5	-13.0	-4.3	6.1	0.0	30.0	30.0	
.0	-29.5	-35.7	-36.8	-26.8	-34.3	-37.6	-38.7	-24.0	-13.7	-13.6	-3.3	-4.5	-6.0	-6.0	-6.0	
.0	-28.3	-28.6	-25.0	-25.5	-28.5	-38.7	-34.9	-20.2	-17.5	-16.4	-7.1	-6.2	-6.0	-6.0	-6.0	

.P.M.   5106
ressure   -0.49

Motec M880	Motec M48
No Kit available - Custom harnessing and sensors made to customer requirements	ECU, Software, Comm Cable, Semi-cust wiring loom for Distr, MAP, CTS, IAT, 6 hrs logging and Wide O2
Fuel/IGN/Boost/RPM trim; IAC; Boost cntrl (opt. speed/gear); traction, launch, speed cntrl; Gear chg, Ignition Cut, Dual Wide O2, Anti-Lag, 3-D fuel/spark Maps, Accel. IGN Adv, Gear Detect	Fuel/IGN/Boost/RPM trim; IAC; Boost cntrl (opt. speed/gear); traction, launch, speed cntrl; Gear chg, Ignition Cut, Wide O2, Anti-Lag, 3-D fuel/spark Maps, Accel. IGN Adv, Gear Detect
na; na; $4160.00 Base ECU Price	
Sequential, Staggered Batch, Batch, TBI, Staged Batch	Sequential, Staggered Batch, Batch, TBI, Staged Batch
6 Direct to ECU - up to 12 Coils with Expander Box	Up to 8 Direct Coils with Expander Box
1-12 seq, 1-8 2-stroke seq, 1-16 batch, R1-3 split, R4+ no split	1-6 cyl 2/4 cycle, 8,10,12 cyl 4 cycle, 1, 2, 3 rotor w/o split
Any volt or freq. device possible as a load input to tune from	Any volt or freq. device possible as a load input to tune from
Torque convertor lockup, traction	Torque convertor lockup, traction
Win95+, 5 MB disc 8MB ram Serial port, floppy Motec cable	Win95+, 5 MB disc 8MB ram Serial port, floppy Motec cable
Both; No	Both; No
147mm X 105mm X 40 mm - 525 Grams	150mm X 100 mm X 36 – 480 Grams
<=40 ranges/<=40 ranges; Yes, user-definable	<=40 ranges/<=40 ranges; Yes, user-definable
User definable up to 21load sites/range; 840 pts	User definable up to 21load sites/range; 840 pts
Yes (doesn't directly fuel)	Yes (doesn't directly fuel)
Yes with pressure sensor, else comp from Batt voltage	Yes with pressure sensor, else comp from Batt voltage
Yes	Yes
Yes, Yes; Yes	Yes, Yes; Yes
No	No
Quick-Lambda & "Lambda-Was" enable adj fuel Map based on predefined lambda goal table and current or logged AFR	Quick-Lambda & "Lambda-Was" enable adj fuel Map based on predefined lambda goal table and current or logged AFR
No	No
No	No
On-site Tuning Available	On-site Tuning available
Upt to 200 Hertz Sample rate; Yes; ECU to 4 MB avail	Up to 20 Hertz Sample Rate; Yes; ECU 512K
4 Digital, 1 RPM, 1 Sync, 8 Analog Voltage, 6 Analog Temperature, 2 Lambda	2 Digital, 1 RPM, 1 Sync, 3 Analog Voltage, 3 Analog Temperature , 1 Lambda
8 Injector, 6 Ignition, 8 Auxiliary Outputs	8 Injector, 2 Ignition, 3 Auxiliary
6 Coils Direct - more using expander box	2 IGN (2 coils max); more direct coils with expander box
1 load input and 1 rpm input minimum;	1 load input and 1 rpm input minimum;
Everything/All Patterns	Everything/All Patterns
User-definable	User-definable
Via 3rd-party dev providing analog voltage signal to ECU; no	Via 3rd-party dev providing analog voltage signal to ECU; no
All	All
Single Distr, Dual Distr, Waste-spark, Coil-on-Plug (w/expander)	Single Distr, Dual Distr, Waste-spark, Coil-on-Plug (w/expander)
Yes	Yes
Yes to trigger inductive coil ignition	Yes to trigger inductive coil ignition
IAC, wastegate boost control, tach, shift light, VTEC, NOS, Drive By Wire, Fully Variable Camshaft Timing, Dual Var Cam Timing	IAC, wastegate boost control, tach, shift light, VTEC, NOS, etc
Yes/Yes (up to 5 stages or 1-stage Nitrous-only Map)	Yes/Yes (up to 3 stages or 1-stage Nitrous-only Map)
Roll-your-own generic/Motec Custom; Mil-Spec avail	Roll-your-own generic/Motec Custom; Mil-Spec avail

Simple Digital Systems EM-xE or EM-xF

ECU, LCD programmer, IAT, CTS or CHT, Crank pos, Fuel trim control, Wiring, IGN amp or waste-coils, LCD prog (laptop not req)
MAP (1-3 Bar), Knock, TPS sensors, O2, Fuel relays, Fast idle solenoids, Injectors, Fuel pumps, RPM switch, Fan relays, Blower relays, A/C solenoids, 15K rpm, Backlit LCD, fuel control-only
8E- $1060, 4E- $940; (4-6 cyl direct-fire 4/6F $1165)
Batch, Staged Batch
Yes
2-6 and 8 cyl, R2 and some marine 2 strokes, (4,6,8 cyl IGN)
MAP, TPS
No
No (dedicated LCD programmer module)
Menu-driven list on small LCD panel; Yes (fuel trim control)
3.875" x 8.5" x 1", 5lbs
38/38 ranges; No (250-RPM increments to 9750)
1; but 64 universal load, 64 fuel enrich-enlean offsets; 204 pts
No
No
No
Yes, No, No
Yes (shipped with unit via injector flow and engine CID)
No

No
No
No
No datalogging
TPS, MAP, IAT/CHT, Knock, Crank pos, O2
Inj(2-8), coils (1-3), Relays(1-4)

3 (waste-spark) with EM-4/6F
RPM, TPS, IAT, CTS; crank pos, knock, O2
Hall
3-Bar sensor
Yes (optional sensor and ECU circuitry); no
1.7-3ohm OR 12-16 ohm (specified in advance)
Single distr w/MSD igniter, optional waste-spark coils
No
Yes (4E requires MSD igniter), No (4F w/direct coils)
Fan, A/C idle-up, VTEC, & supercharg clutch relays, fuel relay
No
Universal 4, 6, or 8 cyl, injector harness specific to application

R.P.M.	5106				.7	-36.3	-41.2	-42.9	-36.3	-34.2	-38.1	-38.8	-23.7	-12.5	-13.0	-4.3	6.1	0.0	30.0	30
Pressure	-0.49				.0	-29.5	-35.7	-36.8	-26.8	-34.3	-37.6	-38.7	-24.0	-13.7	-13.6	-3.3	-4.5	-6.0	-6.0	-6
					.0	-28.3	-28.6	-25.0	-25.5	-28.5	-38.7	-34.9	-20.2	-17.5	-16.4	-7.1	-6.2	-6.0	-6.0	-6

STANDALONE PROGRAMMABLE ENGINE MANAGEMENT SYSTEMS

continued from page 91

to have an accurate approximation of the curve. But engines with wild or variable cams, turbochargers, variable intake plenums or runners, nitrous, and other exotic volumetric-efficiency modifiers may have oddly shaped torque curves that require sophisticated air/fuel and timing curves for good drivability and power.

The resolution of an engine management system refers to the *number of numbers* in the tables being used to approximate the fuel and timing curves over the range of engine speed and loading—that is, the *granularity* of the approximation. Clearly, a 12,000-rpm engine running up to 45-psi boost (a *lot* of dynamic range) requires bigger digital fuel and ignition tables than a 6,000 redline naturally aspirated engine with a mild cam. So, all things being equal, engine management systems with higher resolution are preferable—especially for difficult engines. On the other hand, resolution is like running-shoe size: you need enough, but more than that will not help you go any faster.

Engine management systems with the ability to accurately control fuel injectors with good repeatability to very short injection pulse widths will have the ability to make more power on engines with a big dynamic range because they can produce acceptable idle with the big injectors you'll need for huge power at peak torque.

AIR/FUEL

Some systems provide the ability to define target air/fuel ratios for various segments of the rpm load range, which, combined with wideband O_2 sensors, can be useful in calibrating the system because it can alert you to modifications in pulse width required to achieve the target air/fuel ratio. Or, in some cases, they can actually override the base pulse width to achieve a specified air/fuel ratio. Of course, deciding on the right target air/fuel ratios for various sectors of the speed-loading-rpm graph or matrix may or may not be straightforward and may require experimentation to achieve mean best torque.

Fuel pressure detection and pulse width compensation is very useful, particularly if you're managing an engine with a deadhead (return-less) fuel supply system like the new Ford Focus. This is similar to the battery voltage compensation logic used by virtually all modern engine management systems to correct for increasing laziness in injector opening performance as battery voltage decreases by compensating with longer injection pulse widths. Fortunately, there is a straightforward relationship between rail pressure and fuel delivery per injector squirt, so effective compensation is predictable and accurate. The ECU uses a sensor in the fuel rail to detect pressure and compensates for nonstandard rail pressure with modified injection pulse widths.

Volumetric efficiency (VE) tables specify an engine's cylinder-filling efficiency across the range of rpm and loading. That is, how much air will actually make its way into the cylinder by the time the intake cycle is completed as a percentage of the amount of air filling the cylinder at atmospheric pressure with the engine stopped at bottom dead center (which can be more than 100 percent with forced induction or resonation-tuned racing engines, and will be at or less than 100 percent under other conditions).

Maximum required injection pulse width time is always a function of the injector size, fuel pressure, and maximum power or torque. Minimum pulse width is a function of injector size, pressure, and fuel required to idle. Everything in between is exactly a function that is linear with volumetric efficiency.

Electromotive's VE table, for example, defaults to all zeros (perfectly flat torque curve across the rpm range), and this will actually start and run most engines. The VE table can then be tuned with positive or negative numbers in the various cells to reflect the engine's actual breathing efficiency at various points of speed and loading.

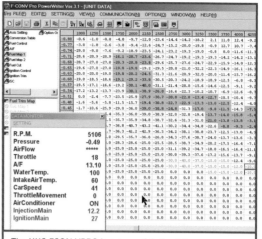

The HKS FCON VPRO is a powerful standalone or piggyback system targeted at professional tuners of Japanese performance vehicles. This HKS screen includes a data monitor window showing current conditions, with the main window showing number one fuel map trim values. Each table cell is a function of rpm and loading, and HKS provides a set of multi-cell modification tools as highlighted on the table. HKS

Many tuners find it more straightforward to directly tune tables of injection pulse width times, but others consider VE table systems more elegant and understandable in their modeling of actual engine behavior with respect to ideal mixture delivery.

CLOSED-LOOP

Most modern EFI systems have the ability to trim fuel injection pulse width at idle and light-cruise to target the ideal (14.7:1) chemical air/fuel ratio (stoichiometric) for cleanest combustion, based on feedback from an exhaust gas oxygen (EGO) sensor measuring residual oxygen in post-combustion exhaust gases that is used to deduce what the charge air/fuel ratio was prior to combustion. The problem is, a standard EGO or O_2 sensor is only accurate at air/fuel ratios very near stoichiometric, and stoichiometric ratios do not work for optimizing power at heavier engine loading.

To receive meaningful air/fuel ratio deductions at the richer mixtures required for best power and torque at heavier loads, the engine management system needs the ability to interface to a more expensive wideband O_2 sensor. To safely trim pulse width at higher, more dangerous levels of power with wideband O_2 feedback, the engine management system needs to have additional smarts.

CALIBRATION, SETUP, PLUG AND PLAY

Properly calibrating an engine from scratch is laborious and difficult, and there is always a risk of damaging a high-output engine if you make mistakes. The problem is, lean mixtures and excessive boost and spark timing can cause abnormal combustion and detonation, which can destroy an engine in a matter of seconds. Learning to tune an engine is thus a bit like a student pilot learning to spin an airplane (which they no longer teach in U.S. flight schools).

The best way to calibrate an engine is *not* to calibrate it, but instead to use a known-good calibration that was built by experts on an identical engine in an identical vehicle, which will plug and play on your vehicle. If identical is impossible, then close is the

		.7	-36.8	-40.7	-43.2	-41.1			-34.2	-38.1	-38.8	-23.7	-12.5	-13.0	-4.3	6.1	0.0	30.0	30.0
R.P.M.	5106	.7	-36.3	-41.2	-42.9	-36.3	-34.3	-37.6	-38.7	-24.0	-13.7	-13.6	-3.3	-4.5	-6.0	-6.0	-6.0
		.0	-29.5	-35.7	-36.8	-26.8	-34.3	-37.6	-38.7	-24.0	-13.7	-13.6	-3.3	-4.5	-6.0	-6.0	-6.0
Pressure	-0.49	.0	-28.3	-28.6	-25.0	-25.5	-28.5	-38.7	-34.9	-20.2	-17.5	-16.4	-7.1	-6.2	-6.0	-6.0	-6.0

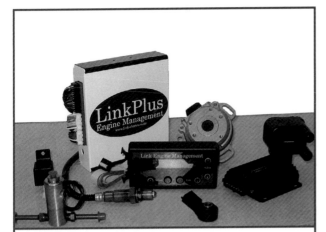

Link Electrosystems' LinkPlus EMS offers plug-and-play capabilities for LS1, Honda, Mazda, Mitsubishi, Nissan, Subaru, and Toyota. Like virtually all systems targeting the plug-and-play market, the ECU hardware has to be powerful, flexible, and highly configurable so that the capabilities needed to handle almost any sensor or actuator are already embedded in the hardware and logic. For example, the 34-output Link2 handles single distributor, dual distributors, waste-spark, or coil-on-plug ignitions. LinkUSA

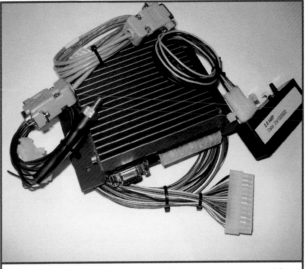

Perfect Power's Pro Race System (PRS) is, as it sounds, targeted for competition use, and is packed with features enabling it to handle almost any configuration of engine from 1 to 12 cylinders and rotary, as well as shifting and shift firmness for automatic transmissions. Perfect Power

next best place to start. A true plug-and-play calibration is gold, and particularly if you are not a pro tuner, the availability of one could be a decisive factor in selecting an engine management system. Library calibrations invoke the phrase caveat emptor. When it comes to calibrations, close is a matter of opinion, and what you don't know *can* hurt you.

Some engine management systems are available with plug-and-play hardware and connectors that enable you to remove a stock ECU, plug in the aftermarket ECU with its adapters, and immediately utilize the original equipment wiring, sensors, and actuators, possibly with the need to replace one or more sensors. Wiring and sensors are expensive, and true plug-and-play hardware compatibility could also be a decisive factor in choosing an engine management system. Of course, you'll still need a plug-and-play calibration or the ability to build your own from scratch or from "close."

When you are learning a new engine management system, the ability to work on the calibration offline is valuable. Calibration wizards that automate the process of building a calibration from a half-dozen or so parameters that is good enough to get the engine started can be valuable, as can automated mapping features such as the ability to change all or part of the fuel or ignition map by a percentage (very useful if you're changing to bigger or smaller injectors).

DATA LOGGING

The ability to take a series of snapshots of engine sensors and ECU status while the engine is running and driving is extremely valuable, because it means that rather than do all tuning in real time while the engine is operating (which requires two people to do safely and well on the test track), you can run the engine through its paces, log the data, and then go back and carefully analyze what is right and wrong about the calibration and improve it at your leisure when the engine is safely stopped. Some engine management systems provide tools that automate the analysis of large amounts of logged data. It may turn out to be convenient if you have the capability to log data in the ECU without needing to have a laptop connected.

INPUT/OUTPUT

Most engine management systems are designed with at least some configurable multipurpose input-output circuitry, and this can be a zero-sum game, For example, the availability of a configurable pulse-width modulated (PWM) output circuit (which is able to open and close, or pulse, a high-speed digital device at a precise frequency) to manage, say, a boost controller might depend on whether you already need the output to drive a fuel injector. The same output might also be capable of driving a tachometer, but not if you need it to interface with a boost controller. When reading a spec sheet on an engine management system, beware: It is extremely important to analyze the *total* number and type of I/O circuits you might need and make sure there will not be conflicts.

SENSORS AND ACTUATORS

Some sensors are required, without which an engine management system will not run an engine at all. And then there are sensors that are optional in the sense that they are not specifically required but then an alternative sensor *is* required. For example, a manifold absolute pressure (MAP) sensor might not be required per se, but a MAP *or* a mass airflow (MAF) sensor *is* required. There are also sensors that are not required in the sense that the engine management system can operate if the sensor is not installed or fails, but are, practically speaking, required and should be considered in a cost analysis of a particular engine management solution. For example, an engine management system might not require a coolant temperature sensor, and would default to 70 degrees Fahrenheit if none were installed, but the engine so equipped probably would not start when cold, and would run horribly rich at full operating temperature.

The flexibility of an engine management system to utilize the sensors and actuators already on your engine could save you a lot of money. Most new aftermarket ECUs are flexible about driving fuel injectors of any resistance, but sometimes they will handle *fewer* low-resistance performance-type peak-and-hold injectors. Many ECUs now have some sort of ability for software-configurability with various types of ignitions, but since the circuitry required to drive multiple coils is expensive, direct-ignition capability is often unbundled

| R.P.M. | 5106 |
| Pressure | -0.49 |

```
.7  -36.8 -40.7 -43.2 -41.1 -30.2 -34.4 -34.8 -22.6  -8.8  -9.3  -2.4   5.9  12.2  30.0  3
.7  -36.3 -41.2 -42.9 -36.3 -34.2 -38.1 -38.8 -23.7 -12.5 -13.0  -4.3   6.1   0.0  30.0  3
.0  -29.5 -35.7 -36.8 -26.8 -34.3 -37.6 -38.7 -24.0 -13.7 -13.6  -3.3  -4.5  -6.0  -6.0  -
.0  -28.3 -28.6 -25.0 -25.5 -28.5 -38.7 -34.9 -20.2 -17.5 -16.4  -7.1  -6.2  -6.0  -6.0  -
```

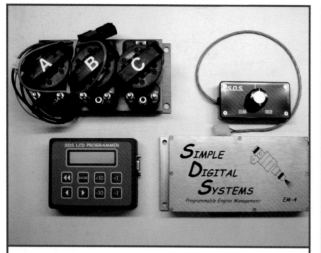

Simple Digital Systems' EM systems were designed for ease-of-use, and arrive complete with a dedicated programmer user interface module for calibration, a rich-lean trim module, optional waste-spark direct-fire coil packs, and much more. The internal data structures are relatively simple, the idea being that an easy-to-program system is likely to yield a higher state of tune than one with a more complex model that requires an unrealistically high investment of tuning time. SDS is famous for having a vast and useful Web site packed with useful information about engine management system construction and development. SDS

as an extra-cost option requiring expansion hardware. A few manufacturers like Electromotive have taken the opposite approach, bundling in direct-fire circuitry and coils (usually waste-fire, in which each coil fires two plugs at once). Bundled direct ignition definitely simplifies ignition compatibility and configuration issues.

NITROUS

Nitrous oxide injection provides a substantial infusion of cheap power, but it essentially involves injecting extra oxygen (linked to nitrogen atoms), which can easily blowtorch your engine if there are fuel delivery or mixture distribution problems. Since port-EFI intake manifolds are not designed to handle wet mixtures of air and fuel, providing required enrichment fuel via extra injectors or orifices is not a good idea unless these are located in every intake port.

Jacking up the fuel pressure to all the injectors when nitrous is flowing to get the required increased fuel delivery is a better idea, though it requires a special fuel pressure regulator and possibly an upgraded fuel pump. The best way to provide the enrichment fuel needed to burn the nitrous is by having the ECU deliver additional fuel via the primary fuel injectors as an offset added to ordinary fuel injection pulse width. This requires an ECU that knows how to do this, and this strategy also requires large enough injectors to handle the total fuel delivery at maximum torque with nitrous on, as well as an ECU that can accurately deliver the fractional pulse width required to *idle* well.

Some ECUs can *stage* the delivery of nitrous (and fuel) or even gradually increase the nitrous stream to improve traction in drag-type conditions. If you're planning to use nitrous, buy an ECU with nitrous capabilities built in so you can manage nitrous along with other EMS tasks rather than handling nitrous as an add-on.

WIRING

The engine management wiring harness is *vital* to reliability, and building a truly robust harness is not cheap. However, faulty wiring is the bane of product support operations, so factory wiring harnesses are typically of extremely high quality, since faulty wiring will cause expensive claims while the vehicle is under warranty and could interfere with emissions functionality (which the U.S. government requires carmakers to warrant for an extended period of time, typically at least 120,000 miles).

If possible, you should use the factory wiring (which requires the availability of plug-and-play adapters and circuitry). The problem of faulty or sub-standard wiring is so severe with aftermarket standalone engine management systems that some premium race-type engine management systems come bundled with a high-quality custom-built wiring harness that is included in the system price (or the buyer must have a pre-approved capability to build custom harnesses).

Less expensive systems may be priced with a generic engine wiring loom equipped with connectors designed (typically) for GM sensors and actuators. The problem is, the generic harness connectors may not fit *your* fuel injectors and engine sensors, and the various segments of the harness may be too short or too long to work really well in your application. This means the harness might have to be extended or shortened at various places or have alternate connectors installed, which requires special tools and expertise to do right. Bottom line: pay close attention to wiring quality or you will live to regret it.

SOURCES—STANDALONE AFTERMARKET ENGINE MANAGEMENT SYSTEMS

Accel/DFI
(888) MR GASKET ext 9999

AEM
(310) 484-2322

A'PEX
(714) 685-5700

Autronic SMC/SM2
(61) 7 4051 6672 (Australia)
(909) 245 9511 (Strader Performance Engineering, CA)

Edelbrock
(310) 781-2222

Electromotive
(703) 331-0100

EFI Technology
(310) 793-2505

FAST Electronics
(662) 224-3495

Haltech
(61) 2 9525 2400 (Australia)

HKS
(310) 491 3300

Holley Fuel Systems
(800) 2-HOLLEY (sales)
(270) 781-9741 (tech support)

Hondata
(310) 782-8278

Link ElectroSystems USA
(949) 646-7461

Motec Systems USA
(714) 897-6804

Perfect Power
(732) 591-2630 (IDA Automotive, USA Distributor)

Simple Digital Systems
(40) 274-0154 (Canada)

STANDALONE PROGRAMMABLE ENGINE MANAGEMENT SYSTEMS

Swapping an EFI engine into a foreign EFI vehicle is one of the most complex and difficult topics covered in this book. Jaguars That Run, a Livermore, California, firm that specializes in Chevrolet TPI and TBI engine swaps has this to say in their manual on TPI/TBI engine swapping.

"The original intention of this manual was to make it possible for 'average shade tree mechanics' to install tuned port injection or throttle body injection engines into older cars and trucks. However, after testing the manual on several 'average shade tree mechanics' and observing their swaps, we learned that they could sometimes make the engines run, but made numerous mistakes in wiring, fuel systems, hose routing, hose selection, cooling, exhaust, emissions-control equipment, air conditioning—just about everything imaginable. A lot of the mistakes could have been dangerous. In other words, the 'average shade tree mechanic' was not qualified to do the TPI or TBI engine swap, or any engine swap."

Author Mike Knell goes on to suggest, "Engine swaps require experience, knowledge, patience, and a willingness to look things up in a book or shop manual. From these experiences, we can only advise most people not to do TPI or TBI engine swaps unless they are experienced mechanics or have experienced mechanics who will carefully inspect the conversion before the vehicle is started, and again before the vehicle is driven. Remember, the purpose of doing an engine swap is to make the vehicle run better, and if you cannot make it better, it is not worth doing."

This book contains a number of engine-swap-project-type vehicles. The BMW M3 into 1993 325-ci is an extremely complex EMS-engine-into-foreign-EMS-vehicle swap, which utilized the vehicle's pre-OBD-II EMS and also involved a turbo conversion. The Toyota 1MZ-FE V-6-into-1991 MR2 Turbo was even more complex, using the engine's OBD-II EMS to control and weave together two partially compatible EMS systems (later converted to aftermarket EMS control).

Projects involving previously carbureted recipient vehicles are usually somewhat less complex in many ways, but nevertheless very serious projects. One involved swapping a turbo-EFI Renault Fuego powerplant into a Lotus Europa, with the addition of an aftermarket EMS and computer-controlled nitrous injection.

The "Jag-rolet" project involved harnessing a powerful Motec M48 aftermarket EMS to manage a supercharged hot rod TPI Corvette V-8 in its new home in a late-model Jaguar sedan. The project required comprehending an extremely complex set of computer-controlled nonengine functions such as automatic load leveling, designing the hardware and wiring to interface such peripherals to the Motec, and programming the Motec to make everything work.

CRITICAL EMS/EFI ENGINE SWAP ISSUES

An EMS/EFI engine swap is not for sissies. Unless you are swapping a complete engine and all related engine systems into a vehicle where this engine is also a factory option, you will almost definitely have to deal with challenges and incompatibilities on a variety of levels:

1. *Any* engine swap involves ordinary physical swap issues: space, motor mounts, clutch and transmission, shifter/linkage, alternator and belt-driven accessories, intake and exhaust systems,

Some things they should have done at the factory, and the 3.0 V-6 MR2 is one. Swapping a 3.0-liter 1994-plus 1MZ-FE Toyota V-6 in place of the 1991 2.0-liter Turbo is at once simpler and more complex than some swaps. Some EMS components (for example, the fuel pump and supply system) work great, but each car has some EMS features that are completely beyond the scope of the other system, an example being the V-6 EMS's desire to see torque converter slip data (the MR2 Turbo is a five-speed-only vehicle).

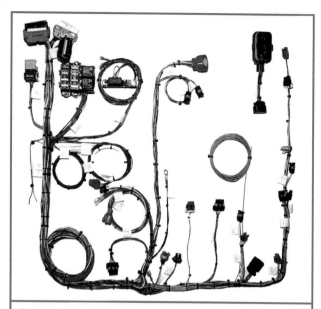

Almost without a doubt, wiring *issues will be the most challenging aspect of an EFI engine swap. Fortunately, outfits like Painless Performance make complete swap harnesses for some common performance swaps, and they also make some semi-custom harnesses you can adapt in more problematic situations.* Painless Performance

cooling system and fans, fuel supply, and so on. Some will be trivial, others extremely challenging.

2. Installing any EFI engine in a carbureted vehicle, involves, in addition to the above, the need to provide a high-pressure fuel supply of sufficient volume, and probably a fuel tank return line. And, of course, you'll need to deal with all the issues surrounding installing and mounting the engine management computer and wiring.

3. Installing a standalone aftermarket EMS on a swap engine in a foreign vehicle involves, in addition to all the above, standard aftermarket EMS installation issues including the need to deal with calibrating the EMS for the specific engine and vehicle. If the vehicle had OEM electronic engine management, you will almost definitely need to deal with managing certain functions only peripherally related to running the engine (gauges, air-conditioning, fan controls, and possibly much more). In order to do this, will you leave the OEM computer in place and alive and happy enough to function where needed? Or is the aftermarket system powerful and programmable enough to manage special vehicle functions such as ride height adjustment, as was required in the Jag-rolet project.

4. When installing an EMS engine in a foreign EMS vehicle, you've got to decide who runs the show: the *engine's* EMS or the *vehicle's* EMS (or perhaps a cooperative effort or a standalone aftermarket system). In addition to handling most of the above-mentioned issues, you're going to have to deal specifically with either: adapting the vehicle's EMS to the engine (perhaps changing sensors and actuators, perhaps changing the calibration, and so on); making the engine's EMS work in the vehicle (opening up a Pandora's box of integrating the foreign EMS with the vehicle's wiring and electrical system and any auxiliary vehicle onboard computers, electronics, and diagnostic wiring); or employing *both* vehicle and engine EMS, integrating functionality of two different control systems.

Whichever strategy you pick, you may also have to deal with issues of:

• Transmission control, including converter lockup, shifting and other transmission issues (particularly if you change from automatic to manual transmission or vice versa; auto and manual ECU electronics or microcode may vary as well)
• Sensors and actuators incompatible with ECU or engine
• Tach driver incompatibility with tach
• Cat and emissions issues, including cat A.I.R. pump and O_2 sensor issues (even if you remove the cat)
• Intake issues (including space for MAF meter and possible MAF turbulence issues)
• Integrated gauges incompatible with EMS or vehicle
• Computer-controlled alternator issues
• Computer-controlled air conditioning issues
• Traction controls
• Electronic throttle
• Handshake between various computers
• Engine position sensor location and timing problems
• Anti-theft, emissions
• In-tank fuel pump and regulator capacity and pressure
• Speedometer incompatibility
• Vehicle (VSS rate, etc.)and so on)
• Fan control incompatibilities
• Coolant temperature control systems (like 1MZ with temp sensors in several places), including cooled coolant from radiator
• Direct-fire versus distributor controls
• Combo-meter dash issues where meter displays from many computers (air bag, cruise, ECU, etcand so on.)
• Multi-mode power steering electronic control problems
• Brake event monitoring
• Knock sensor incompatibilities (if using recipient EMS),
• Altering *length* of stock harness

An engine swap with current state-of-the-art EMS vehicles might involve the future shock of dealing with a fiber optic onboard communications bus and functionality such as active suspension controls and countermeasures, automatic cellphone-based trouble-report assistance, proximity cruise controls, and much more.

THE ACTUAL SWAP PROCESS

1. Get the engine (and transmission) installed and mounted. Obviously, as in any engine swap, you need the right motor mounts that locate the engine with correct geometry so everything fits. You may need to relocate engine components or equipment in the engine compartment, alter engine accessories, or even modify the firewall, radiator bulkhead, radiator, or other components to make the new engine fit in place.

2. Install exhaust system/catalysts. Emissions-controlled engines must have legal exhaust manifolds with properly located oxygen sensors, legal catalytic converter(s), and AIR injection. Vehicles with speed-density EFI engines may need relatively correct exhausts just to keep the engine volumetric efficiency close enough to stock to maintain correct EFI calibration—or a custom PROM chip. Some tight swaps may mandate custom headers, which can be a gray area for California street driving unless the header manufacturer has a California Exemption Order for the modified parts. Clearly, the closer to stock configuration, the better chance a referee will approve the vehicle.

3. On a rear-drive longitudinal engine/trans swap, mount or modify the driveshaft so it bolts to the EFI-compatible transmission, possibly shortening it or replacing one end if the vehicle

The LT1-for-TPI is not exactly a common engine swap, but Fuel Injection Specialties has placed a good number of LT1 powerplants into GM TPI trucks and SUVs, and into non-TPI V-8 and V-6 automobiles. However, if your old TPI Firebird made 200 horsepower, and you've got a hot rod LT4 cranking out 350 or 400 all-motor horses hidden in the closet, we're talking nearly twice the power. You could do this with TPI hotrod parts, but a PROM-based 1993 LT1 ECM will manage the engine, as will almost any TPI computer with a few calibration or auxiliary fuel-air tweaks, and the swap is straightforward if you've done your homework.

rear end is foreign to the transmission.

4. Mount the computer and wiring harness. Except for certain EMS computers with hardening to survive the harsh environment of the engine compartment (moisture, heat, vibration, electromagnetic radiation, etc.), the engine management computer must be mounted in the passenger compartment or trunk where it will stay cool and dry. You may need to lengthen some wiring in order to locate the computer in an optimal place in the vehicle (and may wish to shorten other wiring for a sanitary installation). As long as you use reasonably heavy gauge wire to extend the wiring harness, and as long as you make excellent quality joints (preferably soldering), sensor and actuator readings to and from the computer will not be affected. If you don't get the wiring harness with the EFI engine, you can probably expect to pay $200–800 for the complete harness. New aftermarket wiring harnesses are available for roughly $300 or more, but make sure they have the capability to connect all components you require on your vehicle (such as torque converter lockup, vehicle speed sensing [VSS], and so on).

5. Fit engine accessories. The air injection pump, air conditioning compressor, power steering pump, and alternator accessory locations for the vehicle must fit with the new EFI engine. Engines with MAF sensors may not be compatible with accessories and the radiator in tight engine swap situations. You will probably have to fabricate air-cleaner/MAF ducting from elbows, pipe, OEM ducting, and so on. This takes time, ingenuity, and money.

In speed-density EFI engines, you can mount an open-element air cleaner directly on the throttle body, but you will lose power due to the lack of cold-air inlet. MAF sensors are delicate and require precise aerodynamics to work right, and they complicate the swap air inlet plumbing task. Some engine swappers have had to relocate or replace radiators and fabricate new engine accessory mounts to relocate accessories to be compati-

ble with the EFI engines. You may have difficulty with the stock knock sensor location. Since the shock wave/sound created by detonation travels well through metal, this sensor can be relocated fairly easily. If necessary, Teflon tape wrapped around the sensor threads can be used to reduce the sensitivity of the sensor in the new location.

6. Handle miscellaneous hassles. Some late-model vehicles have vehicle antitheft systems. A coded resister in the key may be compared to data in the computer. Without the right key with the right resister, the vehicle will not start. This can usually be defeated with aftermarket electronics or by changing the calibration.

The vehicle speed sensor (VSS) tells the computer how fast the vehicle is going. Some engine management systems make use of this data in logic to control torque converter lockup, EGR valve, evaporative charcoal canister purge valve, cooling fans, closed loop air/fuel mixture, and idle speed. There is a lot of internal logic in the computer that depends on the VSS, affecting the operation of the above devices. You should use a VSS if you want the engine to run as well as it did stock.

Depending on your selection of transmission, you will need to make sure the overdrive torque converter lockup is wired and operating correctly.

EFI wiring is fairly self-contained and operates fairly independently of the rest of the vehicle's electrical system. The check engine light (and, on some vehicles, the radio volume, electronic speedometer, and other dash gauges) are connected to the engine management computer. Some vehicles have a combo meter or other control module that receives data from several sources including the computer, but does not feed data to the main computer (and may be eliminated in engine swaps).

There is usually some wiring included in the factory injection wiring harness for convenience reasons that has nothing to do with fuel injection but operates appliances like windshield wipers that are simply in proximity to EFI wiring and have been routed in the same wiring loom. There are several wires that must be connected to the rest of the vehicle electrical system. Depending on how much of the factory harness you use, or whether you are using an aftermarket harness, these include fuel pump wiring, torque converter lockup, and power for emissions-control devices. In general, fuel injection systems need constant 12-volt+ power, switched 12-volt+ power, starter 12-volt+, tach or crank/cam trigger, and ignition coil, in addition to the EFI sensors and actuators. You may want to consider an aftermarket EFI wiring harness, since it will only include EFI-specific wiring rather than all engine and related wiring.

You may need to handle throttle linkages, transmission kickdown linkage, and clutch linkage (hopefully you can use a hydraulic clutch, since it will be vastly easier). This is standard swap stuff, except that the EFI throttle body may require different linkages.

If you are using special swap headers, you may have to install an oxygen sensor or additional oxygen sensors in the exhaust system as close as possible to the first exhaust collector or use a heated sensor if you are not able to use the stock EFI exhaust system (weld-in O_2 sensor bosses are available from firms like K&N).

You will definitely have to consider the total electrical power requirements of the vehicle, including all EFI and non-EFI appliances and lights. The alternator must provide sufficient power for worst case (all lights on, air conditioning on, etc.). Remember, the EFI system will not function reliably if the battery voltage gets much below 10 volts and may not run below 8–9 volts.

7. Build a fuel supply system with adequate volume and pressure. Any fuel-injected engine requires (relatively) high-pressure fuel from an EFI-specific electric fuel pump, but, obviously engines with more horsepower require more fuel, and you may decide to raise the fuel pressure above stock to tune the engine for special circumstances, which requires a more powerful pump. If the new engine has a lot more power (and why else would you be swapping in a new engine?), you will need a high-flow in-tank or supplementary inline fuel pump to avoid engine damage from lean mixtures or fuel starvation.

On many EFI engines, pressurized fuel flows into the injector fuel rail, while excess fuel beyond that required to supply the injectors at a specification fuel pressure is released by the pressure regulator and returned to the fuel tank. You may have to fabricate a fuel return line to route this excess fuel back to the tank unless your vehicle already had some variety of fuel injection. Some EFI engines are not designed with a return line from the engine, instead bleeding off excess fuel with a regulator T-ed off the main fuel line in or near the fuel tank.

You may also want to construct a mini tank reserve upstream of the high-pressure pump, fed by a low-pressure pump, in order to keep the engine running during high G-maneuvers or steep hills when the fuel level is low. A fuel return can often be tapped into the fuel filler neck or gauge sender unit. Optimally, the return fuel should dump into the bottom of the tank or spray across the tank to prevent aeration near the fuel pump pickup. This may be difficult to arrange and is not absolutely vital. An OEM fuel tank will probably have reserve and return plus fuel damper built into the tank plus a submersible high-pressure pump. Wrecking yards can supply EFI-type inline external fuel pumps from cars like the old 280Z/ZX. You will need an adequately sized fuel line for sufficient injection pressure and a high-pressure injection in-line fuel filter.

8. Handle cooling. This has nothing in particular to do with the engine management system, though air cleaner/MAF plumbing may interfere with the radiator and fitting. Depending on the engine-vehicle configuration, you may need to increase the size of your radiator or have it re-cored for higher cooling capacity. It is critical that the thermostat and EMS agree on what is normal operating temperature, or the EMS may continue to wet the cylinder walls with cold-running enrichment even after the engine is at normal operating temperature.

9. If you are dealing with an emissions-controlled vehicle for street use (anything newer than about 1967), to be legal you need

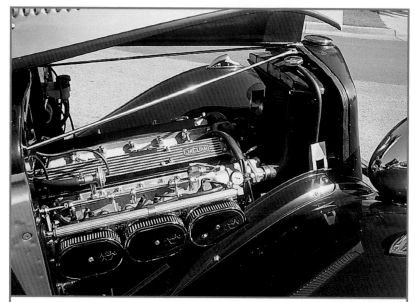

Texas concours d'elegance car collector Dee Howard had his technicians stick a Jaguar 4.2-liter in-line six into this 1940s-vintage Studebaker. The techs then installed a triple-Weber DCOE intake manifold, equipped it with triple TWM throttle bodies incorporating cast-in injector bosses, and installed a Haltech fuel-only engine management system. Sweet. (Though maybe not sweet enough: Howard later decided to convert to carbureted Chevy small-block V-8 power!)

to maintain functionality of all required emissions controls on the new engine. This explicitly refers to all emissions controls applicable to the newer of the engine or vehicle, and all original components (or exempted replacement parts) that can affect emissions—intake manifold, throttle body, exhaust manifolds, air pump, charcoal canister, EGR system, computer and relays, fuel delivery system, air cleaner, air cleaner ducting, vehicle speed sensor (VSS), torque converter lockup wiring, and virtually all wiring. You could fabricate your own wiring, or buy an aftermarket wiring harness, but to be legal, all devices in any way affecting or potentially affecting emissions must work properly.

If the car is not emissions controlled, you do not need original smog equipment, and in this case anything goes as far as speed equipment—such as larger runners, bigger cams, heads, and so forth. The thing is, speed-density systems base fueling on engine speed and manifold pressure, assuming a *fixed* relationship between these parameters (corrected for temperature and altitude) and engine volumetric efficiency—a relationship that goes away completely, for example, when you switch to a cam that doesn't run as much vacuum.

Computer programming on all modern fuel injection systems is set up to provide fuel and spark timing for a particular weight vehicle, with particular drag, particular thermostat, particular induction and exhaust systems, and particular pollution controls. Any changes at all may mean the vehicle engine management is no longer optimal. Radical configuration changes without a PROM change may mean the engine will hardly run at all. EMSs with mass airflow (MAF) sensing actually measure engine airflow into the block, and can compensate for modifications that influence the factory-assumed relationship between manifold pressure, engine speed, and engine airflow—but only to a point (on some systems as little as 15 percent greater airflow than stock). It is extremely important to keep in mind that, ultimately, injectors max out in fuel flow as soon as they are contin-

The builders of this car had the following brainstorm: A Cobra (replica or not) with a huge iron-block 427 Ford V-8 or even a 302 Windsor small-block Ford is packing a lot of front-end weight, and if the goal is handling in a club-race environment, a Ford 2.3-liter SVO Ford four-banger could probably make almost as much power and make up the difference through the turns due to better weight distribution and lighter overall weight. A stock 2.3 Turbo maxed out at 205 horsepower, but a chip change and enough fuel could make something like 355 horsepower on an overboosted 2.3. The Texas-based designers of this Cobra replica swapped in an iron-block, alloy-head SVO 2.3 turbo, added a Freon-cooled intercooler, and cranked the boost to 20 psi. As they say in Dallas, Yeeehhhaaa!

TPI, TBI, AND OTHER GM V-8 ENGINE SWAPPING

GM's small-block V-8 port EFI engines are important enough as performance engines to merit special mention in a discussion of EMS engine swapping.

GM built thousands of tuned port injection (TPI) and throttle body injection (TBI) fuel-injected small-block V-8 engines from 1985 through 1996. Many of these 5.0- and 5.7-liter EFI engines are now in wrecking yards and available for engine swaps at reasonable prices. TPI engines are essentially traditional Chevy small-block V-8 long blocks with the addition of port or throttle body fuel injection components as external bolt-ons. LT1/4 engines are essentially hot rod TPI engines with hotter cams and high-rpm intake runners, reverse-flow cooling, optical ignition triggering, and so forth. As you'd expect, Corvette engines are much more expensive than 305 or 350 TPI Camaro/Firebird motors. ZR-1 32-valve four-cam Corvette engines (assembled by Mercury Marine using special Lotus-designed four-valve-per-cylinder heads) are rare and very expensive—if you can find one.

There are many advantages to TPI fuel injection compared to older carbureted small-block Chevy engines—or even compared to throttle body Injection (TBI) V-8s. Port injection squirts fuel directly at the intake valve, meaning designers don't have to worry about keeping fuel evenly mixed with air in the intake manifold. Therefore, dry manifolds allow greater flexibility of design to make use of ram effects and pulse tuning to increase low- and midrange torque, which is what makes street cars feel fast and responsive.

Port injection inherently produces virtually perfect mixture distribution to all cylinders, meaning all cylinders produce maximum power or economy under precise computer control. Perfect distribution means more efficient, higher compression ratios are available without spark knock. Coordinated ignition spark advance and precise control of part throttle fuel-air mixtures gives fuel-injected engines the ability to run well and lug well at very low rpm. Coupled with automatic overdrive transmissions and locking torque converters, EFI engines are capable of both excellent fuel economy at part throttle and high output at full throttle. The locking converter automatic overdrive transmission typically saves 10 horsepower worth of friction drag at highway speeds at light cruise due to the slower running motor.

The key to drivability is a matched engine-transmission pair, with fuel injection and computer-controlled torque converter lockup. Carbureted engines simply cannot equal the street-able power and economy of EFI. EFI engines always start well in any weather, always idle well, and last longer (typically over 200,000-plus miles) because O_2 sensor-controlled fuel-air mixture feed-back-loop programming makes sure overly rich mixtures don't wash oil off cylinder walls, causing premature wear. A properly done EFI engine swap makes a hot rod into a daily driver instead of a hangar queen—and it's legal in all states.

TPI and TBI, of course, are not the only GM small-block options. LT1/LT4 engines from 1992 to 1997 are very available in wrecking yards or from crate-motor suppliers (including GM), and these engines will bolt up to many traditional small-block

uously open, which is the ultimate flow limit for a particular set of injectors unless you alter fuel pressure.

If you are making radical changes to an engine for off-highway use (or street use on a pre-1967 vehicle), you should consider buying a programmable aftermarket computer, which will provide a friendly user interface with documented access to all features that allow you to alter all engine parameters on the fly via laptop PC. Most aftermarket computers will work fine with TPI-type engine sensors, fueling systems, and the tuned port engine air metering system (speed-density). The newest trend is plug-and-play aftermarket systems that plug directly into factory EFI wiring harnesses and control everything from raw fueling to the check engine light to torque converter lockup.

If you live in California, after an engine swap you may have to have any emissions-controlled vehicle inspected by a referee at the California Bureau of Automotive Repair, which involves both a visual check and a tailpipe sniffer test. You will want all stock systems to work properly.

HELP WITH SWAPPING COMPONENTS

There are a variety of companies that make special equipment to help with EMS engine swapping, including everything from replacement high-volume in-tank pumps to special wiring harnesses, electronic interceptors, special PROMs or calibrations (designed to disable problematic functions like antitheft, MAF metering, emissions equipment), VSS and speedo conversions, tach conversions, and more. Beware of aftermarket simplified wiring harnesses that eliminate essential wiring by fooling the engine management system into thinking, for example, that the car is always in neutral or park, which could hurt drivability, fuel economy, emissions, and even power.

Another approach to achieving huge power while avoiding the cold-blooded spitting, hissing, and other drivability problems you're likely to get when you swap a carbureted side-oiler 427 into a Cobra replica in place of the original 289, is this Norwood Autocraft EMS conversion. By squirting fuel straight at the intake valve at each port from port injectors mounted in TWM DCOE-type throttle bodies mounted on a Weber-patterned 427 manifold, the EFI 427 avoids mixture-distribution problems. Precise Norwood EMS tuning resulted in a car that starts great when cold, idles smoothly (though it's still a bit lumpy), and sloshes the blood deep into your reptilian brain when you mash the go-pedal.

GM V-8 transmissions and motor mounts. LT1 heads and manifolds, however, are *not* strictly interchangeable with traditional small-block blocks (partly because of the reverse-flow cooling and all that it entails)—though the conversion has been done by (brave) machinists who know what they're doing. LS1 engines are a completely different animal and are in no way interchangeable with L98 or LT1 engines. However, properly modified, they are capable of tremendous power, and many are appearing as swaps into older classic carbureted vehicles.

If you have a healthy carbureted engine, you may be considering swapping EFI components onto your engine. This can be done, but there are potential problems: swap-meet EFI components may be missing parts—and you may not know it. Buying additional parts can be terribly expensive! TPI ducting for the MAF sensor and/or remote air cleaner may be incompatible with accessories or radiator and hoses on older engines.

The stock EFI computer will be calibrated for a specific engine configuration, and if your older engine is different in any respect, EFI calibration may no longer be optimal. In addition, 1987 and newer TPI lower-intake manifolds have four bolts that are not compatible with older cylinder heads. The newer TPI EFI manifolds must be modified with welding or grinding to fit older engines. In addition, the distributor drive gear metallurgy on newer EFI distributors will not be compatible with older cam drive gears, resulting in improper wear unless the gear is changed.

Older TPI engines computed engine fueling requirements by directly measuring mass airflow into the engine with a hot wire mass airflow sensor, which is mounted upstream of the throttle body. All air entering the engine must flow through the MAF. Since air mass is equivalent to the air's ability to cool a heated wire, electronic circuitry in the MAF simply measures the current required to keep the hot wire at a fixed temperature.

Post-1990 TPI engines estimate engine airflow by measuring air temperature, manifold pressure, and engine speed, and use these parameters to look up fueling requirements for a specific engine in computer memory. Interestingly, while GM switched from MAF-sensed airflow metering to speed-density manifold-pressure-sensed metering, at the same time Ford did the opposite! GM returned to MAF metering—while retaining the MAP sensor for limp-home mode operation—in 1994.

From the point of view of an engine swap, speed-density EFI is preferable due to the simplified intake plumbing required in the absence of a MAF sensor. If you modify the airflow of the engine, however, the system will not work right unless you switch to a custom PROM chip (still available from firms like Arizona Speed and Marine in Chandler, Arizona). And it may be hard to get a really right PROM unless your engine configuration is identical to something with which the PROM manufacturer is familiar.

Amazingly, MAF-sensed engines work fairly well with the MAF removed entirely. And this is due to excellent limp-home strategies that disregard suspect sensors and make do with historical data and data from other sensors (like TPS in TPI days or MAP sensors after 1993) in order to deduce engine airflow. But they will work better if you find the parts to convert to the late-model computer and speed-density sensors.

GM TPI and TBI engine swaps can result in a powerful and economical street vehicle that operates virtually identically to a GM F-Body or 'Vette. In this case, you will need the stock EFI engine and all accessories and wiring, and you will want to install them with an absolute minimum of modifications. When you are finished, you will have a powerful, modern, efficient powertrain. If you are running off road or using an older nonemissions-controlled engine, you can modify the EFI engine's airflow and fuel delivery with high-performance parts such as bigger TPI runners and fatter cams and other VE enhancements, and recalibrate the GM computer's fueling and spark timing for the new engine configuration with a new PROM chip—or install a completely new aftermarket programmable computer to handle the job.

A TPI engine swap is not a trivial task. It will probably cost more than you think and will require more than average ingenuity. But tuned port injection and the L98 TPI engine represents straightforward, fully compatible fuel injection for all 50 million or so Chevy small blocks and factory V-8-equipped vehicles built between the mid-1950s and the early millennium, which makes it the most successful engine in history!

There is vastly more performance equipment available for this engine than anything else around, and it's capable of achieving any level of performance anyone could possibly require from the small-block V-8. Swapping in a TPI engine is a great way to fuel inject your vehicle or improve the performance of your existing non-TPI EFI engine. Chevy small-block V-8s have been swapped into almost any vehicle you can name. If you take your time, start with a complete engine/transmission and do everything right. The task can be fun, and the result will be a rugged and reliable installation.

This chapter examines the following options for fuel injecting your hot rod.

CARBURETED ENGINES

1. Swap in a complete fuel-injected engine and supporting equipment such as the high-pressure fuel pump. Many existing transmissions currently mated to old carbureted engines will mate to a similar newer fuel-injected engine, which can be virtually identical except for the addition of EFI and electronic engine management. Common examples of this are listed below:

• Ford 5.0- and 5.8 8-liter port EFI V-8s that Ford started installing after 1985, which replace 260-289-302-351W engines used in many '60s1960s, '70s1970s, and '80s 1980s Mustangs and other performance cars.

• Ford 460 port-EFI V-8s that were installed in some Ford trucks after the mid-'80s1980s, and will replace carbureted 429 and 460 V-8s used in '60s1960s- and '70s1970s-vintage musclecars.

• Chevy/GM 305/350 tuned port injection V-8s fitted to GM cars of the '80s 1980s and '90s1990s.

• Chevy/GM 7.4-liter TBI or MPI big-block V-8s fitted to various GM trucks in years after 1986, including the performance 454SS.

• Porsche flat-six 3.6 engines were equipped with electronic port fuel injection starting in the mid-'80s1980s, though some continued with mechanical K-Jetronic injection for a number of years. Vintage Porsche carbureted engines from the '60s 1960s and '70s 1970s can be replaced with later Porsche powerplants and even turbocharged.

• Jaguar 3.8- and 4.2-liter in-line-six powerplants with multiple carbs were upgraded to EFI in the mid-1970s.

Many Ford and GM EFI V-8s will mate to older transmissions as far back as the '50s 1950s in the case of GM, or the '60s 1960s in the case of Ford. Ford's original thin-wall 260 small V-8 came out around 1964. GM's 265 small-block V-8 came out in the mid-'50s1950s. These blocks are similar or identical externally to the modern EFI V-8s of the same family (though the intake manifold will be different). You may be able to find a junkyard EFI V-8 in good shape for a few hundred dollars.

The job then will be to adapt the EFI engine to your car. A number of aftermarket suppliers like Howell, Street and Performance, and others sell simplified wiring harnesses designed for engine/injection swapping.

Generally it is legal to swap in a later model engine (complete with all emissions-control devices) into an earlier vehicle (although in California, you will probably have to have the vehicle and engine inspected at a referee station). It will *not* typically be legal to install an older engine and engine management system in a newer vehicle, though an older short block (or even long block) might not, de facto, cause problems.

A number of examples of EFI-for-carb engine swaps are covered later in this book:

• An an injected 530-ci Cadillac into a GMC motorhome, first using Cadillac electronic controls and later Haltech injection.

• A a hemi-head Renault turbo EFI motor into a Lotus Europa, replacing the stock wedge Renault engine, controlled by DFI electronic injection and ignition controls.

Typically, late-model EFI engines are available from commercial wrecking yards for $1,500–1,750 or less, with older EFI engines in running condition somewhat less. Obviously, engines can be obtained from private parties via newspaper ads or the Internet for substantially less, though caveat emptor is the rule. Make sure you get *all* required parts, including wiring and ECU, which may or may not be included in the asking price. It will be *much* more expensive to later buy additional EFI parts one by one.

2. It may be possible to adapt Port EFI manifolds, fuel rails, throttle bodies, etc., from an equivalent EFI engine to an existing previously carbureted engine using the principles discussed above. This would be an attractive option if your existing engine is in very strong condition, yet similar in camming, porting, displacement, etcand so on., to a factory EFI engine.

The danger is that variables that affect breathing/volumetric efficiency—particularly cam changes—could mean that OEM EFI engine management will no longer provide correct air-fuelair/fuel ratios or timing on your engine. This may still be a good choice, even with a heavily modified engine, as long as you recalibrate or otherwise modify operation of

One practical way to convert from carburetion to EFI and electronic engine management is buying major components from a firm like Accel. This Chevy big block is equipped with a custom large-runner EFI intake and appropriately large plenum. The runners are longer than they look, protruding into the plenum every-which-way, like stacks. In the background, we see a Gen 7 standalone EMS with wiring harness.

Bell Engineering had in mind the best Cobra in the world, and they built the CB1 Cobra replica car from the ground up with many advanced features that Carroll Shelby never dreamed of in the 1960s. The engine is a built twin-turbo 351 Ford with custom intake and turbo system and an Electromotive TEC³ EMS.

the factory engine management system to provide correct air-fuelair/fuel ratios and timing. PROM changes, fuel pressure changes, add-on black boxes and interceptors that change injector pulse width, larger injectors, extra injectors, and other hot rodding techniques are available to rework the stock engine management parameters. In the case of a common engine with common modifications, there may well be plug-and-play options, though none of this comes for free.

OEM EFI parts will be available used for a few hundred dollars and up in a salvage yard. Fuel enrichment will cost from $100– to 200 (PROM replacement or increased fuel pressure via special regulator) on up into the $500–$ to $1000 range for more complex systems. TPI and other manifolds and parts can be polished for an excellent look on a custom car. GM and other car companies sell packages to retrofit TPI to older engines. Incidentally, many European car companies adapted EFI to older carbureted engines. This book illustrates as OEM Bosch L-Jetronic EFI swap onto a Jaguar XKE 4.2 engine in a later chapter, and explains what's involved in swapping GM TPI equipment for the Chevy engine.

3. Bolt on a Holley Pro-Jection throttle body injection system.
These two- and four four-butterfly throttle-body injection systems are relatively inexpensive ($500–1,000), easy to install (since the throttle body with integral injectors and fuel pressure regulator bolts onto a four-barrel manifold exactly like a carb), and are true electronic fuel injection systems with many of EFI's advantages— but not all! These are single-point injection systems in which all fuel is injected near the throttle plate(s). So, like a carbureted engine, the intake manifold must handle a wet mixture, which imposes constraints that would not exist on a dry port-EFI manifold.

Previous versions of the system (Pro-Jection-2 and Pro-Jection-4 still may be found used) were tuned with pots on the ECU. The current Digital Pro-Jection uses a modern digital control unit and requires a laptop for calibration. You'll need some additional relatively inexpensive parts, such as materials to construct a second fuel line to return unused fuel from the throttle body back to the tank (all electronic injection systems require this).

4. Some Use some combination of options 1 or 2 plus an aftermarket programmable ECU and associated parts.
That is, install factory air-supply equipment (EFI manifold, throttle body, etc.), factory fuel-metering equipment (injectors, fuel rail, etc.), with aftermarket computer, wiring, and (probably) engine sensors.

This book covers the installation of aftermarket EFI controls onto a V-6 Gen II MR2, Honda CRX, Honda VTEC del Sol, big-block 1970 Challenger, and the Lotus Europa.

5. Bolt on a turnkey EFI conversion package,
which has everything you'll need to do the job. Such a system may even have an emissions exemption, making it legal on certain or all late-model vehicles. An example is the Edelbrock Pro-Flow system, available for many normally aspirated American V-8s. There are several such packages for the small-block Chevy V-8, small-block Fords, and Chrysler A- and B-block V-8s.

6. A complete roll-your-own add-on fuel injection system, in which you might:
• Modify a carburetor intake manifold to hold port injector bosses (or buy an aftermarket EFI manifold).
• Construct a fuel rail and mountings to hold injectors and provide fuel.

There's almost nothing you can't find out there to convert nearly anything to EFI and electronic engine management. Hilborn developed this EFI conversion for 426 Hemi engines, and although it's computer-controlled, the look is definitely radically stacked mechanical fuel injection.

• Select and install a fuel regulator, filter, supply and return lines, and injectors.
• Build (if you're a machinist) or buy a throttle body (available from TWM in Santa Barbara, Accel, and other suppliers).
• Purchase an aftermarket controller, sensors, and wiring loom as in 5 above.

Such a system will typically include a speed-density ECU and sensors, fuel pump, injectors, fuel-pressure regulator, single-plane manifold plus modifications, and throttle body.

This option is explored in this book in a nitrous-injected, triple turbo Jag XKE, a Chevy 350 twin turbo, and a big-block Mopar 383.

FUEL-INJECTED ENGINES
TBI-to-MPI conversions. In case your engine is already equipped with throttle body injection (for example, certain 1987–95 GM trucks and cars), you can buy a kit of parts to convert to the near-perfect fuel distribution of port fuel injection. Edelbrock's kit provides replacement port-EFI intake manifold with injectors and fuel rail, upgraded fuel pump, injector harness, and other conversion parts, retaining the stock computer, throttle body, and much of the stock engine wiring harness.

CHAPTER 11
INSTALLATION AND STARTUP ISSUES

For a successful, trouble-free EMS installation, you'll need to deal with some or all of the following issues: electrical and wiring; power draw; electrical interference; mounting and connecting the ECU; ignition system compatibility, installing sensors, Vacuum/boost source; laptop computer interface; installing fuel maps; and fuel injectors.

ELECTRICAL WIRING INSTALLATION

Engine management system problems are almost always caused by wiring, peripheral sensors or actuators, virtually never by the ECU itself. I have seen experienced professional EMS experts make false assumptions about wiring integrity that waste countless hours of troubleshooting time.

Good electrical wiring is *essential* to reliable engine management, and you'll need the right tools to do any custom wiring. Engine management gurus like Bob Norwood are constantly amazed by the nightmare wiring that arrives on vehicles entering the shop for troubleshooting—wires twisted together to make connections and sloppily wrapped in electrical tape, poor-quality plastic-shrouded crimp-type connectors "crimped" with a hammer or some other inappropriate tool, multi-wire OEM connectors that

have been separated by force rather than the correct special tool and then held together with conduit ties, and so forth.

It is always false economy to do substandard wiring. Poor wiring *always* fails in the end, usually when it is most inconvenient. And if your luck runs out, bad wiring will fry something expensive in the process.

You definitely need a schematic and a pin-out diagram of the EFI electrical system, which is usually obtainable from the ECU manufacturer. This will be extremely useful when things don't immediately work right on initial startup and you're not sure what's wrong. A schematic and a continuity tester are a godsend when this happens.

When routing the EFI harness, be sensible about running it away from noisy electrical sources—as well as heat or sharp or abrasive components (like braided steel hoses, which can, over time, damage wiring by rubbing against it).

Most engine sensors work by varying resistance based on some physical condition of the engine. Beware of modifying an EFI wiring harness in any manner that would change the resistance of a sensor by adding additional resistance on the way to the ECU. For example, suppose you bought a standard wiring harness

Wiring is one of the most critical areas of any EMS installation, and poor wiring will eat your lunch in troubles immediately or down the line. You need the right tools and the right tricks, but there are entire books about wiring. Norwood Autocraft built a custom wiring harness for this four-cam stroker big-block Chevy with Motec EMS. Note the monster custom throttle bodies and intake plumbing under construction.

that cannot reach a coolant temperature sensor without extension. Use the proper gauge wire if you must extend it, and use a resistance meter to verify that the wiring modifications have not changed the signal as seen at the ECU. In the case of some aftermarket engine electronics, you may discover that any wiring changes void the warranty. However, in most cases, you can have the manufacturer build a special wiring harness to your specifications, which can actually save you money in the long run.

A really good crimping tool can crimp a connector to a wire so well that the wire will tear before the connector releases the wire. But you must have the right crimping tool for a connector and the right connector for a given wire size, and I am talking about a heavy-duty ratchet-type crimping tool. I find that the right tools can actually make doing good wiring *fun*, whereas the wrong tools can make it hell.

Most professional EMS installers never solder anything because if a wire works or moves in the vicinity of a solder-on connector, the strands that make up the wire can gradually break—one by one—where the wire was heated and enters the solder, and the connection can eventually fail. My view is that soldering is definitely preferable to poor-quality plastic-shrouded consumer-grade crimp-on connectors of the type you find at places like Pep Boys. Such things might work OK for a trailer hitch, but *not* for a computer EMS wiring harness. The problem is, it is hard to have the right wiring tool for every situation if you are not an EMS pro. You will not want to run out to buy obscure crimping tools on a case-by-case basis when you are wiring something (and the right tool may be expensive and difficult or impossible to find at the last minute). If you are doing an ambitious EMS wiring job, buy the right tools in advance and pay what it costs to get good quality tools. In an emergency, unless you have access to the right high-quality crimp-on connectors and the right crimping tool(s), soldering is more reliable.

Give some thought to layout of the system wiring harness before installing ECU or sensors. It is often helpful to connect the sensors and ECU to the harness and drape the wiring in place and get a feel for what will work the best.

POWER DRAW

All ECUs have the ability to correct injection pulse width for the state of charge of a battery, which affects the speed with which injectors open. However, most ECUs have a threshold below which they will not start the engine. This is usually around 8 volts. At the same time, a low battery or inadequate electrical charging system can effectively reduce the maximum injector flow rate due to the slower opening time. The fuel pump maximum pumping volume declines with voltage (just as it *increases* with voltage when devices like the Kenne Bell Boost-a-Pump jack up the voltage to super-normal levels to run the pump faster and harder during boosted engine conditions).

An engine management system will consume a certain amount of power that would not be required on a carbureted engine, depending on engine speed and injector resistance. Electronic ignition systems require power, as do electric fuel pumps—all of which are in addition to the power requirements of accessories such as air conditioning, headlights, wipers, and so on. You must add up all the electrical requirements of the vehicle under worst conditions and compare them to the output of the vehicle's charging system (and the power capacity of certain critical wires delivering electrical power to high-draw appliances). With everything on at night, your battery might be slowly draining. Eventually, the engine would stop. In some cases you may have to install a larger alternator as part of an EFI conversion or engine swap.

Most ECUs have built-in fuel pump relays to power a high-pressure electric fuel pump, but you can probably increase the capacity of your pump by running a heavy-gauge wire directly to the battery and using the ECU's pump wire to trigger the heavy-duty relay (which will itself consume power). When using more than one fuel pump, do not depend on the ECU to directly energize the fuel pump—use it to trigger a relay.

If you're not sure about the pumping capacity of your fuel pump whatever the voltage, test it by jumpering the fuel pump to the battery with the fuel return line diverted into a closed container of known size, and measure the time it takes to fill. Using the formulas described in the section of this book on actuators (of which the fuel pump is one) you can determine the fuel pump capacity at rail pressure. Make sure it is more than the best-case horsepower for the worst-case brake-specific fuel consumption for the engine (described elsewhere in this book), including nitrous fueling, if applicable. If you're running forced induction, you will have to pump up the fuel pressure regulator to full boost pressure to determine the fuel pump flow at the higher pressure of boosted conditions.

ELECTRICAL INTERFERENCE

High-output ignition systems with extremely high voltages (particularly with wire-core plug wires) produce electromagnetic radiation (radio waves) that can interfere with the proper operation of an engine management system. Most ECUs require resistor-core plug wires and mounting inside the vehicle or in the trunk, away from the heat, moisture, and electrical interference of the engine compartment. Even if the ECU is approved for engine compartment, mount the ECU inside the vehicle if at all possible. Electronic components do not like heat, and heat can reduce their life, even if it does not kill components. Electromotive's TEC[1] and TEC[2] ECUs were approved for mounting in the engine compartment; the TEC[3] is not. I wonder why.

ECU

It is essential to provide excellent constant and switched voltage to the ECU itself. It is best to route constant power and ground directly from the battery. Switched power must be connected to a source that is not turned off during cranking (or drained to less than 8 or 9 volts).

If the ECU provides cranking enrichment, make sure it is triggered by a power source that is only active during cranking, or the system will run rich.

When mounting the ECU—not in the engine compartment, of course—be aware that the ECU case is often used as a heat sink. Mount the ECU in a cool place away from the exhaust system and heater where dry cool air will help to cool it, and where heat can transfer directly from the case into the bulkhead to which it is bolted. *Never* mount the ECU on the floor of the vehicle, which can easily become flooded and permanently ruin the ECU's electronics. Finally, make sure the ECU's user-interface port will be accessible for connecting the communications cable to configure and calibrate the system once the ECU is in place.

IGNITION SYSTEM COMPATIBILITY

Engine management ECUs require electrical signals to determine how fast the engine is turning and—for sequential injection or spark control—to determine crankshaft position. Many ECUs can use a low-voltage signal from the negative side of the coil as a trigger to determine engine speed. Others may utilize a crank ref trigger or cam sync trigger sensor—or a combination of coil, crank, and sync.

If you are using a capacitive discharge ignition, you must use a special tach output signal as an ECU engine-speed trigger (unless you are not using a spark trigger at all), since the multiple-spark capability of a capacitive system can damage your ECU. Most ECUs will work with common performance-car OEM ignitions, common aftermarket ignitions, standard points ignitions, and a variety of special ignitions. In some cases, the ECU may require dip-switch settings or factory hard-wiring to properly determine trigger threshold, but most are now configurable using EMS software. Check with the manufacturer of the ECU and the ignition (if aftermarket) for compatibility. On the Motec M48, for example, you'll need to configure the ignition as one of: rise trigger; fall trigger; Mazda three-rotor; Mazda Series 5 or Series 4; Ford SAW 4, 6, or, 8 cylinder; Ford SAW 10 cylinder; and Ford SAW special.

You'll also have to configure the engine triggering system used to fire ignition and injectors and so on, meaning you'll need to know the number of reference teeth or flying magnets on the crank angle and cam angle sensors and the number of missing or additional teeth used to sync top deal center on the trigger wheel.

Bear in mind that the bottom line of programmable ignition is a good hot spark at the plug electrode. Make sure to select plugs that are the right heat range for your application. If yours is a turbo conversion or a massive boost pump-up, you probably need colder plugs than stock, and you should probably set the gap smaller than stock, even with a high-powered ignition. Check with the coil/coil-driver vendor, but many pros assume the gap on a hot rod turbo application should be no larger than 0.029.

Whatever you do for ignition, you're going to need a way to drive a tachometer. If you are using the stock ignition module or coil-driver, there is probably a tach driver signal you can continue to use. If not (say, if the stock ECU provides this signal but you're changed to aftermarket), the aftermarket ECU probably has a tach signal that will work, or one that can be inverted with a little wiring and a resister. If this is an engine-swap situation to a powerplant with a different number of cylinders, you may find your ECU can provide a pulse width–modulated signal of programmable duration (see the MR6 project in this book).

INSTALLING SENSORS

You will probably be locating or installing air- and coolant-temperature sensors, a throttle-position sensor (TPS), a MAP sensor, one or more oxygen sensors, and possibly a crank or cam sensor.

Heat soak from surrounding metal can be a problem on some air-temperature sensors. Try to locate the sensor in such a way that engine heat does not soak the body of the sensor, giving a falsely high reading (which will probably lean out the mixture). Make sure the air temperature sensor you're using is fast enough to react to the rapid changes in temperature that occur with turbo-supercharged and intercooled induction systems so that the air temperature sensor can be safely installed downstream of an intercooler where it belongs to react intelligently to temperature changes as it should.

Oxygen sensors do not work well until they heat up to about 600 degrees Fahrenheit. It is essential to locate a standard O_2 sensor as close as possible to the exhaust ports, but in such a way that it measures exhaust from multiple cylinders (if possible, all cylinders). Heated oxygen sensors provide their own heat by internal electrical heating and can be located farther from the exhaust port, for example in a header collector. A heated O_2 sensor has the added advantages of beginning to work sooner after startup and not cooling off too much to work properly during prolonged idling.

On ECUs that do not support multiple O_2 sensors, you might want to install two oxygen sensors on a V-configuration engine through an A-B toggle switch for switching which sensor gets used if you are concerned about possible variations in exhaust gas oxygen content from cylinder to cylinder. If the engine develops poor fuel distribution due to dirty injectors or begins to misfire due to poor spark-plug condition, one or two cylinders emitting way too much or too little oxygen into the exhaust can cause the ECU to adopt air/fuel strategies that are wrong for all cylinders.

Throttle position sensors come in many varieties. Even though most modern aftermarket engine management systems can learn TPS electrical characteristics at idle and wide-open throttle, be careful if you're using an existing original equipment TPS with an aftermarket ECU. Not all linear variable resistors are the same. Fortunately, most programmable ECUs have the ability to display what percentage of throttle opening they think the sensor is reading. Check the TPS with closed and open throttle to make sure it's working properly as seen at the ECU, and make sure physical throttle positions in between translate into plausible readings on the EMS virtual-dash engine data page.

Many ECUs do not pay much attention to the TPS voltage as such, but are rather more concerned with the *changes* to voltage (delta) on speed-density systems that indicate *transitions* from one engine speed to another. However, absolute throttle position is one factor that the ECU probably uses to trigger closed-loop fuel trim and idle speed stabilization logic. However, even if the ECU is not programmable for absolute throttle position, do not automatically assume there is a problem if the sensor's full range is not available to the ECU.

Make sure the TPS turns in the same direction as the throttle shaft, and *never* use the TPS as a throttle stop. If the ECU can program itself for TPS voltage or frequency, make sure that the throttle stop is backed off all the way when you install and clock (if clockable) the TPS so the TPS does not prevent the throttle from closing all the way should that become desirable. Once the TPS is clocked, adjust up the throttle stop to the proper position and then go through the ECU-TPS learning procedure. If the ECU must read a specified voltage for idle, with the engine off, key on, throttle closed, and EMS user-interface at the virtual dash displaying sensor values, clock the TPS until the TPS reads zero throttle position.

ECUs that provide spark control require some kind of engine position sensor. Exact engine sync position relative to the power stroke for a certain cylinder can only be provided by the camshaft, since a crank sensor has no way of knowing whether a given position represents compression or exhaust stoke on a four-cycle engine. The camshaft/distributor tells the exact position in the engine cycle, but if you are using a flying-magnet-type system with only one position event per combustion event (versus, say, a 36- or 60-tooth gear read optically or magnetically at crank speed), the ECU cannot react as quickly to micro-changes in engine speed. Any cam-speed position information can be affected by slop in the cam belt. Most ECUs that need to know engine position do require both a crank sensor and use the cam/distributor purely to sync the crank position to compression on the number one cylinder.

VACUUM/BOOST SOURCE

Speed-density engine management systems require accurate manifold pressure to provide proper fueling. This is particularly critical at idle. Some wild-cammed engines are not suitable for speed-density EFI because the manifold pressure at idle is unstable. Engines like this should use throttle-angle (Alpha-N) EFI control, or a system like the Electromotive MAP-blend system that switches over to TPS or a percentage of TPS and manifold absolute

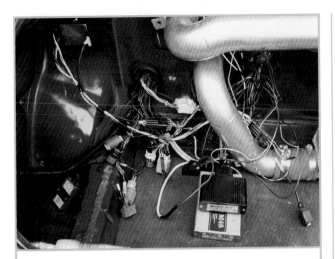

The greatest challenge will not be to build from-scratch wiring, but partially disassembling stock wiring and combining it with new wiring for aftermarket electronics. This in-progress wiring from the MR6 project in this book involved several varieties of Toyota MR6 wiring connectors, Motec connectors, GM Weatherpak connectors, Bosch connectors, and more. You must have the right tools (especially to disassemble factory connectors without destroying them) and patience.

pressure (MAP) sensor data at idle to keep a wandering MAP signal from causing the ECU to chase its tail at idle.

A problem could occur if you're running multiple throttle bodies, since the manifold pressure reference signal from one port would only indicate the vacuum of this one port and might fluctuate rapidly at idle, confusing the ECU. In this case, it is essential to construct a vacuum accumulator/manifold of 2-ci volume with vacuum lines that are as short as possible running to reference ports on all intake runners. You would then reference the MAP sensor to the accumulator. Such a device will dampen fluctuations yet is small enough in volume to react quickly to changes in manifold pressure.

You should always endeavor to have the shortest possible manifold pressure reference line to a MAP sensor, or even mount it directly on the manifold with an O-ring as in some OEM applications. The larger the volume of the reference, the lazier the MAP sensor data will be in relation to actual manifold pressure, which could cause problems with acceleration and deceleration enrichment and other calibrations. Metal or hard plastic reference lines can help because they will come up to full pressure slightly faster than softer rubber vacuum hoses that will expand a little before they reach full pressure—a bit like a balloon.

BOOSTED APPLICATIONS

A few aftermarket engine management systems are not suitable for running turbos or superchargers without a lot of work and heartache. An example would be any EMS that uses throttle position to estimate engine loading (since the same throttle position could yield vastly different engine VE depending on the status of compressor speed on a turbo application). The point this raises is that it helps to know where you are going with a project before you decide what your EMS strategy will be. Though it is a MAP-based speed-density system, an Edelbrock Pro-Flow small-block Chevy system is not suitable for forced induction if you decide later that you need more power.

Some engine management systems are not well suited for managing nitrous oxide injection, while others have highly sophisticated nitrous capabilities. If you are planning to boost the engine with nitrous, get an EMS with nitrous capabilities so you can provide fuel enrichment using the electronic injectors, and plumb the intake manifold with *port* nitrous jets. The closer you inject fuel and nitrous to the intake valves, the less likelihood of distribution problems, and the less chance of turning the intake plenum into a pipe bomb in case of a backfire. You should never inject fuel upstream of an intercooler for the same reason.

Do not make the mistake of trying to inject nitrous oxide and fuel (a fogger nozzle, for example) at a single point. Port EFI manifolds are *not* designed to handle wet mixtures, and you could burn down the engine (see the Honda CRX project in this book). It may also be tempting to inject nitrous at one place near the throttle body, but resist the temptation: You could have uneven nitrous distributions depending on exactly where you inject the nitrous upstream of the throttle body, and if too much nitrous goes to one or two cylinders, you could still have a lean burn-down situation even if the nitrous enrichment fuel is perfectly distributed through the port electronic injectors.

PERSONAL COMPUTERS/INTERFACE

The vast majority of programmable engine management systems are designed to use a personal computer as a user interface for tuning and monitoring the EFI system. You will find it helpful to have a rudimentary knowledge of the Windows 95 or later operating system, although most EMS user interfaces provide a fairly complete set of self-contained commands that will do the job to create and modify fuel maps, and to manage them on floppy or hard disk.

It is important to have a laptop computer so you can calibrate or troubleshoot the vehicle while driving. If it's all you have, with a lot of trouble you might be able to use a desktop system to modify tuning off-line between test drives in a feedback-based iterative process, particularly if the EMS can datalog sensor data to the ECU while you drive for later review.

Never attempt to tune and drive at the same time; get a friend to drive the vehicle while you tune. Many expert tuners claim that some road tuning is required even if the raw fuel map was created on a dyno with an air/fuel-ratio meter or gas analyzer. It is definitely not safe to drive and tune at the same time, so either drive the car while logging data and use the datalogs to tune while the car is stopped, or be smart and get someone else to drive while you tune.

Edelbrock Pro-Flow systems, SDS's EMS, and some of the interceptor/piggyback systems that modify the actions of the factory computer are available with their own user interface—usually with a keypad and an LCD display—which means you do not *need* a PC (though some of these can *use* a laptop if you have one or don't want to buy the proprietary interface device). However, given the nontrivial cost of a programmable engine management system, the superior friendliness and usability of a big-screen laptop user interface, and the ease of borrowing or renting a laptop, laptop concerns should not be an issue to most people.

Some programmable EFI systems support an optional mixture-timing-boost trim module that enables a tuner to vary the air/fuel mixture by an adjustable percentage rich or lean by turning a knob on the interface. This allows you to easily change the mixture and instantly "feel" or analyze the effects. Some provide an across-the-board trim that simply modifies the "final" injection pulse width, timing, or maximum boost calculation by the selected percentage.

Others have provided a separate adjustment for, say, idle versus wide-open-throttle. Once you know what percentage change is

Fuel Line Size Chart

Horsepower	Rigid Line Size	-AN Hose Size	Pump GPH at System Pressure	Pump GPH Free Flowing
200	5/16"	-04	17	46
350	3/8"	-06	29	81
550	1/2"	-08	46	127
800	5/8"	-10	67	184
1200	3/4"	-12	100	276

Injector Static Flow Rate, In C.C. of Fuel/Minute, Toyota 3S-GTE, MR2 Turbo

1	723.0
2	718.3
3	720.4
4	721.0
AVG:	720.7

TOTAL SYS. LBS./HR.-STATIC:	274.54
CC FUEL/MIN/H.P.-STATIC:	6.78
B.S.F.C./@ 80%/@ EST. H.P.:	0.52
DUTY CYCLE AT EST. B.H.P.:	0.85

Horsepower At 80% Duty Cycle at listed Brake Specific Fuel Consumption
@ .65: 337.9
@ .55: 399.3
@ .50: 439.3

COMMENTS: SYSTEM BALANCE IS AT 0.7%

RC Engineering

Like everything electrical, GM Weatherpak connectors require the right tools to assemble. You slide on the rubber sleeve, strip the wire, crimp on the metal connector, crimp the connector around the rubber sleeve, and insert the prong into the main plastic connector until you feel it click. Removal requires a special tang to release the connector, though there are circular tools with many tangs for various types of connectors. If you don't have the right tools, you'll be tearing out your hair before the connector comes apart.

required at a certain speed and loading to improve the calibration, you can program the change permanently into the EMS. Unless you are quite experienced, you will probably find this method more reliable and easier than trying to directly alter the raw fuel map—even with the help of a wideband air/fuel-ratio meter!

If you have a good AFR meter, it's a simple matter to set the engine speed and loading at a particular breakpoint and trim the module until the air/fuel ratio is where you want it (or where torque is optimized on a dyno). You can then stop, look at the percentage trim it took to optimize the mixture (or timing, or maximum boost), then modify the injection pulse width or VE-fueling at the speed-loading point (and smooth the points surrounding the chosen test breakpoint, since virtually all engine management systems *average* pulse width or timing with surrounding data to arrive at the final fuel or timing calculation).

To repeat, if you do not smooth the data in surrounding speed-density cells, you will *not* get the pulse width or timing you expected due to averaging. Most engine management systems will display the speed-density cell from which an engine is being managed as well as the current injection pulse width. If the value in the current cell is not the same as the actual pulse width being applied, there is either averaging going on with other cells, or some kind of warm up enrichment or closed-loop O_2-based trim going on. Bottom line, don't let anyone convince you that you don't need the rich-lean module.

I recommend you practice navigating through the menus (screens) of the laptop EMS user interface before attempting to tune the car and read the software manual several times, *memorizing* the features and commands. Things can happen very fast while tuning, and some can damage an engine. You do *not* want to be fumbling around reading the manual while you tune. Tuning a high-per-

formance vehicle involves a certain amount of risk. Lean mixture or detonation under high power can damage your engine.

Always save *all* your fuel maps so you can recover to an earlier fuel curve if you reach a point where things are not working well on the current map, or if you accidentally erase something (some systems like the Motec M48 virtually force you to do this).

Keep a detailed log of everything you did and change only one thing at a time. One good way to do this might be with a hand-held recorder, as long as background noise is not unacceptably loud.

FUEL AND TIMING MAPS

When installing an EMS you will need a map close enough to optimal to start the engine. Professional engine builders and tuners have collectively spent a huge amount of time building precise fuel (and spark) maps for various engines. It is always a complex and painstaking process to build a map from scratch if maximum power and efficiency are vital and if you don't have access to an exhaust-gas analyzer, exhaust-gas-temperature sensors, and a chassis or engine dyno.

Some pro tuners are capable of programming a fuel/spark map for an engine to an acceptable state of tune just by driving the vehicle on the streets and working with a PC or other interface. However, since engines will produce best torque within a fairly broad range of air/fuel mixture, it is difficult and time consuming to fully optimize both power and efficiency, particularly where it is not feasible to operate in closed-loop or at stoichiometric mixtures at less than wide-open throttle.

The good news is that many fuel maps already exist for common engines and modifications, not to mention plenty of uncommon engines with uncommon modifications. Keep in mind, any changes from the original engine configuration or even the vehicle configuration could potentially cause a prebuilt map to be less than optimal on your vehicle. What's more, an otherwise identical vehicle and engine configuration will require some rescaling of the prebuilt map if the injectors or fuel pressure are different from the engine on which the calibration was built. But it is always better to start with an optimized, preprogrammed map from a similar or identical engine if you can find one.

If this is *not* possible, today many engine management systems have software that will automatically build a startup cali-

The weld-in or epoxy-in injector boss can be critical for building a custom port-EFI engine management system, for example, if you need to convert a wet intake manifold to port injection or if you're adding additional injectors somewhere in the intake system. MSD Fuel

If you are building a custom or heavily modified exhaust, it is likely you'll need one of these weld-in O^2 sensor bosses. Painless Performance

bration based on a set of parameters you enter about your engine, vehicle, fuel, and ignition systems.

As a last resort, use the closest map you can find and then rework it for your engine. Even if the calibration is not that close, if it was a good calibration for a real engine, it will be better than nothing.

FUEL INJECTORS

Fuel injectors are usually *not* included in a standard aftermarket EMS kit, because they are not cheap and because they must match the horsepower and dynamic range of a particular engine application. As discussed in detail in the chapter on actuators, it is essential to select and install injectors that provide a small enough amount of fuel to idle properly without pulse width much below 1.5 to 2.0 milliseconds (below which the injection events are not precisely repeatable) and still provide adequate high-end fueling without going static, in which the injectors stay open constantly (which can overheat them) and can no longer meter fuel.

Injectors are rated in terms of resistance to current flow. Low-resistance internal coils, typically used in high-flow injectors draw more power and require heavier-duty drivers. Such injectors are usually a peak-and-hold design, with a higher current used to open the injector fast, and then a reduction to the minimal power needed to hold it open while reducing heat. Make sure your ECU has the capability to reliably power the injectors you plan to use.

Many ECUs have the ability to fire multiple injectors per injector driver without overloading the circuitry (it all depends on the resistance of the injectors, the current draw, and the capacity of the drivers). This means that you can potentially use two or more injectors per cylinder. It also means you might be able to add more injectors to handle more cylinders than the system was originally designed to handle if you're using batch-fire injection—just by wiring in more injectors onto existing drivers.

Some ECUs have the ability to *stage* injectors, providing idle and low-end fueling by pulsing one primary injector per cylinder, while providing wide-open-throttle fueling by pulsing one or more *secondary* injectors per cylinder once fuel requirements reach a certain threshold. This solution provides far greater dynamic range in the case of a forced induction engine or other motor with greatly differing idle and peak-power fuel requirements. Haltech ECUs have always been examples of an EMS with the ability to stage injectors based on engine loading so all loading points beyond a specified manifold pressure receive fuel from all injectors.

There are a number of aftermarket devices that provide fuel enrichment for factory EFI systems by staging one or more additional injectors (sometimes as many as four, six, or eight) located usually at a single point upstream of the throttle body. This solution can work reasonably well if executed carefully on the right engine with the right intake manifold. But theoretical problems are several: You are probably forcing a dry manifold to handle a wet mixture, which can cause distribution problems. And since you are probably providing enrichment for a power-adder conversion, lean cylinders can quickly lead to disaster. The extra injector(s) must pulse at a rate that provides correct fueling and at the same time provides consistent enrichment to all cylinders.

Some tuners have installed a standalone aftermarket programmable EMS to help an OEM injection engine with fuel delivery by providing fuel enrichment under turbo boost. Bob Norwood, of Norwood Autocraft in Dallas, applied a fuel-only Haltech ECU to control two gigantic alcohol injectors when building an Acura NSX turbo—successfully fueling 10 pounds of boost on an engine with a 10.5:1 compression ratio. Bell Engineering Group of San Antonio installed a fuel-only Haltech ECU and four additional injectors when building a super-high-output turbo system for a Mazda Miata. In both cases, the Haltech unit was not activated until the engine entered turbo boost, and the Haltech did not use temperature or TPS sensors since the factory injection system provides cold enrichment, acceleration enrichment, and so forth.

Electromotive TEC[3] Pre-installation Checklist

1. TEC[3] Computer
2. DFU(s)
3. Resistor Core Spark Plug Wires
4. TEC[3] Wiring Harness with Power Harness
5. Windows-based PC
6. Serial Connector Cable (DB9) for PC
7. Crank Position Sensor (Magnetic Sensor)
8. 60 (-2) Tooth Crank Trigger Wheel or 120 (-4) Tooth Cam Trigger Wheel
9. Coolant Temperature Sensor (CLT)
10. Manifold Air Temperature Sensor (MAT)
11. Manifold Air Pressure Sensor (MAP)
12. Throttle Position Sensor (TPS)
13. Exhaust Gas Oxygen Sensor (EGO)
14. Idle Air Control Motor (IAC)
15. Knock Sensor (KNK)
16. Fuel Rail(s) and Fuel Pressure Regulator
17. High-pressure Electric Fuel Pump
18. Fuel Injectors
19. Fuel Injector Wiring Harness
20. Throttle
21. Wire Terminal Crimping Tool
22. Shrink Tubing
23. Assorted Wire Crimp Terminals
24. Drill
25. Bolts for DFU(s) and TEC[3] ECU
26. Soldering Gun

CHAPTER 12

DESIGNING, MODIFYING, AND BUILDING INTAKE MANIFOLDS

Intake manifold design really matters. A great intake manifold design can provide really substantial performance advantages over a less optimal one. Clever designs have taken advantage of high air-flow characteristics, inertial effects, and pressure-wave tuning to achieve air density in cylinders that substantially exceeds atmospheric pressure *with as much a 10 psi positive pressure at the intake valve.*

Sooner or later, you may want to fabricate a custom intake manifold for a fuel-injected engine. Important intake manifold design goals typically include the following:
- Low resistance to airflow
- High air velocity for a given flow rate
- Excellent fuel and air distribution among the various cylinders
- Runner, plenum, and air intake sizes that take advantages of ram and tuning effects
- Overall intake volumes that combine good response with smoothing

A wet intake manifold—designed for handling both air and fuel flow for carburetion or throttle body injection systems—has substantially less freedom of design for optimizing airflow, because the overriding priority is always to keep the air and fuel well mixed and to distribute fuel originating at a single point as evenly as possible to all the cylinders. Single-carb or TBI manifolds use many tricks and kludges to correct air- and fuel-flow problems, and it is seldom feasible to build such a manifold with any method other than mold building and casting. Computational fluid dynamics is one of the more complex types of computer modeling and simulation, and there is still a lot of trial and error in manifold design. If you need a wet intake manifold, buy one.

On the other hand, building a tunnel-ram manifold for a port EFI powerplant is quite feasible, and construction may be rather simple on inline engines once you know the correct runner size and length for your application, as well as the ideal plenum volume. Of course, V-type engine manifolds are more difficult to fabricate, but, again, are simple compared to single-point wet manifold designs. This is because there is usually a way to design the runners so every one is perfectly symmetrical with the others in such a way that all runners have even lengths and take-off points, and all merge into the plenum in precisely the same way.

TYPES OF DESIGNS

Intake manifolds may consist of hollow *runners* that mate to the intake ports on a cylinder head. They may or may not merge together into one or more larger chambers called *plenums*, into which air enters through a *throttle body* or *carburetor*, which is connected to the air cleaner and atmosphere (or perhaps to a supercharger or turbo compressor) via an *intake ram tube*.

Whether wet or dry, there are several different classes of intake manifolds.

Dual-plane manifolds divide the cylinders into two groups, which are separated into a divided plenum or two smaller plenums. This type of manifold produces better low-end torque and better throttle response than most other manifolds.

Single-plane manifolds route all intake runners into a single common plenum, which smooths out induction pulses better than the dual plane design and typically provides better peak power at the expense of low-end torque. Higher-revving engines should use a single-plane intake design.

Tunnel-ram manifolds are a special case of the single-plane design, except that all intake runners are straight and meet symmetrically at a common plenum. This design is, of course, still good for top-end power, with the added benefit of intrinsically excellent fuel distribution. The tunnel-ram typically has a fairly large plenum, which tends to reduce the signal strength of induction (at a carburetor venturi) and hurt throttle response. Designers typically work hard with tuning effects to improve the responsiveness for adequate streetability.

Individual-runner manifolds eliminate the tunnel-ram plenum and equip each runner with its own throttle. There is a tremendous amount of pulsing through each throttle, which can actually improve low- and midrange performance due to increased peak runner velocities. However, with no plenum to dampen the induction pulses, high-rpm tuning is more problematic. There tends to be a lot of fuel standoff above each throttle on carbureted individual-runner manifolds, however.

Dual-resonance (Varioram) manifolds are basically a cross between a dual-plane and a tunnel-ram manifold. The system uses a large plenum divided into two smaller halves by a wall fitted with a large throttle plate. The throttle plate can close to route intake air from half the runners through a smaller dual-plane plenum that functions as a collector for matched sets of cylinders that Y together into a ram tube extending out to the throttle. When the plenum throttle opens at higher rpm, the dual plenums become a large single tunnel-ram plenum collecting all runners into a single-plane plenum. Both the dual-plane and tunnel-ram configurations use advanced tuning effects to optimize airflow, inertia, and resonation effects for improved performance in either mode of operation (see sidebar).

No matter which type of manifold is in use, swirl resulting from air moving past the throttle plate should be dampened out by a restrictor to avoid biasing airflow to certain cylinders. Similarly, there must be no sharp corners or edges in any of these manifold types in order to avoid turbulence or eddies that might bias volumetric efficiency to particular cylinders.

INTAKE (AND EXHAUST) MANIFOLD TUNING

The intake and exhaust systems of a running engine constitute a highly complex system of high- and low-pressure waves resonating and bouncing around inside the plumbing. Such waves are caused by the inertia of the air and the opening and closing of the valves at a certain frequency. If an engine designer can arrange to have high-pressure waves approach an intake valve just as it opens or closes (or to have *low*-pressure waves approach an *exhaust* valve from the exhaust system as exhaust valves open), volumetric efficiency will increase and the negative effects of valve overlap will decline.

A millennium-vintage problem: the unmodifiable plastic intake manifold. It's cheap to build, and it insulates charged air from engine compartment heat. This Ecotec manifold also has long, narrow runners optimized for low-end torque. GM

Pressure waves can actually move from the exhaust system through the exhaust valve and combustion chamber out through the intake valve into the intake manifold. Intake and exhaust tuning must harness these effects that force a little more charge into the engine as the intake valve closes or fight reversion when the intake starts to open. It is unfortunate but true that tuning that is optimal for one engine speed usually degrades performance at some other part of the power band.

Engineers use two main models to predict the effects of manifold tuning on engine volumetric efficiency. The first approximates a plenum-runner manifold with a quarter-wave organ pipe combined with a Helmholtz resonator, assuming that either component can operate either independently or in combination as a system. And also assuming that there are no more than four runners per plenum, and that the cylinders sharing a plenum must be even-firing, with pulses entering the plenum at even intervals. The second models manifold behavior as a function of wave motion within the manifold, varying according to rpm.

In either case, the intake manifold (and the intersections and terminations of its runners, plenum(s), intake ram tubes, and so on), valves, cylinders, and pistons constitute a complex system containing resonant cavities formed by cylinders with open ports and manifold branches leading to closed ports. Elements of the system change impedance (resistance to the net movement of alternating *gas flow*) where a pipe alters a cross-sectional area.

WHAT HAPPENS IN TUNED INTAKE AND EXHAUST SYSTEMS?

The intake system on a four-stroke automotive engine is designed to get as much charge air/fuel mixture into the cylinders as possible. One way to help with this is by tuning the lengths of the pipes.

When a piston is moving down and the intake valve is open, atmospheric pressure forces air into the cylinder at very high speed. When the intake valve suddenly closes, the column of intake air still has velocity and inertia, producing a shock wave as the air smashes into the valve, sending a pressure pulse backward into the intake manifold at the speed of sound. At certain frequencies of valve opening and closing events, the air column in the runner will start to resound or echo in a coordinated way otherwise known as resonance, which can cause energy from reflected pressure waves to combine with the force of static air pressure to push more air into the cylinder than would happen from air pressure alone. Designing intake pipes, plenum sizes, and runner size and length to take advance of positive and negative inertial and resonation effects to supercharge the volumetric efficiency of an engine is known as intake tuning.

RESONATION EFFECTS

• The exhaust valve opens and an acoustic pulse (+) goes down the pipe at the speed of sound (which varies according to exhaust temperature).

• The wave arrives at the end of the pipe and enters the atmosphere, triggering a negative pressure wave that travels back up the pipe. If the pipe length, rpm, cam overlap, and duration are tuned correctly, the low-pressure wave reaches the combustion chamber just before the exhaust valve closes.

• This helps suck out the last of the exhaust gas.

• The negative pressure waves continue on through the combustion chamber (valve overlap happening) and out the intake valve, where it helps to get the sluggish intake charge started into the combustion chamber.

• The negative wave travels up the intake port and into the tuned intake manifold runner where it again meets the atmosphere.

• At this point the powerful wave is once again reflected back as a positive pressure wave traveling back down the intake runner. If the length is correct, the remaining energy is expended, propelling extra charge into the combustion chamber just as the intake valve closes.

• Power increases are commensurate with multi-psi positive boost pressure.

• At all other rpm but the target "sweet" engine speed the effect is negative—subtracting volumetric efficiency—and making for a very peaky motor.

• Engineers use box structures in the intake system to dampen strong pressure waves and flatten the torque curve.

The air space in the intake-system box acts as a spring, and the air in the intake tube acts as a weight. At a target frequency the air in the tube vibrates back and forth without actually departing the tube. The volume of the box and the length and diameter of the tube determine the frequency of the system. Strong pressure waves excite the box into resonance at a matching impedance, dampening the pressure waves and flattening the torque curve.

This is a complex situation. Another manifestation of intake tuning is as follows:

When an intake valve is open on an engine, air is being pushed by atmospheric (or boost) pressure into the combustion chamber at very high velocity (sometimes at supersonic speed). The piston creates a negative pressure wave as it recedes down the bore that travels through the intake valve until it hits the plenum, where it is inverted and returns to the intake valve as a positive pressure wave that can help supercharge air into the cylinder if it arrives at the right time while the valve is still even partly open.

When the intake valve suddenly closes, incoming air slams to a stop against the valve and inertia causes it to stack up on itself, forming a region of high pressure that makes its way up the intake runner away from the cylinder in a high-pressure wave. When the pressure wave reaches the end of the intake runner, a negative pressure wave will be produced that bounces back down the intake runner, crashes into the now-closed valve and is reflected back a second time as a negative pressure wave that again moves up the runner toward the plenum at the speed of sound (1,250–1,300 feet per second in hot intake air). At the plenum the wave is again inverted—this time to a positive pressure wave—which returns again to the intake valve where it can help boost air into the cylinder, given the right engine speed.

A well-tuned intake will have a high-pressure wave at the intake valve as it's opening and again at the same time the engine is in its overlap period with both intake and exhaust valves open. If the exhaust is tuned to the same rpm range, there will be low pressure in the exhaust at the same time due to scavenging and negative pressure waves returning from the exhaust system to the vicinity of the exhaust valve.

With the intake above atmospheric pressure and the cylinder much lower than atmospheric, air blasts into the cylinder, causing a rapid drop in pressure that moves away from the cylinder through the intake. On the heels of this, as the piston accelerates down the bore, a second negative wave moves through the intake valve into the manifold, one right after the other. The various pressure waves can create as much as 10 psi of positive boost pressure at the intake valve, which can raise volumetric efficiency above 100 percent in high-rpm racing engines.

How long would an intake runner have to be to take advantage of pressure-wave tuning? Assume the engine is turning over at 5,000 rpm. On a four-stroke engine, each intake valve opens once every 720 degrees of crankshaft revolution. Assume it stays open for 250 degrees. This means that there are 470 degrees between the intake valve closing event and the next opening event. It will take the engine 0.012 seconds per revolution at 5,000 rpm, and since 470 degrees is about 1.31 revolutions, each intake valve will remain closed for 0.0156 seconds. At the speed of sound—1,300 feet per second—the pressure wave would travel about 20 feet in this time, which is 10 feet up and 10 feet back.

Unfortunately, the tuning effect is only true within a narrow range—around 5,000 rpm—meaning the pressure wave will be out of sync at, say, 3,250 or 4,250 rpm. Variable intake systems like Porsche's Varioram can create more than one sweet spot, but peaky power is just the name of the game in pressure-wave tuning. The other problem is that 10-foot runners are way too long to fit in a car.

But any even divisor of 10 feet—say 2.5 feet, or 30 inches—allows the pressure wave to resonate up and down the pipe four times before the intake valve opens again, arriving at the valve at the right time.

Intake aerodynamics are complex and tricky and offer trade-offs. High-velocity intake air is a good thing, because intake air moving into the cylinders at high speed has more momentum and increases turbulence that helps the air and fuel mix and increases the speed of combustion and helps fight detonation. But if you increase the air velocity by forcing intake air to enter the engine through a smaller-diameter intake runner, at higher engine speeds with high-CFM airflow, small runners will produce a pressure drop that degrades volumetric efficiency. So it is better to have large-diameter runners at peak-power engine speed.

Some four-valve engines are designed with primary-secondary runner systems that use one (sometimes smaller diameter) primary runner at lower engine speeds when super-high charge velocities will improve cylinder filling. A butterfly valve blocks the secondary (sometimes larger diameter) runner at lower speeds, then opens up at higher engine speeds to reduce the pressure drop at peak power, increasing maximum horsepower.

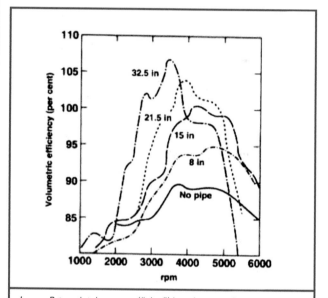

Jaguar D-type intake runner ("pipe") length comparison plots volumetric efficiency for various engine speeds with 32.5, 21.5, 15, 8, and zero inch pipes, indicating the effects of inertia and resonation effects on cylinder filling. Clearly, the two longest pipes achieved a supercharging effect below 4,000 rpm, while the 15-inch pipe was better above 4,000. The 8-inch and zero pipe were suboptimal to at least one of the longer pipes under all conditions.

RUNNER LENGTH

Longer, narrower intake runners are better for optimizing low-rpm performance because they have a lower resonant frequency, and the smaller diameter helps increase the air velocity and enhances inertial cylinder-filling effects at lower engine speeds. Shorter, wider runners are better for high rpm because they have a higher resonant frequency, and the larger diameter is less restrictive to all-out airflow.

Grape Ape Racing (grapeaperacing.com) provides a workable method of computing intake runner length using wave motion theory, which is designed to provide reasonable results compared to known-good tuned manifolds.

Begin by computing an effective cam duration (ECD), which typically involves subtracting 20 to 30 degrees from the advertised cam duration. Use 30 degrees for solid-cam drag motors, 20 degrees for less radical powerplants.

researchers in this example were working with a Suzuki motorcycle engine with head ports estimated at 13 centimeters in length. This length had to be subtracted from the total runner length to derive the intake manifold portion of the runner length. At 10,500 rpm, using the second set of pressure waves to reduce volume, improve response, and save space, the total runner length is 28.5 centimeters. So the intake manifold runners would be 28.5 – 13 = 15.5 centimeters.

Manifold designers should terminate manifold runners in the plenum with trumpeted stacks to reduce the loss of air entering the runners.

Runner Area

In contrast to tuning runner length (which affects power in a narrow rpm range), tuning the diameter or area of the runners affects power at all engine speeds. If the port is too small it will restrict top-end flow. If too large, cylinder-filling velocity will suffer, and this will hurt low-end power. As a rule, the larger the runner, the less strength pressure waves will have.

Ecotec turbo dragger powerplant with GM Racing custom manifold. When GM Racing decided to build this super-duty Ecotec powerplant, the plastic intake had to go, and the name of the game for super-high boosted power in the 600–900 horse range is shorter, larger runners. The shorter runners and space-frame chassis permit flexibility of manifold design, and the shorter-than-stock runners simplify geometry compared to the long plastic runners of the stock 2.0-liter Ecotec. The manifold is a rather simple construction, with throttle body and head flanges, four straight pipes, and a round plenum. Runner length and diameter and plenum volume are critical to optimal power, as is radius-ing the various piping that enters and exits the plenum.

ECD = 720 - Adv. duration - 30
For a 305-degree duration race cam:
ECD = 720 - 305 - 30
The ECD of that cam would be 385.

The formula for optimum intake runner length (L) is:
L = [(ECD x 0.25 x V x 2) ÷ (rpm x RV)] - (D ÷ 2)
Where:
ECD = Effective Cam Duration
V = Pressure wave speed (400–1250 ft/s)
RV = Reflective Value – the utilized pressure wave set
D = Runner Diameter (inches)
If an engine with a 305-degree race cam needed to be tuned to 7,000 rpm using the second set of pressure waves (RV=2) and had a 1.5-inch diameter intake runner, the optimum runner length formula would look like this:
L = [(445 x 0.25 x 1300 x 2)÷ (7000 x 2)] - 0.75
19.91 inches would be the optimal runner length.

The equation produced the above graphs, verifying the requirement for longer runners at lower engine speeds to produce tuning effects as well as indicating the effect of using various reflective values.

Note: Runner length includes the length of the intake passage from the manifold plenum all the way through the head intake port to the intake valve, which includes the set-length passage from the manifold gasket to the valve in the cylinder head. The

Since the intake valve is the most restrictive part of an intake system, intake runners should be sized to optimize airflow through the valve area. Equipped with a camshaft that is a good match, most respectable cylinder heads will flow air through the valve area equivalent to that of an unrestricted port with about 80 percent the area of the valve(s), while some excellent racing heads can be ported to flow 85 percent of the valve area. For example, a 2.02-inch valve with 3.20 square inches of area will flow the same as an open tube with about 2.56 square inches of area (this is 80 percent of 3.20). Therefore, the port area should be about 2.56 square inches in the area of the port leading directly up to the valve.

INTAKE PORT TAPER
For best cylinder filling, it helps to have a high flow velocity at the back of the valve, because faster air exerts less pressure (this is the Bernoulli effect). To increase velocity, the intake port can be tapered. To be effective, there should be a 1.7–2.5 percent decrease in intake runner area per inch of runner approaching the valve, which represents a 1–1.5 *degree* taper.

For example, let's say you want a 2 percent increase in taper per inch of runner for the 2.02-inch valve discussed above, which has a port area of 2.56 square inches near the valve. A 10-inch runner with a 2 percent taper and final area of 2.56 inches near the valve should start out with 3.12 square inches of area 10 inches away from where the port joins the plenum. As taper approaches 2.5 percent, you are at the limit of what taper can do to help airflow.

PLENUM VOLUME

The purpose of a plenum is to permit charge gases to slow down and gain density, and to help isolate the resonant pulses of the individual runners from a desirable overall smooth flow of air through the throttle. An intake manifold plenum can be effective over a fairly wide range of volume, and general design rules work pretty well in calculating optimal volume. This lack of sensitivity is fortunate, since there is no simple answer to the plenum volume requirement for a given application or rpm range. But general rules work fairly well. The tradeoffs involve supplying sufficient plenum volume to deliver plenty of airflow versus keeping the volume low enough to keep throttle response from becoming sluggish.

The rule for plenums without tuned runners and with airflow restrictors is to design an intake plenum with one to two times the engine displacement in order to provide a large amount of available reserve air. In this case, air volume acts as a low-pass filter (capacitor) between the engine and the restrictor, smoothing the flow through the restrictor. Without a plenum, the engine would choke the air through the restrictor with each intake stroke of a piston. Using this method with the above example predicts a plenum of 1.2 liters.

The rules change when a Helmholtz resonator is being used. The Helmholtz plenum makes a dense charge by use of pressure waves, using the same physics as a tuned-port intake runner.

For engines operating at a maximum rpm in the 5,000 to 6,000 range, Grape Ape Racing recommends the following guidelines for plenum volume: V-8s sharing one large plenum do not work well as far as resonation-effects tuning, but the optimal tuning size would be 40 to 50 percent of total cylinder displacement. For a four-cylinder, 50 to 60 percent of displacement works well. For three cylinders or a six with twin plenums, each plenum should be 65 to 80 percent of the displacement of the three cylinders being fed. For engines operating closer to 7,000 to 7,500 rpm, reduce plenum volume by 10 to 15 percent. For a power boost at 2,500 to 3,500, the plenum should be 30 percent larger.

Using a tuned plenum and tuned intake pipes requires a plenum volume of approximately 300 cc in the previous Suzuki example, which applies specifically to engines that use tuned intake pipes.

Intake Ram Pipe Diameter

The velocity in the plenum intake ram pipe should always remain at less than 180 feet per second at maximum rpm and horsepower.

Grape Ape Racing's formula to compute the diameter of a plenum intake pipe is:

$$D = SQR\ (CID \times VE \times rpm) \div (V \times 1130)$$

Where:

D =   Pipe diameter
SQR =   Square root of
CID =   Cubic inch displacement
VE =   Volumetric Efficiency
V =   Velocity in ft/sec

(for liters, change cubic inch displacement to liters and the constant to 18.5.)

Example

153-ci four-cylinder engine with 85 percent VE, revving to 6,000 rpm and 185 ft/sec:

$$D= SQR\ (153 \times 0.85 \times 6000) \div (180 \times 1130) = 1.96\ inches$$

If you are using more than one throttle bore to feed the cylinders, the ram pipes associated with each bore should be half of the total required. To get the diameter, use the formula for the area of a circle–Area = π r²–in reverse as: Diameter = 2 x SQR (Area ÷ π).

Where SQR = Square root

A dual-bore ram pipe for the above example would replace a single 3.02 02–square- inch 1.96-inch diameter pipe with twin bores each of 1.39 39-inch diameter.

Intake Ram Pipe Length

Grape Ape Racing suggests adjusting the length of the intake ram pipe last, using a pipe that can be easily adjusted in length for dyno or airflow testing

Start with a 13-inch-long ram pipe to provide airflow enhancement at about 6,000 rpm. For each 1,000-rpm change in speed from 6,000, add or remove 1.7 inches, shortening the pipe for higher rpm. The inlet of a ram pipe should have at least a 1/2-inch radius on the intake for smooth flow.

On the dyno, experiment with 1/2 inch adjustment in either direction to see what this does for *both peak and average power and torque.* Grape Ape suggests tuning the intake ram pipe about 1,000 rpm lower than the intake runner length tuning speed for increasing average power. Obviously, a target speed of 8,000 rpm does not require a very long ram pipe.

HOW TO FABRICATE AN INTAKE MANIFOLD

(This information was provided by Simple Digital Systems)

This section details the complete fabrication process on a custom intake manifold for a developmental Mazda RX7, but the basic design can be easily adapted for almost any engine design. Required tools include a drill press, lathe, bandsaw, hole saws, files, and a TIG welder. SDS constructed this manifold from mild steel for ease of assembly, but in most ways aluminum is easier to fabricate and is the material of choice for intake manifolds.

The plenum type design uses a single throttle body attached to a plenum with separate intake runners for each port. Port type injectors are mounted in each runner.

The main flange is constructed from 1/4-inch steel plate to reduce warpage during welding. An intake manifold gasket can be used as a guide to trace the flange outline and mark the port and bolt holes, and the flange plate can be cut out using a bandsaw or plasma-arc torch. Bolt holes can be drilled on a drill press. Intake runner tube holes can be made with a hole saw or plasma arc if they are not round. Try to design the port holes so the ID of the tubing used for the runners is the same as that for the finished port size in your head. If this is not possible, you can probably open up smaller runners with porting or using the extrude-hone process. The runners should be spigotted inside the flange hole for ease of jigging.

Runners are made from 0.050 to 0.065 tubing. Runner lengths can be adjusted within the space constraints to help boost torque within the desired range. Short runners are good for high-rpm torque as used in racing. Long runners are more applicable for street use at lower rpm. For most street engines, try to keep the length from the end of the runner to the valve at least 9 inches long and preferably longer. Available space usually limits this dimension when using straight runners, so curved runners using 90-degree mandrel bent tubing can sometimes be used to increase the runner length.

As applied to engines with oval or square ports, you will have to form the tubing into the port shape, which is considerably more complicated. Tubing should be cut off at precise lengths with a tubing cutter and carefully deburred. For maximum airflow, tapered runners with velocity stacks inside the plenum can be made if you can do this type of work. While very time consuming, this design shows a 20–25 percent increase in flow over straight tubing runners.

Depending on engine displacement and the throttle body used, the intake plenum is typically constructed from 3-inch diameter 0.065 wall tubing for engines under 1,600 cc or 4-inch diameter for most other engines. Holes must be cut in the plenum tubing with a hole saw for the intake runner tubes and throttle-body tube. The throttle body may also be placed directly on one end of the plenum in place of an end plate, depending on component layout. The RX7 manifold uses 4-inch tubing. The plenum should be made at least 2 inches longer than the dimension of the front-to-rear outside intake ports to allow better airflow to the outer ports. This is a good time to plan and drill any manifold vacuum ports for brake boosters, MAP sensors, and so on.

Build end plates from 0.050–0.066-inch plate stock to cap the plenum ends. Cut them out with a bandsaw or plasma arc. Make them about the same size as the plenum tubing OD. If you plan to mount the throttle body on one end of the plenum, this end cap should be made from 3/16–1/4-inch plate to allow for tapping threads to hold the throttle body. If the throttle body will be mounted on a ram tube, use 1/4 plate for the TB flange. Tap directly into the plate with fine threads for maximum strength.

Throttle body options are numerous. Select a throttle body that can flow the desired amount of air and fits within the space limitations. Think about the suitability of the throttle body for plumbing in an air cleaner, intercooler, or turbocharger. Too large a throttle body for street use will often mean a very sensitive tip in throttle response, which can hurt drivability. There is also no need for a giant 3.5-inch throttle body if your turbo plumbing is only 2.25 inches. Acquiring a throttle body with a potentiometer-type TPS already installed will save time and trouble. For many applications, Ford 5.0-liter and 4.6-liter V-8 throttle bodies work well. They are cheap, available, and relatively compact. Ford Motorsport and Edelbrock make larger sizes in 5-mm increments. The 4.6-liter throttle bodies are compact and have a minimal amount of extra clutter. All Ford throttle bodies have a potentiometer-type TPS sensor. Ford 65-mm 4.0-liter Explorer throttle bodies are easily available used or from a Ford dealer. These are relatively inexpensive and include minimal garbage on the outside. Put some thought into a throttle linkage at this point also.

SDS recommends fabricating injector bosses out of 1-inch bar stock, and sealing the vacuum side of the injector with a 5/8-ID, 3/4-OD O-ring slid over the nose of the injector body. You can pull off the stock O-ring and pintle cap. Bore the bar stock to 0.640–0.650, straight through. This allows a slight air gap between the boss and the injector to reduce heat transfer and fuel boiling. Machine a 0.740 counterbore, 0.040 deep at the end for O-ring retention and sealing. Cut off the bosses at 45 degrees so that they are about 1.35–1.5 inches long. This is a

good entry angle for many injectors into the runner and is an easy angle for sawing.

The fuel rail is typically made from 3/4-inch square tubing with a 0.050 to 0.065 wall. This shape allows much easier drilling and jigging. One-quarter-inch holes can be drilled at the port spacing interval. Over the top of these holes will be welded the upper O-ring bosses machined from 3/4-inch bar stock. These will usually be about 1/2-inch long and have a 0.540- to 0.545-inch hole drilled for the O-ring to seal on. Carefully chamfer the entry side for easier O-ring fitment. End caps for the rail are made from 1/4-inch plate and tapped for 1/8-inch NPT fittings. For a more detailed description on building a fuel rail, go to the article on our tech page.

Once all of the pieces are cut out and deburred, if possible clamp the head flange to a scrap cylinder head to reduce warpage during welding. Carefully jig the plenum and runners to the flange and measure for straightness. Tack weld the pieces lightly and recheck for straightness. You should still be able to move things around a bit at this stage. Tack the opposite side of each joint and re-check again. Once you are satisfied, TIG weld all of the joints. Weld on your end caps and TB flange. Let everything air cool.

Now cut holes in the runners for the injectors. Scribe a line across the runners where you want the center of the injectors to be. Now intersect each mark on the runner with another line down the center of each tube. Punch mark and drill a 1/4-inch hole. Now drill straight through with a 1/2-inch drill. Once you pierce the tubing, slowly lean the drill over at 45 degrees to oval the hole. The injector bosses must be positioned over these oval holes; make sure before starting that you have enough room around the boss and flange so you will be able to get the welder in close proximity for welding.

Assemble the injectors into the rail and slide the bosses over the injectors. Carefully align the assembly so the injectors are dead center through each injector hole in the runners. Clamp in position and recheck. Lightly and quickly tack each boss in position. As soon as this is completed, either pull the injectors out or water quench each tack to avoid heat damage to the injectors (*very* easy to do). Once you are satisfied that everything is straight, remove the injectors if they're still there and finish welding the bosses. Be aware that the bosses must be very straight and at identical depth for proper sealing.

You can weld bolt-down tabs onto the fuel rail and weld attachment points for the bolts onto the manifold. A couple of long, 1/4-inch bolts usually suffice. The manifold can now be thoroughly cleaned with soap and hot water, then lacquer thinner. A quality engine enamel or powdercoating will provide a nice-looking finish.

CHAPTER 13
EMS TUNING 101

HOW PROGRAMMABLE DIGITAL FUEL AND TIMING SYSTEMS WORK

Programmable engine management systems permit (and require) a user to calibrate the basic raw fuel and ignition curves for an engine by defining tables of numbers that specify spark timing advance and base fuel delivery for pulse width–modulated (PWM) electronic fuel injection as a function of engine speed and airflow. Such tables are typically represented to the user (tuner, calibrator) in graphical or tabular format via user interface/editing software running on a Windows computer (usually a battery-powered laptop).

This PC-based engine management software simultaneously controls communications-driver software that has the ability to interact, via serial data cable, with the onboard electronic control unit (ECU) that actually manages the spark and fuel requirements for an engine. The software allows the user to move data back and forth between ECU memory and the laptop for tuning purposes.

Typically, programmable engine management systems provide both the ability to interactively modify many operating parameters and table cells in the ECU while the engine is running, as well as the ability to upload the entire set of all engine management tables into a Windows PC for subsequent offline calibration and eventual download back to the same or alternative ECU(s).

A programmable engine management system is, therefore, comprised of the onboard ECU, firmware, and calibration data, plus a Windows laptop computer (or, in a few cases, an alternate dedicated interface module), laptop engine management software, and data-link hardware. For full functionality of the system, *all* components must be present and operational. Once the ECU is calibrated, it is fully self-contained and will run the engine without additional components. However, without the laptop present and connected, the ECU is in no sense programmable.

Although user-interface software of various engine management systems may present internal engine management tables to the tuner with its own unique graphical appearance—and although the underlying resolution or granularity of the tables may, in fact, actually vary—behind the scenes the basic engine management operating concepts are exactly the same: When it is approaching time for the digital microprocessor managing an engine to schedule the next fuel injection or spark event, the processor first checks engine sensor data to get a snapshot of how fast the engine is turning and how much air it is ingesting.

The ECU then performs a series of look-ups to the appropriate entry for the current speed and airflow in the various fuel and timing tables to determine when to open and close the injector(s) to spray a pulse of fuel into one or more cylinders and how many degrees before top dead center to trigger an external igniter or coil to fire the next spark plug. If engine speed or loading fall somewhere in between two specified points in an engine management table, the ECU will provide base fuel or spark or corrections by applying a weighed average of the closest several points on either side of the actual value.

Regardless of the exact means used, the process of calibrating an engine management system on a new engine has the ultimate goal of optimizing torque, fuel economy, or emissions at each and every breakpoint of engine speed and loading defined in the fuel and ignition tables (there are usually between 24 and 800 such points).

In theory, this can be a laborious process, but it is one that can frequently be expedited with the use of automated map-builder computer modeling algorithms, pre-existing calibrations from similar engines, intelligent extrapolation and interpolation between rpm ranges and load points, and even self-learning routines that make use of wideband O_2 sensors and dynamometers. Nevertheless, it is worth keeping in mind that OEM carmakers frequently spend months or even *years* calibrating the engine management systems for a particular engine in a particular vehicle.

John Carmack's twin-turbo Ferrari F50 on the Norwood Dynojet, getting ready for a Car and Driver *magazine testing regime. Bob Norwood tweaked the Ferrari's Bosch Motronic system at lower power settings, but when it came time to crank up the boost to higher levels on the $750,000 vehicle's 10.5:1 compression 60-valve V-12, the Ferrari converted to Motec EMS for ultimate control.*

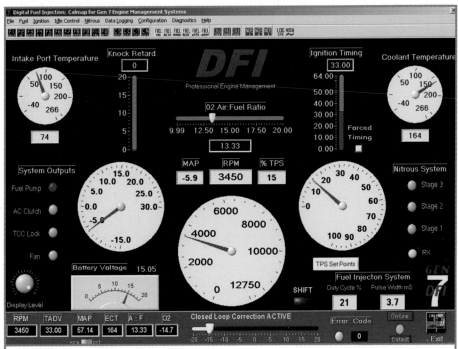

Laptops are great for displaying EMS sensor status in a highly readable format like this "virtual dash" screen from Accel's Gen 7 version of CalMap EMS software. There is a lot of great data on this page, and it's very readable. In some cases, you can set alarms to warn of dangerous conditions.

The magnitude of this task is ameliorated by the *sensitivity* of gasoline in combustion—the spread between the air/fuel ratio producing lean best torque (LBT) and rich best torque (RBT)—that is, the *range* of air/fuel ratios over which the engine produces best torque. Best torque is *not* a single point but rather a *range*. In a normally aspirated powerplant that does not require surplus fuel for combustion cooling and knock control, the spread between RBT and LBT at wide-open throttle is typically 11.5 to 13.3, with mean best torque in the 12.0 to 12.5 range. This spread narrows at high engine speeds as highest possible flame speed becomes increasingly critical to achieving complete combustion in the time available.

In addition to providing users the facility to define and edit engine management tables from scratch, programmable engine management systems provide the user with a routine to upload complete pre-constructed library maps stored on disk in a laptop to the onboard ECU. Similarly, a user can upload entire ECU-resident maps into the Windows PC for viewing, analysis, modification, or export to other engine management ECUs. Factory OBD-II ECUs provide this same type of capability (though it is password protected by a security seed). Older factory systems typically required a PROM change to install a new calibration.

Virtually all engine management systems now give a tuner the ability to dynamically modify a wide range of fuel, timing, and enrichment maps, on the fly, *while the engine is running*. A tuner does this by editing a graphical or tabular representation of a given map on the laptop screen. The tuner then launches an immediate ECU update, at which point the modified data is uploaded to the ECU at high speed to replace the appropriate parameter or table item for directing engine management operations for the next combustion event.

Many programmable engine management systems allow a tuner to manipulate a mixture trim device consisting of one or more linear potentiometer dials in order to dynamically enlean or enrich fueling—usually across the board or sometimes just at idle or wide-open throttle—by a percentage determined by the position of the dials. This is analogous to turning the mixture-adjustment screw on a carburetor with a screwdriver. Such trim can be a valuable tool in making an instant determination (via air/fuel-ratio meter, exhaust gas temperature gauge, dyno display, or other tools) of the empirical effect of changes to fuel, timing, or maximum boost and determining the percentage change that is required to fix a problem area when reprogramming the ECU's internal tables.

In addition to calibrating base fuel and ignition tables, programmable EFI systems all offer the ability to define special enrichment during acceleration (when the engine suddenly changes speed, upsetting the internal equilibrium of air and fuel in the intake system) and cold-running/startup (when the engine temperature has not yet reached normal operating ranges and the heavier fractions of gasoline do not vaporize well). Most systems also allow configuration of fuel or ignition corrections for changes in air temperature.

Systems like Accel's Gen 7 break down cold-running enrichment into cranking enrichment, after-start enrichment, and warm-up enrichment. DFI and some other systems provide ignition control, nitrous control, knock sensor ignition retard, and idle speed correction via stepper-motor-controlled throttle air bypass. These are all controlled via PC screen menus. Some sophisticated engine management systems provide compensation for variations in fuel-rail pressure and individual-cylinder variations in volumetric efficiency, temperature, and injector flow.

The majority of engine management systems permit a tuner to directly define fuel injector open time (pulse width) in cells of one or more speed-density fuel tables as time in milliseconds (and tenths or hundredths of milliseconds). However, several sophisticated engine management systems provide a facility that is at once elegant, yet less straightforward to some people, utilizing a concept called the VE table.

In a VE table EMS, the tuner begins by specifying a maximum injection time at full engine loading with a specific injector size (occurs at peak torque rpm) and may then define and tweak the minimum injection time for the engine at idle if it varies from the linear interpolation between peak pulse at full load and a zero pulse at zero load. These two points define pulse width at all loading points in between, with engine speed assumed not to be a factor due to a perfectly flat torque curve at all engine speeds.

However, since the torque curve will *not* be perfectly flat across the range of operation on most engines, the VE-table tuner must provide for offsets to the raw computed linear pulse width values for speed-load sites with sub- or super-normal volumetric efficiency.

BEFORE TUNING: SKILLS AND TOOLS REQUIRED TO GENERATE A SCRATCH CALIBRATION

"While we believe that our system is the simplest on the market to install and program, it is still beyond some people's capabilities. This is not a magic bullet that will solve all of your problems without any work. Have realistic expectations and goals for your project. We get many people contacting us saying that they intend to have a streetable 10-second car that gets great fuel economy and idles like a watch. This probably will not happen in reality. Evaluate your skills and knowledge. . . .

"After installation, you must program the system properly. Even though we supply very complete instruction manuals, if you have no idea about the concepts of rich and lean, basic EFI theory or ignition timing requirements then you will probably not be able to program the system properly. We suggest that you explore our Tech Page and Manuals online thoroughly to gain as much understanding of basic theory and the system as possible before you decide to buy.

"If things seem way over your head after this, our system may not be for you. You will need patience and understanding to be successful with our system. There is a learning curve with all systems, including SDS. Be prepared to spend some time understanding the various parameters and what effect they have."

An alternative to the laptop is a dedicated programmer module like this user interface from SDS or a similar unit from Edelbrock and one or two other firms. You can accomplish most of the functionality of a laptop, but you are peering through a tiny window into the EMS environment. Of course, once the system is calibrated, the user interface is of no consequence when driving.

Simple Digital Systems

To tune a running engine that has an unfamiliar engine management system without making things worse or risking damage to the engine while you learn, you're going to need to gain a working familiarity with the engine management software that enables you to quickly grasp what is happening and then react quickly. You definitely do not want to be trying to read the manual to find out what to do when detonation or thermal loading is threatening a meltdown on your new hot rod powerplant. And for that, you're going to need patience, judgment, and experience with the software and ECU data structures, as well as familiarity with the datalogging software you will *definitely* need in order to calibrate power-adder engines in boosted-power ranges.

The ability to work fast without making stupid mistakes is important, because running engines do not necessarily remain in a steady state: They heat-soak in all sorts of interesting ways that can affect the tuning and power. The oil gets thinner. Resistance to detonation degrades. Engines in a dyno cell pollute the air with hydrocarbons, carbon monoxide, and heat, which affects intake air temperature, volumetric efficiency, intercooler efficiency, combustion quality, and other running parameters. The ability to react quickly is important for reasonable tuning efficiency. Once an engine warms up out of a certain range on the warm-up enrichment map, for example, it will probably be hours before it cools off enough to give you another crack at calibrating the colder start-up temperature ranges.

If you're buying dynamometer time for tuning, slow tuning abilities can be costly. Slow reaction time can even be dangerous. A poor state of tune can damage the engine if it takes the tuner too long to recognize and figure out how to correct the problem, or if the correction is in the wrong direction. As Ray Kurzweil points out in *The Age of Spiritual Machines*, human beings are lightning-fast at recognizing a pattern and taking specific preplanned action (like realizing a kid has run in front of your car and hitting the brakes), but we think quite

slowly when faced with new situations that call for reasoning through a problem.

To calibrate most programmable engine management systems, you will first need to have at least some familiarity with the operating environment of a Windows laptop PC. In the first place, you're going to have to install the EMS software that you'll be using to interface to the onboard engine management system in order to calibrate engine control tables and parameters. Later, you will need to use the laptop keyboard to enter configuration data and to use the arrow keys or pointing device (mouse, trackpoint, and so on) to change the shape or height of fuel and ignition graphs, and so on.

Once the software is up and running, you'll need sufficient familiarity with a given engine management software environment to be able to dynamically track (while the engine is running) which portions of the main fuel, main ignition, cold-running, and transient enrichment maps are being used by the engine management system to manage the engine. Virtually all engine management systems provide some kind of engine data page that can be used to monitor engine sensor values and actuator commands (such as current injection pulse width and ignition timing), and determine the map table cells and corrections currently being used in the engine management calculations.

The main fuel, main ignition, engine temperature compensation and other tables typically provide some sort of highlighting, bolding, pointer, or other indication of which cell or graph is the current one in a particular table. Many systems have some sort of hot key that a tuner can press to instantly jump to the currently active map or map cell for immediate editing.

So you're going to have to practice working with the user interface of the EMS software and calibrating the parameters and tables.

The trouble is, even if you understand the software environment, you still have to have tuning skills and diagnostic tools to do the job well. Professional tuners persistently complain of computer-savvy customers with editing access to programmable engine management systems and *little or no tuning experience*.

These customers insist on tampering with a good calibration in an effort to make more power and succeed only in degrading the state of tune. They then must bring their vehicles back to the pro tuner to be fixed. This sort of unintentional de-tuning is so common it's virtually epidemic, and it's not always rank amateurs who are involved.

A tuner with *a great deal* of experience with a particular type and brand of engine with specific types of modifications and a specific brand of engine management system may be able to do a reasonable calibration based on past knowledge. Otherwise, you *must* have the right diagnostic tools. Here's the thing: *It is not possible for most people to tune an engine well by seat-of-the-pants, "butt-dyno," trial-and-error methods.*

A fact of life is that the *less* experienced you are, the *more* you need sophisticated tuning tools like fast, wideband AFR meters, multi-gas analyzers, exhaust gas temperature sensors, and chassis dynamometers to do an adequate calibration. This is even truer for engines with nitrous or forced-induction power-adders, but it's also true of the simple stuff, such as tuning well for a good idle: How do you tell the difference between a rich mixture that's producing a rough-running engine and a lean mixture that is causing the engine to misfire? Many people would not know the difference unless they could see rich black smoke. Ironically, the more you need diagnostic tools, the less likely you probably are to have them.

Experienced or not, you'll need to be very organized, patient, and logical, as well as familiar with scientific methods of experimentation (for example, changing *one*—and only one—variable at a time so you establish true cause-and-effects in an unambiguous way). Like a scientist doing an experiment, you'll need to keep good notes for later analysis and tracking.

On occasion, you'll need the patient ability to shut down the motor, sit quietly, and think logically. What are the possible reasons an event may have happened? What are the implications of the event? What are the possible actions? Are there any risks? What are the possible outcomes of various actions? Which is the highest priority action to try first?

To effectively calibrate an engine with power-adders without damaging it, you'll need a strategic tuning plan *firmly rooted in good modeling theory*. Particularly on a power-adder engine, you *must* start with a timing map *known* to be conservative (on the retarded side of the scale, but not ridiculously so), and a higher-load fuel map *known* to be safely on the rich side. This will require either a calibration from an identical or nearly identical engine, or *calculated* maps built with modeling software or hand calculations.

Even then, you'll need to sneak up on boosted conditions, datalogging quick forays deeper and deeper into boosted territory, analyzing the data, and correcting encountered speed-loading points *and points beyond them at higher boost* you have not yet actually encountered when timing and air/fuel ratios are headed in the wrong direction (one thing you can count on is that higher levels of boost will *not* want less fuel and more timing).

Even with plausible conservative maps, when you complete the easier idle and light cruise tuning and are ready to tune for high specific power on a boosted engine, it is essential to arrange for availability of a fast, wideband AFR meter, preferably EGT feedback, and a dyno or strap-on inertial accelerometer like the G-Tech.

If at all possible, there should be a good way to correlate the wideband AFR data to logged sensor data from the EMS (the Motec M48 EMS used in several projects in this book will log wideband O_2 data). Things are happening too fast at high boost to make corrections, and when an acceleration run is over, it is very difficult to figure out what was happening when without a datalog.

Normally aspirated street powerplants seldom generate enough thermal and mechanical loading to hurt themselves with poor tuning unless the compression ratio is very high and you are running the vehicle at top speed and wide-open throttle for a long time. Not so with boosted powerplants at higher loading, when small mistakes can damage the engine catastrophically.

Incorrect spark timing and lean fuel mixtures can induce immediate severe knock that can, in turn, overheat the combustion chamber and result in uncontrolled pre-ignition. Boosted and nitrous-injected engines can break ring lands almost immediately if the engine knocks due to incorrect calibration parameters (and *will* break the apex seals on a rotary engine like the RX-7 *immediately*; as they say, a rotary engine will knock *once*, and then be broken).

But there is more to think about than the fuel and ignition maps. It will not help to get the ignition timing and injection pulse width *perfect* if required fuel exceeds the capacity of the fuel supply system such that fuel rail pressure drops, or if the pulse width required to get a certain weight of fuel in the cylinders exceeds available time for injection at higher rpm and the injectors go static (constantly open).

Nitrous is a special case, and nitrous tuning is both easier and more difficult than other types of calibration. Of course, no one should ever begin nitrous tuning until normally aspirated tuning is *perfect*. On one hand, nitrous is eminently capable of causing catastrophic engine damage when air/fuel mixtures or timing are wrong. Nitrous injection can torch valves in seconds or less, crack pistons, and break rods or even crankshafts.

On the other hand, nitrous is so dangerous that nitrous system manufacturers these days virtually always sell foolproof fail-safe cookbook nitrous solutions that require zero tuning and are virtually incapable of damaging the engine if installed correctly (though they *require* a stock-quality nonboosted calibration and adequate fuel supply). Bottom line: If you do not have experience calibrating engine management systems, do *not* attempt to learn on a boosted or nitrous motor. Boosted and nitrous motors absolutely cannot be safely tuned with trial-and-error methods; it's simply too dangerous.

If you are exploring strange new worlds with your tuning project, pushing mechanical and thermal limits that are unknown and difficult to predict, *things can and will break*. Ask yourself if you have the financial resources to deal with engine damage should it occur for whatever reason. If your resources are severely limited, get help from an expert (preferably an automotive engineer) to compute what is a highly *safe* state of tune for various critical parts, and do not go beyond it for any reason.

If you cannot afford engine damage, remember that more expensive engine management systems usually offer improved logic, capabilities, and alarms that make it safer to push the limits. An example of this would be the ability of an engine management system to receive, process, and take action based on data from exhaust gas temperature (EGT) probes—one of the most reliable early-warning signs of impending engine damage from lean mixtures thermal loading—to push back the throttle or institute other countermeasures faster than a human could react to prevent damage. Keep in mind the old saying, "You can pay me now or you can pay me later."

If you do not have much tuning experience, you may be able to benefit from self-tuning ECUs like Electromotive's TEC[3].

VIRTUAL CARBURETORS

Some old-timers and bike tuners have a lot of experience tuning carbureted engines, and they know a lot about how people tuned engines before there were fast exhaust gas analyzers, wideband air/fuel-ratio meters, or affordable chassis dynos. This can be dangerous.

It is certainly true that a good basic knowledge of how and why engines take fuel and spark timing at various speed and loading ranges is important. It is also true a lot of carbureted engines weren't operating at a very accurate and high state of tune. The mixture distribution between cylinders was often terrible, with some single carb cylinders running much leaner than others and tuning efforts directed toward making sure the leanest cylinders were running safely rich.

What's more, the specific power (horsepower per cubic inch) of musclecar-era performance engines was vastly less than recent sport-compact engines, and even the gross horsepower was often less. There are plenty of 1.8-liter turbocharged Hondas making more horsepower than a 1970 426 Hemi elephant engine.

Not only were older engines stressed less than today's super-high-output powerplants, but—fortunately or unfortunately—points ignitions and carburetors targeted by manufacturers for specific engines were, practically speaking, limited in their adjustments. People would adjust idle-air-screws to set air/fuel mixture at idle, and they *might* swap main jets around a little to tune for best acceleration near wide-open throttle, but they were less likely to make radical changes to basic components like venturis that greatly affect midrange air/fuel ratios. Where they made changes, these were likely to be incremental, in ways that affected one particular area of the timing or fuel curve *and that left alone the basic shape or slope of a fuel or timing curve.*

There were never many carbureted turbo motors (for really good reasons), and without the requirement to deal with boosted engines, hot rodders tended to follow a cookbook approach to performance tuning for specific normally aspirated performance engines. For, example, "a stock compression 350 Chevy motor with a Com Cams x1 grind, x2-inch tube headers, and a Holley x3 carb needs Holley x4 main jets, x5 power valve, with the accelerator pump rod in the x6 position, and you need x7 centrifugal advance springs in your distributor and x8-degrees total timing."

Jet-swapping is smelly, time-consuming, and costs actual cash-money for the precisely machined jets (unlike key strokes on a computer, which are free), so tuners were less likely to make the kinds of radical changes that would completely screw up fuel delivery. Bottom line, in the old days it took real work and money to royally screw up physical engine management systems, so it didn't happen as much as it does today when computers have greatly increased the efficiency with which people can radically screw up electronic engine management system calibrations.

One trend in engine management modifications is what I like to call the "virtual carburetor," that is, modifying a stock engine management system for power-adder conversions to fix bad OEM behavior with a potentiometer-equipped electronic interceptor box that gives a tuner the semi-automated option of tuning selected subsets of the fuel and timing curves by turning a knob—and without the challenge of building a scratch calibration equal to the factory's. Companies like Techlusion do a great job of providing the ability to make incremental changes targeted at specific areas of injection and timing, keeping it simple and protecting people from themselves—just like in the old days.

HOW TO HIRE SOMEONE ELSE TO DO THE CALIBRATION/TUNING

Many performance shops pride themselves on their engine management problem-solving abilities. They love it (perhaps secretly) when they get a chance to be a hero, fixing a problem created by a customer or (even better) a competing performance shop, especially if it involves selling expensive new parts or equipment. "We'll have to change out the Framajigger for a Whatsahoosits; a Framajigger isn't worth a damn and could never work on your 1912 John Deere Super-60-powered trike. Squirrel Performance should never have installed it, but since they're under 60 counts of fraud and a death sentence for incompetence, you get what you pay for."

Performance tuners can be very good at looking you in the eye and conveying absolute certainty about their ability to achieve low-earth-orbit with a 1912 John-Deere-powered trike, even if they aren't actually sure what one is. Many performance shops have a very hard time turning down targets of opportunity that are outside their ordinary specialization and expertise.

The deadly combination of pride and greed can lead to shops taking on tuning projects that, by definition, require on-the-job training (or dusting off old expertise on systems they haven't seen in a long time). Some shops welcome the chance to get paid to learn a new engine management system, which could result in unnecessary parts and equipment changes. The worst outcome would be if a shop damaged your engine or vehicle working on a system with a steep learning curve.

The following are some rules of thumb that can help you to avoid this sort of heartache:

• I would not use a tuning shop that lacked extensive experience with your specific engine management system. Preferably, they should be a dealer.

• Unless the tuner were a wizard who I *knew* to be a wizard, I would probably want references from satisfied customers. Other customers can verify satisfaction with a tuner's technical expertise, and customers will also have opinions about support capabilities, financial policies, and fairness.

• I would not normally use a tuning shop for engine management calibration that did not have its own in-house chassis dynamometer. Under exceptional circumstance, this rule might be modified if a tuner had a close working arrangement with a dyno shop that did have a dynamometer. Yes, it is possible to do a good tuning job with a strap-on (no, not *that* kind of strap-on!) accelerometer-type pseudo-dyno fifth-wheel device (that is, G-Tech) and a portable wideband air/fuel-ratio meter (or an engine management system that supports a wideband O_2 sensor), or, at a minimum, with a wideband AFR meter and a bunch or deserted road. However, I am more comfortable with a shop that has made the investment in a $20,000-plus dynamometer than one that has not. What's more, a shop that owns the dyno does not have to buy dyno time, so you're likely to get more extensive dyno tuning and probably a better calibration than a situation where your tuner is on the clock.

• Possession of an *engine* dyno (for example, a Superflow) test cell is a leading indicator of great seriousness. You can drive a car onto a chassis dyno and rip off a pull or two in a few minutes; an engine dyno is more complicated. It was somewhat easier in the days of carbureted distributor-ignition engines but attaching a fuel-injected powerplant to an engine dyno and connecting or fabricating everything it takes to make it work outside the engine bay of a car typically takes *hours*. However, when they calibrate engines in Detroit or Munich or Japan or wherever, they start with radically expensive water-brake or eddy-current engine

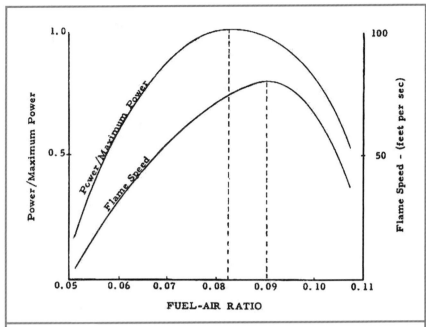

Here's the heart of the matter in fuel calibration: Power and flame speed vary greatly with the air/fuel ratio. In this graph we see typical power and flame speed versus fuel-air (not air-fuel) ratio curves for constant-speed operation. Peak power with gasoline fuel occurs just rich of rich best torque, at an air/fuel ratio of just above 11:1. The cooling heat of vaporization of such rich mixtures along with the high flame speed help fight detonation.

dynos that enable engineers to hold the engine as long as necessary at any precise combination of speed and loading (even combinations not achievable at sea level on the road with street tires) to set best torque at precisely the perfect air/fuel ratio.

• A tuner with a lot of experience is, of course, extremely desirable: How many scratch calibrations has he done? Assuming he has a dyno, how many dynoruns or pulls does he have in his archives? And how many of these involved calibrating the onboard computer on the dyno (there should be *many* dyno pulls in such a case). Ask to see the run notes associated with dynoruns. Does he keep good, scientific documentation of the tuning progress? Does he keep records as equipment and settings change (possession of such can be very useful in the future on a similar vehicle)? Lots of calibration dyno runs on either an engine or chassis dyno, and lots of good anal-retentive documentation are *very* good signs. I do want to add that I know some excellent supertuners who do *not* keep very good documentation; in part, because they are so confident and experienced that they don't need it as much anymore. However, many shops with Dynojet chassis dynos are beginning to do a better job of documenting tuning sessions than they did at first.

• Desirable additional diagnostic equipment for a tuner should include fast, wideband air/fuel-ratio meters (for example, a Horiba), exhaust gas analyzers, OBD-II diagnostic analyzers, oscilloscopes, and so forth. I absolutely would not consider using a tuner who lacked a Horiba, Motec meter, or similar.

• How long has he been in business? The longer, the better. Not only is this an indication of experience, it is an indication of someone who knows how to run a successful business, and tuners that go out of business can't offer good product support.

• Is the owner still doing the heavy lifting on calibrations? What other responsibilities does he have? Engine calibrations can be risky, so most tuner-owners let their grease monkeys handle the

installations and take care of the tricky stuff on the dyno themselves. I personally think an owner-tuner is preferable to an employee working on your calibration.

• A history of racing indicates a tuner has—or used to have—some money. *Winning* experience indicates he knows what he's doing, is careful and scientific, and has (or used to have) some money. Someone who calibrates race cars that win can tune my engine any time.

• Have there been many articles about the shop or the owner in various performance car magazines? This can be a good thing, but there are plenty of articles these days that magazine people refer to internally as advertorial, an unhappy compound of the words advertising and editorial. Sometimes such articles are also called blow-job stories, particularly when they are puff-pieces about advertisers. Most good tuners and performance-equipment suppliers who want magazine PR know how to get it, and it's out there. Many technical articles about tuners and shops are honest and fairly complete. But some articles may not tell the whole story, especially about advertisers. And there are articles out there about people who have more expertise about how to get PR than they do about calibrating engines. *Magazine PR is not proof that the tuner is someone you want working on your car!*

WHAT IF A PRO TUNER BREAKS YOUR ENGINE?

I interviewed several tuners on this question.

Calibrating some super-high-output engines is inherently risky, and when you are doing an R&D engine project, pushing the limits of strange new worlds, experimenting to find out how much power is *possible* or how much torque engine components can stand, something may break and this is really not the fault of the tuner. You wanted more power, and neither you nor he knew for sure the mechanical or thermal limits of critical parts. You are basically hiring the tuner to do the job to the best of his ability, and you should understand that you may have to replace some parts in an R&D effort. That's why it's called R&D.

A tuner is *not* responsible if mechanical parts on your vehicle break, unless he breaks parts through negligence by over-boosting or over-revving the engine. Broken rods or pistons, blown head gaskets, or other damage unrelated to tuning are not the fault of the tuner.

On the other hand, a professional tuner is expected to have a high degree of expertise, and not to destroy parts and equipment through negligence or carelessness. According to Dallas supertuner Bob Norwood, a professional performance tuner should *never* hurt your engine with lean mixtures or wrong ignition timing. A pro tuner should *not* detonate your engine.

But.

In the real world, a tuner is unlikely to have insurance that would cover engine damage from R&D or any other reason. The personal resources of the tuner are certainly limited. If a tuner destroys your high-dollar engine through negligence, he might be unable to fix it for free even if he is so inclined. Discuss this issue ahead of time and come to an agreement. If it would be catastrophic if your engine breaks, make sure the tuner understands this. A tuner might work differently—more carefully, farther from the ragged edge—if it is essential to avoid damage versus if the highest priority were to search for the absolute limits of deliverable maximum power.

The bottom line is that the reason factory engines often have the capability to be hot rodded to higher power levels is that automakers must warrant them and therefore always design them to run at power levels a conservatively safe distance from mechanical and thermal damage. Actually, in the distant past, back in the 1960s, there were a few vehicles sold by automakers for street or track use *without warranty* for racing use with radical hairy engines like Chrysler's 426-ci Race Hemi, Ford's SOHC Cammer 427, and a few others. Today all streetable vehicles have to pass emissions and have warranties, and no one sells anything without a warranty.

Basically, if something breaks on the dyno while tuning and it is clearly the tuner's fault through egregious negligence, the tuner may want to fix it to protect his reputation. If this happened essentially through dishonesty or fraud (he sold you junk that was bound to break or to break the engine in your application), you could sue him, and maybe you'd win. And if you won, maybe there'd be something to take. But those are big ifs.

The best insurance is to find a good, experienced tuner with a good reputation, tools, and financial resources. Pay his high prices and expect him to do his best for you.

And if something goes wrong that is beyond his and your control, tell yourself, "That's racing."

ENGINE TUNING THEORY

To tune an engine effectively, you will need to understand the theory of volumetric efficiency, burn rate, spark advance, air/fuel ratios, temperature, air-pressure, detonation, and fuels.

VOLUMETRIC EFFICIENCY AND ENGINE FUEL REQUIREMENTS

A piston engine is, in a sense, a self-propelled air pump that breathes in a quantity of air through the throttle related to the displacement of the engine, and exhales nitrogen and combustion products through the exhaust. But even at wide-open throttle (WOT), the amount of air that enters a spark-ignition piston engine varies considerably depending on—stuff. There are many factors that affect how much air *actually* enters the engine's cylinders under various operating conditions (almost always less than the static displacement of the engine except with highly optimized performance engines utilizing resonation-effects tuned intake and exhaust systems, or on turbo or supercharged engines, which may easily operate at 100 percent or more of static displacement, also known as the *volumetric efficiency* or VE).

Torque is the instantaneous twisting force that an internal-combustion engine is able to impart to a crankshaft, averaged over the combustion cycle by a flywheel. Over time, torque can be used to do work such as accelerating a car. Peak torque occurs at the engine speed and loading at which an engine (with power-adders, if present) is most efficient at ingesting air into the cylinders. Therefore, peak torque is also peak *volumetric efficiency*, or VE. Peak torque requires the largest amount of fuel per combustion cycle, hence, the longest injector open-time or pulse width. Even though engine breathing is less efficient at speeds above peak torque and therefore each putt is less powerful, more combustion events are occurring per time, and the engine will make its peak power at faster speeds than peak torque. Due to the definition of power as a function of torque/5,250, these two measurements are always the same at 5,250 rpm. A properly tuned engine will always use the most fuel per time at peak power.

On modern normally aspirated street engines, engineers strive to design engines to have as high a VE (torque) as possible across the range of engine operating speeds. One hundred percent VE is the amount of air that would be in the cylinder of an engine at bottom dead center at rest, minus the combustion chamber clearance volume (the displacement of the engine). It is extremely difficult or impossible to achieve 100 percent VE when the engine is running, because it takes a certain amount of time for air to rush into the cylinders, and there is a limited amount of time (eight or nine thousandths of a second at 7,000 rpm), and there are restrictions in the way. Most street piston engines cannot achieve anything like 100 percent VE, settling instead for maybe 70 to 90 percent, but engineers work to build a high, flat torque/VE curve across the rpm range.

A number of factors conspire to prevent a full 100 VE in normally aspirated street engines. There is normally at least a slight pressure drop through a throttle body or carburetor, even at wide-open throttle (particularly with a carb due to the need for a restricting venturi). All carburetors and fuel injection throttle bodies restrict airflow—producing a pressure drop—when the throttle blade is partially closed (and sometimes even when it's wide open). Intake ports and valves offer at least some restriction at some engine speeds. The exhaust stroke does not expel all burned gases because some exhaust is trapped in the clearance volume. The exhaust valves and exhaust pipes offer some restriction as well. The camshaft profile of an engine has a huge effect on the VE of an engine at various speeds and loading.

If a normally aspirated engine had a perfectly flat torque curve across the rpm range, injection pulse width would not vary at all with engine speed, but would be a function only of engine loading. If VE is entirely independent of engine speed, then the amount of air entering a cylinder on the intake stroke will depend entirely on the amount of air pressure in the intake manifold, hence, injection pulse width is a linear function varying precisely with the voltage from a manifold absolute pressure sensor, with peak VE and torque at the highest achievable manifold pressure. Since many modern street engines *do* have a pretty flat torque curve, a calibration in which pulse width varies in a linear fashion from zero at zero manifold pressure to maximum power pulse at full achievable manifold pressure is a good approximation of an ideal steady-state air/fuel table optimized for peak power.

Given the varying VE of an engine, the amount of air in the cylinder at any given breakpoint is not perfectly predictable. Again, you can assume that the maximum VE occurs at the point of peak torque. This will be the point of maximum injection pulse width, with pulse width falling off both above and below this. Maximum power will occur at a higher rpm than peak torque where the engine is making more power strokes per time increment, although the power strokes are less efficient at speeds above the torque peak and will therefore require less fuel to be injected per power stroke.

In some cases, engine torque and VE vary considerably with rpm. Normally aspirated engines optimized to run efficiently at very high speed with resonation-effects intake-runner and exhaust tuning at the expense of very weak low-rpm torque capabilities are one example. Some such engines have achieved as much as 10 psi positive pressure at the intake valve from inertial and resonation effects. Another example of varying VE is

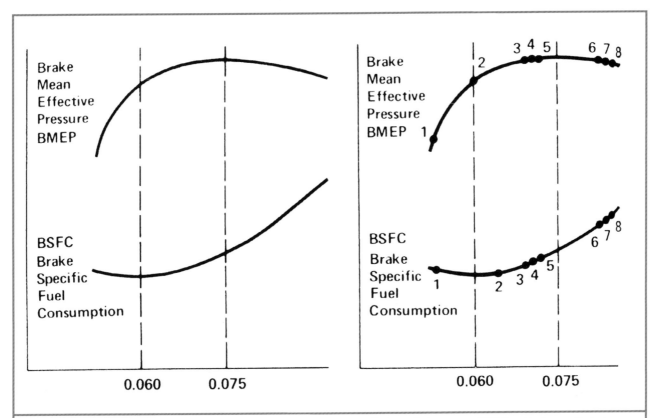

Another way to look at the calibration task is through the lens of peak break mean effective cylinder pressure (BMEP) and brake specific fuel consumption (BSFC), which do not occur at the same air/fuel ratio. In an engine operating with perfect distribution, for best power, a fuel-air ratio of 0.075–0.080 produces the lowest fuel consumption. For best fuel economy, a fuel-air ratio of 0.060 is optimal, though produces less power. The right graph charts the fact that single-point (carb or TBI) wet manifolds produce suboptimal fuel distribution. Cylinders 6, 7, and 8 are too rich, producing less power and wasting fuel. Cylinders 1 and 2 are too lean. To prevent detonation in 1 and 2, all would have to be enriched to bring 1 and 2 to the flat part of the curve, meaning only 1 and 2 would produce peak power with minimal fuel consumption for that output. Holley

turbocharged engines optimized for high-speed operation that cannot make boost at lower rpm. However, even if the torque curve is not flat, given the relatively broad range of air/fuel mixtures at which an engine will still produce peak torque or near-peak torque, a straight-line approximation of injection pulse may still produce relatively good performance *even with no VE corrections.* Aftermarket EMS supplier Electromotive points out that "most production engines (even modern DOHC engines) can be made to run acceptably well with *no* volumetric efficiency table offsets to the raw fuel curve, even when fitted with individual throttle-per-cylinder setups."

As an engine wears out, its VE decreases—but this may not occur evenly across all speeds and loading ranges or between various cylinders. The newest original-equipment factory engines vary less than 1.5 percent in power, but historically, factory-built engines have varied greatly in VE due to significant variations in cam lobe profiles and other machining, with perhaps 5 percent very fast (high VE), 15 percent very slow, and the rest somewhere in the middle.

If the load-measuring device for an engine management system is a mass airflow (MAF) sensor (output voltage or frequency *highly* dependent on engine speed), then the basic fuel calculation is based on the MAF voltage divided by engine rpm. Note that MAF-based fuel calculations require no temperature correction for air density, since the MAF reading already reflects air density effects on mass airflow into the engine.

FUEL BURN RATE

Normal combustion in a spark-ignition engine occurs after the spark plug ignites the compressed air/fuel mixture, starting a flame front that spreads out in a wave from the plug in all directions, something like dry grass burning in a field. It takes time for the flame to spread in the air/fuel mixture, exactly as it takes time for a field of dry grass to burn away from where the fire was started. During normal combustion, the flame front in a combustion chamber moves across the chamber until it reaches the other side, burning smoothly and evenly.

Special cameras recording the combustion event through a porthole in a cylinder reveal that flame-front speeds for gasoline-air mixtures vary from 20 feet per second to more than 150 feet per second, depending on air/fuel ratio, density, compression ratio, turbulence among the charge gases, and combustion chamber design. Flame speed is fastest at rich mixtures near 11.1:1 AFR, falling off dramatically in both the rich and lean directions from this point (especially in the rich direction). The slower the flame front speed, the greater the chance of abnormal combustion.

High-compression engines make more power because they are more efficient at getting energy from the air/fuel charge mixture. Normally, combustion chamber pressure rise is 3.5 to 4 times the initial compression pressure. The difference in pressure gain in an 8.5:1 compression engine versus a 10:1 compression engine can easily be 50 percent, as much as 250 psi. A typical

160 psi compression pressure rises quickly to over 600 psi following combustion. Pressure rise rates can exceed 20 psi per degree of crankshaft rotation. Cylinder pressures are highest at wide-open throttle and are lowest at idle and light cruise, when intake manifold vacuum is very high and engine VE very low. Combustion proceeds faster under higher compression.

Turbulence is carefully designed into combustion chambers. Some swirl will produce greater flame speed, more efficient combustion, and even help resist knock by spreading the flame front more quickly. Too much swirl will produce undesirable disturbance of the quench layer around the combustion chamber, increasing heat loss into the cylinder walls.

The actual performance of fuel in an engine is closely related to the release of energy in the combustion process. This is affected by dissociation of the reactants and products of combustion (splitting of molecules into smaller molecules or single atoms or ions) and heat transfer through the combustion chamber. Increased heat in the combustion chamber makes power, but heat can also be pro-knock. At wide-open throttle at high speed, heat transfer amounts to about 15 percent of the fuel's thermal energy. This heat loss decreases with increased brake mean effective pressure in the cylinder. The heat transfer is maximized at stoichiometric air/fuel ratios, decreasing with the higher fuel flow rates of rich mixtures. Heat transfer into the block and water jacket also depends on gas turbulence near the combustion chamber surfaces, and it increases with knock and other violent combustion conditions.

Given that engines induct a fairly constant mass of air for a given speed, manifold pressure, camshaft timing, and intake temperature, the need to improve spark-ignition engine power for racing has led to two main strategies in an effort to increase the amount of fuel that can be burned in the combustion chamber. The first has been to increase the mass of oxygen intake by supercharging or supplementary oxygen injection, mostly in the form of nitrous oxide. The second has been to search for fuels with high specific energy.

In addition to the octane number requirement of an engine in order to avoid detonation, high-performance engines have other performance requirements for fuels, including volatility, flame speed, weight, and volume.

SPARK ADVANCE

Spark advance, which is optimally timed to achieve best torque by producing peak cylinder pressure at about 15 degrees ATDC, increases octane requirements by a half to three-quarters of an octane number per degree of advance. Spark advance increases cylinder pressure and allows more time for detonation to occur.

Engine speed range and fuel burn characteristics affect ignition timing requirements. As an engine turns faster, the spark plug must fire at an earlier crank position to allow time for a given air/fuel mixture to ignite and achieve a high burn rate and maximum cylinder pressure by the time the piston is positioned to produce best torque, which is dependent not only on engine speed but on mixture flame speed, which, in turn, is dependent not only on the type of fuel but on operating conditions that change dynamically, such as air/fuel mixture.

Therefore, an independent variable affecting the need for spark advance as rpm increases is the need to modify ignition timing corresponding to engine loading and consequent volumetric efficiency variations, which demand varying mixtures. Throttle position, for example, affects cylinder filling, resulting in corresponding variations in optimal air/fuel mixture requirements.

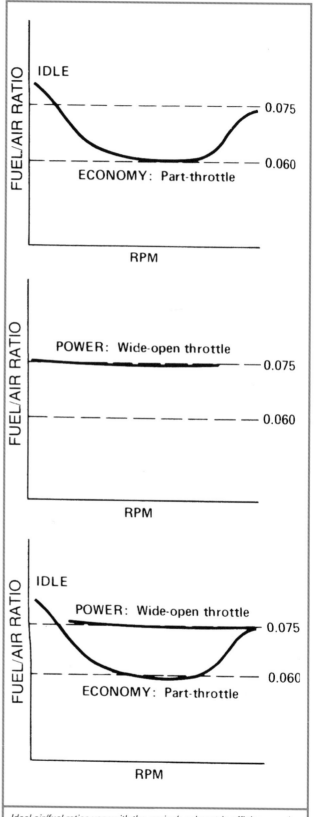

Ideal air/fuel ratios vary with the engine's volumetric efficiency and the necessity in modern automotive engines to produce maximum power at full throttle, maximum economy at part throttle and idle, and good drivability in all ranges. Holley

An important factor that affects VE—and potentially flame speed—is valve timing. Remember, a denser mixture burns more quickly, and a leaner mixture requires more time to burn. Valve timing has a great effect on the speeds at which an engine develops its best power and torque. Adding more lift and intake/exhaust valve opening overlap allows the engine to breathe more efficiently at high speeds. However, the engine may be hard to start, idle badly, bog on off-idle acceleration, and produce bad low speed torque. This occurs for several reasons. Increased valve overlap allows some exhaust gases still in the cylinder at higher than atmospheric pressure to rush into the intake manifold exactly like EGR, diluting the inlet charge—which continues to occur until rpm increases to the point where the overlap interval is so short that reverse pulsing is insignificant. But big cams result in gross exhaust gas dilution of the air/fuel mixture at idle, which consequently burns slowly and requires a lot of spark advance and a mixture as rich as 11.5 to 1 to counteract the lumpy uneven idle resulting from partial burning and misfires on some cycles. Valve overlap also hurts idle and low-speed performance by lowering manifold vacuum. Since the lower atmospheric pressure of high vacuum tends to keep fuel vaporized better, racing cams with low vacuum may have distribution problems and a wandering air/fuel mixture at idle, which may require an overall richer mixture in order to keep the motor from stalling, particularly with wet manifolds. Carbureted vehicles with hot cams may not have enough signal available to pull sufficient mixture through the idle system, leaning out the mixture, requiring a tuner to increase the idle throttle setting (which may put the off-idle slot/port in the wrong position, causing an off-idle bog). Obviously, this is not a problem with fuel injection.

Short-cam engines may run at stoichiometric mixtures at idle for cleanest exhaust emissions. Some late-model high-performance engines like that of the Acura NSX make use of a dual-mode cam profile, in which one set of cam lobes is used for low speed conditions and another for high rpm.

Coming off idle, a big-cam engine may require mixtures nearly as rich as at idle to eliminate surging, starting at 12.5–13.0 gasoline air/fuel mixtures and leaning out with speed or loading. Mild cams will permit 14–15 gasoline mixtures in off idle and slow cruise.

With medium speeds and loading, the bad effects of big cams diminish, resulting in less charge dilution, allowing the engine to happily burn gasoline mixtures of 14 to 15:1 and higher. At the leaner end, additional spark advance is required to counteract slow burning of lean mixtures. Hot cams may produce problems for carbureted vehicles when changed engine vacuum causes the power valve to open at the wrong time. Changed vacuum could affect speed-density fuel injection systems but would have no effect on MAF-sensed EFI. Where air/fuel mixtures are inconsistent or poorly atomized, flammability suffers, affecting many other variables.

With a big cam, the spark advance at full throttle can be aggressive and quick; low VE at low rpm results in slow combustion and exhaust dilution, lowering combustion temperatures and reducing the tendency to knock. Part-throttle advance on big-cam engines can also be aggressive due to these same flame speed reductions resulting from exhaust dilution of the inlet charge due to valve overlap.

Turbulence and swirl are extremely important factors in flame speed—more important, within limits, than mixture strength or exact fuel composition. Automotive engineers have long made use of induction systems and combustion chamber geometry to induce swirl or turbulence to enhance flame speed

and, consequently, anti-knock characteristics of an engine. Wedge head engines, with a large squish area, have long been known to induce turbulence or swirl as intake gases are forced out of the squish area as the piston approaches top dead center.

In the 1970s, automotive engineers began to de-tune engines to meet emissions standards that were increasingly tough. They began to retard the ignition timing at idle, for example, sometimes locking out vacuum advance in lower gears or during normal operating temperature, allowing more advance if the engine was cold or overheating. Since oxides of nitrogen are formed when free nitrogen combines with oxygen at high temperature and pressure, retarded spark reduces NOx emissions by lowering peak combustion temperature and pressure. This strategy also reduces hydrocarbon emissions. However, retarded spark combustion is less efficient, causing poorer fuel economy, reduced power, and higher heating of the engine block as heat energy escapes through the cylinder walls into the coolant. The cooling system is stressed as it struggles to remove the greater waste heat during retarded spark conditions, and fuel economy is hurt since some of the fuel is still burning as it blows out the exhaust valve, necessitating richer idle and main jetting to get decent off-idle performance. If the mixture becomes too lean, higher combustion temperatures will defeat the purpose of ignition retard, producing more NOx. Inefficiency of combustion under these conditions also requires the throttle to be held open farther for a reasonable idle speed, which, combined with the higher operating temperatures, can lead to dieseling (not a problem in fuel-injected engines, which immediately cut off fuel flow when the key is switched off). By removing pollutants from exhaust gas, three-way catalysts tend to allow more ignition advance at idle and part throttle. Undesirable products of combustion include formaldehyde, NOx, CO_2, and fragments of hydrocarbons.

In any case, various high-performance fuels vary in burn characteristics, particularly flammability, flame speed, emissions, and so on, and all affect spark timing requirements. Gasoline engines converted to run on propane, natural gas, or alcohol require a different timing curve due to variations in combustion flame speed of the air/fuel mixture.

AIR/FUEL RATIOS

Air/fuel ratio has a major impact on engine octane number requirement (ONR), increasing octane requirements by +2 per one increase in ratio (say from 9:1 to 10:1). Ideally, air/fuel ratio should vary not only according to loading but also according to the amount of air present in a particular cylinder at a particular time (cylinder VE). Richer air/fuel ratios combat knock by the intercooling effect of the cooling heat of vaporization of liquid fuels and a set of related factors. The volatility of fuels affects not only octane number requirement but drivability in general.

The chemically ideal air/fuel mixture (by weight), at which all air and gasoline are consumed in combustion occurs with 14.68 parts air and 1 part fuel, which is usually rounded to 14.7. This ratio is referred to as stoichiometric. Stoichiometric mixtures vary according to fuel, from a low of nitromethane at 1.7:1, to methanol's 6.45:1, ethanol's 9:1, up to gasoline at 14.6:1 and beyond to natural gas and propane, which are in the range of 15.5–16.5:1. Mixtures with a greater percentage of air than stoichiometric are called lean mixtures and occur as higher numbers. Richer mixtures, in which there is an excess of fuel, are represented by smaller numbers. Mixtures, by the way, are often expressed as a percentage of stoichiometric, usually referred to as lambda. Therefore, the stoichiometric AFR is 1.0 lambda; the best-power ratio for gasoline (12.2:1) is 0.83 lambda.

At high loading and wide-open throttle, richer mixtures give better power by making sure that all air molecules in the combustion chamber have fuel present to burn. At wide-open throttle, where the objective is maximum power, all four-cycle gasoline engines require mixtures that fall between lean and rich best torque, in the 11.5 to 13.3 gasoline range. Since this best torque mixture spread narrows at higher speeds, a good goal for naturally aspirated engines is 12.0 to 12.5, perhaps richer if fuel is being used for combustion cooling in a turbo/supercharged engine.

Typical mixtures giving best drivability are in the range of 13.0 to 14.5 gasoline-air mixtures, depending on speed and loading. At higher engine speeds, reverse pulsing through a carb in engines with racing cams tends to enrich the mixture as reversion gases pass through the venturi twice. Naturally, this is not a problem with fuel injection.

EFI engines are not susceptible to hot weather percolation, in which fuel boils in the carb in hot weather when a hot engine is shut down, flooding into the manifold. This is somewhat dependent on the distillation curve of the fuel, which varies according to the weather for which it is intended. Hot weather percolation can be aggravated by high vapor pressures produced by fuel boiling in the fuel pump or fuel lines, forcing additional fuel past the float needle. Similarly, vapor lock—produced by mechanical fuel pumps supplying a mixture of gas and vapor to a carb and eventually uncovering the main jets as fuel level drops—is not a problem with EFI.

The main difference between computer-controlled engines and earlier modes of control is that with EFI engines, variations in speed and loading and other engine parameters translate into spark advance and fueling pulse width values that at any particular point on the spark or fuel map can be completely independent of other nearby points (and could vary quite abruptly, if necessary). This was impossible with older mechanical control systems. Computer-controlled engines eliminate compromises in fueling and spark control. Multi-port injection eliminates distribution and inconsistency problems with handling wet mixtures in the intake manifold associated with less than one carb per cylinder, often resulting in improved cold running, improved throttle response under all conditions, and improved fuel economy without drivability problems.

Computer engine management with individual cylinder-adjustable electronic port fuel injection really shines on extremely high-output engines with high effective compression ratios. Dallas supercar builder Bob Norwood points out that for peak power, any engine must be treated as a gang of single cylinder engines, each of which should be optimized for peak power.

On NASCAR engines—restricted to single-point carburetion—this has been done by custom modification of the compression ratio, rocker arms, and cam lobes for individual cylinders. But using sequential programmable EFI with fuel and ignition calibration on an individual cylinder basis, whatever the volumetric efficiency of each "single-cylinder engine," optimal timing and mixture can be provided by computer.

Individual cylinder calibration is good for 50 horsepower on a typical high-output small-block Chevy V-8, says Norwood. Not only does individual cylinder calibration improve power, but it saves engines from knock damage while allowing a higher state of tune. By contrast, common practice on street carbureted engines or batch fuel injection is to tune to the leanest cylinder (read best-flowing). A Chevy V-8 with its dog-leg ports typically varies 10 percent in airflow between best and worst cylinder, and it is typical that V-type motors respond better than

in-line motors to individual cylinder calibration. On a dyno, best and worst cylinders that are within 100 degrees EGT of each other is considered excellent. Leaning out an engine on a dyno or track will frequently make more power until suddenly one or two cylinders die. "You always burn one or two cylinders," says Norwood. "In the meantime, the other slackers are running rich." Some advanced engine management systems have implemented EGT-feedback closed-loop systems that equalize EGT from all cylinders at a target level with individual-cylinder calibration, simultaneously providing best power across the board. With this system, every cylinder can reach maximum power at very high compression ratios with the best high-octane gasoline.

Testing by Smoky Yunick for *Circle Track* showed that the small-block Chevy test engine was prone to detonation on three of the eight cylinders. This was thought to be due to unequal distribution or coolant circulation that caused those cylinders to run hotter. "Now we could have cut these three cylinders to 14:1 compression," writes Smoky, "and maybe run the other five cylinders at 15:1 and gained some power. Don't know, but I do know this, you are gonna get a safer, more powerful engine if you adjust the compression, timing, and distribution to each cylinder."

TEMPERATURE

Inlet air temperature increases octane requirements by 0.5 octane number per 10 degree increase. Temperature affects fuel performance in several ways. Colder air is denser than hotter air, raising cylinder pressure. Colder air inhibits fuel vaporization. But hotter air directly raises combustion temperatures, which increases the possibility of knock.

EFI systems normally have sensors to read the temperature of inlet air and adjust the injection pulse width and spark timing to compensate for density changes and fuel volatility limitations. Engines will make noticeably more power on a cold day because the cold, dense air increases engine volumetric efficiency, filling the cylinders with more molecules of air. This is bad news for a carb, which has no way of compensating (other than manual jet changes—one size per 40 degrees temperature change). Piston-engine airplane pilots must always know the ambient air temperature in order to estimate take-off distance (much shorter on colder days), and aircraft fuel systems (including carbureted systems) allow dynamic manual adjustments of air/fuel mixture. As you'd expect, racing automotive engine designs always endeavor to keep inlet air as cold as possible. Even stock street cars often make use of cold-air inlets, since each 11 degrees Fahrenheit increase reduces air density 1 percent.

Cold engines require enrichment to counteract the fact that only the lightest fractions of liquid fuel may vaporize at colder temperatures while the rest exists as globules or drops of fuel that are not mixed well with air. The actual vaporization/distillation curve of various gasolines and fuels differs, depending on the purpose for which the fuel is designed, and gaseous fuels like propane and methane do not require cold running enrichment and have no intercooling effect once in gaseous form. But in a cold engine, most of a fuel like gasoline—in liquid form—will be wasted.

Liquid fuels do not vaporize well in cold air (though the oil companies do change their gasoline formulation in an effort to counteract this by increasing the vapor pressure in cold weather, which can lead to a rash of vapor lock in sudden warm spells in winter). Choking systems in carbureted vehicles and cold-start fuel enrichment systems on EFI vehicles are designed to produce a rich-enough mixture to run the vehicle even when much of the

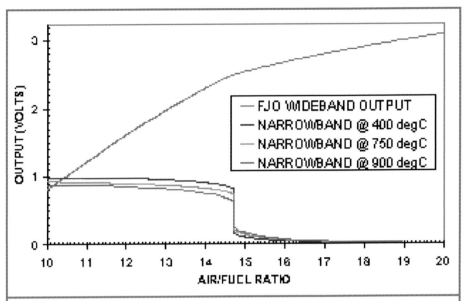

A wideband or UEGO is not the same as any old O$_2$ sensor, heated or not. Standard O$_2$ sensors—heated or not—have little ability to distinguish changes to air/fuel ratio except as a rich-lean indicator very near 14.68:1, the stoichiometric air/fuel ratio. What's more, the temperature of the standard-type sensor away from stoich has an effect on the voltage output that is greater than the effect of large air/fuel ratio changes. Standard sensors are useless for power tuning and should be left to do their work to produce low emissions at idle and light cruise.

fuel exists not as burnable vapor, mixed with air, but as drops of nonburnable liquid fuel suspended in the air or on the manifold walls and combustion chambers. Most air pollution is produced by cold vehicles burning mixtures as rich as 3 or 4:1 (even 1:1 during cranking). Engine management systems sense coolant temperature in order to provide cold-start enrichments (cold prime, cranking, after start, and warm-up). Electronic injection systems usually spray fuel straight at the intake valve into the swirling turbulent high velocity air that exists there, which greatly improves atomization and vaporization. In general, the higher the pressure of injection, the better. Experiments have shown that pressure as high as 300 psi produces excellent shear of fuel droplets for the best quality atomization. EFI does not normally need the exhaust-gas heating that carbs require to provide acceptable warm-up operations, though EFI manifolds may use coolant heating of the manifold to increase vaporization at idle in very cold weather.

Unlike carbs, which are susceptible to fuel starvation if ice forms in the carb as vaporizing fuel removes heat from the air and from metal parts of the carb in cold weather, EFI systems will deliver fuel even if ice forms in the throttle body. In any case, EFI is not as susceptible to icing since fuel vaporizes (stealing heat) at the hot intake valve, not at the throttle body. EFI systems don't require a choke stove to heat inlet air for cold-start driving like carbureted vehicles.

Increases in coolant temperature increase octane requirements by roughly one octane number per 10 degree increase from 160 to 180 degrees Fahrenheit.

AIR PRESSURE

Increasing altitude reduces octane number requirements by about 1.5 octane numbers per 1,000 feet above sea level. Supercharging, which provides a denser breathing atmosphere, has the opposite effect—increasing ONR due to the increased

density, higher effective compression ratio, as well as the tendency of turbo-superchargers to heat inlet air. Air density varies with temperature, altitude, and weather conditions. For a given pressure, hot air, with greater molecular motion, is less dense. Air at higher elevations is less dense, as is air with a higher relative humidity. Air is less dense in warm weather, and air that is heated for any reason on the way into the engine tends to become less dense and thus affect the weight of fuel required to achieve a correct air/fuel ratio. Intake system layout can have a large effect on the volumetric efficiency of the engine by affecting the density of the air the engine is breathing. Air cleaners that suck in hot engine compartment air will reduce the engine's output and should be modified to breathe fresh, cold air from outside for maximum performance. Intake manifolds that heat the air will produce less dense air, although a properly designed, heated intake manifold will very quickly be cooled by intake air at high speed and will probably improve distribution at part throttle and idle.

Turbo-superchargers, which output heated, turbulent air, are known to enhance good air/fuel mixing, and to enhance atomization.

OCTANE REQUIREMENTS

A fuel's octane rating represents its ability to resist detonation and preignition. Factors influencing an engine's octane requirements are listed below:
- Effective compression ratio
- Atmospheric pressure
- Absolute humidity
- Air temperature
- Fuel characteristics
- Air/fuel ratio
- Variations in mixture distribution among an engine's cylinders
- Oil characteristics
- Spark timing
- Spark timing advance curve
- Variations in optimal timing between individual cylinders
- Intake manifold temperature
- Head and cylinder coolant temperature
- Condition of coolant and additives
- Type of transmission
- Combustion chamber hot spots.

When an engine knocks or detonates, combustion begins normally with the flame front burning smoothly through the air/fuel mixture. But under some circumstances, as pressure and temperatures rise as combustion proceeds, at a certain point, remaining end gases explode violently all at once rather than burning evenly. This is detonation, also referred to by mechanics as knock or spark knock. Detonation produces high-pressure shock waves

Lambda Table

Load\RPM	0	500	1000	1500	2000	2500	3000	3500	4000	4500	5000	5500	6000	6500	7000	7500	8000	8500	9000
333	0.85	0.85	0.85	0.85	0.85	0.85	0.85	0.85	0.85	0.85	0.85	0.85	0.85	0.85	0.85	0.85	0.85	0.85	0.85
300	0.85	0.85	0.85	0.85	0.85	0.85	0.85	0.85	0.85	0.85	0.85	0.85	0.85	0.85	0.85	0.85	0.85	0.85	0.85
267	0.86	0.86	0.86	0.86	0.86	0.86	0.86	0.86	0.86	0.86	0.86	0.86	0.86	0.86	0.86	0.86	0.86	0.86	0.86
233	0.87	0.87	0.87	0.87	0.87	0.87	0.87	0.87	0.87	0.87	0.87	0.87	0.87	0.87	0.87	0.87	0.87	0.87	0.87
200	0.88	0.88	0.88	0.88	0.88	0.88	0.88	0.88	0.88	0.88	0.88	0.88	0.88	0.88	0.88	0.88	0.88	0.88	0.88
167	0.88	0.88	0.88	0.88	0.88	0.88	0.88	0.88	0.88	0.88	0.88	0.88	0.88	0.88	0.88	0.88	0.88	0.88	0.88
133	0.9	0.9	0.9	0.9	0.9	0.9	0.9	0.9	0.9	0.9	0.9	0.9	0.9	0.9	0.9	0.9	0.9	0.9	0.9
100	0.9	0.9	0.9	0.9	0.9	0.9	0.9	0.9	0.9	0.9	0.9	0.9	0.9	0.9	0.9	0.9	0.9	0.9	0.9
67	0.9	0.9	0.9	0.9	0.9	0.9	0.9	0.9	0.9	0.9	0.9	0.9	0.9	0.9	0.9	0.9	0.9	0.9	0.9
33	0.9	0.9	0.9	0.9	0.9	0.9	0.9	0.9	0.9	0.9	0.9	0.9	0.9	0.9	0.9	0.9	0.9	0.9	0.9
0	0.9	0.9	0.9	0.9	0.9	0.9	0.9	0.9	0.9	0.9	0.9	0.9	0.9	0.9	0.9	0.9	0.9	0.9	0.9

in the combustion chamber that can accelerate wear of an engine or actually cause catastrophic failure. Preignition is another form of abnormal combustion in which the air/fuel mixture is ignited by something other than the spark plug, including glowing combustion chamber deposits, sharp edges or burrs on the head or block, or an overheated spark-plug electrode. Heavy, prolonged knock can generate hot spots that cause surface ignition, which is the most damaging side-effect of knock. Surface ignition that occurs prior to the plug firing is called preignition, and surface ignition occurring after the plug fires is called post ignition. Preignition causes ignition timing to be lost, and the upward movement of the piston on compression stroke is opposed by the too-early high combustion pressures, resulting in power loss, engine roughness, and severe heating of the piston crown. It can lead to knock or vice versa. Dieseling or run-on on carbureted engine is usually caused by compression ignition of the air/fuel mixture by high temperatures, but can be caused by surface ignition.

Apparently identical vehicles coming off the same assembly line can have octane requirements that vary by as much as 10 octane numbers. The octane number requirement (ONR) for an engine in a particular vehicle is usually established by making a series of wide-open-throttle accelerations at standard spark timing using primary reference fuels with successively lower octane ratings until a fuel is found that produces trace knock.

The single most important internal engine characteristic that requires specific fuel characteristics is compression ratio, which generally increases the ONR +3 to +5 per one ratio increase (in the 8–11:1 compression ratio range). High compression ratios squash the inlet air/fuel mixture into a more compact, dense mass, resulting in a faster burn rate, more heating, less heat loss into the combustion chamber surfaces, and consequent higher cylinder pressure. Turbochargers and superchargers produce *effective* compression ratios far above the nominal compression ratio by pumping additional mixture into the cylinder under pressure. Either way, the result is increased density of air and fuel molecules that burn faster and produce more pressure against the piston. Another result is an increased tendency for the remaining gases to spontaneously explode or knock as heat and pressure rise.

Keep in mind, high peak cylinder pressures and temperatures resulting from high compression can also produce more NOx pollutants. Lower compression ratios raise the fuel requirements at idle because there is more clearance volume in the combustion chamber that dilutes the intake charge. And because fuel is still burning longer as the piston descends, lower compression ratios raise the exhaust temperature and increase stress on the cooling system.

Until 1970, high-performance cars often had compression ratios of up to 11 or 12 to one, easily handled with vintage high octane gasolines readily available in the 98–99 ((R+M)/2) range. By 1972, engines were running compression ratios of 8–8.5:1. In the 1980s and 1990s, compression ratios in computer-controlled fuel injected vehicles were again showing up in the 9.0–11:1 area based on fuel injection's ability to support higher compression ratios without detonation, coupled with the precise air/fuel control and catalysts required to keep emissions low. Race car engines typically run even higher compression ratios. In air-unlimited engines, maximum compression ratios with gasoline run in the 14–17:1 range.

Ratios above 14:1 demand not only extremely high-octane fuel (which might or might not be gasoline), but experienced racers use tricks like uniform coolant temperature around all cylinders, low coolant temperature, and reverse-flow cooling. Extremely high compression ratios also require excellent fuel distribution to all cylinders, retarded timing under maximum power, very rich mixtures, and probably individual cylinder optimization of spark timing, air/fuel ratio, and volumetric efficiency.

MAKE SURE YOUR ENGINE IS HEALTHY AND TUNABLE

Prior to performance tuning, a performance engine must not burn oil, which will tend to act as a pro-knock agent. The engine must have low oil consumption past rings, turbo shaft seals, valve stems, and so on. Obviously, the engine should not be leaking water internally (which would turn oil white), nor should there be excessive scale in the engine cooling jacket.

The motor itself must be mechanically sound, with adequate oil pressure under all circumstances. Most powerful EMSs can monitor oil pressure and sound alarms and take countermeasures to stop the engine if oil pressure drops dangerously low. There are labs with the ability to analyze used engine oil to check for abnormal metal content and to analyze the source of abnormal metals, which can be very useful in evaluating internal engine health.

If this is a super-duty engine with power-adders, the pistons should be thermal-coated, or the engine equipped with oil-squirters, or the rods equipped with an orifice between the big end at an angle toward the piston wrist pin designed to provide a pressurized cooling spray of oil against the undersides of the pistons to cool it.

The engine management system and sensors must be working correctly.

The external coolant system must have integrity, with the radiator working efficiently. The fans and shroud must be working efficiently, and if the car is on a stationary chassis dyno, there *must* be good, powerful fans to route cool air toward the vehicle's own cooling system. Particularly if the engine exhaust is not scavenged to the outdoors, cooling fans in a dyno cell or shop must also be adequate to keep exhaust gases from feeding back into the engine intake, which will reduce the oxygen content of air and increase hydrocarbons and carbon monoxide. Besides killing human brain cells, inadequate ventilation will hurt uncorrected performance by heat-soaking the air (perhaps unevenly in such a way that the dyno's weather station might not compensate correctly for temperature changes, which would produce inaccurate horsepower and torque readings).

ANALYZE COMBUSTION CHAMBERS

A tuner should have a good concept of combustion chamber geometry. The purpose of ignition advance is to account for the lag time following spark ignition for combustion pressure to peak at 15 degrees after top dead center. Combustion chambers with a large volume, and combustion chambers with a relatively long distance (or obstructed or nondirect pathway) from the plug to the farthest extent of end gases in the chamber require more time to burn, and more ignition timing advance. The longer the burn, the greater the possibility of remaining unburned gases exploding before normal conclusion of the burn, which is called knock or detonation. Engines with bigger bores and lower effective compression (sum of cranking compression and boost) ratios almost always require more timing advance. Four-valve pentroof combustion chambers tend to be very effective in supporting fast, efficient combustion and resisting detonation. Traditional wedge-head combustion chambers almost always have excellent squish, where charge gases are squished away from certain areas as the rising piston comes into very close proximity with the head in the final stages of the compression stroke, producing combustion chamber turbulence that enhances air/fuel mixing and spreads the flame front faster throughout the chambers.

Engines with combustion-chamber and piston-crown thermal coatings enhance horsepower and speed up combustion by keeping heat in the combustion chamber and out of the engine cooling jacket. Heat makes power by producing higher gas pressure against the pistons. Thermal coatings also resist detonation and pre-ignition by keeping metal hot spots from forming on piston crowns or head chambers.

Not being designed for racing, many street cars—particularly turbocharged cars—are not actually capable of withstanding abnormally lengthy episodes of thermal loading, but thermal coatings can help out here. The idea is that even though a really fast sports car like the Turbo Supra making full boost does not have the ability to sustain continuous full throttle at maximum speed without thermal loading overheating and eventually killing the motor, such a vehicle will reach illegal and dangerous highway speeds long before the engine is damaged.

A car on a continuous-loading chassis dyno will often reach cooling system and other thermal limits faster than you'd think. A tuner is wise to know whether or not an engine's combustion chambers have been coated. OEM turbo engines typically use piston-cooling oil squirters to protect the piston crowns from thermal damage, but this strategy does not keep heat from stressing the cooling system and the engine oil, nor does it keep the heat *in* the combustion chambers to make power (and then energetically spool turbochargers)

LIBRARY STARTUP MAPS

Startup maps are basically generic fuel curves designed to approximate actual requirements of a similar engine well enough so the engine can be started and run so it can be fine-tuned via computer. The engine displacement is less important than the cam specs of an engine. In other words, a mild-cammed 350-ci Chevrolet small block is closer in fuel requirements to a mild-cammed Chevy 400-ci small block than two 350s are to each other—if one is mild cammed and the other is wild cammed.

The following is a typical list of startup maps you might find for an aftermarket EMS.

MAP name	Engine Application
STD4	Most 4-cylinder engines
STD6	Most 6-cylinder
STD8	Most 8-cylinder
STD12	Most 12-cylinder
MOD4	Highly modified 4-cylinder
MOD6	Highly modified 6-cylinder
MODS	Highly modified 8-cylinder
TURB04	Most turbocharged 4-cylinder
TURB06	Most turbocharged 6-cylinder
TURB08	Most turbocharged 8-cylinder
MC4	High-rpm 4-cylinder (10,000 rpm+)

COMPUTING TARGET AIR/FUEL RATIOS, PULSE WIDTH, AND TIMING

Can we provide a handy formula for programming an ECU offline that will give excellent results for all load-speed combinations immediately upon startup? It is relatively easy to calculate or estimate the volume of a cylinder and therefore the weight of air that enters the cylinder under ideal operating conditions. It is also fairly easy to make a good guess at how long an injector should stay open to spray in an amount of fuel that is a particular fraction of the weight of the air (such as the 14.7:1 stoichiometric or chemically correct air/fuel ratio).

Of course, a complicating factor in designing a fuel curve is that even if you know how much air is entering the cylinders at a given time, it is not necessarily clear what air/fuel mixture is optimal. Most hot rodders want the maximum power possible at wide-open throttle, using the least possible fuel to accomplish this (lean best torque). But under some circumstances, it may be desirable to run at rich best torque (or even richer) in order to design in a safety factor to help prevent detonation with bad gas.

And what about throttle response versus efficiency at part throttle? Are you willing to sacrifice some idle quality for fuel economy or reduced emissions? Does the overlap of your cam dilute the idle mixture so you need a particularly rich mixture for an acceptable idle? These kinds of questions complicate tuning even further.

Additionally, there's the real-world performance of the circuitry that activates the injectors and the physical response of the injectors themselves, which can vary from injector to injector.

Given the complexity of designing a theoretically correct fuel map, most tuners don't even try. If possible, they use a pre-existing map from a similar vehicle or engine. Otherwise, they build a startup map designed purely to be rich enough to get the vehicle running and warmed up. This is made easier by the fact that warmed-up fuel-injected engines will run with mixtures as rich as 6.0:1 on up to lean mixtures near 22.0:1, and even cold engines are fairly flexible about air/fuel mixtures that will run the motor.

With the engine running, ideally on a dyno, tuners adjust the engine at each breakpoint combination of speed and load, and, using test equipment and a mixture trim module, set ignition timing and injection pulse width to achieve low emissions, best torque (lean or rich), or a specific air/fuel mixture—or some combination of all three. Then they road test the vehicle and fine tune it under actual driving conditions, also fine tuning the enrichment maps.

If your ECU has closed-loop capability and is running in a speed-throttle position range in which closed-loop operation is activated, your ECU may be able to tell you what the mixture would have been if you were running open-loop based on what amount of mixture correction was required to achieve a stoichiometric air/fuel mixture. You can then make corrections to the raw fuel map based on this information.

TARGET AIR/FUEL RATIOS

Gasoline, being a stew of hydrocarbons of varying structure, is not a single homogenous molecule that burns with oxygen in a precisely predictable fashion. Nonetheless, given unlimited time for combustion and a perfectly mixed batch of fuel and air, it takes between 14.6 and 14.7 pounds of air to burn a pound of gasoline in a reaction that produces 100 percent water and carbon dioxide (in this ideal reaction, the nitrogen in air is purely along for the ride). So an air/fuel ratio of 14.7:1 is the chemically correct or stoichiometric ratio, sometimes abbreviated by the Greek letter *lambda*, as in lambda of 1.0. For example, lambda of 0.9 translates as 0.9 x 14.7, or 13.23 air/fuel ratio.

In reality, in a high-speed piston engine, there is limited time for combustion, the air and fuel are *not* perfectly mixed, and tiny amounts of nitrogen will be burned by hot combustion temperatures and pressures. And inevitably some gasoline molecules will not be burned and some oxygen will not find any fuel to burn. Best power always occurs when there is at least a small fuel surplus above stoichiometric because air (oxygen) is the scarce commodity in spark-ignition piston engines. The best power strategy is therefore always to make sure there is enough fuel available so that virtually every molecule of oxygen finding its way into the engine will be able to burn fuel. The best economy strategy is to make sure there is a small oxygen surplus so that every bit of fuel will get burned. The best emissions strategy is complicated by the fact that lowest hydrocarbon (HC) emissions occur, as you'd expect, with an air surplus, but such lean mixtures produce hotter combustion, which burns more nitrogen and increases oxides of nitrogen, known as NOx. The best overall emissions strategy, therefore, turns out to be to target 1.0 lambda, the stoichiometric ratio, which does not minimize either HC or NOx, but minimizes the sum of both.

Let's first consider target air/fuel ratios for steady-state operation, when an engine is operating continuously at a particular speed and power output and the various engine fuel, air, and cooling systems have reached equilibrium.

Assuming the engine is not knock-limited and therefore does not require additional surplus fuel to cool combustion via heat-of-vaporization effects, for best torque with optimal fuel economy at wide-open throttle, aim for a 12.5:1 air/fuel ratio midway between lean and mean best torque (12.8–12.2 AFR). If you're willing to sacrifice a little torque, mixtures as lean as 0.92 lambda (13.5:1) on mild street engines that are not turbocharged will produce excellent peak-power fuel economy at the cost of about 4 percent power. Bob Norwood typically saves up a calibration by removing 2–3 percent timing and adding the same percentage of fuel.

On turbocharged or other boosted powerplants that *are* knock limited, at a very minimum you are aiming for mean best torque of 0.82 lambda (12.2 AFR), if not rich best torque of 0.8 lambda (11.76 AFR. This has the side benefit at very high speed

of increasing flame speed (which increases all the way to 0.75 lambda, at which point a healthy flame front will move through this air/fuel mixture at 80 feet per second).

Under very high vacuum (deceleration conditions), run 15.5:1 or 16.0:1 AFRs (1.05–1.09 lambda) or even fuel cut on deceleration to prevent lean exhaust system backfiring.

Mild-cammed engines with tuned runners and little valve overlap operating at full operating temperatures will idle at 14.7:1, but such a 1.0 lambda mixture is fairly close to producing misfires, and most engines produce a better, smoother idle and less stalling with richer air/fuel ratios. It is common practice to set idle lambda at 0.9 (within the correction range of the oxygen sensor) and allow the closed loop system to pull mixtures leaner to stoichiometric at idle and light cruise under O_2 sensor control. If, for any reason, the O_2 sensor fails, the engine will only idle better. Big-cammed engines will *not* idle well at stoichiometric mixtures, needing at least 13.0–13.5:1 mixtures (0.88–0.92 lambda) to idle as smoothly as possible, even at full operating temperatures.

Subject to engine VE distortions, for best drivability, air/fuel mixtures should smoothly increase fuel in a linear fashion from light cruise at low rpm toward heavier loadings and peak-torque speeds. Build in smooth increases in mixture as the engine increases in speed toward best torque and as loading increases.

The lambda table used by the Motec M48 on the MR6 project in this book as tuned by Bob Norwood sets target lambda at 0.82 at one atmosphere of boost, decreasing to 0.9 lambda at atmospheric pressure, and remaining at 0.9 lambda as manifold pressure decreases to higher ranges of vacuum.

During transient operation—when an engine is starting, warming up, accelerating, or decelerating rapidly—ordinary target air/fuel mixtures are no longer applicable; instead, the engine management system will apply corrections to the base air/fuel ratios to supply additional fuel.

MANUALLY COMPUTING TARGET PULSE WIDTHS

But let's not give up, let's actually compute some target injection pulse widths. Keep in mind that injectors are not accurate much below about 1.7 milliseconds (a millisecond is 0.001 of a second). In addition, some injection systems will not open the injectors more than 16 milliseconds under any conditions. Also keep in mind that the amount of time available for injection is determined by the speed of the engine. At some point, the available time for injection and the required pulse width crisscross, and the injectors become static, that is, open all the time. When in this realm, you have lost control of fuel flow in relation to engine fuel requirement changes, and additional engine speed will *decrease* the total fuel injected and available per power stroke per cylinder. Since this would most likely occur under conditions of high loading at high speed, additional loading could lead to disastrously lean mixtures. Check to make sure that the injection pulse width is less than 56,000 divided by rpm. For example, at 6,000 rpm, the available time for injection is 56,000/6,000, which is 9.3 milliseconds, so a specified pulse width greater than 9.3 milliseconds would drive the injectors static. Most OEM fuel injection systems are designed to never require an injector duty cycle greater than about 80 percent, and research by fuel injection expert Russ Collins indicates there is good reason for this. According to Collins, injectors can begin to fibrillate at such duty cycles, becoming nonlinear in fuel delivery, potentially causing lean-out problems. In practice, I've seen many engines run great to pulse widths of 100 percent or above, but it is *not* a good idea.

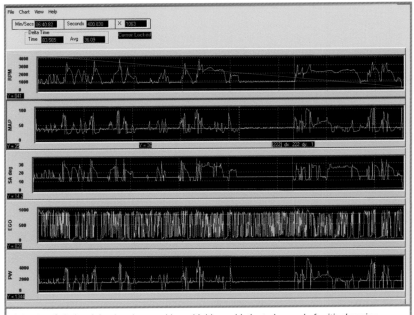

Injection Solutions' datalogging provides a highly sophisticated record of critical engine parameters. The data shown here as a function of time are injection pulse width, exhaust gas oxygen (air/fuel ratio), spark timing, manifold pressure, and rpm.

$0.002261428 \div 12.5$ (target air/fuel ratio, peak torque) = 0.000180914

$0.000180914 \div 2 = 0.000090457$ (pounds fuel per injection, batch fire)

30 pounds per hour injectors ÷ 3,600,000 = 0.000008333 = pounds fuel per millisecond

$0.000090457 \div 0.000008333 = 10.85$ ms pulse width at peak torque with 30 pounds per hour injectors on 454-ci engine at 90 percent VE

Remember, most EFI regulators vary fuel pressure to keep it constant in relation to manifold pressure. If necessary, convert fuel volume to weight (using formulas in the injector selection section). Convert fuel flow in weight per hour to flow per second by dividing by 3,600. For example, 24 pounds per hour divided by 3,600 equals 0.0066667 pounds per second.

Now, divide the fuel flow in weight per time by the fuel weight calculated to achieve the desired mixture. This will tell you the fraction of a second you need to open an injector in order to inject the right amount of fuel. Injectors do not open instantly, but assume this is figured into the pulsed flow rate. Convert this time to thousandths of a second, if necessary, always rounding upward. Don't forget to divide this in two if your injection system uses two batch fires per power stroke. Perform the necessary math if you are using fewer throttle-body injectors. Make sure the pulse width is less than the lesser of 16 milliseconds or 56,000 divided by rpm, but greater than 1.7 milliseconds.

Most EFI system software gives you the ability to set multiple speed-load points at once and to modify multiple points by a given percentage or to a given slope. You should not have to manually set all 256 or 512 breakpoints in the startup map.

Maybe you don't want to mess with theoretical mixtures. Given the flexibility of engines to run with widely varying air/fuel mixtures, in the old days Haltech simply recommended building a startup map by setting light-load bars for all rpm ranges to 2.5 milliseconds, sloping pulse width smoothly upward to 7.5 milliseconds in the higher manifold pressure or wider throttle angle bars.

These days, of course, in many cases, a wizard will do most of the math for you.

ACTUAL EMS CALIBRATION

Preliminary Pre-start Checklist
• Make sure there is a working fire extinguisher within easy reach.
• Verify good ground connections to the engine, cam covers, and intake.
• Verify switched and constant power to critical sensors and actuators.
• Connect a laptop with the appropriate (latest revision) EMS software to the electronic control unit (ECU), and—with the key on, engine stopped—verify that the ECU, sensors, and actuators are communicating with each other, with plausible data from all sensors via the engine data page screen:
– Engine temperature gauge (coolant or head temperature) on a cold engine should read ambient temperature.
– Intake air temperature (IAT) should read ambient temperature (same as engine temperature).

If you don't have a VE chart, but know the peak torque speed for the engine, assume 80 percent VE for a street engine at peak torque, 90 to 100 percent for racing engines, and 100 percent or more for supercharged engines. This figure can be used to compute peak fuel injection requirements, and is identical to what Electromotive calls user-adjustable pulse width, the required pulse width when the MAP sensor is reading full scale (though some forced-induction engines cannot achieve or safely operate at full scale on a multi-bar MAP sensor).

To compute fuel injection requirements, first compute the weight of air in the cylinders. Convert static displacement to displacement at volumetric efficiency (80–100 percent). Divide displacement in cubic inches by cubic inches per cubic foot (1,728), which gives you engine displacement in cubic feet. Multiply this by pounds of air per cubic foot (0.07651) at standard temperature and pressure, which gives you pounds of air that enter the engine per two revolutions. Divide this by the number of cylinders to get air weight per power stroke at each cylinder (injector). Divide this figure by the target air/fuel ratio to get pounds of fuel per power stroke. Divide this number by the number of injections per power stroke (usually one for sequential injection, two for batch-fire port injection), which gives you the weight of fuel per injection. Compute the pounds of fuel injected per millisecond of injector pulse by dividing the pounds per hour of fuel flow by 3,600,00.

And, finally, divide the weight of fuel per injection by the weight of fuel per millisecond of injector pulse to get the number of milliseconds of pulse width to set calibration. Here's an example:

454-ci engine x 0.9 VE = 408.6-ci air drawn into engine at peak torque

$408.6 \div 1728 = 0.236458333$ cubic feet of air drawn into engine

$0.236458333 \times 0.07651$ (pounds per cubic foot air) = 0.018091427 pounds of air entering engine per two revolutions

$0.018091427 \div 78 = 0.002261428$ (pounds per power stroke at each cylinder)

– MAP (manifold absolute pressure) sensor should read atmospheric pressure 1 bar (100 kilopascals).

– The O_2 sensor should heat up (you should *always* use an electrically heated sensor), meaning the sensor body should feel warm after a little while. If the O_2 sensor is out of the exhaust system, you should see the tip get red hot shortly after you energize it by turning the key to *on*. If you are using a wideband O_2 sensor, the laptop should display a 15.9 air/fuel ratio once the wideband sensor has heated and begun functioning.

– The throttle position sensor (TPS) should move from 0 to 100 percent when you open the throttle; if not, virtually all EMSs now have a calibration facility that allows you to mark the closed and fully open TPS voltages as 0 and 100 percent throttle.

• If the ECU has injector test functions like a Motec M48, test each injector, listening for a rapid clicking noise from every injector.

• Make sure the spark plugs are new, of the right heat range, and gapped correctly for the engine in its current configuration. Turbo conversion engines usually run a much narrower gap than stock (that is, 0.029 or less, versus, say, 0.044)

• Cranking an engine when the injectors are washing the cylinder walls with raw fuel while there is no functioning spark to light off the engine is not something you want to happen. If the ECU has spark test functions, pull the high-tension coil plug wire and set the end near a grounded metal surface of the engine, run the test, and watch for rapid sparking. If the engine has direct-fire capability, you may need to unscrew the spark plugs,

re-insert each in the coil-on-plug connector, and ground the plugs. In some cases—like the Motec M4-8 with direct-fire expander module—you may need to redefine the ignition as single-coil and then jumper spark to each individual low-voltage coil wire at the ECU connector to watch for a spark at each plug. If there are no spark test ECU routines, disable the injectors—or make sure there is no fuel pressure—and crank the engine to test the ignition as above.

• Check the firing order *one more time*—for plugs, direct-fire plugs, and injectors if it's SEFI. Make sure you have the official engine firing order and know how the engine manufacturer numbers the cylinders. With a distributor ignition engine stopped at top dead center at number-one cylinder on the crankshaft timing mark on the compression stroke (both valves *closed*), make sure the rotor is pointing at the number-one distributor lead and that the successive distributor wires in the direction of distributor rotation are routed to the correct cylinders in the firing order. A direct-fire ignition module is normally hardwired with certain pins on the connectors designated first to fire, second to fire, and so forth. Make sure the wiring is right.

• Attach a pressure gauge to the fuel rail. Jumper the fuel pump to prime the system and check for pressure of 39–45 psi with no leaks. De-energize the fuel pump and watch the rail pressure closely. The fuel pressure regulator should hold steady pressure for a long time, meaning minutes rather than seconds (the shop manual for the car may have a design pressure drop per time specification). If pressure drops rapidly (seconds rather than minutes), the regulator is defective or stuck partly open with dirt or

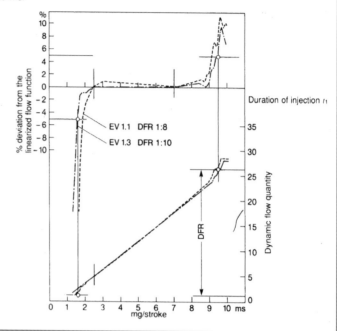

Characteristic curves.
DFR (dynamic flow range): Variation ratio between minimum and maximum injected fuel quantity at maximum 5% deviation from the linearized flow function. Periodic time T = 10 ms.

This chart effectively illustrates the performance of Bosch fuel injectors at various pulse widths. For a workable calibration, it is critical to stay within the range of excellent repeatability for the injectors in use, since the variation ratio between minimum and maximum injected goes beyond the critical 5-percent limit at very short and very long pulse widths. Bosch

debris, *or one or more injectors are stuck open.* Injectors that have sat for a long time in a wrecking yard are particularly susceptible to sticking. When you're done testing, duct tape the pressure gauge in place on the windshield or other location where it can be seen after the engine starts and eventually in early test driving or on the chassis dyno.

• If there is any possibility one or more injectors are leaking, it is critical to *remove the spark plugs and injector wiring and crank the engine* to purge any gasoline from the combustion chambers—and then fix the problem by replacing or cleaning the injector(s). When a piston is stopped very near before top dead center and a leaking injector is filling up the combustion chamber, the crankshaft will have extreme leverage against a combustion chamber full of gasoline because the crankshaft moves many degrees of rotation to produce the tiniest piston movement. *This type of Hydro-lock may not just stop the engine dead in its tracks; it can break rods and pistons and smash broken rods though the block.*

• Attempting to start an engine without an rpm or crank trigger signal can be a really bad idea. Disconnect the injector wiring so the engine will not start (unplug the injector harness or unplug all the individual injector connectors), and crank the engine. Verify that the rpm display on the engine data page of the laptop shows a plausible non-zero cranking rpm.

• Once you definitely have a correct rpm, connect a timing light to the number one cylinder and verify that the spark timing as shown by the timing light matches the laptop's reported timing on the engine data page. If it does not, if the engine uses an adjustable distributor ignition, the distributor might be indexed wrong, or the ECU might be incorrectly configured for crank trigger offset. Rotate the distributor until the actual timing matches the ECU's target timing. If the distributor indexing cannot be adjusted or if the engine uses direct-fire ignition, reconfigure the crank trigger angle in the EMS software until actual timing matches the ECU's current timing.

• It is *critical* to verify that the fuel injectors are sized correctly for the engine so they are small enough to idle well and large enough to fuel the engine at maximum torque and power, preferably at less than 80 percent duty cycle. Injectors that are too big will *never* idle well, and injectors that are too small could kill a super-high-output, boosted, or nitrous engine with lean mixtures and detonation. Verify that the fuel pump is big enough to deliver enough fuel at maximum power at the required pressure.

• Is your application beyond the envelope of the EMS in a way that is at all questionable?

• Analyze potential risks and remember Murphy's Law (if something *can* go wrong, it will). Do you have high enough octane fuel to protect the engine from detonation while you are calibrating it (unleaded racing gasoline is available up to about 114 octane, leaded to about 118)? Too much octane never hurt anyone, but too little has killed many an engine. Are the plugs cold enough and gapped narrow enough (a turbo conversion should use 1–2 ranges colder than stock and a maximum of 0.030 plug gap)? What is a realistic redline (even with super-duty parts)? Too much rpm will try to stretch rod bolts when the rod tries to pull the piston away from TDC on the intake stroke when there is no pressure helping move the piston, and rod stretch is typically much harder on the rod that boosted combustion pressure. How will you know if the engine is knocking? It can be difficult or impossible to hear. What could make the engine knock? *Nothing*—if you're running sane levels of boost, enough octane, enough fuel, safe coolant temperatures, cold enough plugs, and enough ignition retard. Is there a strategy in place to prevent boost overrun and overboost? Do you have warnings in place (gauges, audible, warning lights, ECU warnings, helper(s) watching gauges, etc.) in case of dangerous lean mixtures, knock, fuel pressure drop, injectors at 100 percent, low oil pressure, overheating, overboost, leaks, etc.? Are you checking the plugs between test runs for signs of thermal overloading or other problems? Should you be running EGT probes in the headers to watch for serious combustion over-temperatures? Have you taken the easy way out on *anything* that might matter? For example, is your aftermarket knock sensor in the best place or just the easiest? And last, but not least, how close to the hairy edge can you afford to push it? Can you afford to have the engine fail?

INITIAL STARTUP CALIBRATION

You will need a calibration sufficiently accurate to get the engine started—at which point the engine can be accurately tuned using feedback methods—but many aftermarket engine management systems have now automated the process of building a safe and workable startup map. Many systems are equipped with modeling software that will construct a custom startup map for your engine good enough to get it running the first time—based on a set of configuration parameters you enter via the laptop user interface (number of cylinders, estimated horsepower, injector size, and so on). Some are *very* good at building a workable map that will actually run and drive the vehicle quite well.

If there is no auto-map software for your system, select the closest library calibration that's available for your EMS on CD or at the vendor's Web site, and download it into the ECU. A map from an identical engine would be ideal, but there are so many factors affecting what makes the optimal calibration that even minor modifications could render an otherwise perfect map suboptimal. What's more, even a really great map developed by experts in white coats on a million-dollar research dyno from an otherwise identical engine will be wrong if the injectors used in the calibration were a different size (and there is virtually always a *range* of injector sizes that will work for a given engine). Even if the injector sizes were the same, the perfect calibration might not work perfectly if, for example, the fuel rail pressure was different at the dyno. You simply cannot assume any calibration is perfect without experimental evidence proving this is the case. It is disheartening how many engines are driving around that could have free power and better fuel economy from a more optimal calibration. In any case, a good candidate startup map *must* be from an engine with the same number of cylinders, with relatively similar VE characteristics (including power-adders or the lack thereof, cam timing, and head-flow characteristics), displacement, and similar-sized injectors. If the injector size is wrong, there is probably a way of rescaling the map's injection pulse width to match the variation in injector flow.

You will almost definitely have to enter some configuration data that defines certain global operating characteristics required to match the EMS with the engine (number of cylinders, ignition type, firing order, and so on).

At this point, the library or auto-map calibration will probably start the engine. If not, the EMS may have a trim module with one or more dials you can use to provide across-the-board enrichment or enleanment, usually in the plus or minus 10 to 30 percent range. Swinging the trim pot(s) while cranking may help get the engine running, particularly if the problem was a lean mixture. A trim module can be very useful during the initial warm-up.

Once the engine is running, try to *keep* it running until fully warm. If the engine is a brand-new, 0-miles powerplant, make sure to look for leaks and perform all the other maiden-voyage checks required to make sure a new engine doesn't get damaged from low oil pressure, lack of coolant, or incorrect assembly.

When the engine starts, the manifold pressure (MAP) sensor readout should instantly switch from atmospheric pressure to a high-vacuum low-flow airflow rate much closer to 0 kilopascals if the system uses a MAP sensor to estimate engine airflow. As the engine warms up, the engine temperature sensor readout should begin to climb from ambient temperature toward normal operating temperature in the 180 to 200 range. Make sure the EMS is set up for Fahrenheit or Celsius, depending on which you're more comfortable using. The heated O_2 sensor readout should begin to display a plausible air/fuel ratio, which may be displayed as lambda, the ratio of the air/fuel ratio divided by the 14.68:1 chemically ideal stoichiometric air/fuel ratio. Thus, an air/fuel ratio of 13.5:1 is about 0.92 lambda. Once the O_2 sensor is hot and alive, when the engine is warming up and in open-loop mode (no automatic mixture trim), adjusting the mixture-trim module should cause the O_2 sensor readout to vary with changes in mixture trim in the expected direction. Beware being misled by a narrow-band O_2 sensor; such sensors are increasingly inaccurate as mixtures vary from stoichiometric. Even wideband sensors (which a few EMSs now support for closed-loop mode) can be fooled if mixtures are far enough away from optimal to cause misfiring (rich-mixture misfiring can fool the O_2 sensor with big gusts of unburned oxygen).

When the engine reaches about 140 degrees Fahrenheit, if configured to allow closed-loop mixture trim, the EMS will start feedback mixture trim based on data from the O_2 sensor. At this point, the EMS will have some means of indicating that it is in closed-loop mode. For example, a readout may begin displaying the amount of fuel the engine is adding or subtracting to achieve stoichiometric ratios. Some EMSs have air/fuel ratio tables that allow a tuner to specify nonstoichometric target mixtures for closed-loop operations in various speed and loading segments of the engine operating range.

The trick with closed-loop operations is to tune the raw fuel curve of the engine so the engine runs great under all circumstances even if closed loop is turned off. Under no circumstances should uncorrected mixtures ever be leaner than stoichiometric. If the O_2 sensor fails, the system will default into open-loop mode, which you always want to be perfect or on the rich side. The base fuel maps should be constructed so the closed-loop system is always either idle or *pulling fuel*, never working to *add* fuel to the base map. Keep in mind that the O_2 sensor is not *in* the combustion chamber but rather is located downstream somewhere in the header(s) or exhaust system, and there is a varying degree of lag in the O_2 sensor reading, depending on engine speed and exhaust velocity. A properly calibrated EMS should *fail safe*, which, in this case, means fail *rich*. Typically, OEM EMS closed-loop systems are clamped to limit the maximum fuel trim to a rather small percentage; if the system is adding or subtracting huge amounts of fuel, something is probably really wrong. Later on, we will talk about using exhaust gas temperature readings to tune safely at high power levels, but even EGTs have lag time, and combustion temperatures are typically much hotter in the chambers than even a short distance downstream in the exhaust. Bottom line, the base fuel map should err on the rich side, but never by more than about 5 percent.

Whether or not you plan to run closed loop once the EMS is optimized, you need wideband capability to dial it in. Keep in mind, narrowband O_2 sensors are not capable of providing accurate readings very far from 14.68:1, that is, below 14 or much above 15 (though they will immediately and clearly warn of a too-lean condition without the driver having to think through what it means). The automatic start-up mapping facility may produce a map that will start and operate the vehicle with decent drivability, but air/fuel ratios and timing will almost definitely not be optimal. Therefore, you're going to need wideband O_2 capability to find the "free" power and efficiency in your engine. It is a fact of life that virtually all pro tuners work with wideband O_2 sensor data in all tuning operations, and many Dynojet chassis dynos are now equipped with wideband capabilities that log air/fuel ratio information and graph it along with power and torque. Alternately, some of the newer EMSs have a wideband option, which may allow the EMS to read and log data from a five-wire O_2 sensor (which typically costs hundreds of dollars), though not necessarily to use for closed-loop feedback operations. Narrowband O_2 sensors deliver a voltage between nearly 0 and nearly 1 volt (a few are frequency-based). Wideband sensors typically deliver a 0–5 volt signal that is *not* compatible. More rarely, some EMS systems can now work with a conditioning processor that converts wideband O_2 data to the 0–1 volt signal needed for actual closed-loop trim operations (an example is Link's UEGOlink). Another option is employing a standalone wideband air/fuel-ratio meter that is not part of the EMS but instead indicates the air/fuel ratio in real time on a digital display or analog gauge needle. Wideband sensing of any variety greatly increases the accuracy of tuning under all conditions, and is accurate in detecting air/fuel ratios from about 10:1 all the way to above 16:1. The trouble is, standalone AFR meters cannot be correlated with datalogging information from the EMS, but they

are very good, nonetheless. Most ambitiously, some EMSs now have an auto-mapping capability that enables the ECU to work with a wideband sensor to build correction tables that trim injection pulse width to achieve the values in a target air/fuel ratio table (assuming you know the ideal AFRs, which frequently vary according to the engine or engine type). Such systems are good for driving around town and allowing the system to tune itself at light and midrange power settings, but I would *not* trust such a system to find optimal AFRs for high-boost, high-power operations. You're going to want to sneak up on dangerous high-power air/fuel ratios manually.

VE TABLES

VE tables are useful for engines that are not equipped with mass airflow sensors, and which rely on manifold absolute pressure and engine speed to *estimate* engine airflow. VE tables correct for the fact that only engines with a perfectly flat torque curve have a perfectly linear relationship between manifold pressure and engine airflow across the rpm range.

VE tables add a kind of elegance to engine management system design that does not exist on EMSs that simply look up fuel injection pulse width or timing advance in a table of rpm and manifold pressure cells. Although *any* speed-density table of timing or pulse width data is going to correlate roughly with engine VE, a pure VE table describing the engine's breathing efficiency across the range of operation is uncomplicated by competing purposes such as the desire to optimize some portions of the operating range for efficiency or low emissions and others for maximum power or the need to warp fuel delivery or timing to cool combustion or fight detonation in certain areas of the map. Any of these, for example, could cause a speed-density-pulse width table to depart from perfectly describing VE. An engine management system in possession of reliable engine VE information in effect knows actual mass airflow into the engine and can use this information in a variety of ways to optimize engine management in various types of self-learning and targeted optimization strategies.

The engine management system's volumetric efficiency table is a critical chart of numbers that represents the engine's breathing efficiency across the range of engine speeds and loading. The VE table essentially tells the engine management system what percentage of the engine's displacement will make it into the cylinders at each and every possible combination of rpm and manifold pressure. Manifold pressure is itself a function of engine breathing capabilities, throttle angle, and the operating status of installed power-adders. VE is dependent on the basic engine design, including head design, camshaft profile, power-adders, and engine speed.

Engine VE typically varies somewhat from 100 percent in a positive or negative direction (usually negative), depending on what the engine is doing and how it is designed. Once the system knows how much air there is in the cylinders, computing how much fuel the engine needs is simple.

There are an infinite number of possible engine speed and loading numbers, so various engine management systems settle for VE tables of a fixed number of cells (such as 16x16 or 20x40), with points falling in between the resolution of the table calculated as a weighted average of the closest defined cells (sometimes referred to as breakpoints). If an engine is breathing an amount of air equal to 100 percent of its displacement, the VE is 100 percent. Depending on the table, this number may be represented as a percentage (1.00), or a plus-or-minus offset (0.00), or even as the fuel injection pulse width (for a certain size injector) that

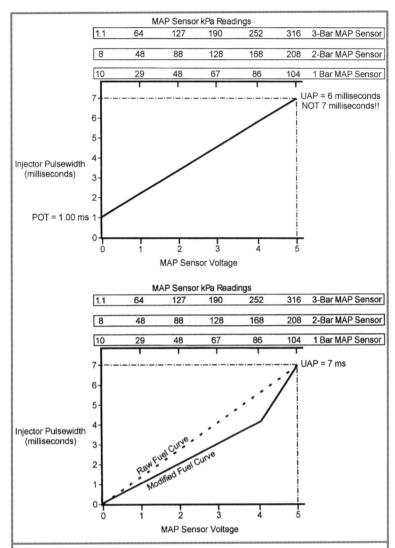

The scope of fuel calibration is reflected in these Electromotive graphs. The company's TEC[3] logic defines raw injection pulse width in terms of a line defined on the left by the minimum "Pulse width Offset Time" as MAP voltage approaches 0, and on the right by the maximum "User Adjustable Pulse width," the maximum possible injector open time at maximum MAP voltage. The bottom graph shows how this raw fuel curve can be modified by adjusting a separate VE table to reflect changes in maximum torque with engine speed on engines where engine VE varies significantly from a flat torque curve. Electromotive

air/fuel ratios to achieve varying purposes at various engine loadings (for example, maximum power at full throttle, maximum efficiency at light cruise).

Logically, a VE table looks something like a big spreadsheet with one floating point number in each cell or box, each of which represents a particular combination of engine speed and loading. The good news is that, once defined, the VE table of an engine never changes (except as the engine begins to wear out or if you modify engine breathing with new hot rodding equipment). If you have a VE table, setting it accurately is a convenient tool a little like the concept of projecting the map of the world onto a flat surface so you can measure distances and draw borders with a straightedge.

From one point of view, a VE table projects or normalizes the convoluted surface of a 3-D graph describing engine airflow (therefore, torque) onto a flat surface, so all other calibration tasks can assume the torque curve to be flat. If torque is flat (or there is a VE table), a straight line drawn between maximum injection at peak torque rpm and minimum injection at idle rpm accurately intersects all other points, so all you have to do to tune the engine is to define these two points.

From another point of view, if you know the mass of air entering an engine, fuel injection pulse width is one simple calculation for best power, another simple calculation for best economy. If you've got a good VE table, tinkering with a few values at the extremes defines everything else. If you have a lambda table, getting it right will make everything else easier. If you have a pulse width table, the surface will look a lot like the surface of the engine VE table except that it will be warped higher in the richer high-power areas in relation to other light-cruise sectors of the map where you're probably tuning for leaner lambda 1.0 mixtures. In any case, if your EMS has a VE-table architecture, getting the VE table right turns your MAP sensor into an accurate meter of actual engine breathing rather than a tool that's not entirely dependable and that you'll struggle with to compensate for with other means.

So how do you build a VE table? Some experts recommend setting acceleration enrichment before messing with the VE table on an EMS with a specific VE table like the Electromotive TEC[3]. However, this presupposes that the main fuel table is reasonably correct. In this procedure, we stick with the default transient enrichment settings and focus on getting the VE right first in steady-state running.

In the first place, the engine should be at normal operating temperature, with a plausible, conservative base timing map. The trick is to work first on defining the main VE table that will greatly affect fuel delivery by entering accurate values in as many cells as possible to optimize basic fuel delivery before moving on to optimize spark timing and other parameters. Keep in mind that some cells of a VE table are not reachable (for example, extremely high levels of boost at, say, 500 rpm), though you should enter plausible values *everywhere*, since the values in unreachable cells may be averaged into the closest bordering cells that *are* reachable, since most EMSs maintain a moving average across the several closest cells to the current rpm and manifold pressure.

is correct for that mass of air equal to 100 percent of displacement divided by the number of cylinders.

The table of optimal spark timing advance values will also correlate highly to the VE of the engine and take into account time per degree of engine rotation at various speeds, since denser charge mixtures at points of higher volumetric efficiency burn faster and require less spark timing. Again, keep in mind that if extra fuel is being used to cool combustion under boosted conditions as an antidetonation countermeasure, or if spark retard is similarly being used to fight detonation, the numbers in such a table will depart from engine VE in these areas. Which is why some EMSs have a pure VE table one step removed from the actual base injection pulse width or timing. As mentioned above, the numbers in a main fuel table will also not correlate perfectly with engine VE if injection pulse width is optimized for varying

One good way to build a VE table is to use the O_2 sensor to target identical air/fuel ratios in all speed-loading points, which is made easier by the fact that many EMSs now have a target AFR table. If you have a wideband O_2 sensor, you can set all cells in the target lambda (air/fuel ratio) table to 12.8:1, which should be safely rich enough to set VE in cells referencing high power and torque settings on all but the most radical high-boost powerplants (but not so rich as to harm idle quality or produce surging at light cruise). If you are using a narrowband sensor, it will not have the authority to accurately distinguish air/fuel ratios this rich. Some people suggest that on a normally aspirated engine without too much compression and power, you can try setting up target ratios of 13.2 (0.89 lambda) in all cells and run the process, but standard O_2 sensors are not particularly accurate in this range so far away from lambda 1.0. Under no circumstances should you attempt to set VE at the more dangerous high-boost VE cells with just narrowband data.

Turn off the parameter that deactivates closed-loop auto correction in conjunction with the target AFR table, but make sure the O_2 sensor readout stays active.

It is a good idea to get some idea of which cells are being used by the EMS to fuel the engine by using a cell highlight if there is one to mark the speed-density cells as you operate the engine. This will enable you to concentrate on correcting the appropriate cells first.

Working cell by cell on the road with a tuner riding shotgun with the laptop and a driver with one foot on the gas and one foot on the brake (hopefully on a deserted country road or test track) working to achieve target rpm and engine loadings (or, even better, using a load-holding engine- or chassis-dyno), adjust as many cells as you can reach in the VE table to bring the actual values in the target air/fuel table back to the identical target 12.8:1 (or 13.2) air/fuel ratio. The trick is to work methodically, staying away from sudden acceleration, stabilizing the vehicle at the various breakpoints before moving to the next.

If you have a fresh, accurate wideband sensor, a tankful of racing gasoline and a known-conservative startup map, it may be feasible to turn on auto-correct closed-loop operation and drive around, noting the amount of fuel the system is having to subtract or add (hopefully not much) to achieve the target ratios, fixing the VE table as you go.

Keep in mind that the trick is changing not just one cell, but smoothing in the surrounding cells to form a smooth slope or graph. If you do not change surrounding cells along with the target cell, you will probably not get the full magnitude of

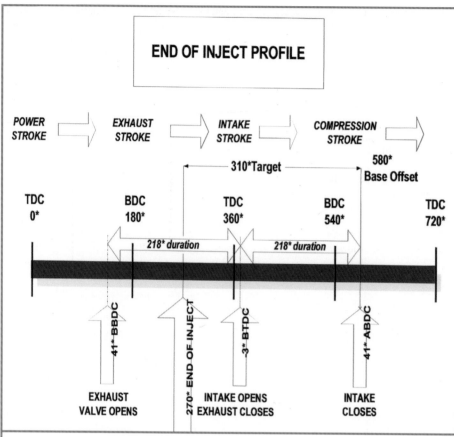

END OF INJECT PROFILE

With sequential fuel injection you will probably have the option to define injection phasing, in particular, the engine position in the 720-degree four-cycle by which injection must conclude (or start). Injection timing can typically be adjusted at various rpm points and optionally at various efficiency points. Adjusting the injection timing allows the fuel to be injected at the optimum point in the engine cycle so the fuel is used most efficiently. Engine power, emissions, economy, and idle stability may be optimized by adjusting the injection timing.

changes reflected in actual injection pulse width, since the EMS will average out sharp peaks and valleys. So change one breakpoint a modest amount, then switch over to graph mode and pull the surrounding cells up or down as required to make a smooth slope. If this is not enough, do another iteration. What will *not* work well is to yank one cell to ridiculous heights before getting around to other nearby cells.

On some systems you can datalog the actual air/fuel ratios as you drive the vehicle through the range of speeds and loading, and have the system tell you the pulse width required to achieve entries in the target ratio table—or even do the actual pulse width compensation. Using datalogging this way can be efficient because a relatively quick test run can be made to encounter *many* cells in the air/fuel or VE table, which can then be fixed following the run, offline if you like. If is also safer, because a brief foray into each cell is enough, with no requirement to linger in high-power cells while you make sense of what is going on and modify the VE or injection pulse width. However you do it, work your way through the table, adjusting VE to bring the actual AFR values in the fuel table back to 12.8:1 (or 13.2:1 if narrowband). Beware of the most dangerous high-boost cells on a 2- or 3-bar boosted engine.

Note: Some highly sophisticated EMSs do *not* use a special VE table. Instead, they use VE-type tables of pulse width and timing advance data to manage the engine, and you will have to optimize each entry for best power, efficiency, or lowest emissions. Even if you do have a VE table, you probably still want to experiment to optimize the target air/fuel ratios, since on a modified engine, it is difficult or impossible to predict in advance what the optimal air/fuel ratios will be. For that, the trick is to use a dynamometer or an inertial accelerometer device like the G-Tech. That said, unless you're patient and experienced, the above method of targeting fairly well understood target ratios with the help of a wideband sensor and varying degrees of auto-correction could produce better results.

OPTIMIZING THE FUEL MAP

There is simply no way to fully optimize an engine management system without a dynamometer and wideband air/fuel-ratio meter or fast exhaust gas analyzer. The combination of target AFR tables and closed-loop operations using wideband O_2 and knock sensors can produce good drivability and compensate for funky base maps. But to find 100 percent of the free power where you need it and best economy where you want it, you'll need a dynamometer and wideband AFR capabilities. You will no longer be searching for a particular air/fuel ratio, but instead looking for mean or even lean best torque. You may decide to use exhaust gas temperature probes as a safety measure to detect dangerously rising EGT before it is too late. If you do not have a dyno, you will definitely need a good AFR readout for tuning. If you are *extremely* patient and methodical, air/fuel ratio information and a stopwatch can yield good results. I would *not* try this; in lieu of a dyno I'd spend the $140–250 for an accelerometer device like the G-Tech that essentially turns your vehicle into an inertial dyno. If not, it is virtually guaranteed that your map will be suboptimal. To get it really right you *need* scientific backup. *Seat of the pants will* not *cut it,* and do not believe anyone who says otherwise.

However, an engine calibrated on a dynamometer of almost any kind will almost definitely run suboptimally rich on the street and track. To produce smooth performance under all conditions, when you are done with a dyno calibration, you'll need to road test and recalibrate to eliminate hesitation, stumbling, and surging in ordinary driving, as well as hunting at idle. An accelerometer device is invaluable for street tuning, as is a wideband readout. But be on the lookout for black exhaust smoke and lazy response, which indicates rich mixtures, or intake backfiring, coughing, and misfiring, which indicates lean mixtures. And *always* listen for engine-damaging detonation, meaning never tune with the stereo blaring and the air conditioning at full roar.

To avoid detonation and lean-mixture meltdowns, always work from rich to lean. Start by tuning the engine at *idle* in *open-loop* mode but with the O_2 readout turned on. The target idle AFR for engines with mild, streetable cams is 14.68:1, also known as stoichiometric, but you probably want to tune *slightly* richer and let the closed-loop system pull fuel under actual operations. This ratio should produce a good idle (though probably *not* the absolute fastest idle, which occurs rich of stoichometric), and it is most compatible with catalytic converters and narrow-band closed-loop O_2 feedback operations.

Even with port-EFI, engines with big cams may need richer AFRs in the 13s to idle well due to overlap-induced reversion that can blow charge the mixture back into the intake runners. If you are not using timed sequential fuel injection, this can really help, and if you are, you may find that changing the phasing of the end of injection helps to minimize the impact of idle-spoiling reversion effects.

Engines with truly radical cams (with a lot of duration and overlap) may hunt at idle even with rich air/fuel ratios due to an inherently unstable and wandering manifold pressure causing the system to chase its own tail, so to speak. If such an engine still has a MAF sensor, MAF accuracy and operation could be compromised by reversion and turbulence in the inlet tube near the MAF, and variations in airflow could, again, feed back and make it hard for the engine to idle well. Throttle position, however, will *not* change when an engine is idling, so an EMS is likely to achieve a better idle with difficult engines by disregarding the MAP signal and using throttle position sensor (TPS) data as an indicator of engine loading instead of manifold absolute pressure (MAP). Alpha-N systems manage an engine using tables of pulse width for rpm versus throttle position. Alpha-N works well on some engines, particularly on light vehicles like performance motorcycles, but it does not work well on engines that require significant changes to fueling based on minor loading changes when there might be *zero* change in throttle position. For this reason, some systems like the Electromotive TEC[3] and the Motec M4-8 provide the option to blend TPS and MAP data so the engine gets fueled at idle and light loads based on throttle position, but transitions to MAP-based fueling at higher levels of engine load (particularly during boosted operation when manifold pressure rises above atmospheric pressure). ECUs set up to estimate engine load based on throttle position may need a fairly steep increase in pulse width at light loads, particularly on systems with large throttle areas in which there is a large change in airflow for a small throttle angle change.

When you have a good idle, turn on the air conditioning, which will noticeably load the engine harder. Some engines have an EMS-actuated dashpot that immediately kicks in when the engine is really cold or when the air conditioning is energized to initiate a fast idle. You might want to consider using it with an aftermarket programmable EMS or just wiring it to the air conditioning clutch circuit to come on when the air conditioning compressor comes on. Although most EMSs now have idle speed stabilization logic that can regulate an IAC stepper-motor to bypass air around the throttle plate to the varying degree required to maintain a target idle speed under all conditions, the IAC system may not be fast enough to deal with air conditioning–induced load changes at idle seamlessly without an annoying momentary drop in rpm when the air conditioning kicks on.

If you're lucky, the air conditioning loading will cause the EMS to switch to another speed-loading cell, which can be optimized for the best idle with the air conditioning on (perhaps with a little more timing or fuel to instantly stabilize the idle). It is for this sort of reason that sophisticated EMS systems like the Motec provide for user-defined nonlinear granularity of rpm and loading ranges. You can create narrow ranges where required to allow the system to provide special fueling or timing in response to subtle changes in engine speed or loading. Most modern EMSs now have sophisticated IAC logic that enables you to command the rate at which the system reexamines idle speed before it can take corrective action, as well as the magnitude of the initial step taken in a corrective direction based on how wrong the idle is. Reprogramming the IAC system can help a lot. Some systems have special functions to handle air conditioning problems.

Move to higher rpm ranges, still with zero load, stabilizing at a target speed, and set the injection pulse width slightly rich of stoichometric. Do *not* get involved at this time in optimizing acceleration enrichment. Stabilize the engine speed steady at one place and hold it there while you trim fuel until the AFR readout shows no more than 5 percent rich of 14.68 or 1.0 lambda.

When you've got the zero-load cells set, smooth any jagged transition to adjoining cells in the fuel table, find a helper, and go for a drive on a safe deserted road or test track. Try to drive the car at a steady state at as many speed-loading cells as possible. It will help the driver if the laptop can be setup with the display magnified so it's easy to read from a distance and positioned so that the driver can use the screen to home in and make a certain cell current—and then back off while the tuner riding shotgun modifies this and surrounding cells. The target AFR for light cruising is 14.5–14.9 or 0.99–1.01 lambda, though you may be able to successfully tune an engine with very mild camming for good fuel economy at light loading with air/fuel mixtures approaching 16:1.

Again, a mixture trim module can be very useful for instantly testing percentages of enrichment or enleanment. You or the driver can just turn the trim pot(s) to optimize performance or home in on a certain AFR. Then you can note the actual pulse width fueling the car and the actual speed-loading cell, or note the percentage of change being made, and then back off and make the changes to the target and nearby cells.

At medium engine loading, you'll need to target richer mixtures in the neighborhood of 13:1.

As the midrange starts to get fleshed out, bring up the graphing function and work ahead on the graph to keep higher throttle settings in line with how the midrange is shaping out and consistently richer than what is proving to work in the midrange. Again, the trick is to make brief forays into high-loading conditions, watching the AFR and the rate of acceleration and drivability, fleshing out the active cells and cells in adjacent areas. Some people find it much easier to tune tables in graphical format, using the mouse and arrow keys. Others find it easier to edit numbers in the table, perhaps flipping back and forth between tabular and graphical format (there will usually be a hot key for this purpose). However you do it, getting an occasional look at a 3-D graph of the fuel or VE table is very valuable, because you'll begin to get a sense of the overall surface of the map, and you'll also be able to see any anomalies in the surface that probably don't belong there. You'll probably find that only a relatively small portion of the fuel map is used much and that you have to work to get the engine running very far off the beaten path, so to speak. If you started with a plausible startup map and have already mapped out a main VE table, what we're doing here is just fine tuning, but always stay on the rich side.

For normally aspirated street vehicles or boosted engines that are not severely knock-limited, mean best torque should fall in the 12.4:1 range at the highest power settings, but boosted engines may require more fuel cooling to fight knock and meltdown. If, as recommended, you are running really high-octane unleaded racing gasoline to minimize the effect of lean mixture mistakes, runaway boost, or poor tuning technique, knock should not be a problem at this time, particularly with a conservative timing map, and, preferably, a good EMS knock-sensing and retard EMS strategy or possibly an add-on device like the J&S Safeguard (which will actually retard individual cylinders only to the degree needed to stop knocking *in that cylinder*). At some point, you will have to turn off the knock-retard program in order to optimize basic safe timing and safe-up timing, fuel, and boost to work with the gasoline octane you'll be running, but not quite yet.

Bob Norwood points out that super-duty engines always make the most power as you lean them out right before they burn down. Really experienced, brave tuners like Norwood working carefully for every last horsepower on a race car will sometimes run high-power AFRs leaner than 12.4:1 (say as much as 13:1),

but such mixtures are more dangerous in really high-output engines, and—as Norwood, who has set a variety of land-speed records at the mile-high Bonneville Salt Flats, can testify—such tuning makes an engine more sensitive to variations in altitude and relative humidity. An engine is producing its most powerful putts at peak-torque rpm, ingesting the most air per power stroke, producing the highest cylinder pressures. The higher stress of increased peak-horsepower rpm will definitely produce greater stress on mechanical parts like connecting rods, and peak horsepower will result in more thermal loading to the cooling system and so forth (which can increase the tendency for detonation). But peak torque makes the most cylinder pressure, and it requires slightly richer mixtures than peak power. Many tuners recommend optimizing the high-power calibration by tuning from a known over-rich setting and then *carefully* leaning the mixture until torque falls off slightly, then moving in the rich direction until you achieve peak torque.

If you develop the fuel map on a chassis or engine dyno, you will almost definitely need to lean out the map for best performance in the real world of either competition or street performance. For one thing, dynos cannot accurately create real-world conditions like underhood airflow at speed or ram-air effects. What's more, dynos capable of holding the engine indefinitely at a certain speed and loading provide extremely high thermal loading that demands safe-rich mixtures that may be inappropriate for achieving optimal performance on the street under less severe conditions. Also, EMS fueling strategies typically involve averaging in nearby cells to a fueling calculation, so rich mixtures required at cells that are essentially unreachable except on the dyno can adversely affect the AFR on the street at nearby areas of the fuel map that *are* reachable.

High-power tuning is when exhaust gas temperature (EGT) data can be useful to warn you of danger before it is too late. You can take it to the bank that a piston engine will make peak power at somewhere between 11.8 and 13.0:1 AFR. However, the EGT at the peak power AFR is less predictable and depends on many factors, including ignition timing, cams, pistons, headers, etc. EGT at peak power might be anything between, say, 1,250 and 1,800 degrees, or even a bit hotter. Typical peak EGT on a performance automotive street engine would be more likely in the range of 1,350–1,550 degrees. Peak EGT typically occurs at about 75 degrees Fahrenheit lean of best power, and temperature falls off from there in both the lean and rich directions.

EGT is commonly used on piston-engine aircraft running essentially at steady state to lean the air/fuel ratio for peak power or best economy, and the procedure is always specific to a particular engine type in a particular aircraft. But pilots typically begin leaning from a particular rich mixture (based on a particular flow in gallons per hour on the fuel flow meter) used to provide a high-power climb out. From there they gradually lean the mixture for best economy in cruise flight manually until the EGT reaches a peak temperature (typically measured at the turbine inlet), and then continue leaning until EGT is 50 degrees cooler than peak.

Continuous EGT above 1,550 degrees would be a cause for concern on many engines. If you make a note of the EGT on an engine working at a known optimal peak-power AFR, then certainly a rapidly rising EGT moving out of the expected range could warn you that something is dangerously wrong in time to take countermeasures. As an example, detonation will rapidly increase combustion temperature and EGT, but this wouldn't affect the exhaust gas oxygen content as seen on a wideband AFR sensor. A pyrometer has the additional advantage that it is not

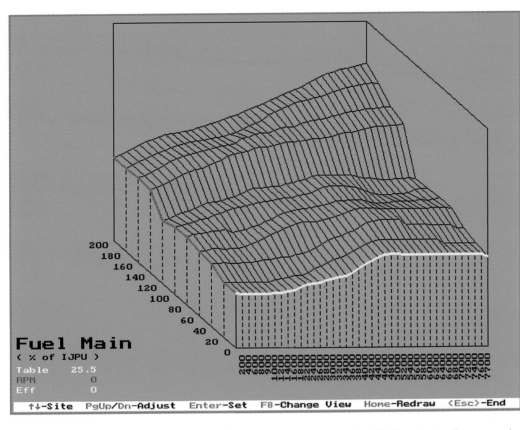

```
            200
          180
        160
      140
    120
  100
 80
60
40
20
0
```

Fuel Main
(% of IJPU)

Table	25.5
RPM	0
Eff	0

↑↓-Site PgUp/Dn-Adjust Enter-Set F8-Change View Home-Redraw <Esc>-End

Motec's 3D Main Fuel Map plots injection pulse as a percentage of the maximum defined injection pulse (IJPU) as a function of engine rpm and loading in kilopascals (in this case 200 is the maximum due to the 2-bar MAP sensor). Most EMSs allow you to view and modify the fuel curves as either tables or graphs, most of which are very similar to this chart. The surface of the chart should roughly correspond to engine volumetric efficiency—keeping in mind that tuners almost always optimize pulse width at wide-open throttle for power, at light throttle for fuel economy.

damaged by leaded fuel, which neither wideband nor narrow-band O_2 sensors can tolerate for an extended time.

Spark plug condition can tell you a lot about steady-state combustion at peak torque or peak power (see sidebar). Check the spark plugs after hard acceleration to get a snapshot of air/fuel mixture. The way to do this is to abruptly shut off the engine at full load (making sure you don't lock the steering if you're actually driving rather than on a dyno) and immediately check the plugs. Overly light or white plugs indicate a lean mixture, while dark, sooty plugs imply too rich a mixture. When things are right, normal plugs should be light brown immediately after a hard run.

A wideband closed-loop system can be very useful to speed up tuning and smooth out variations in air/fuel ratio at midrange and high-power settings once you're in the neighborhood of optimal. Once you've built up a number of islands of correct fueling (*not* just single cells) and have some idea of the optimal AFR, you can probably use an interpolation function to build a smooth linear surface connecting the known-good islands. Now set the target AFR table cells in the vicinity with the known-accurate AFR values, and then turn on wideband closed-loop fuel correction and run the engine. Note the amount of correction the EMS is applying to reach the target AFR at various cells, then correct the fuel table to minimize the amount of auto-correction (the main fuel table should always be constructed so auto-correction is in the direction of pulling out fuel).

Some interceptor devices present a user interface which, in effect, is that of a programmable standalone ECU, in which case calibration is the identical process (only easier, since you are starting with a known-excellent factory calibration and working to change critical areas—usually to add fuel and remove timing for power-adders).

If you are working with an entirely independent add-on auxiliary fuel or timing device such as an additional injector con-

troller (AIC) or timing boost-retard processor, the first thing is to verify that it is installed correctly and functioning properly. The AIC will be triggered by a combination of loading and rpm, so test the unit by running the engine within a trigger rpm range with the engine lightly loaded, and pump up the manifold pressure reference line with a hand pump to the configured threshold triggering pressure, using an injector tester to verify that all auxiliary injectors begin operating when they're supposed to when pressure rises to the trigger point. Alternately, bleed off fuel pressure to the addition injector(s), and using a mechanic's stethoscope, listen to hear an additional injector clicking rapidly. As a last resort, test the AIC with at least one of the additional injectors spraying into a safe container so you can actually *see* it operating (this could be very dangerous if you are not extremely careful). Before actual loaded testing on the road or dyno, you must also verify that the fuel-supply system is sufficient to fuel the primary and additional injector without a pressure drop.

When you actually begin testing the AIC with a load, the trick, as always, is to work from rich to lean, or from a conservative timing setting in the direction of more timing advance if you are working with a boost retard device. Since a turbocharger adds approximately 10 percent torque per pound of boost, you can figure needing 10 percent more fuel per-psi boost. If possible, check out the main fuel map pulse width in the target speed-loading cells that will be boosted, and set the AIC to deliver at least 10 percent extra fuel per-psi boost, correcting for any differences in injector fuel-flow capacity. If you do not know the stock pulse width, there are pulse width meters that can be used to measure injection pulse width. Otherwise, estimate the additional horsepower or torque you expect to generate per-psi boost and apply the correct formula to the amount of fuel required per horsepower per cylinder at a 12.0–12.5 AFR. Follow the instructions on the AIC, and test with careful forays into boosted territory, using a wideband O_2 sensor, and preferably a knock sensor to help protect the engine from knock or lean-mixture damage.

OPTIMIZING SPARK ADVANCE

As EMS-builder Electromotive says in the TEC^3 user manual, "Perhaps the most important step in tuning an engine is establishing the required ignition advance. An engine with too much timing will detonate, regardless of how much fuel is thrown at it. An engine with too little timing will perform poorly and overheat the exhaust in short order." Late ignition timing can make an engine run roughly, and it will definitely increase the exhaust gas temperature (since the mixture continues burning later in the combustion cycle). If the exhaust headers are glowing, you know there is too little timing; if the engine is knocking, timing late may be too advanced.

Optimal spark timing varies, with the overall goal being to light off the charge mixture in time so that peak combustion pressure occurs when the piston is at 15 degrees after top dead center on the power stroke. A critical overarching requirement, however, is also to avoid engine damage by delaying light-off to the degree required to avoid having unburned portions of the charge explode as combustion pressure and heat build during the burn (as the piston may still be compressing the mixture). The optimal amount of timing advance varies inversely with volumetric efficiency because denser mixtures burn faster and require less lead time to achieve the 15-degree peak.

Therefore, engines need more advance at low-load, narrow throttle settings when VE is poor. At the same time, when the engine is running at high speed, a combustion event of fixed length in time consumes more degrees of crankshaft rotation, requiring more timing advance. Combustion speed and crank degrees per millisecond are thus independent variables. VE is poor at idle, but engine speed is slow, so 8–20 degrees of timing typically works well. As the crank quickly accelerates off-idle, a huge infusion of spark advance produces the best torque and responsiveness, with some wild-cammed engines with very poor low-end VE requiring as much as 28 degrees total advance. Engines cranking to start turning 200–300 rpm require moderate spark advance with nearly as much timing as idle.

As the throttle opens and the engine speeds up toward peak torque-peak VE, combustion speeds up along with VE and thus requires less spark advance. Peak timing usually occurs by 3,000 rpm on most engines. On the other hand, high engine speeds at very light loading require huge amounts of advance—as much as 40 degrees or so. If the engine is turbocharged, the onset of boost increases the effective compression ratio, and combustion speeds up dramatically, requiring less timing, and simultaneously increases the risk of detonation. Maximum timing at 1.0-bar boost is typically in the range of 23 degrees at any engine speed, and 2.0-bar boost might only want 22 or even 21 degrees of timing advance. All things being equal, bigger engines need *more* timing advance because it takes the flame longer to burn its way across a large bore than a smaller one. Engines with small combustion chambers need *less* timing (opposite reason), and engines with pop-up pistons or other irregularities that slow down flame travel in the combustion chamber need *more* timing.

Spark advance maps plot spark timing as a function of engine rpm and engine loading (normally manifold pressure or mass airflow). Unlike conventional distributors with mechanical springs and advance weights and vacuum-advance diaphragms, which are mechanical devices incapable of providing timing advance that varies except as a smooth curve or straight line,

computer-controlled timing can dance all over the place if that's what's optimal. It can certainly approximate a complex 3-D surface with various peaks and valleys if that's what produces the best performance on a complex engine with, say, multiple power-adders and numerous accessories and so on.

Conventional distributors smoothly increase timing with engine speed. Though two springs of varying strength may be used to bring in some timing advance quickly with initial speed increases, then bring in more as the rotating weights overcome the force of the second spring. High vacuum at light loading was used to add timing. It would be a simple matter for EMS designers to provide two separate timing tables—one for rpm advance, a second for load retard—but one additive timing table makes it simpler to calibrate, and, besides, the real trick is an exhaustive search for power on a dyno while the engine is running. Who cares why you need more or less timing?

A modern EMS with knock-sensor/spark-retard logic should be calibrated with enough timing to provide peak torque across the board using design-octane fuel, but may rely on knock-sensor-feedback logic to pull timing to protect the engine from occasional light knock in case of a poor tank of fuel or if the engine is unusually hot (or if anything else *unusual* is happening that makes the engine especially knock prone). However, tuners should work rigorously to calibrate timing conservatively enough that the knock sensor normally *never* becomes active—by eliminating excessive timing advance in any area of the timing map where the knock sensor sounds off in order to clamp maximum timing within a normally safe range. Setting maximum timing by advancing timing as far as possible to the ragged edge of knock-sensor activation will probably yield best fuel economy though not necessarily best torque. (It was once common to set timing on older carbureted engines with distributor-points ignition by advancing the distributor in actual road testing to the ragged edge of incipient spark knock.) Over-advanced timing may also cause the engine to surge at cruise.

Aftermarket mechanical hot rod valvetrain parts may generate enough background noise to falsely trigger the knock sensor. If this is the case, keep in mind that any knock sensor should ideally be selected to match the frequency of the detonation characteristics of a particular engine, which might not be true of a particular generic aftermarket sensor. There are a lot of knock sensors, and another one tailored for a different frequency might work better. For example, an original-equipment sensor selected specifically for your engine—or, if your engine has been modified and you are getting false positives, perhaps specifically a *non*-OEM sensor). Normally, a small four-cylinder engine should have a knock sensor designed for a small four-banger and so on.

A possible alternative to changing sensors would be to experiment with changing the location of the sensor in the block or head(s) or even mounting it with Teflon tape or other insulating material around the threads to decrease sensitivity. Some aftermarket knock-sensor control systems have the ability to tune or condition the EMS or knock-controller response to various types of knock-sensor signals to narrow the response to what is actually detonation as opposed to engine noise. However, if the engine is simply too noisy for a knock sensor and you have the bucks and like experimenting, you *could* install in-cylinder pressure transducers, some of which are bundled with special spark plugs to detect knock and provide a readout of torque or brake mean effective pressure (BMEP). Measuring cylinder pressure is actually much more accurate than listening for knock-like vibrations using a knock sensor.

Anything that affects engine VE or efficiency (displacement, heads, cams, power-adders, compression) will change the shape of the optimal spark advance map. And it is worth the effort to optimize timing, because there is free power in optimized timing—sometimes a lot. To optimize spark timing throughout full-throttle acceleration runs, you need a dynamometer (or a dragstrip-type highway and a datalogging accelerometer like the G-Tech). As with fueling, it is much better to start with an optimized map from a similar or identical engine, but, remember, engine wear effects VE, and carbon buildup in combustion chambers raises the compression ratio, which increases combustion flame speed.

Brand new engines increase in VE for as much as 20,000 miles as the rings wear in, but eventually VE heads slowly south, so no engine is truly identical to another. As with fuel calibration, it is much better to start conservatively and feel your way toward maximum safe power. The default EMS timing map should be a good, conservative place to start on the retarded side, and timing is too advanced when the engine exhibits trace spark knock, but it's sometimes hard to hear knock, and by the time you do, it could be too late. A knock-sensing and retard interceptor like the J&S Safeguard will protect the engine but show you graphically when the device is fighting knock with retard, enabling you to search for optimal advanced timing. In the case of fuel, conservative means rich. In the case of timing, it means *less* spark advance (temporarily retarding spark timing is the strategy used by knock-sensor engine control logic).

KEEP IN MIND:

• Emissions considerations will affect optimal timing and may conflict with maximum performance or responsiveness.
• Maximum safe timing can be influenced by the *gear*, and some EMSs have special maps to trim spark for certain gears.
• You might want to safe-up your engine by pulling timing and adding more fuel at peak power or above if you occasionally like to rage along at full throttle in top gear and want to save the engine at the cost of a little power. This is a typical OEM strategy, which is why tuners sometimes *remove* fuel and *add* timing to make a little more power on a stock engine.
• At light cruise, as much as 44 degrees timing advance could enhance fuel economy.
• A small-block wedge-head V-8 like the traditional small-block Chevy with traditional heads wants about 38 degrees maximum total timing, though fast-burn aluminum heads can lower this to, say, 34 degrees.
• OEM engineers sometime use timing (or even cam phasing) to smooth over transitions from gear to gear on an automatic-transmission vehicle. Honda now does some of this to smooth out the transition from the low- to high-speed cam on VTEC engines. Some tuners have dialed in more than 40 degrees of timing in the midrange to match the torque converter flash point, again pulling timing in speed-loading cells where the converter lugs down.

When you are done optimizing the timing map, it can pay dividends to return to the fuel map, particularly if you find the spark map was way wrong. Because air/fuel ratio affects the tendency to detonation and the flame speed of combustion, experienced supertuners like Bob Norwood move back and forth between the fuel and timing maps as they home in on an optimal calibration. Timing and fuel interrelate because they both have an effect on the tendency of an engine to knock, and because air/fuel mixture affects flame speed, which affects ideal timing, air/fuel mixtures burn faster with increasing richness all the way to an 11.1:1 air-gasoline mixture (which is richer than rich best torque, RBT).

For example, let's say you started tuning a turbocharged engine at 11:1 AFR, then began to take out fuel in the direction of 12.2:1 to make power. Flame speed would be slowing down as you leaned the mixture. You'd now have more power at 12.2:1, but if you had optimized timing correctly for 11:1, assuming the fuel octane was high enough, you'd then discover that you could make more power by advancing the timing to make up for the slower flame speed of 12.2 combustion. On the other hand, if the engine required fuel to fight detonation at 11:1, you might need to retard timing at 11:1 *before* pulling fuel because there'd be higher cylinder pressures at the higher power 12:2:2 AFR levels and less surplus vaporizing fuel available to fight knock. Which is one great reason to experiment with really high octane fuel: you can experiment with more combinations of spark and fuel than you could otherwise safely reach.

COLD-START AND WARM-UP

Gasoline is a stew of various hydrocarbons of various weight and chemical composition. The heavier components, in particular, do not vaporize well when the air or internal engine surfaces are cold. Engine management systems, therefore, must deliver extra fuel to start the engine when it is cold and while it is warming up to keep it from stalling. Cold-start tables will arrive from the EMS vendor with default data, but—unlike, say, battery-voltage compensation tables (user-modifiable, but why?)—cold-start tables will almost definitely need to be reworked to perform really well on your engine with good drivability and no stalling. This will be especially true if you are running really large injectors, really large displacement, unusual intake manifold designs, injectors with unusual geometry, injectors that are farther than usual from the intake valve, or radical, unusual combination of hot rod parts.

Most engine management systems can now provide the following:
• **Cold-running enrichment** as a percentage of the ordinary fuel pulse width calculation, with the percentage declining with increases in coolant or cylinder-head temperature until the engine reaches maybe 140 degrees Fahrenheit.
• **Cold-air enrichment** based on intake air temperature (which may disappear if the intake system heat-soaks). Note: it is critical to locate the intake air temperature (IAT) sensor in a place where it is measuring the intake air temperature, not the temperature of the intake manifold or air-intake tube.
• **Special priming enrichment** while the engine is cold and cranking to deliver the super-rich mixtures needed to fire up fast when the engine is stone-cold. On early EFI engines, this almost always involved energizing an additional cold-start injector (a spray valve, typically aimed at the throttle body), though increasingly, the priming enrichment is delivered through the main port injectors.
• **After-start enrichment,** which typically continues for a fixed number of engine revolutions (say 250) to prevent stalling.
• **Idle-air control enrichment,** which typically might increase idle speed when the engine is very cold.
• **Hot temperature enrichment** to help protect a hot engine via the cooling effect of fuel evaporating. Also called the heat of vaporization, meaning the heat required to vaporize something.
• **Air-temp enrichment**
  Some EMSs can provide the following:
• **Temperature-based increases in acceleration enrichment,** as a percentage.
• **Temperature-based increases in spark timing** to compensate for the decreased flame speed when a lot of raw fuel is floating around the combustion chambers in tiny droplets rather than fully vaporized.

- **Temperature-based limitations to maximum boost** to prevent engine damage while the oil is still thick.
- **One-second temperature-based start enrichment**
- **One-second constant enrichment**
- **Fixed one-second starting pulse width**

Many cold-running tables are designed as a final percentage modification to the ordinary fuel and timing calculations, which is why it is essential to get the EMS working really well at normal temperature before you mess with the cold tables. However, once the engine is running great under ordinary conditions, it is really simple to wait until the engine is fully cold, bring up the warm-up map as a graph, start the engine, and adjust the map up or down at the current temperature (an arrow or something similar will highlight the current coolant temperature) until the engine runs smoothest, then repeat the process as the engine warms up and the temperature increases.

A cranking engine will typically be affected by both warm-up and cranking enrichment, and initially it can be hard to tell if you need more cranking fuel and less warm-up or the reverse. Immediately after starting, it can be hard to tell if you need more or less after-start enrichment or if you should be modifying the main warm-up map. One clue would be to watch what happens when the cranking and after-start enrichments disappear. Does optimal warm-up enrichment suddenly decrease or increase?

A typical engine might need 10 to 15 percent enrichment immediately after a cold start in order to achieve a 12–12.5:1 AFR on a mild street engine (perhaps as low as 11:1 on more radical engines), and this requirement will decline to nothing somewhere at or above an engine temperature of 140 to 180 degrees Fahrenheit, with more enrichment required in really cold climates. When cranking in really cold winter weather, the engine may need *grossly rich* priming AFRs in the neighborhood of 4:1 to start the engine fast. Beware a sudden cold snap or, say, a trip to a mountain ski area from palm-lined Los Angeles when you have never had a chance tune below 50 or 60 degrees. You might be unable to start the engine without a laptop. Though in a pinch you could probably start the engine by spraying starting ether, carb cleaner, WD-40 (or perhaps even gasoline, somehow, if you're ingenious) into the manifold plenum through an unplugged vacuum reference line.

The engine must be stone cold to set cold cranking and warm-up enrichment accurately, which means you will want to attack the problem first thing in the morning after the engine has cold-soaked over night. You really want the coolant temperature at 70 degrees or lower to begin the process.

It may work best to adjust the throttle plate to a reasonable setting for cold-running idle and later work to program the IAC for an ideal cold idle speed after the mixture is right. As always, a wideband O_2 readout may be useful, though the goal of cold-running enrichment is purely to eliminate stalling and improve drivability, you never need more fuel than what is obviously needed for this purpose. A mixture-trim module could be very useful when you are working to optimize the AFR for fastest cold idle.

Start the engine and try dialing in greater or lesser cold enrichment percentages until the engine runs best. As always, modest trim changes are best, since a few percent can make a large difference in how the engine runs. The EMS will not have an infinite number of temperature cells. Stay on top of it as the engine warms up and moves to a higher temperature range. You may see the engine change speed or change audibly when the engine moves to a new discrete segment of the warm-up map.

If an engine with reasonable timing will not start, it is either because there is too little enrichment or because the engine is flooded. A flooded engine will typically smell strongly of gasoline. To confirm flooding, you can remove the plugs to check if they are wet. If the engine will not start when cold and the plugs are not flooded, there is probably not enough fuel.

ACCELERATION ENRICHMENT

Some people compare acceleration enrichment on a port-EFI system to the accelerator pump typically used to squirt raw fuel down the throat of a traditional Holley four-barrel carb when the throttle rotated. Though EFI typically needs less acceleration enrichment than a carb because bogs in air velocity that affect a carb's mechanical ability to get fuel in air have no effect on EFI. The accelerator pump was something of a Band-Aid heavily related to air velocity and problems with the fuel equilibrium in a wet air/fuel manifold.

Acceleration enrichment (or enleanment) is critical to responsive drivability, and all EMSs will arrive with plausible default data designed in. However, there is almost always room for improvement, and sophisticated EMSs offer a large and powerful set of acceleration enrichment functions to make sure there is never a problem with transitional enrichment that cannot be handled in some way.

Transitional enrichment is the final major system to optimize and should be attacked when everything else is in good shape. If you mess with it before the main fuel tables are right, it could confuse the process of calibrating normal operation. Transient enrichment is normally used to keep the engine from stumbling when you suddenly open the throttle (not normally a problem on port-EFI systems), and to make sure that the engine is as responsive as possible when you suddenly stab the throttle.

Most EFI systems can now provide acceleration enrichment based on increases in throttle angle and/or sudden changes in manifold pressure. Providing enrichment based on falling MAP is somewhat analogous to the way a Holley carb used a power valve to fix a lean hesitation and surge when there was a large drop in vacuum (for example, if you started up a steep hill) prior to a corresponding increase in throttle angle.

The main components to acceleration enrichment are basically attack (sensitivity) and sustain, and there the EMS may provide separate rules for enrichment at low rpm (when the vehicle can probably accelerate very hard) and at higher rpm. Attack is the percentage of enrichment that should be commanded while the throttle angle is increasing at or above a threshold rate. Sustain specifies how fast to phase out the enrichment. Electromotive points out that the typical duration of required transient enrichment on acceleration is around one second.

There is no diagnostic tool that will tell you when the accel is right. Inertial chassis dynos like the Dynojet will measure increases in torque and power as you roll on the throttle. But the Dynojet is not very good at dealing with partial throttle situations because it is difficult to distinguish the human factor in banging down the gas a certain amount, versus how the engine responds when you bang it down at a certain rate (one imagines robots increasing the throttle an exact amount at an exact rate of increase).

However, if you have the time and the dyno, the Dynojet could possibly help you optimize accel enrichment. In any case, sooner or later, you'll need to hit the road, and the trick here is basically a lot of patience and intuition, and plenty of trial and error with a laptop attached and available when you discover a problem. If you are working on the dyno, it may be easier to see

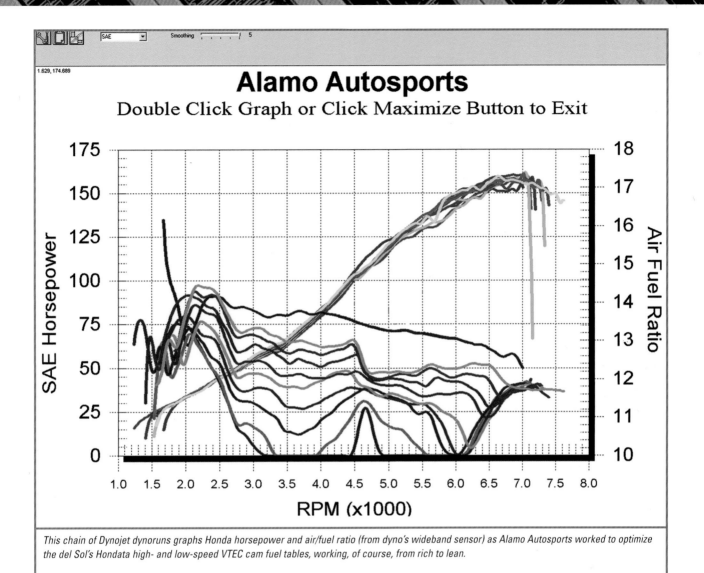

Alamo Autosports
Double Click Graph or Click Maximize Button to Exit

This chain of Dynojet dynoruns graphs Honda horsepower and air/fuel ratio (from dyno's wideband sensor) as Alamo Autosports worked to optimize the del Sol's Hondata high- and low-speed VTEC cam fuel tables, working, of course, from rich to lean.

rich, black smoke from too much accel. When you are road testing the calibration, it could be very useful to have a helper in a chase vehicle watching for black smoke on enrichment (and possibly exhaust flames, particularly on deceleration).

Adjust the acceleration increase, sensitivity, or whatever it is called upward if there is an instant bog, and add to the sustain if there is a slightly delayed bog. You may also be able to clamp acceleration enrichment to a percent of total pulse width for various loading or rpm ranges if some conditions result in overenrichment.

Most engines will not require any acceleration enrichment over 4,000 rpm and will require more enrichment the lower the rpm. Ideally, your EMS will provide multiple cells to define acceleration enrichment at various ranges of loading or rpm, enabling you to require decreasing enrichment with increased engine speed.

If you reach 100 percent increase (or maximum acceleration enrichment) and the engine still lean-coughs, the base fuel map is too lean somewhere, or the engine data configuration is wrong, or the ECU hardware is wrong for your application. Engines with very large throttle area (such as one throttle blade per cylinder) will greatly increase manifold pressure with a slight turn in throt-

tle angle, and might require MAP-based acceleration enrichment, less throttle area, or possibly some method of multiplying the change in throttle angle at lower rpm (for example, a gear-driven, variable-rate-of-gain TPS).

IDLE SPEED CONTROL

IAC units are stepper motors under computer control that work to stabilize engine rpm at idle. If the IAC is configured wrong, it may do nothing (for example, if the throttle plate is already opened enough to force rpm above the target IAC idle speed), or it may hunt. Hunting occurs when the computer is slow to react to a change in idle speed, but then overreacts, and idle speed can surge up and down forever, failing to stabilize. Some EMSs now have the ability to configure high and low thresholds for corrective action. You should also be able to program the magnitude of the corrective action, possibly depending on how wrong idle speed is and how fast it got so wrong. Work with the IAC parameters until idle stays where you want it.

A LOT OF WORK

If you are thinking by now that building a really good calibration sounds like a lot of work, you are correct. While some great

Ultimate control. It doesn't get more ultimate than an engine dyno, where you pull the engine from the vehicle and mate it to a computer-controlled load-holding water brake or eddy-current dynamometer with extensive instrumentation.

tuners like Norwood can build a pretty good scratch calibration in a few hours on a dyno, they are also likely to spend significant time test driving the vehicle with a laptop connected to fix problems, and they may well return to the shop for more dyno time where optimal performance is critical.

Beyond this, they will also need some cool mornings to optimize the warm-up maps, and they cannot really know if the winter-weather calibration is truly right until the really cold weather rolls around. Automakers often spend months or even years building the perfect calibration that will be mass produced in hundreds or thousands of vehicles. Tuners and their customers certainly cannot afford to do this. But an individual hot rodder or club racer who does it for fun can take the time to do it right and stay with it until the calibration is really right. If you are patient and scientific and willing to buy or rent diagnostic equipment, if you keep notes of what you have tried and the results and attack one variable at a time, if you take pride in your work, you will home in on a *great* calibration.

ADVANCED STUFF: INDIVIDUAL CYLINDER INJECTOR AND TIMING TRIM

Some engine management systems provide a set of advanced features that can be useful if you want to optimize engine performance under all circumstances. Examples of such features include the following:
• Individual-cylinder fuel trim
• Individual-cylinder timing advance trim
• End-of-injection phasing (crank angle)
• VE correction to target air/fuel ratios
• Engine coolant temperature VE correction
• VE correction coefficient versus estimated intake port temperature (density *decreases* with temperature, but hotter air pumps more easily)
• Knock-sensor functions (threshold definition, rate of knock-

induced advance retard, rate of advance increase on recovery, maximum retard allowed)

Many of these advanced features are useful only within the lab-type environment of an engine dynamometer equipped with substantial diagnostic equipment, or on an exhaustively instrumented race-type vehicle with a large datalogging capability. As an example, individual-cylinder trims can be useful to correct for minor variations between cylinders in volumetric efficiency, thermal loading, and so forth. Even on a well-designed and built engine, some cylinders may run a little hotter or colder, some cylinders may have slightly more tendency to knock, some cylinders may have slightly different intake or exhaust runners that can affect VE under certain circumstances.

In the days of high-output single-carbureted V-8 engines with points ignitions, fuel distribution could be terrible and certain cylinders almost always had better cooling than others. Unfortunately, there was no choice but to tune the carb rich enough to avoid dangerous mixtures and detonation in the leanest cylinders and to set timing to avoid detonation in the hottest cylinders. Modern port-EFI engines obviously avoid most problems with mixture distribution, though testing and matching injectors for perfect flow definitely makes sense if you are gearing up to optimize individual-cylinder performance. There can also be minor differences in cooling and VE that are not practical to eliminate, but which can at least be optimized: an eight-cylinder engine is actually a gang of eight single-cylinder engines.

To make any sense, individual-cylinder tuning requires individual-cylinder instrumentation. Accurate, fresh EGT probes on each cylinder are critical. Wideband O_2 probes cost as much as $1,000 each, so it's clearly not practical for most people to run individual-cylinder AFR probes. In-cylinder pressure transducers are *extremely* effective in measuring the effect of individual-cylinder tuning strategies, and there are spark plugs with built-in pressure probes. You clearly need the ability to measure small changes

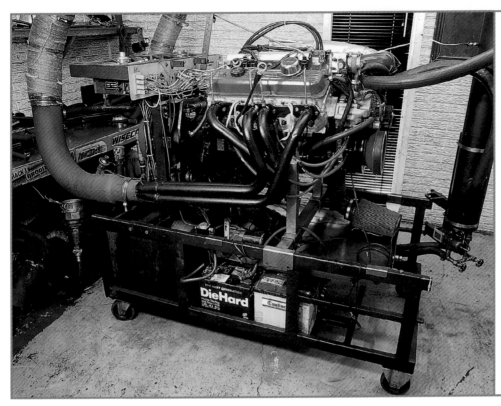

This supercharged and intercooled Pontiac LT1 powerplant did its thing on Kim Barr Racing's Superflow dyno. Note the EGT probes in each exhaust port. Exhaustive calibration is a matter of manually holding a load at a wide selection of rpm and loading cells within the EMSs main fuel and main timing tables. Unlike an inertial chassis dyno's 5- to 15-second dyno pulls, an engine calibrated this way on an engine dyno has proven itself in the hell of continuous loading at high power.

in torque, which are, after all, the bottom line. You clearly want a dynamometer with load-holding capability; a race-type engine full-instrumented with torque-measuring strain gauges on the transmission input shaft, or an EMS with the ability to estimate micro-changes in individual-cylinder torque by looking at tiny differences in acceleration and deceleration of the crankshaft. Such an EMS can effectively turn an engine and EMS into its own dyno, but this is almost science-fiction-type stuff within the repertoire of only the most sophisticated automotive engineers. Practically speaking, aftermarket tuners and racers—if they do this at all—are going to take on individual-cylinder tuning using EGT probes on a load-holding engine dyno.

The trick is to search the rpm and loading range, allowing the EGT probes to stabilize (good, fresh probes are important), and checking for cold or hot cylinders. Since most EMSs only permit individual-cylinder trim as a percentage (*not* a complete, separate fuel map for each cylinder), you're looking for the worst differentials, at speed-density points that matter. Now, you work to trim any individual slackers that are running too cold or too hot with individual fuel trim, watching the dyno for increases or decreases in torque. When you think you've fixed the problem as well as it can be fixed, log a full-throttle pull to redline, and compare with *before* results to make sure that you didn't make things worse somewhere else. When you've got the fuel perfect, if the engine is knock-limited by particular cylinders (a J&S Safeguard individual-cylinder knock-retard module is a great tool to identify individual cylinders that are more knock-prone), you can experiment with removing timing only on the ones that need the most help.

INEQUALITY FOR ALL: INDIVIDUAL-CYLINDER ENGINE MANAGEMENT

What is a V-8 engine, really? Eight single-cylinder engines hooked up in a line, doing their best to work together. These cylinders are not identical engines, however. This is why the most sophisticated aftermarket engine management systems, like the Motec M48, the Race System from EFI Technologies, and others offer not only sequential injection, but individual cylinder calibration of injector open time (pulse width), injection start time (phasing), and spark timing. Such systems produce higher output and cleaner power on otherwise identical engines.

Good fuel atomization is critical to efficient combustion, and really good atomization will produce more power for a given amount of fuel. Engines are run at richer than stoichiometric air/fuel mixtures at peak torque/power in order to increase the odds that every molecule of air that makes its way into the engine finds some fuel to burn. Excellent atomization makes sure that there are fewer localized islands of air in the air/fuel mixture without fuel, reducing the need for richer overall air/fuel mixtures. A really well-mixed homogeneous air/fuel mixture produces higher flame speed during the burn event, which implies more horsepower and less required ignition timing. If injectors spray on a runner wall or against a closed intake valve, fuel will pool and run into the engine in droplets. Injectors should aim at the back of the valves with minimum contact with port walls and fire precisely in sequence at the correct time.

Variations in the intake system, cylinder head, and exhaust system, even with equal-length runners, cause airflow to resonate slightly differently for different cylinders, resulting in slightly varying volumetric efficiency. "Even with identical header length," says EFI expert Graham Western, "some exhaust tubes are more forgiving than others. On true racing engines, fuel pulse width trim of 1 to 2 percent correction is required to eliminate mixture variations due to this slight breathing differential."

While batch injection (all injectors firing at once) is reasonably efficient on lower-rpm multicylinder engines with little pulsing in the intake, on high-rpm, wildly pulsing competition motors, sequential phasing is more effective, particularly at peak torque where tuned intake and exhaust systems are most efficient. On racing engines, where charge density of atomization can be improved by taking advantage of the intercooling high heat of vaporization by injecting farther upstream, spraying at the wrong time can result in fuel being blown back out the stacks or directly through into the exhaust system on valve overlap, producing distribution problems or wasted fuel and high emissions.

In order to eliminate standoff distribution problems, it is critical to inject fuel when the intake valve is open so that fuel does not hit the closed valve and bounce back along with the decelerating, bouncing air column. On a plenum-stack turbo engine with compressed air entry at one end, depending on the resonant frequency, standoff fuel can be blown to the stacks farthest away. Individual cylinder pulse width or injection-phasing adjustment can correct for distribution problems related to resonant differences in the airflow to each cylinder (particularly when the injectors are open a lot)—realizing as much as 8 percent more power. According to Western, a 60 to 70 percent injector duty cycle usually makes best power. Larger injectors get all fuel in while the valve is open, but the tradeoff with short injection time is poorer vaporization and cooling.

Western says you always see power increases over batch fire injection at mid- to high-range torque; a six-cylinder truck that tested at 280 horsepower with batch-fire injection will pick up 20 ft-lbs of torque at peak torque and 12 to 16 horsepower at peak power with sequential injection. A 700-horsepower racing engine might make up to 5 percent more horsepower due to general phasing correctness. Individual spark trim is usually associated with detonation-limited engines. Retarding timing on just the offending cylinders (typically 2 to 3 degrees) lowers peak combustion temperature and cylinder pressure, helping fight detonation. Where exhaust gas temperatures are too high on turbo motors, tuners may run a little extra fuel or spark timing in the cylinder causing problems—or increase overall compression ratios. Raising compression is preferable to advanced timing. The increased expansion ratio of gases from a smaller combustion chamber results in greater loss of temperature than retarding timing, and the increased compression ratios improve peak and average cylinder pressures instead of simply peak cylinder pressures.

AIR/FUEL OPTIMIZATION PROCEDURES ON A PISTON-ENGINE AIRCRAFT

Modern turbocharged aviation piston engines, which typically run at 75 percent power settings more or less constantly, and which must run at high takeoff power settings for long minutes or tens of minutes at a time on takeoff and climb-out, typically leave airports full-rich, which would typically be 100 to 150

degrees rich of peak EGT (75 degrees rich of peak typically makes maximum power). The EGT gauge is critical to the leaning procedure used to arrive at an optimal power and AFR setting for cruise flight and illustrates something about the meaning of safe. Obviously, if you burn down an aviation engine, you cannot simply coast over to the side of the road.

Engines with poor mixture distribution are sensitive to leaner mixtures and will run roughly on the lean side of peak EGT. Many aviation engines, which typically still use carburetion of constant mechanical fuel injection, vary in AFR by 8 to 12 percent among the various cylinders But on well-balanced aviation engines, an excellent procedure for arriving at mixtures producing high levels of safe power is to begin the leaning procedure with the engine running more than 100 degrees rich of peak EGT, with manifold pressure slightly down from full throttle. The pilot or flight engineer then carefully leans the turbocharged engine until EGT peaks (which would often be in the neighborhood of 1,500 degrees Fahrenheit. At this point, the procedure is to continue leaning until EGT cools to 40 to 60 degrees below peak EGT (this would typically be only 15 to 30 degrees lean of peak on a normally aspirated engine).

Consulting an aviation EGT chart for a common piston engine, we see that EGT falls off fairly evenly on both the rich and lean side of peak EGT, whereas cylinder head temperatures on an air-cooled engine were 35 to 40 degrees cooler on the *lean* side of peak EGT, and fuel flow was markedly lower.

However, power is down on the lean side, right? Well, yes, assuming you continue to run the same manifold pressure. If a plane requires more power in cruise flight, in this procedure, the pilot increases the throttle to raise manifold pressure to the required power setting with a few more inches of manifold pressure. High-performance aircraft engines typically have constant-speed propellers that use hydraulics to vary the angle of attack of the prop to maintain a particular manually adjusted prop speed Within a particular envelope, such engines will not increase in speed with more throttle.

This procedure requires good mixture distribution, or the engine will run a little rough due to uneven mixture distribution. But the implication of this procedure is interesting for EFI automotive engines with excellent mixture distribution, running high octane gasoline. Leaner mixtures stress the cooling system less, and the EGT is *not* higher. Power might be down a mite, but more boost handles that, at less fuel flow. Something to think about, because a piston aviation engine pulling against a constant-speed prop is a lot like a piston automotive engine pulling against a steady-load engine dynamometer.

All EFI systems are ultimately controlled by an onboard computer that, as we've already discussed, has been referred to by various automakers as an electronic control unit (ECU), electronic control module (ECM), or powertrain control module (PCM).

Automotive ECUs seldom fail. They are extremely rugged and are built to withstand the harsh automotive environment of vibration, freezing winters, baking-hot summers in the sun (or, sometimes, under-hood temperatures), high humidity, electrical interference, and other factors and still perform with the precision required to calculate engine fueling, spark, and other management tasks in millisecond time frames. The ECU must perform perfectly on a -10 degrees Fahrenheit morning in Buffalo, as well as a baking-hot south Texas summer afternoon just after a thunderstorm, in 100 percent humidity. It must perform perfectly in a four-wheel-drive vehicle, jouncing and vibrating through extremely rough off-road conditions.

Automotive engineers rarely put the ECU in the engine compartment, preferring to locate it inside the vehicle or trunk where the cooler, drier environment is kinder to electronic components. Why tempt fate any more than you have to? However, it is possible to design an ECU that will withstand the incredibly harsh engine-compartment environment in which underhood temperatures can reach hundreds of degrees, where the electromagnetic and radio interference close to high-voltage ignition components can induce killing voltage in normal nearby wiring, and where vibration can shake apart sensitive electronic components—not to mention mechanical parts. Some Corvettes, for example, are equipped with computers that live in the engine compartment.

Mil-spec (military specification) equipment—designed in some cases to withstand the electromagnetic pulse of a nuclear blast—is shielded from radio interference by the enclosure, and protected from humidity and vibration by a bath of hardened plastic resin that is poured into the enclosure, completely surrounding, strengthening, reinforcing, protecting all components. Mil-spec cabling and harnesses use fail-safe connectors and attachment systems. However, such systems are difficult to justify if the simple alternative exists of locating sensitive components out of harm's way inside the vehicle's passenger compartment or trunk. But on race cars and nuclear reentry vehicles, you go mil-spec. There are a few hardened automotive ECUs available in the aftermarket, including race units from companies like Motec, EFI Technology, and Zytek, which can sell for many thousands of dollars. All electronic control units are very rugged, but it is only common sense when designing an EFI system to mount your ECU and wiring so the ECU stays happy.

Assuming the engine itself is healthy, in most cases, if there's a problem with an engine management system, the problem is *not* with the computer but with the data it is receiving from its sensors—or some related electrical or physical air/fuel system problem. Low fuel pressure, clogged injectors, defective sensors, leaky vacuum hoses or fittings, and electrical system shorts or incorrect wiring can drive you crazy when trying to troubleshoot an EFI system.

It is important to go about troubleshooting in a methodical, scientific way, *with the right tools*, checking one thing at a time, substituting one part at a time, checking the results, then moving on to the next step. If you accurately follow a troubleshooting algorithm, it is inevitable that you will isolate the problem.

A chassis dynamometer like this bolt-on Dynapack tells you if tuning or equipment changes made things better or worse. There is just no substitute—well, except for an engine dyno. An engine that has not been tuned on a dyno with air/fuel ratio data will virtually always have "free power" waiting to be unleashed.

It is important to realize that troubleshooting a modern computerized EMS is a little like playing chess with a hidden opponent. The computer itself—with its own bag of tricks—will be doing its best to analyze or work around or compensate for problems. Of course, the ECU's view of the world is completely through the sensors. A problem like a small vacuum leak can fool some ECUs into assuming a lean condition exists, causing it to erroneously enrich the mixture. One bad injector—stuck either closed or open—can throw off the average air/fuel mixture as perceived by the computer based on a single sensor measuring exhaust gas oxygen coming from a group of cylinders. If there's only a single exhaust sensor, located on one bank of a V-configuration engine, the effect of one malfunctioning cylinder will be magnified.

Incorrect fuel pressure will result in the wrong amount of fuel squirting from working injectors. This will show up immediately in systems operating in open-loop mode. However, an ECU operating in closed-loop mode at idle or light cruise will attempt to compensate for incorrect fuel pressure. The fuel pressure problem may thus be masked at idle, only to show up disastrously at full throttle, which is still almost always required to work in open-loop (predetermined injection pulse width).

A bad EGR valve can cause the ECU to believe the mixture is lean. A plugged fuel filter may only show up during very heavy loading or high rpm, or intermittently at other times, causing a closed-loop computer to constantly chase correct mixtures. Many OEM computers maintain internal lists of correction factors designed to compensate for changes in engine VE due to aging or other such factors.

Modern ECUs also make use of tactics that cause them to disregard suspect sensors under certain circumstances (typically called limp-home mode or something similar). An ECU may outsmart itself in attempting to correct for unusual circumstances, possibly making things worse. When the computer attempts to compensate for an inaccurately perceived lean condition by enriching the mixture at idle, the engine may run roughly and produce a rotten-egg smell at the exhaust.

Throttle position sensors that are broken or out of adjustment can cause the engine to stumble at certain throttle positions or on sudden acceleration.

If the EFI system incorporates knock sensing, ECU strategies to prevent engine damage by knocking may mask other problems. Bad gas, plugged injectors, new platinum spark plugs, or oil introduced into the mixture by turbochargers or positive crankcase ventilation (PCV) systems can all cause pinging or detonation. Since the ECU may retard timing to prevent detonation, the problem may be masked. What the user sees is lack of power.

In virtually all cases, EFI problems can be traced to the computer somehow receiving inaccurate data or some other physical problem outside the ECU. And for efficient troubleshooting, to start with, you need the right tools.

EMS TOOLS
Accelerometer Performance Meters

To optimize the performance of an engine management system on a particularly vehicle, you need a way to accurately measure power and torque *before* and *after* tuning changes or VE modifications.

Once you know a vehicle's *exact* weight (including driver and payload, fuel and fluids, etc.), accelerometer performance meters can do an excellent job converting acceleration data into precise torque and horsepower and complete data acquisition. For as little as $150 or so, an enthusiast can record acceleration runs and

analyze shift points, delivered horsepower, and reaction times. You still won't necessarily know what percentage of crankshaft power is lost to drivetrain and rolling losses (you wouldn't on a chassis dyno either), though some performance meters like the Vericom will report various types of friction. Either way, like an inertial chassis dyno, an accelerometer performance meter is great for recording *before* and *after* balls-out acceleration runs to determine the effect of performance modifications.

The ubiquitous G-TECH/Pro Performance Meter Competition is a sophisticated automotive tuning tool designed to be used in place of an inertial chassis dynamometer. The G-Tech Pro measures reaction time; 0 to 60 feet time; 0 to 330 feet time; 0 to 60 miles per hour time; 1/8 mile E.T. and speed; 1,000 feet time; quarter-mile E.T. and speed; horsepower; torque; rpm; braking distance; handling Gs; accelerating and braking Gs; rpm versus time graph; horsepower and torque versus rpm graph; speed versus time graph; speed versus distance; and Gs versus time graph.

Similarly, Vericom's meter provides braking performance; speed; longitudinal G-force; drag factor; time; lateral G-force; braking force; distance; vertical G-force; coefficient of friction; horsepower (corrected standard); gradient; dynamic friction; torque; static friction; rpm; delta velocity; slip friction; and gear ratio (revolutions per foot), in English or metric units.

It doesn't take too many dynoruns before you coulda hadda G-TECH (or Vericom).

Code Access Keys

Typically these devices look like a cross between an ignition key and a paperclip. Their job is to plug into the driver-compartment diagnostic connector of specific vehicles in order to induce the engine management system to flash out malfunction codes using the check engine light. A code access key is convenient and inexpensive; on the other hand they don't do anything you can't do with slightly more effort using a jumper wire or a paper clip. Unlike a paperclip, code keys are specific to a particular vehicle or class of vehicle.

Computers

You're almost definitely going to need a laptop PC to tune and modify engine management systems. Yes, there are a few systems with their own proprietary calibration modules, but they are the exception, and in some cases a laptop will also provide a user interface and one with more power.

Fortunately, there are a lot of used laptops out there, and you don't necessarily need a very powerful one, especially if you only use it to calibrate engines. Although the Windows 95 and later operating systems have almost totally supplanted MS-DOS (the Mac OS is no factor at all in industrial operating systems supporting functions like engine-tuning software), many of the engine-calibration packages are still available in versions that will run under DOS. Because DOS requires *far* less computer power to run than any version of Windows 95 and later, if you are watching your pennies and have no other use for a laptop, you will find that old gutless laptops without enough CPU power or memory to run Windows will run DOS-based packages like a racehorse, and old gutless laptops can be found nearly for free.

That said, if you can afford it, a good laptop that runs Windows 2000 Professional or Windows XP will be far more robust and less subject to crashing than Windows 95/98 or even 2000 (even Microsoft admits this). If Windows 2000 or XP is specifically *not* supported for some calibration software package, OK, maybe there's a good reason, but otherwise keep in mind that a laptop that crashes in the middle of updating some factory engine management systems can permanently damage the ECM.

Fluke's 190C Color ScopeMeter provides continuous capture of waveform sample points, with 27,500 points in deep memory in up to a 48-hour time span. A metering oscilloscope can be invaluable for diagnosing difficult problems with almost any engine management or automotive electrical system, and many people find they are much less complicated to use than they expected. Fluke

What really matters in a calibration laptop is the ability to survive abuse. Every pro tuner I know has trashed multiple laptop computers in the harsh environment of road testing and dyno tuning. It happens: Laptops slide off fenders and roofs and smash on the concrete. Laptops get slammed in doors, they get overheated and cooked, they get sprayed with brake cleaner, their communications cables and recharging power cords get ripped from them. People have *crashed cars* because a laptop started to leap off the passenger seat during a high-G maneuver and the driver saved the laptop at the cost of the car.

There are mil-spec laptops and computers where you wear the keyboard on your belt and the system unit in a fanny pack. There are laptops designed to survive a nuclear electromagnetic pulse. There are also foam pads, Velcro, and big, industrial-strength cable-ties and there are turnkey laptop protector solutions. It is really worth thinking about how to protect your automotive laptop from an early demise.

DIAGNOSTIC SOFTWARE FOR LAPTOPS/PDAS

With the right cabling and hardware interface, a Windows laptop or PDA can become a diagnostic device to gather data from the engine management system while you drive.

An example is EFILive's GM OBD-II Scanner. When EFILive connects to a vehicle, it performs a series of automated requests to determine the status of the vehicle's PCM. The program can retrieve and display: interface cable type and com port properties, OBD-II protocol detected, VIN, PCM number, number of generic and enhanced parameters supported by the PCM, number and type of modules connected to the class-II data bus, system readiness tests, EPA test results, noncontinuously monitored

systems' test results (manufacturer specific), oxygen sensor test results, MIL status, diagnostic trouble codes (if present), history data if present, and customer-specific data associated with the detected VIN.

EFILive V-6 uses the latest Microsoft GUI technology as seen in native Windows XP applications. The virtual dashboard allows the user to configure tabular data, gauges, and scrolling charts in the same window. The gauges may be customized to different shapes, sizes, and colors, including large, crisp digital displays that can be read up to 10 meters away. Each gauge has its own audible and visible alarms. Gauges can be customized with your own bitmap images to give them a genuine look and feel. EFILive will interact with transparent portions of a custom bitmap to display visual alarms. When things go wrong, EFILive is designed to provide comprehensive help. EFILive presents the error message, what went wrong, and steps you can take to rectify the problem. If the problem cannot be resolved easily by the user, the vendor provides backup support.

Diacom's laptop package records and displays the onboard computer in real time, views up to 30 functions simultaneously, and automatically scans for system trouble codes. Diacom includes interface cables and operator's manuals. According to Diacom's Web site, the package requires an IBM PC, AT, PS/2 or compatible PC or laptop, a 5.25-inch or 3.5-inch floppy drive, hard disk (optional but recommended for vehicle data storage), one parallel printer port, one serial port (when using optional remote trigger module), at least 512K of RAM, CGA, and EGA- or VGA-compatible CRT or LCD screen and controller.

Obviously, if you drop your laptop or PDA, it will probably break—unlike some rugged professional scan tools designed for a shop environment. For this reason, some people will tell you the laptop/PDA solution is better for the sophisticated enthusiast than the dedicated professional.

DIAGNOSTIC CHARTS AND ALGORITHMS

A diagnostic chart or algorithm can be hardcopy- or computer-based. It can be vehicle/engine specific or generic. A good diagnostic chart or algorithm is guaranteed to solve the problem if you follow the algorithm and do the testing correctly. The chart/algorithm also functions as a kind of checklist of possible problems and considerations. Airplane pilots know the value of a checklist, and the FAA requires that pilots use them before start, before takeoff, before landing, and at other times when forgetting something could be catastrophic. If you follow the checklist, you will not forget to put the landing gear down, and if you follow a problem-solving checklist, you will not decide to rip the engine out of the car for a rebuild because you forgot that something like low fuel pressure could cause the same symptoms you're experiencing. No matter how experienced you are, checklists are a great way to avoid embarrassing and expensive diagnostic mistakes. Many shop manuals, aftermarket EMS installation and user guides have diagnostic charts. This book includes a generic chart for your reference.

DIAL-BACK TIMING LIGHTS

New factory engine management systems typically do not permit adjustment of engine position sensors or distributor clocking, but sooner or later you're going to need a way to verify crank or cam trigger position, or verify that spark timing is what the computer thinks it is. Or you may need to verify that one or more plugs are firing—at all, much less when. A good timing light enables the strobe to trigger not just from a high-voltage source like a plug wire, but from the low-voltage circuits that fire direct coils. A timing light's dial-back capability will also enable you to set cam timing, to precisely set adjustable cam sprockets on both banks of a four-cam powerplant, and to strobe other cam-speed events that occur far from TDC. The dial-back capability simply enables the tuner to dial-in delay to the strobe flash with a calibrated dial that provides a readout of the correction required to dial-in a particular event to top dead center.

For example: To determine the advance you want at 2,500 rpm, use a dial-back timing light. Set the amount of advance you want—say 35 degrees—on a dial-back timing light. Connect the dial-back timing light to the number one cylinder. Some EMSs may require you to disconnect a wiring jumper to ascertain that timing is guaranteed to be at a certain advance at a specific rpm. Run the engine at 2,500 rpm. Rotate the camshaft or distributor position sensor until TDC timing mark is centered at TDC. You will now have the amount of advance you dialed into the timing light. Tighten any standoffs and verify that timing has not shifted.

DIGITAL MULTIMETERS

A digital multimeter can be an essential tool for troubleshooting automotive engine management and electrical systems. The Fluke 88 Deluxe Automotive Meter with rpm has been around a long time and is typical of an excellent automotive-oriented multimeter. The Fluke 88 provides volts, ohms, amps, continuity and diode testing, milliseconds injection pulse width measurements, rpm measurements with inductive pickup, minimum/maximum/average recording, frequency and duty-cycle measurements (for MAF meters, IACs, and so forth), low ohms function, 10 megohm input impedance, parameter change alert, smoothing, and a sleep mode. The Fluke 88 provides leads with interchangeable test probes and wide-jaw alligator clips.

DYNAMOMETERS

A dynamometer will not hurt your engine or vehicle, but it will enable you to optimize performance in a scientific manner. A few baseline dyno pulls on a Dynojet 248 chassis dynamometer will probably cost around $50 and take a few minutes. For R&D, you may want to buy dyno time by the hour or the day.

Professional automotive engineers at places like General Motors use super-expensive dynamometers that cost in the six- or seven-figure range to calibrate and test engines, to measure the power and torque output under varying conditions, and to record emissions data while the vehicle is operating under tightly controlled conditions. Compared to data acquisition from road testing or competition telemetry, dynamometers provide a controlled environment that makes it vastly easier to evaluate what is going on in an engine and optimize tuning for particular types of operation.

There are many different types of dynos, but automotive engine dynos are designed to connect to the flywheel or flex-plate of an engine that has been removed from a vehicle, while chassis dynos are designed to interface with the driving wheels of a vehicle as either a rolling road or as devices that bolt to the lug nuts of a vehicle in place of the driving wheels. Typically, a vehicle drives onto a set of rotating drums (one large roller under each driving wheel or two smaller rollers) and is fastened in place with tie-down devices, which allow the vehicle to operate its powertrain against the rollers while remaining stationary.

The rollers are attached to various sensors that measure speed, loading, time, and other essential information that can be crunched in a computer to analyze performance. Inertial dynos measure the rate at which an engine or vehicle can accelerate a set of heavy drums (Dynojet 248 chassis dynamometers usually have two 1,600-pound drums). Although chassis dynos have been around since the 1930s, Dynojet Research revolutionized the performance aftermarket in the early 1990s for bikes, cars, and light trucks with an affordable, easy-to-use inertial chassis dyno costing less than $30,000. Before the Dynojet, most claimed performance increases for aftermarket parts were optimistic guesses, and street-car enthusiasts typically had no idea whether a performance modification had made things better, done nothing, or was even performance-negative.

Inertial dynos measure horsepower and torque by measuring the rate of acceleration of an engine at wide-open throttle working to accelerate a heavy mass of known weight and rolling resistance. Most inertial dynos are chassis dynamometers, but there are inertial engine dynos that attach to the flywheel of a powerplant (Dyno-mite, for example). Whether of the engine or chassis type, inertial dynamometers are most effective working at full throttle, and have no way to hold a precise, steady engine load at a given engine rpm (some inertial dynamometers also have optional water brakes or eddy current power converters). For that, you probably need a water brake, oil brake, eddy-current brake, generator, or other active power-absorbing dynamometer technology that can hold a precise braking force on the flywheel of an engine.

The big disadvantage of an engine dyno is that you have to remove the engine from a vehicle before you can work with it on the engine dyno. Compared to the old days of self-contained carbureted engines, there may be a discouraging amount of interfacing work on EFI engines to make all cooling, fuel, and exhaust systems work in the dyno environment. On the other hand, once a powerplant is *on* an engine dyno, it is probably *much* easier to work on than it is in the engine compartment of the vehicle. For example, if a head gasket blows, replacing it is probably trivial on an engine dyno compared to the hours of R&R work required in a tight engine compartment.

Inertial chassis dynamometers like the Dynojet typically are unable to provide precise holding at a *particular rpm or load*, but they are very repeatable and accurate, very simple to use, and easier on the vehicle and engine than brake-type dynos. The passive inertial mass of the inertial dynamometer will often be less than the weight of the vehicle itself, and the dynorun is typically over in 5 to 15 seconds, with little time for mechanical or thermal forces to become critical.

Whatever unknown translation factor is required to convert from wheel horsepower on a particular vehicle to SAE crankshaft horsepower (the weight of all components in the powertrain as well as tire and rolling resistance will affect measured horsepower), inertial dynos are extremely good for providing *before* and *after* comparison testing of aftermarket performance parts. On the other hand, installing heavier wheels can change the inertia of the drivetrain enough to show a horsepower gain (although, indeed, a lighter drivetrain probably means the vehicle *will* accelerate faster). The equation used to convert acceleration to horsepower is Horsepower = [(Mass x Acceleration x Drum Circumference) ÷ (Time x 550)]. Torque can be derived from horsepower with the formula Torque = [(Horsepower ÷ 5250) ÷ Engine Speed]. Dynojets capture engine speed from an inductive pickup attached to a plug wire or direct-fire coil.

ELECTRONIC AND ELECTRICAL COMPONENTS

Where do you go for PROMs, computer cables, wiring, relays, switches, and so forth? Radio Shack, Digikey, or Newark Electronics.

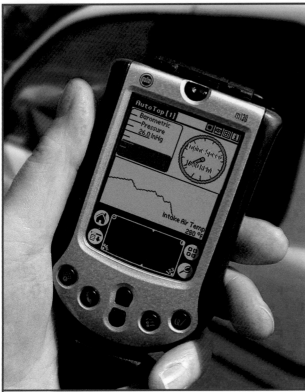

A scan tool that taps into the diagnostic port of a factory EMS is extremely valuable for viewing and logging factory engine data from sensors while the engine is running. This Palm-based Auto Tap system provides a lot of data in a small full-color user interface, but it is very lightweight and convenient for use while driving. Note barometric pressure, engine rpm, and moving graphs of mass airflow rate and intake air temperature. Auto Tap

ELECTRONIC STETHOSCOPES

For $100–200 you can purchase an electronic stethoscope designed to pinpoint noise and locate bad bearings, bushings, dirty fuel injectors, wind and air leaks, and noisy valves and filters. Typically, a flexible shaft reaches tight areas and an ultrasensitive microphone and amplifier provide the full range of sounds needed for diagnosis, with seven sound level control settings from 60 dB to 120 dB.

FUEL PRESSURE GAUGES

Fuel pressure is critical to proper operation of electronic fuel injection. Without an accurate, high-pressure gauge, you are guessing about the fuel rail pressure. You need a good gauge with a selection of fittings for common NPT, flair fittings, and air conditioning–type Schrader valves. Yes, the water-pressure gauges from hardware stores are less expensive, but they are not reliable, and who wants fuel spraying into a hot engine compartment at 40–60 psi? An alternative on some aftermarket engine management systems would be to install a pressure sensor on the fuel rail and record and datalog the data to a virtual dash via EMS software, with a warning alarm set if pressure drops below a critical level.

IGNITION AND SPARK TOOLS

You *think* the plugs are firing fine. You *think* the spark is hot and true. But removing a plug to watch it fire is a major operation on some cars. While nothing substitutes for actually watching the ignition fire the actual plug that will light the fire, spark-gap testers are quick, simple, and cheap. If you've ever experienced a full-bore high-voltage shock trying to hold a plug against a ground with your hand, you probably already understand the value of this tool.

INJECTOR TESTER

Yes, you can often *hear* injectors clicking when they are working. But if the injector is *not* clicking is it because the injector is stuck or because the wiring is bad? An injector tester will provide the answer. It should have adapters for all possible injector connectors. If it can read out the injection pulse width, so much the better. Some powerful automotive multimeters like the Fluke 88 can read out injection pulse width.

PROM TOOLS

If you recalibrate a lot of PROM-based OEM engine management systems, you will need a PROM burner, which allows you to download a calibration to the burner from a laptop or other PC. Batronix's (www.batronix.com) USB chip programmer is a new-generation programming device with a USB connection to the PC, giving it special conductive properties. This PROM burner is designed to be much faster than a parallel burner, yet the device is efficient enough that power is supplied through the USB port with no network adapter required. Via charge pumps, the device automatically generates all voltages required for programming chips—between 3 and 25 volts—out of the voltage supplied through the USB port. The housing is only 2.5 centimeters high. You can use this type of programming device to read, burn, copy, and compare all commonly used memory chips. The vendor provides frequent software updates on their Web site.

Emulators typically convert a factory EMS to dynamic programmability. Diablosport's Romulator is an electronic module that emulates the ECU's read- only memory (ROM). The Romulator actually replaces the ECU's program memory with its own internal RAM, which enables the user to monitor and change parameters. A knowledgeable tuner could use a general-purpose binary editor like WINHEX to modify anything in the PROM image, but Diablosport's EMS-calibration software package is designed to provide user-friendly graphical access to specific tables and parameters in the PROM for specific engine management systems and vehicles (see the chapter on modifying factory engine management systems). The Romulator simply plugs into the same port the chip module employs and is used in conjunction with a PC.

SCAN TOOLS

A scan tool is designed to provide a window into the factory EMS computer, though there are now some good packages that convert a laptop or PDA into a scan tool. A good aftermarket scan tool can typically provide some or all of the following factory scan tool functions:
1. Display and clear check engine error codes
2. Sensor display: airflow sensor, throttle position sensor, rpm, injector pulse width, battery/system voltage, coolant temperature, barometer, O_2 sensor, spark advance, ISC stepper position, EGR temperature
3. Switch position: idle, park/neutral, air conditioning request, air conditioning control, power steering
4. Turn on/off each of the actuators: injectors, fuel pump relay, purge solenoid, fuel pressure solenoid, wastegate solenoid
5. Datalogging: fuel trims, knock sensor, air volume, acceleration enrichment

Scan tools and digital multimeters have some overlap in functionality. Scan tools:
- provide easy access to trouble codes
- provide easy checking of temperature sensors
- are good at finding intermittent wiring problems (wiggle test)
- monitor events while driving
- record reading changes while driving for later playback
- can sometimes measure pulse width
- show command issued by ECU, but can't show if executed (pre-OBD-II)

SERVICE MANUALS

Modern engines and vehicles are very complex, and the factory shop manuals are virtually always better than anything you can find from Chilton, Haynes, or other manual publishers. Of course, they are not as cheap, and—oddly enough—you usually can't get a factory manual from a dealer but have to order it from an outside supplier.

SOLDERING, CRIMPING, CIRCUITRY, AND WIRING TOOLS

Bad wiring is the cause of a huge number of EMS problems. Let me repeat: *Bad wiring is the cause of a huge number of EMS problems!* If you do any automotive wiring at all, you *need* a good crimping tool. This means a ratcheting crimping tool with the right jaw for all common size metal connectors. They cost $20–100 and are infinitely superior to those wimpy things that cost $5 or $10 and are supposed to crimp those plastic-cased wire connectors you find at places like Pep Boys. In the first place, you should *not* be using plastic-cased connectors but rather good metal connectors covered with heat-shrink tubing. You can find good electrical tools at Howard Electronics, Inc. (Xytronics desoldering/soldering station with temperature control), Pace (eBay), Metcal Gear (eBay), Radio Shack, Paladin Tools, Altex and Fry's Electronis.

TEMPERATURE INSTRUMENTS

A cooling system thermometer can be useful if you are struggling with flaky sensors or if you're attempting to locate a heat-soaked air- or coolant-temperature sensor that measures temperature more representative of the actual air or coolant temperature.

VACUUM PUMP

MAP sensors, wastegate actuators, oil pressure sensors, fuel pressure regulators. A hand pump with a gauge readout and a selection of plumbing to interface the tool to various types of sensors and vacuum/boost-operated devices can be really valuable to operate these types of devices without the engine running to make sure they are functioning properly. How many hours of wasted time does it take to justify this tool? Not many, especially if measuring the response of a wastegate actuator or a MAP sensor keeps you from destroying your engine.

VOM METERS (VOLT-OHM-MILLIAMMETER) AND DIGITAL MULTIMETERS

These devices are commonly used for sensor testing of O_2, MAP, MAT, TPS, Knock, CTS, and IAC. You might decide to acquire a digital multimeter in place of a scan tool. Compared to a scan tool, the digital multimeter:
- tests same sensors as scan tool (with more effort)
- performs tests without influence by wiring or ECM problems
- tests sensors that scan tools cannot test (IAC, cam position, and crank position)
- tests injector pulse width and resistance
- tests traditional electrical components (alternator, regulator, starter, coil, plug wires)
- finds bad connections and high-resistance circuits
- finds average or maximum readings using recording/averaging feature without having to watch the meter

WIDEBAND AIR/FUEL-RATIO METERS, SENSORS, AND EXHAUST GAS ANALYZERS

Too many people make the mistake of buying an air/fuel ratio meter that uses a one-wire, two-wire, three-wire, or four-wire sensor. These are not useful for calibrating an EMS except at idle or light cruise on a system that will be running a stoichiometric air/fuel ratio of 14.7:1.

The problem is that output of a standard O_2 sensor (with fewer than five wires) is binary, which means it can only tell you if the engine is running rich or lean or 14.7:1, not how much. This type of sensor is extremely sensitive near 14.7:1, meaning that small changes in exhaust gas oxygen content (fairly good evidence of a surplus or deficit of fuel compared to oxygen in the engine charge mixture) produce large changes in the voltage output of the O_2 sensor. But as soon as the air/fuel ratio moves a bit further away from stoichiometric in either direction, the minute changes in exhaust gas oxygen content are effectively out of the authority of the sensor, which barely changes voltage output at all rich of, say 13.5 or leaner than 15.0.

Once the mixture goes a little bit fat, cheap air/fuel-ratio meters based on standard O_2 sensors will peg to the rich side, even if the faceplate of the meter is graduated all the way from 10.0 to 20:1. Likewise, if the meter goes lean, the meter will peg to the lean side. In reality, such a meter might as well be labeled *Rich, Stoich, Lean*. Of course, such a meter is great for tuning idle and light-cruise mixtures, but then many programmable EMSs can tell you the same thing and can trim the mixture to 14.7 in closed-loop mode.

There are two wideband sensors commonly available from Bosch and NTK/NGK, and they both have five wires. Although simple, switch-type OEM sensors require only a voltmeter to detect the switchpoint in exhaust gas oxygen content as air/fuel mixtures near stoichiometric change from rich to lean, five-wire sensors require sophisticated closed-loop controllers to measure the air/fuel ratio. One difference between ordinary O_2 sensors (heated or not) and a five-wire wideband sensor is that the switchpoint zone can be moved around to different air/fuel ratios in the wideband, meaning five-wire sensors are in effect equipped with an electronically controllable target air/fuel ratio in their reaction cavity.

Low-cost AFR meters are available from Lambdaboy and Powertrain Electronics. The Lambdaboy user interface is available through a proprietary display or a PDA. Lambdaboy claims a range of 8–25 AFR, with response time of 0.10 seconds. The unit does require air calibration for sensor drift compensation. Claimed accuracy is 1 percent at 14.7, 2 percent at full-scale. Display resolution is 1 AFR on eight sunlight-viewable LEDs and 0.1 AFR on PC and display. Standard analog output is 1.8 to 2.8 volts, 8 to 24 AFR, with optional 0 to 1 volt, 8 to 24 AFR. 0- to 1-volt simulated EGO sensor output (so you can run without your stock sensor).

TROUBLESHOOTING SENSORS

Car companies have been working hard to increase the self-diagnostic capabilities of original-equipment electronic fuel injection

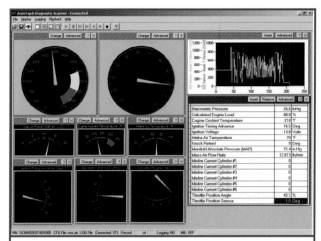

A Palm-based scan package is nice, but there is nothing like the large color screen of a laptop for providing the complete picture of what the EMS thinks is happening in the engine and vehicle. Note the "Calculated Engine Load" virtual meter and the individual-cylinder misfire displays. OBD-II monitors can be really useful for onboard diagnostics. *Auto Tap*

engine management systems, but these systems can still seem inscrutable when you have no idea what's wrong.

Ten or fifteen years ago, on an EMS like GM's final PROM-based system that managed a TPI or early LT1 performance V-8, Sutton Engineering reported that 80 percent of drivability problems didn't report a trouble code in the computer. This is *much* less likely to be true on 1994 to 1996 and later factory OBD-II vehicles, which are greatly improved in their ability to detect and comprehend engine system troubles. Intermittent problems, such as loss of ECU ground or a bad sensor, may just not show up in the diagnostics trouble codes.

OK, when something is wrong, we won't blame the computer because the problem is much more likely to be that the computer is not receiving good data from the sensors. But how do we look at sensor data to decide if there is a sensor problem?

A paperclip or wire jumper will cause the check engine light to flash out trouble codes on a dash panel, but you really need either a scan tool or a digital multimeter (DMM) for more comprehensive troubleshooting.

Some engine management systems require an analog multimeter or a digital multimeter (DMM) with analog simulation so you can see trouble codes spelled out as pulses. The scan tool's ability to function while you drive is nice, but it reflects what the ECU *thinks* is happening, not necessarily what is actually occurring. A scan tool measures physical pulse width at the injector—rather than *assuming* actual pulse width by analyzing the situation at the ECU—which can point out a bad injector driver. It can also be used to compare PROM chips.

Exhaust Gas Oxygen Sensor

The exhaust gas oxygen sensor is very important because many factory ECUs process data from this sensor with the highest priority. A functioning O_2 sensor offers a good snapshot of how the engine is running (although it too can be fooled). It can be a good idea to routinely replace the sensor with a new one before proceeding. It's relatively inexpensive and a good, new sensor is a valuable aid in diagnosing other problems, such as other bad sensors. The O_2 sensor is used in feedback mode by

the ECU to trim the air/fuel mixture and thus directly affects injector on-time.

The voltage range fed by the sensor to the ECU is 0.1 to 0.9 volts, with 0.5 being the stoichiometric or theoretically ideal mixture. The sensor must be heated to 600 degrees Fahrenheit to work (and may have supplemental electrical heating to bring it up to temperature within 10 or 20 seconds of start). The oxygen sensor bounces from rich to lean every few seconds by design, but it is possible to check these oscillations with a scan tool or digital multimeter (DMM). The sensor voltage should jump from rich to lean every few seconds. If this doesn't happen, or if the sensor voltage stays in the 0.3- to 0.6-volt range, or stays lean, the sensor is bad.

Some DMMs have an averaging feature that allows you to read the idle mixture, also telling the maximum high and maximum low, giving you some idea of the sensor's condition. A scan tool lets you look at the block learn counts that will tell you if the ECM is drastically altering the air/fuel mixture. As an example, a GM ECM shows 128 as the normal block learn count. Above 138, the ECU is interpreting sensor data to imply lean running and will add fuel, while below 118, the reverse is true.

The oxygen sensor must remain connected to the ECU in normal closed-loop operation during measurement. You can scrape insulation off the sensor wire or possibly make use of a breakout box connected between the sensor and wiring to connect the meter. Remember, we are talking about using a *digital* multimeter. An *analog* multimeter can damage the sensor by drawing too much current. If you're using a two-wire oxygen sensor, you can disregard the ground wire, according to Lucas. Single-wire sensors are available that work fine in place of the two-wire variety, leaving the ground wire disconnected. Check a shop manual for the wire function in multiple-wire sensors. (Some Chrysler oxygen sensors use four wires!)

Since low voltage corresponds to a lean mixture, failing sensors seldom erroneously indicate a rich condition. Sometimes an O_2 sensor will intermittently quit, which shows up as rich running, with the check light occasionally coming on. This could show up in an ECU trouble code, indicating that the O_2 sensor seemed to be staying lean too long (voltage went below 0.5 and stayed there for a long time).

Oxygen sensors usually last for 10,000 to 40,000 or more miles. Evidence suggests that some sensors are much better quality than others. Some experts suggest using a Bosch or Lucas aftermarket sensor if you don't want to pay the money for an OEM unit.

The following procedure checks for possible wiring problems between the O_2 sensor and the ECU:

1. Disconnect the O_2 sensor connector.
2. Tap into the sensor wire with a nail or similar object.
3. Leave a meter hooked up to the sensor.
4. Hold the nail in one hand while touching the positive battery terminal with the other (safe with battery voltages).
5. The ECU will see a high voltage, interpret this as a rich condition, and lean out the mixture, causing the motor to stumble and possibly die.
6. The meter connected to the O_2 sensor will indicate voltage dropping to 0.1 volts or below.
7. Switching your hand to the minus or negative battery terminal should cause the ECU to enrich things considerably, indicating 0.8 volts or above on the meter.

When you drive the system lean, a new sensor will indicate below 0.1 volts, while an old tired sensor may not go below 0.2 volts. The reverse is true at the rich end of the scale.

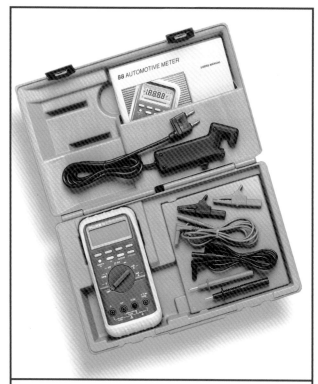

The Fluke 88 is a powerful tool for EMS testing. It provides volts, ohms, amps, continuity, diode testing, milliseconds pulse width, rpm (with inductive pickup), frequency and duty-cycle measurements, min-max-average recording, and more. Fluke

To verify a bad ECU causing a problem with closed-loop mode, or a bad oxygen sensor circuit, connect a scan tool to the ALDL and with the ignition on, engine off, disconnect the sensor lead, and ground the lead. The scan tool should indicate 0.2 volts or below.

If you have a DMM, with engine off and ignition on, check the voltage at the ECU lead, which should be 0.3 volt or below. Verify that the ECU is very well grounded!

TEMPERATURE SENSORS

When an EFI engine is cold, before the oxygen sensor reaches operating temperature, the ECU must rely on temperature and other sensors to get the engine running. Indeed, basic fuel injection pulse width is computed based on engine speed and manifold pressure or mass airflow, corrected for temperature. The oxygen sensor is important for tuning or trimming the basic injection pulse width, but the key word is *trim*. The ECU logic allows it to go only so far in making oxygen sensor corrections. A bad temperature sensor can result in a situation in which the car is running rich, and the ECU knows it is rich (it is, in fact, storing a trouble code that says the mixture is rich), yet it cannot correct sufficiently.

Suspect a bad temperature sensor in this case, but before hauling out the diagnostic tools, first check to make sure that a low coolant level isn't causing the temperature sensor to read inaccurately.

You can test a thermistor-type coolant-temperature sensor by checking the resistance across the two leads (one usually being a ground routed to the ECU, which often supplies its own ground to sensitive sensors) with the connector removed using a DMM (a scan tool makes it much easier if the sensor is hidden in some

inaccessible place). The coolant temperature sensor is usually located on the intake manifold, sometimes on the water pump or head. Check for corrosion on the leads; anything affecting resistance through the sensor wiring would change the sensor reading. Compare resistance to the nearby chart, then verify the figures with a known-good thermometer.

The MAT (manifold air temperature) and IAT (inlet air temperature) sensors, if fitted, can be tested the same way. Look for these sensors in the manifold or air cleaner ducting. If the car cold-soaks overnight, all temperature sensors should read the same. When reconnecting the sensors, make sure the connectors are seated correctly and that none of the connector pins push out of the connectors (some can!). Electronics supply houses sell conductive chemicals that help to prevent corrosion on connectors.

Temperature Sensor Resistance

GM and Chrysler Temperature (deg F.)	Resistance (ohms)	Ford Temperature (deg F.)	Resistance (ohms)
210	185	248	1,180
160	450	230	1,550
100	800	212	2,070
70	3,400	194	2,800
40	7,500	176	3,840
20	13,500	158	5,370
0	25,000	140	7,700
-40	100,700	122	10,970
		104	16,150
		86	24,270
		68	37,300
		50	58,750

Note: At 210 degrees Fahrenheit, the GM and Chrysler systems should be producing stoichiometric mixtures (14.7:1). At -40 degrees Fahrenheit, they should be producing 1.5:1.

MASS AIRFLOW SENSORS

There are two kinds of MAF sensors, those that output varying direct current (DC) voltage like most sensors, and those that output a varying frequency. For varying frequency, you'll need a scan tool. MAF failures typically begin with intermittent symptoms that are followed by total failure. For example, you're driving down the highway and feel a momentary loss of power, usually accompanied by the service engine light flashing on and off. If the MAF does fail completely, the light will stay on, though the car should still be drivable in limp-home mode.

Tapping on the MAF with the engine running may indicate a failing MAF if the engine stumbles. If the engine is running rich or lean and disconnecting the MAF helps, this indicates a failure.

Jumper wires can enable you to connect a DMM to the MAF to read voltage. MAFs usually have three wires: ground, 12 volts in, and sensor output. You can connect the DMM to the ground and sensor output, which typically yields about 2.5 volts on GM systems. Switching to DC volt frequency allows you to look for big frequency changes in response to small changes in engine speed. You can blow through the MAF with something like a hair dryer, tapping the MAF to see if it's affected. Road testing is possible with the scan tool or DMM with extended test leads.

A scan tool will read grams per second of air, which, for example, might be 4–7 at idle on a V-6 engine, the complete range for the engine varying from 3–150 grams per second. A failed MAF will show up on the scan tool as the default output by the ECU. DMMs show frequency variance from 32–150 hertz. The idea is to look for no output or suspiciously rated output. Remember, the scan tool displays what the ECU thinks is happening at the sensor rather than what is actually happening.

The following results come from testing on a 3.8-liter turbo V-6 Buick:

Mass Airflow Sensor Testing Values

MAF output	DC volts	Hertz
Idle	2.599 (avg.)	45
5 psi boost (Second second gear)	2.613	118
10 psi boost (Second second gear)	2.620	125
17 psi boost (Second second gear)	2.635	142

MAP AND BP SENSORS

The MAP (manifold absolute pressure) sensor outputs a value based on absolute pressure, not relative to the altitude the vehicle happens to be at right now. Before diagnosing the MAP sensor, check the hose feeding the sensor for any leaks. Like the MAF, MAP sensors come in two breeds, those that output varying voltage (GM) and those that output a square-wave 5 volts DC that varies in frequency (Ford), requiring a DMM that can measure frequency in hertz. It can be hard to get documentation on what the output of a MAF should be; consider measuring a known-good vehicle first and then comparing it to the one in question.

TROUBLESHOOTING HALL EFFECT SENSORS

Wiring shorts, loose or corroded connectors, or arcing damage to internal sensor circuitry can cause problems with the Hall Effect sensor. Most aftermarket engine management systems will immediately detect Hall Effect faults such as missing crank REF or cam SYNC and display the problem via laptop interface, while OEM systems will set the malfunction indicator light (MIL). To troubleshoot original-equipment Hall Effect sensors, Wells Manufacturing (which builds a variety of automotive sensors, actuators, and engine management components) recommends using a self-powered indicator-light type of Hall Effect tester that flashes when the sensor's output signal changes.

"If the sensor is being tested on the vehicle," says Wells, "simply plug the tester into the sensor, then crank the engine and observe the indicator light. It should wink on and off while the engine is cranking. If the sensor is being tested off the vehicle, plug the tester into the sensor connector and observe the indicator light. It should remain on steadily. Then insert a metal blade into the sensor's magnetic window (air gap). The indicator should go off while the blade is in the window, then wink back on when the blade is removed. No change would indicate a faulty sensor."

Wells points out that on-car sensor problems can be caused by several things:

The sensor must have the proper reference voltage (VRef) power supply from the computer. If the sensor is not receiving the required voltage, it won't work. Measure the supply voltage between the sensor's power supply wire and ground (use the engine block for a ground, not the sensor's ground circuit wire). If you don't see the specified voltage, check the sensor's wiring harness for loose or corroded connectors.

The sensor's ground circuit must be in good condition. A poor ground connection will have the same effect on the sensor's operation as a bad VRef supply. Measure the voltage between the sensor's ground wire and the engine block. If you see more than 0.1 volts, the sensor has a bad ground connection.

If the sensor has power and ground, the next thing to check would be sensor output. The voltage output signal should switch back and forth from maximum to near zero as the engine is cranked. No change would indicate a faulty sensor. Watching the sensor's voltage output on an oscilloscope is a good way to spot problems that might escape normal diagnosis. You should see a nice, sharp square-wave pattern that goes from maximum to near zero (or vice versa) every time the sensor switches on and off. The sync pulse signal from a crank/cam position sensor should also stand out and be readily apparent in the waveform. If you see rounded corners, spikes, excessive noise or variations in amplitude from one pulse to the next, the erratic operation of the sensor may be causing the computer to miss signals or pick up false signals.

Wells Manufacturing recommends testing an electronic ignition system with Hall Effect pickup in the distributor by inserting a steel feeler gauge, pocket knife blade, or similar object, into the sensor window with the ignition turned on to see if this fires the ignition coil, making sure the spark has a proper path to ground via a spark-gap tester. For you old-timers, blocking the magnetic window in a Hall Effect sensor should have the same effect as opening the points in a breaker point ignition system.

A Hall Effect sensor *must* be aligned correctly with the interrupter ring or shutter blade to generate a clean signal, and any contact between moving parts and the sensor might cause idle problems or fatal damage. A worn, stretched, or jumped timing chain or belt can affect the timing of the sensor's output (which is one reason why using any cam component to provide precision timing information is riskier than using the crank). Failure of a crank position sensor usually causes the engine to die immediately since the EMS no longer has a reference for ignition and fuel injection.

On older Chrysler Hall Effect ignition systems, the shutter blades under the distributor rotor *must* be properly grounded to produce a clean signal. Wells recommends using an ohmmeter to check continuity between the shutter blades and distributor shaft. "If the blades are not grounded, the sensor will check out okay but won't produce a good signal when the engine is cranked with the rotor in place," says Wells. "On Ford applications with Thick Film Integrated (TFI) ignition distributors, the Hall Effect signal is called the 'Profile Ignition Pick-up' (PIP) signal. Corroded connectors between the Hall Effect PIP unit in the distributor and the TFI module on the distributor housing are common." Replacing the sensor on these applications requires R&R of the distributor shaft.

TROUBLESHOOTING THROTTLE POSITION SENSORS

With the engine stopped and the ignition turned off, unplug the electrical connector and attach an ohmmeter between the center terminal and one of the outer two terminals on the TPS. Manually open the throttle slowly to operate the TPS. The ohmmeter reading should increase or decrease smoothly as you operate the TPS through its complete range of travel.

Connect the ohmmeter between the center terminal and the other outer terminal as above. Again, the ohmmeter reading should increase or decrease smoothly as you work the sensor. Momentary infinity (open), full continuity (0 ohms), or erratic readings indicate a defective or shorted sensor.

Check the reference voltage from the computer with a digital voltmeter or oscilloscope to verify the sensor's output signal. If the sensor is not receiving a full 5 volts from the ECU, it will be unable to generate the correct range of return signals. Less that 5 volts on the VRef indicates a possible wiring or computer problem (assuming the battery is not excessively low). Verify good continuity between the TPS ground wire terminal and ground with the key on. Verify that the voltage reading indicates less than 0.1

volts. Voltage output between the signal wire and ground with the key on should vary from less than 1 volt at idle up to nearly 5 volts at wide-open throttle.

A scan tool or laptop computer can also be used on most OEM and aftermarket engine management systems to observe the TPS output voltage as seen by the onboard computer system. Note that since the scan tool displays the TPS output voltage as a digital reading, the display may not update quickly enough for you to detect a momentary skip spot in TPS output as you open and close the throttle.

TROUBLESHOOTING O₂ SENSORS

According to Wells Manufacturing, which builds replacement OE-type O_2 sensors, oxygen sensor performance diminishes with age as contaminants accumulate on the sensor tip and gradually reduce its ability to produce voltage. "The sensor becomes sluggish and takes longer to react to oxygen changes in the exhaust causing emissions and fuel consumption to go up. The effect is most noticeable on engines with multi-port fuel injection (MFI) because the fuel ratio changes more rapidly than on the throttle body or feedback carburetor applications. If the O_2 sensor fails, the computer may go into open loop operation." Contaminants include lead, silicone (from internal coolant leaks or inappropriate types of RTV sealant), phosphorus (from oil burning), and sulfur.

The operation of the O_2 sensor, according to Wells, should be checked any time there's an emissions or engine-performance problem. Wells recommends that the O_2 sensor be checked when the spark plugs are replaced, or if the vehicle has a maintenance reminder light calling for a check of the O_2 sensor. Reading O_2 sensor output voltage with a scan tool or digital voltmeter while the engine is running and warmed up can tell you if the sensor is producing a signal, but the digital numbers will jump around too much to enable an accurate determination if the sensor is good. A high-impedance analog voltmeter is somewhat better for viewing O_2 sensor voltage transitions, but may respond too slowly to be conclusive on systems with higher voltage transition rates. However, a digital storage oscilloscope (DSO) will display the sensor's voltage output as a wavy line that shows both amplitude (minimum and maximum voltage) as well as the frequency of transition rate from rich to lean. A good O_2 sensor will produce an oscillating waveform at idle that transitions from minimum voltage (0.3 volts or less) to maximum (0.8 volts or higher).

Wells suggests that making the fuel mixture artificially rich by feeding propane into the intake manifold will cause the sensor to respond almost immediately (within 100 milliseconds) and go to maximum (above 0.8 volts) output. Creating a lean mixture by opening a vacuum line should cause the sensor's output to drop to its minimum (0.3 volts or less) value. If the O_2 sensor voltage does not change back and forth quickly enough, it needs to be replaced. A sluggish O_2 sensor may not set an MIL code, so don't assume the O_2 sensor is OK just because there is no code.

Alternately, Wells suggests, the O_2 sensor can be checked by removing it from the vehicle, connecting a digital voltmeter to the sensor, and using a propane torch to heat the sensor element. With a nonheated zirconia O_2 sensor, connect the positive test lead to the sensor's wire lead and the negative test lead to the sensor housing or shell (ground). Hold the sensor with pliers, then use the propane torch to heat the sensing element. The flame consumes most of the oxygen, so the sensor should generate a rich voltage signal of about 0.9 volts within 60 seconds or less. Now remove the flame while observing the voltmeter. If the sensor is good, the sensor reading should drop to less than 0.1 volt within three seconds.

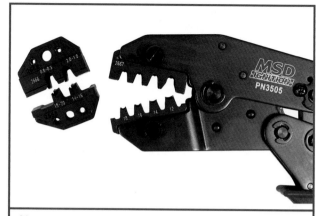

Of course you're going to need a standard set of mechanics tools for tuning and modifying engine management systems, but since wiring is so important, it is worth having a really good ratchet-type crimp tool with the correct slots for a handling a variety of crimping jobs. This MSD tool has multiple, interchangeable jaws with a variety of connector gripping slots. Such a tool can (and should) cost a few bucks, but remember that a broken or shorted wire could destroy your vehicle.

To test a sensor's heating circuit, Wells recommends connecting an ohmmeter across the sensor's two heater wires. Set the meter on the low scale. If the heater circuit is OK, the ohmmeter will register a resistance reading. The precise reading is unimportant as long as the meter registers other than an open circuit.

Titania O_2 sensors on Jeep and Toyota engine management systems change resistance rather than generating a voltage like zirconia sensors. However, a titania O_2 sensor can also be tested with an ohmmeter and propane torch. Connect an ohmmeter across the black and gray sensor leads, setting the meter on the 200K scale. Heat the tip of the sensor with the torch as above, and note the ohmmeter reading. A good sensor should show resistance that varies with flame temperature, and good sensor should return to infinity after removing the flame.

TROUBLESHOOTING ENGINE TEMPERATURE SENSORS

A coolant temperature sensor (CTS) is critical in triggering many engine functions, which means a faulty sensor or wiring can cause a variety of warm-up performance problems and emissions failures.

Symptoms that might be caused by a bad coolant sensor include the following:
• EMS fails to launch closed-loop feedback mode once the engine is fully warm
• Poor cold idle due to excessively rich fuel mixture, no early fuel evaporation (EFE), or lack of heated air
• Stalling due to excessively rich mixture, retarded timing, or slow idle speed
• Cold hesitation or stumble due to lack of EFE or excessively early EGR
• Poor fuel mileage due to rich mixture, open loop, spark retarded

Keep in mind that such problems may not indicate sensor failure but might occur due to wiring faults or loose or corroded connections. An engine thermostat that is too cold for the application can also upset cold-running EMS strategies, as can low coolant due to incomplete filling or cooling system leaks. The sensor element must be in direct contact with liquid coolant to generate accurate data.

The easiest way to test a coolant temperature sensor is with a sweep test with an ohmmeter. To conduct a sweep test:

1. The engine must be cold.
2. With the key off, disconnect the wiring connector from the thermister.
3. Attach an ohmmeter across the sensor leads and note the resistance.
4. Reattach the CTS connector. Run the engine for 2 minutes and then shut it down.
5. Repeat steps 2 and 3.
6. There should be a difference of at least 200 ohms between the two readings.
7. If not, the sensor is defective or possibly coated with cooling system sludge that insulates it from changes in engine temperature.
8. Try cleaning the sensor element and then retest.
9. Good shop manuals will contain a chart specifying coolant sensor resistance for various engine temperatures. Verify that the reading meet specs.
10. Alternately, check the sensor's operation signal as the engine warms up with a voltmeter: Measure the CTS reference voltage (normally 5 volts) and returned voltage (3–4 volts when cold, dropping to less than 2 volts at normal operating temperature).
11. No change in returned signal voltage (or lack of a return signal) indicates a faulty sensor.
12. Incorrect VRef indicates a wiring problem.
13. Alternately, observe the sensor's waveform with a digital storage oscilloscope (DSO).
14. If an EMS provides live sensor data through its diagnostic connector, you can also read the coolant sensor's output directly with a scan tool (usually in degrees Celsius or Fahrenheit).
15. Aftermarket programmable engine management systems will display engine temperature sensor readings along with other sensor values on an engine data page via laptop computer.

Sample trouble codes indicating a coolant sensor circuit problem:

General Motors: Codes 14 (shorted) and 15 (open)
Ford: Codes 21, 51, and 61
Chrysler: Codes 17 and 22
OBD-II: P0117, P0118, P0125, P1114, P1115, and P1620

TROUBLESHOOTING MAP SENSORS

A faulty MAP sensor, shorts or opens in the wiring circuit, or vacuum leaks in the sensor hose or intake manifold can cause drivability symptoms that include rough or erratic idle, black exhaust smoke (rich fuel condition resulting in high hydrocarbon emissions), stalling, hard starting, hesitation, engine misfires, pinging, poor fuel economy, and bad general engine performance.

There are various ways to check out a MAP sensor that you suspect has problems. It is trivial to use a laptop attached to most programmable engine management systems to display the MAP sensor data and watch how it changes as you load and unload the engine. A scan tool can check for MAP sensor trouble codes on OEM systems. A digital storage oscilloscope can observe the sensor's output as a waveform. A multimeter can compare the sensor's output voltage or frequency to specifications in a manual. And the following two relatively simple procedures courtesy of Wells Manufacturing will quickly tell you whether or not a MAP sensor is responding to changes in intake vacuum.

General Motors and Chrysler MAP sensors can be tested on a vehicle using a digital voltmeter (DVOM) and two jumper wires. The following test verifies that the MAP sensor is responding to changes in engine vacuum. If the reading does not change, it means the sensor is faulty or the vacuum hose is plugged or leaking.

- Disconnect the MAP sensor's electrical connector.
- Connect one jumper wire between the connector and the MAP sensor's terminal A.
- Connect another jumper wire from the connector to terminal C.
- Connect the positive lead on the DVOM to terminal B (the sensor's output terminal) and the negative DVOM test lead to a good engine ground.
- Turn the ignition key *on* and observe the voltage. If the reading falls in the voltage range of 4 to 5 volts (2 to 3 volts for turbocharged engines) at sea level, the sensor is functioning properly at this point.
- Be sure the vacuum hose between the MAP sensor and engine is in good condition and does not leak. Then start the engine and let it idle. An idling engine will produce a large amount of intake vacuum, which should pull the MAP sensor's voltage down to a low reading of 1 to 2 volts (note: readings will vary with altitude).
- You can also do this test with the key on, engine off by applying vacuum to the MAP sensor's hose with a hand-held vacuum-pressure pump.
- On Ford applications, you'll need a multimeter that can display frequency required to check the sensor's output, or you can use an ordinary diagnostic tachometer (tachs display a frequency signal) as follows:
- Set the tachometer to the four-cylinder scale (regardless of how many cylinders the engine has).
- Connect one tachometer lead to the middle terminal on the MAP sensor and the other tachometer test lead to ground.
- Connect the two jumper cables the same as before, attaching each end terminal on the sensor to its respective wire in the wiring connector.
- If you want to measure engine vacuum so you can correlate it to a specific frequency reading, connect a vacuum gauge to a source of manifold vacuum on the engine, or tee the gauge into the MAP sensor hose.
- Turn the ignition *on* and note the initial reading. The reading on the tachometer should be about 454 to 464 at sea level, which corresponds to a frequency output of 152 to 155 hertz.
- Start the engine and check the reading again. If the MAP sensor is functioning properly, the reading should drop to about 290 to 330 on the tachometer, which corresponds to a frequency output of about 93 to 98 hertz. No change would indicate a defective sensor or leaky or plugged vacuum hose.

TROUBLESHOOTING VAF METERS

There are special testers for troubleshooting VAF sensors, but you don't always need one to check sensor operation. By watching the VAF sensor's output with an analog voltmeter or ohmmeter, or better yet an oscilloscope, you can look for a change in the sensor's output as airflow changes. One simple check is to look for a voltage change as you slowly push the airflow flap all the way open. A good sensor should produce a smooth and gradual transition in resistance (ignition off) or voltage readings (ignition on) all the way from full closed to full open. If you see any sudden jerks in the movement of the needle (analog ohmmeter or voltmeter) or dips or blips in the scope trace (similar to sweeping a TPS), the VAF sensor needs to be replaced.

Changes in the sensor's voltage output should also produce a corresponding change in fuel injector duration when the engine is running. Injector duration should increase as the VAF flap is pushed open.

On Ford EFI systems, you can use a breakout box and voltmeter to check VAF sensor voltage readings. Pushing the flap open should cause a steady and even increase in the sensor's out-

continued on page 167

Troubleshooting and Diagnostics

STARTING PROBLEMS

A healthy engine with fuel, air, and spark MUST run. If it's not running properly, there MUST be a problem with one of more of these parameters.

TROUBLE: ENGINE WILL NOT START
Possibilities:
Injectors are not operating because:
  ECU is not receiving 12 volt switched power.
  ECU is not receiving 12 volt constant power.
  Insufficient ground.
  Ignition system is not compatible with ECU.
  Bad ECU.
     Check connections, fuses, relays, plug wires.
ECU not receiving an rpm signal from the ignition system
     Check wiring for proper connection, ignition compatibility, ignition system operation.
Voltage drops below 9 volts during cranking.
     Check battery voltage, connections.

Injectors are operating but:
  Fuel mixture is too lean.
  Fuel mixture is too rich (spark plugs will be wet with fuel).
     Check plug condition and adjust fuel delivery bars as required.

Injectors do not have proper fuel pressure.
     Check pump operation, fuel rail pressure.

Engine is not receiving ignition spark or Engine ignition timing is incorrect.
     Check ignition system function, and timing.

Engine must have a small amount of air to start.
     Engine is not receiving enough air to start.
     Check throttle-stop setting, IAC system configuration parameters.

NO-START Algorithm

1. Check for spark; if spark exists, go to 2.
   1.1 On fuel-only EFI system with standalone ignition: Disconnect EFI tachometer wire from ignition system; if still no spark, repair ignition; otherwise go to 1.2.
   1.15 On fuel-spark EMS, check for RPM reading as viewed from a laptop or scan tool. Verify crank sensor alignment and air gap with the trigger wheel. Verify crank sensor resistance (typically 600-700 ohms***)
   1.2 (Electrical problem in EFI system) Disconnect ECU; if spark, substitute spare ECU known to be good and recheck; otherwise go to 1.3.
   1.3 (Bad EMS wiring harness) Check EMS wiring harness.

2. Check fuel pump—with ignition on, pump should audibly run a few seconds, then stop. If pump runs, go to 3; otherwise go to 2.1
   2.1 (Bad fuel pump or electrical problem in ECU) Check switched and continuous power to ECU, check all grounds. If bad, fix; otherwise go to 2.15.
   2.15 Check pump fuse and all other fuses. If blown, go to 2.3; otherwise go to 2.2.
   2.2 (Fuse OK) Check pump power and ground for several seconds after activating ignition. If power OK, replace fuel pump; otherwise check wiring harness and if good, substitute replacement ECU.
   2.3 (Fuse blown) Replace the fuse (make sure it's the correct

rated fuse!). If fuse again blows, go to 2.4
   2.4 (Fuse blows again) Check wiring harness, repair if necessary; otherwise go to 2.5.
   2.5 Disconnect fuel pump, injectors, and other sensor connectors from the ECU and try new fuse. If the fuse blows again, substitute new ECU;
otherwise go to 2.6.
   2.6 Reconnect EFI components one by one until the fuse blows, and then replace the faulty component.

3. (Fuel pump runs, engine will not start)
Check fuel flow direction at pressure regulator and pump. If correct, check fuel rail pressure with a high-pressure gauge (upstream of regulator, if there is one or more!) for ten seconds after energizing the EMS. The fuel pressure should increase to the rating of the fuel pressure regulator (typically 39-43 psi with no manifold pressure). If pressure is OK, go to 4; if pressure is low or there is no pressure, go to 3.2; otherwise go to 3.1.
   3.1 (High pressure) Check return line for blockage. If OK, replace regulator.
   3.2 (Low pressure) There is either air in the line or the regulator has malfunctioned. Trying running the pump for a for a minute or two by jumpering +12V to the hot side of the fuel pump. If pressure increases with air bleeding, continue. Otherwise, clamp off return line, then retry. If pressure builds to a high pressure, replace regulator; otherwise go to 3.3.
   3.3 Check supply line from pump to injectors and regulator and fix if blocked (blocked fuel filter, pinched hose, blocked fuel feed screen at tank, low fuel, and so on). If OK, replace fuel pump.

4. (Pressure in fuel rail OK)
With ignition on, crank engine with coil wire disconnected. Fuel should pulse in 0.01 second squirts from one or more injectors (listen for a loud clicking from the injectors; this will be easier with a mechanic's stethoscope or with a screwdriver tip against the injector and the other placed against the ear). If fuel pulsing is OK, EMS is OK (replace coil wire, check ignition timing, firing order of the plugs, check at least +9V battery voltage to the ECU during cranking, and so forth); if no fuel is pulsing, go to 4.1; if continuous fuel flow from one or more injectors, go to 5.
   4.1 (No fuel pulsing). Verify valid Crank Ref or Tach signal to ECU. If signal is OK, check +12V power to injectors with a test light with the ignition turned on to one terminal on one injector. If +12V is present on one terminal, check for continuity between the other terminal and the injector driver/channel of the ECU, which pull to ground to fire the injector. If you have an oscilloscope, look for a +12V square wave at the injector connector. If not, the ECU could be defective, and this would be a great time to substitute a known-good ECU (with the correct Map installed). Verify in the EMS software that the injection pulsewidth table is not zero-ed out completely, resulting in NO INJECTION. If the electrical tests show no problem but the injectors are not clicking, consider that one or more injectors might be stuck fully closed or even fully open. An injector stuck closed will not click or flow any fuel; an injector stuck open will rapidly flood the cylinder, and can hydro-lock the cylinder and destroy pistons or even put a rod through the block JUST CRANKING THE ENGINE!! Try substituting one or more known-good injectors into the fuel rail. Otherwise go to 4.2.
   4.2 (No tach signal or crank reference). Check wiring harness and connectors/pins to all components, then check ECU. Check ignition components that generate the tach signal or the crank trigger mechanism that generates the Crank Ref signal.

5.(Continuous fuel from one or more injectors)
Disconnect electrical connections on injectors, recheck for spray. If no injector sprays, check wiring harness, connections, and ECU; otherwise go to replace damaged injector(s). **Note:** Stuck-open

injectors can hydro-lock the engine and cause catastrophic damage! If you suspect excessive flooding might be a problem, immediately remove the spark plugs before cranking for further troubleshooting efforts.

6. (Insufficient air)
Crank engine with the throttle manually opened a small amount. If the engine starts, adjust IAC motor settings for proper start-up.

TROUBLE: ENGINE STARTS, THEN DIES (FULLY WARM)
Possibilities:
   Tach/Crank Ref source problem
   Sensor problems
   Intake air leaks
      Check sensors, tach signal source, integrity of intake system.

Start-Stall Algorithm:

6. Check critical engine connections (Crank Position, tach lead to tach source for fuel-only batch-fire injection (coil, tach driver), lead to coolant temperature sensor, ground connection, MAP/MAF sensors, and so on). Correct problem if required; otherwise go to 6.1.
   6.1 Check resistance of tach-to-ECU or Crank Ref circuit. If required, replace any wiring/connectors; otherwise go to 6.2.
   6.2 Check other sensors for legitimate electrical values based on known physical conditions (temperature vs. legitimate coolant sensor resistance, air pressure in manifold vs. MAP signal, and so forth). Replace sensors if bad; otherwise go to 6.3.
   6.3 Check for major air leaks in intake system (throttle body, runners, vacuum hoses and ports, and so on).
   6.4 Check for other ignition problems.

Trouble: Difficult Cold Start, Poor Operation When Cold
Possibilities:
   Bad coolant temperature sensor
   Warmup map needs more enrichment at lower temperatures
   Zero-rpm map needs more duration on cranking bars
   Cranking primer needs more duration
      Check coolant sensor, maps for proper adjustments.

Difficult Cold-Start Algorithm:

7. Disconnect coolant temperature sensor, check resistance of the sensor to ground (need chart of resistance per degrees temperature) with ohmmeter (in the case of thermistor-type sensor). Verify that Zener diode-type sensor is good or substitute known good sensor. If bad, replace sensor and retry running; otherwise go to 7.1.
   7.1 (Good coolant sensor) If the engine is not hard to start, adjust the prime fuel injection pulse width under warm conditions; otherwise go to 7.2.
   7.2 Verify operation of startup enrichment devices (extra injector or injectors, thermo-time switch, fast idle equipment, and so on). If bad, replace; otherwise go to 7.3.
   7.3 (Incorrect startup enrichment on programmable system) Reprogram the warmup enrichment percentage(s), verifying that the values are correct when the system is again cold (which may take some time).
   7.4 Engine flooded (use "clear flood" mode if it exists).
   7.5 Low battery voltage (below 8.5volts) when cranking. Try starting with additional voltage or by powering the ECU with a separate battery.
   7.6 Unusual fuel pump condition (vapor-locked, overheated motor, and so forth).

Trouble: Difficult Hot Start
Possibilities:
   Engine flooded.
   Air temperature sensor (if installed) is heat soaking.
   Fuel lines are heat soaking.
   Fuel pump problem.
      Check air temperature sensor mounting location, fuel line routing (shield from engine/exhaust heat).

Trouble: Fuel Pump Does Not Operate
Possibilities:
   ECU is not receiving power or ground.
   Fuel pump relay is not operating. Pump is not receiving power or ground.
      Check connections, relays.

Trouble: Tach Reading Is Incorrect on Fuel-only Batch-fire EMS
Possibilities:
Number of cylinders or other data entered incorrectly to ECU.
Incompatible ignition type.
      Check ID page, ignition compatibility chart.

RUNNING PROBLEMS

Drivability Improvement to Engine That Runs
8. Verify correct prime engine sensors (MAP, MAF, tach signal, and others). Replace if bad;
otherwise go to 8.1.
   8.1 Check fuel pressure setting during running conditions with gauge connected to fuel rail.
   8.2 Check for air and vacuum leaks.
   8.3 Check injectors for correct operation. Injectors should match flow at all duty cycles as closely as possible (for racing, injectors must be individually cleaned and matched by experts).

Fuel Economy Improvement For Engine That Runs
Possibilities:
   TPS poorly calibrated.
   Wrong air-fuel mixture.
   Acceleration enrichment problem.
   Leaky injectors.
   Other engine problem.
      Check TPS, mixture, acceleration enrichment maps, injector balance test.

Fuel Economy Improvement Algorithm
9. Check TPS for calibration (incorrect setting could cause undue acceleration enrichment).
   9.1 Check air-fuel mixture at idle and cruise (14.7:1 is correct stoichiometric; smooth, lean idle is best). Check content of exhaust gases for carbon monoxide (CO), hydrocarbons (HC), and so forth.
   9.2 Check acceleration enrichment percentage and sustain. Back off the enrichment until the vehicle begins to bog slightly, then enrich slightly. Adjust sustain according to manufacturer's recommendations, then try backing off until vehicle bogs, and enrich slightly again. Poor economy is probably not due to faulty acceleration enrichment if drivability is good. Black fuel smoke under heavy acceleration is a good indicator that acceleration enrichment is excessive.
   9.25 Check for leaky injector(s).
   9.3 (EFI system good) Check distributor curve, engine timing, vacuum advance, and other functions.

Trouble: Engine Runs Rich
Possibilities:
- ECU adjustments set too high.
- TPS calibrated incorrectly.
- Leaky injector(s).
- Excessive fuel pressure.
- Nonfunctional temperature sensor.
- Bad ECU.

Trouble: Engine Runs Lean
Possibilities:
- ECU pulsewidth adjustments set too low.
- TPS bad or incorrectly adjusted.
- Low fuel pressure.
- Low switched or continuous power to ECU.
- Bad ECU.

Trouble: Engine Revs To An Rpm And Develops A Misfire
Possibilities:
Incompatible ignition
   Check rpm signal on engine data page.
Fuel mixture changes drastically from one rpm range to the next.
Fuel mixture is too rich.
Fuel mixture is too lean.
   Check Fuel delivery maps for proper adjustments.
Engine is running out of fuel
   Check Fuel pressure.

Trouble: Slow Throttle Response
Possibilities:
- Fuel delivery maps too rich.
- Ignition timing retarded.
   Check fuel maps, ignition timing.

Trouble: Engine Coughs Between Shifts
Possibilities:
- High-rpm light throttle areas of maps too rich.
- High-rpm light throttle areas of maps too lean.
- Accelerator pump values too rich.
- Accelerator pump values too lean.
   Check accelerator pump maps, fuel delivery maps.

Trouble: Engine Coughs Under Rapid Throttle Movement
Possibilities:
- Accelerator pump enrichment too lean.
- Accelerator pump enrichment extremely rich.
- Fuel maps too lean at part throttle.
- Ignition timing retarded.
   Check accelerator pump maps, fuel maps, ignition timing.

Trouble: Engine Misfires Under Cornering
Possibilities:
- Improper fuel pressure.
- Fuel pickup problems.
   Check fuel pressure.

Trouble: Engine Performance Will Not Repeat From One Test To The Next
Possibilities:
- Fuel pickup /pressure problems.
- Crank Ref or ignition tach signal problem (on fuel-only EMS)
- Fuel maps adjusted erratically.
- Engine warmup map adjusted erratically.
- Poor or erratic manifold vacuum signal (speed-density units).
- Battery voltage fluctuating.
   Check fuel pressure, tach signal on engine data page, fuel maps for proper adjustment, vacuum signal, battery voltage.

Trouble: Engine Surges Under Light Cruise Conditions
Possibilities:
- Fuel mixtures too rich.
- Fuel cut on deceleration feature is on.

Check fuel maps, fuel cut problem.

Trouble: Poor Or Erratic Idle
Possibilities:
- Throttle plates are not synchronized (multiple butterfly systems).
- Fuel maps adjusted erratically in idle area.
- Poor manifold vacuum signal.
   Check throttle synchronization, fuel maps, manifold vacuum.

Trouble: Engine Misfires Under Hard Acceleration
Possibilities:
- Fuel maps too lean.
- Fuel maps too rich.
- Improper fuel pressure.
- Ignition system failing.
- Battery voltage falling.
   Check fuel maps, fuel pressure ignition system, battery voltage.

Trouble: Engine Lacks Performance/Power
Possibilities:
- Full-throttle settings are too rich (black smoke from exhaust).
- Full-throttle settings are too lean (high exhaust gas temperatures).
   Check fuel maps.

Trouble: Problems at Low-, Medium-, and High-Load
Possibilities:
Air-Related Load Problems
   Check air-related load problems arise from improperly sized throttles. Make sure that a throttle can flow enough air for an engine. Stock throttles on heavily modified engines will typically cause upper-rpm performance problems.

Fuel-Related Load Problems
Fuel pressure dropping with load because:
   Fuel pump cannot keep up with the engine's fuel needs
   Fuel pressure regulator is not functioning properly.
   Check fuel pressure under load (should increase by 1psi with every 1psi of boost on turbo or supercharged engines (unless a rising rate regulator is installed, in which case it should increase 4-8 psi per psi boost)

Injectors Undersized
   Check injector flow rate.

Tuning issues
EGO Sensor Correction Problems
   Check tuning recommendations, O2 sensor tuning recommendations

Spark-Related Load Problems
Calibration.
Secondary Ignition Failure
Improper Wiring.
   Check coil or direct-fire coil units are wired correctly before proceeding.
   Check tuning (software timing problems can seem like spark-related problems). An engine that is detonating has too much timing in the Timing Advance Table, or is not receiving enough fuel. An engine that is not performing well may have too little timing, or too much timing. Try adding timing until light detonation is detected, then back off a bit (but do not attempt to run a rotary into detonation!)
   Check for secondary Ignition failure (often blamed for load-based ignition problems), including the spark plugs and spark plug wires.
   Check the coil screws tightness
   Check Direct-Fire Coil Units for proper ground operation (If misfiring occurs under load, loose coil screws or ungrounded Direct-fire Coil Units are often the culprit).

continued from page 163

put from 0.25 volts when the flap is closed to about 4.5 volts with the flap fully open. The reference voltage to the airflow sensor from the computer should be 5 volts.

Ford's trouble codes that apply to the VAF include: code 26 indicates a VAF reading out of range, code 56 indicates sensor input too high, code 66 is sensor input too low, and code 76 indicates no sensor change during the goose test.

Most manufacturers also give specific resistance specs for the various VAF terminals in a shop manual. Bosch, for example, lists the following for some of its applications:

Terminals 6 and 9	200–400 ohms
Terminals 6 and 8	130–260 ohms
Terminals 8 and 9	70–140 ohms
Terminals 6 and 7	40–300 ohms
Terminals 7 and 8	100–500 ohms
Terminals 6 and 27	2,800 ohms max at 68 degrees

Fahrenheit

On a Mazda 323, the resistance reading between VAF sensor terminals E2 and VS should be 20 ohms with the flap closed, and 1,000 ohms with the flap open. Readings that are out of range would indicate a bad sensor.

TROUBLESHOOTING KARMAN-VORTEX SENSORS

Both Toyota/Lexus and Mitsubishi Karman-Vortex airflow sensors put out two signals: an airflow frequency signal and an air-temperature or barometric-pressure voltage signal. The Karman-Vortex signal should be a square-wave signal that flips back and forth from 0 to 5 volts. The frequency of the signal will increase (narrower pulse width) as airflow increases. Frequency should increase smoothly and steadily with rpm. Drivability problems such as surging, hesitation, stalling, and elevated emissions may indicate a sensor failure. Most sensor problems are caused by a loose or corroded wiring connector.

In early Mitsubishi applications, the sensor is mounted inside the air cleaner. A poorly fitting air filter may prevent the air cleaner lid from fully seating against the sensor and cause problems. Similar problems can be caused by air leaks in the intake plumbing or manifold.

Codes 31 and 32 indicate no signal from the airflow sensor on the Toyota and Lexus applications. Code 12 refers to a fault in the airflow sensor circuit on Mitsubishi applications.

TROUBLESHOOTING MAF SENSORS

MAF problems will normally cause the ECU to set a trouble code on OEM engine management systems. An engine with a bad MAF sensor may be hard to start or stall after starting. It may hesitate under load, surge, idle rough, or run excessively rich or lean. The engine may also hiccup when the throttle suddenly changes position. However, keep in mind that low engine compression, low vacuum, low fuel pressure, leaky or dirty injectors, ignition misfire, and excessive exhaust backpressure (plugged converter) can produce similar drivability symptoms.

MAF sensors can be tested either on or off the vehicle in a variety of ways. You can use a MAF sensor tester and tachometer to check the sensor's response. If testing on the vehicle, unplug the wiring harness connector from the sensor and connect the tester and tachometer. Start the engine and watch the readings. They should change as the throttle is opened and closed. No change would indicate a bad sensor. The same hookup can be used to test the MAF sensor off the vehicle. When you blow through the sensor, the readings should change if the sensor is detecting the change in airflow.

Another check is to read the sensor's voltage or frequency output on the vehicle. With Bosch hot-wire MAF sensors, the output voltage can be read directly with a digital voltmeter by back-probing the brown-and-white output wire to terminal B6 on the PCM. The voltage reading should be around 2.5 volts. If out of range, or if the sensor's voltage output fails to increase when the throttle is opened with the engine running, the sensor may be defective. Check the orange and black feed wire for 12 volts, and the black wire for a good ground.

Power to the MAF sensor is provided through a pair of relays (one for power, one for the burn-off cleaning cycle), so check the relays too if the MAF sensor appears to be dead or sluggish. If the sensor works but is slow to respond to changes in airflow, the problem may be a contaminated sensing element caused by a failure in the self-cleaning circuit or relay. With GM Delco MAF sensors, attach a digital voltmeter to the appropriate MAF sensor output terminal. With the engine idling, the sensor should output a steady 2.5 volts. Tap lightly on the sensor and note the meter reading. A good sensor should show no change. If the meter reading jumps or the engine momentarily misfires, the sensor is bad and needs to be replaced. You can also check for heat-related problems by heating the sensor with a hair dryer and repeating the test.

This same test can also be done using a meter that reads frequency. The older AC Delco MAF sensors (like a 2.8-liter V-6) should show a steady reading of 32 hertz at idle to about 75 hertz at 3,500 rpm. The later model units (like those on a 3800 V-6 with the Hitachi MAF sensor) should read about 2.9 kilohertz at idle and 5.0 kilohertz at 3,500 rpm. If tapping on the MAF sensor produces a sudden change in the frequency signal, it's time for a new sensor.

On GM hot-film MAFs, you can also use a scan tool to read the sensor's output in grams per second (gps), which corresponds to frequency. The reading should go from 4–8 grams per second at idle to as much as 100–240 grams per second at wide-open throttle.

Like throttle position sensors, there should be smooth linear transition in sensor output as engine speed and load change. If the readings jump all over the place, the computer won't be able to deliver the right air/fuel mixture, and drivability and emissions will suffer. So you should also check the sensor's output at various speeds to see that its output changes appropriately.

Another way to observe the sensor's output is to look at its waveform on an oscilloscope. The waveform should be square and show a gradual increase in frequency as engine speed and load increase. Any skips or sudden jumps or excessive noise in the pattern would tell you the sensor needs to be replaced.

Yet another way to check the MAF sensor is to see what effect it has on injector timing. Using an oscilloscope or multimeter that reads milliseconds, connect the test probe to any injector ground terminal (one injector terminal is the supply voltage and the other is the ground circuit to the computer that controls injector timing). Then look at the duration of the injector pulses at idle (or while cranking the engine if the engine won't start). Injector timing varies depending on the application, but if the mass airflow sensor is not producing a signal, injector timing will be about four times longer than normal (possibly making the fuel mixture too rich to start). You can also use millisecond readings to confirm fuel enrichment when the throttle is opened during acceleration, fuel leaning during light load cruising and injector shut-down during deceleration. Under light load cruise, for example, you should see about 2.5 to 2.8 milliseconds duration.

EMS TROUBLESHOOTING

TROUBLESHOOTING KNOCK SENSORS

For this test you need a strobe gun so you can physically see the timing change via the cam belt or the timing mark on the crank pulley wheel.

• Inspect the KS connector and its pins for good condition and ensure that there is no damage to the cable form like cuts, splits, burns, and so on.

• Check that the connector plug fits home on the KS and is not loose and that the pins make good contact. Spray the contacts with WD-40.

• Using the strobe gun, attach the probe of the inductive pick to the HT lead of cylinder number one (follow manufacturer's instructions if you don't know how to do this).

• Turn on the engine and allow to idle.

• Gently tap the engine block close to the number-one cylinder with the handle of a screwdriver.

• The timing should be seen to retard.

TESTING A VEHICLE SPEED SENSOR

The VSS can be tested using a scan tool or a lab scope; however, vehicle manufacturers' test procedures vary widely. A road test with the scan tool or lab scope hooked up is often the most effective way to test a VSS. Compare your scan tool readings with the manufacturer's specifications. When using the lab scope, you may see either a square-wave pattern if the VSS has an integrated buffer circuit or an AC sine wave if the buffering and conversion are done within the computer. In some cases, you can even count VSS cycles with a DVOM. Use the analog bar because the digital display usually won't be fast enough or accurate enough to give reliable readings.

If the readings on your scan tool or lab scope match the manufacturer's specifications, then the VSS is working properly and the problem is most likely in the computer. If the readings don't match the specs, check the sensor wiring harness and connectors for damage, corrosion, and high resistance. If the wiring and connections are good, the sensor is bad and must be replaced.

One final note on VSS, changing to a tire size that differs from original equipment may cause problems, especially if the change is excessive and the buffer has not been recalibrated. The speedometer readings will be wrong and there is a chance other vehicle systems may be affected because the VSS signals will no longer show actual vehicle speed.

TROUBLESHOOTING IACS

Wells Manufacturing, which builds a wide variety of EFI sensors and actuators, recommends the following methods of diagnosing IAC problems.

A common condition is an idle-air-bypass valve that's fully extended (closed). This is often a symptom of an air leak downstream of the throttle, such as a leaky throttle body base gasket, intake manifold gasket, vacuum circuits, injector O-rings, etc. The computer has closed the bypass circuit in a vain attempt to compensate for the unmetered air leak that is affecting idle speed.

Incorrect idle speed (too high) also can be caused by a shorted air conditioning compressor clutch wire or defective power steering pressure-sensor circuit. A code 35 on a GM application indicates a problem in the idle-air control circuit. To troubleshoot the problem, disconnect the IAC motor, then start the engine to see if the idle speed increases considerably. Turn the engine off, reconnect the IAC and start the engine again. This time the idle speed should return to approximately the previous rpm. If it does, the problem is not the IAC circuit or motor. Check for vacuum leaks or other problems that would affect idle speed.

If the idle speed does not change when the IAC is unplugged or does not decrease after reconnecting the unit, use a test light to check the IAC wiring circuits while the key is on. The test light should glow when connected between terminals A and B and terminals C and D if the PCM and wiring are OK (indicating the problem is in the IAC motor). If the test light fails to glow on either circuit, the fault is in the wiring or PCM.

On Ford central fuel injection (CFI) throttle-body injection applications, a solenoid or vacuum diaphragm is used to open the throttle linkage along with an idle-air bypass valve. Ford also uses idle-air bypass on its multi-point injection applications. Codes 12, 13, 16, 17, and 19 all indicate idle speed is out of spec (too high or too low). Codes 47 and 48 indicate a fuel-mixture problem that could be caused by an air leak. The diagnostic procedure when any of these codes are found is to turn the engine off, unplug the ISC bypass air-solenoid connector, then restart the engine to see if the idle rpm drops. No change would indicate a problem in the solenoid, wiring or EEC processor.

A Ford ISC solenoid can be checked by measuring its resistance. A reading of between 7.0 to 13.0 ohms would be normal. Also check for shorts between both ISC solenoid terminals and the case. If the ISC checks out OK, check for battery voltage between the ISC connector terminals while the key is on. Voltage should also vary when the engine is running. No voltage indicates a wiring or EEC processor problem.

With Chrysler, a code 25 means there's a problem in the AIS motor-driver circuit. The AIS driver circuit can be checked by a scan tool actuator test. When the actuator display shows a code 03, the AIS motor should cycle one step open and one step closed every four seconds. You can remove the AIS from the throttle body to see if the valve pintle is moving in and out or simply listen for the motor to buzz.

Using a scan tool to run the engine-running test mode #70 (which checks the throttle-body minimum airflow), depressing and holding the proper scan tool button should force closed the AIS bypass circuit. At the same time, idle speed should increase to about 1,300 to 1,500 rpm. If it does not match specifications, minimum airflow through the throttle body is incorrect.

On late-model vehicles with OBD-II, the following codes may indicate a problem with the idle-air-control system:

P0506—idle too low

P0507—idle too high

P1508—idle-air-control stuck closed

P1509—idle-air-control stuck open

P1599—engine stall detected

When installing a new GM IAC or Chrysler AIS motor, the pintle must not extend more than a certain distance from the motor housing. The specs vary, so check the manual. Chrysler says 1 inch (24.50 millimeters) is the limit, while some GM specs allow up to 28 millimeters on some units and 32 millimeters on others. If the pintle is overextended, it can be retracted by either pushing it in (GM) or by connecting it to its wiring harness and using actuator test #03 to move it in (Chrysler). You can also use a digital storage oscilloscope (DSOs) to view IAC wave forms.

CHAPTER 15
EMISSIONS AND ONBOARD DIAGNOSTICS

EMISSIONS-CONTROL DEVICES

Air pollution has been a major problem in heavily populated areas of the United States since the late 1950s and early 1960s when the first major smog formed from the photochemical action of sunlight on air pollution in southern California. The state of California has required air pollution control devices of one form or another on cars since 1961, when the state began requiring car companies to scavenge crankcase vapors rather than vent them into the atmosphere.

California laws dealing with automotive emissions became stricter after 1965 and 1972, with the other 49 states eventually following California's lead. California currently has laws requiring all owners of light trucks, cars, and RVs newer than 30 years old to put their vehicles through emissions testing to certify that HC, NOx, and CO emissions are within specs for the particular vehicle and engine.

The law also requires a visual antitampering check designed to make sure no one has removed emissions-control devices or made other modifications that might affect emissions (such as changing to a different carburetor, modifying the ignition, changing heads, or modifying the exhaust system, the camshaft, etc.— that is, almost any hot rodding mods that affect engine volumetric efficiency. Aftermarket manufacturers that build products that are not simple OE-type replacement parts—and might affect emissions—must get an exemption order from the state of California. The exemption order stipulates that aftermarket manufacturers follow a structured procedure at a certified lab designed to prove that vehicles' emissions are within a 10 percent tolerance of original vehicle emissions—in order for the vehicle to be street legal. Both manufacturers and end users have been subject to fines of $2,500 for each violation (in contrast to the old days when the government more or less penalized the installer/manufacturer).

New federal laws took effect in 1991 that put the rest of the country in sync with California on emissions requirements (other than the fact that performance parts for 49-state vehicles do not require an explicit EO; manufacturers must self-test the parts). In fact, 49-state manufacturers are supposed to emissions-test their products on a kind of honor system in which they are expected to obtain and maintain evidence that their products are legal, if the government should request such evidence.

As a practical matter, there are still states and cities that don't yet have an emissions testing procedure in place (depending on the administration in Washington, they have been under a degree of federal pressure to implement one). Many manufacturers are now seeking California EOs for performance parts, since California is a huge market for car parts, and since a California EO automatically provides compliance with federal laws.

By 2000, air pollution had become a worldwide problem. The most polluted city in the world was no longer Los Angeles, which has benefited from air pollution control measures, but Mexico City, which is terribly overpopulated and has virtually no air pollution measures—mostly because air pollution controls cost money and may have an adverse economic impact. Add-on pollution-control devices for an engine cost money. The

The standard OBDII connector

Pin 1 - Proprietary

Pin 2 - J1850 Bus+

Pin 3 - Proprietary

Pin 4 - Chassis Ground

Pin 5 - Signal Ground

Pin 6 - CAN High (J-2284)

Pin 7 - ISO 9141-2 (K Line)

Pin 8 - Proprietary

Pin 9 - Proprietary

Pin 10 - J1850 Bus-

Pin 11 - Proprietary

Pin 12 - Proprietary

Pin 13 - Proprietary

Pin 14 - CAN Low (J-2284)

Pin 15 - ISO 9141-2 (L Line)

Pin 16 - Battery Power

The standard OBD-II connector is the pathway into all 1996-plus and some 1994-plus engine management systems. This 16-pin connector provides really valuable diagnostic data, and it also provides a marvelously simple access to reflash the EMS calibration—if you've got the proprietary security seed to gain access. Fortunately, there are brilliant hackers who are also car guys, hence, products like LS1edit exist to recalibrate performance engines like the LS1/6 'Vette. OBDII Diagnostic Secrets Revealed, Peter David

situation had become so critical in Mexico City that on bad days, only half the cars in the city could drive (which in itself has an economic impact). Newspapers regularly showed people wearing surgical masks or gas masks just to go outside. Los Angeles and California had experimented with laws requiring a certain percentage of cars sold there to be zero emissions (electric), though the government had backed away from requiring a specific percentage by 2003.

Smog is a result of the photochemical action of sunlight on unburned hydrocarbons (HC) and oxides of nitrogen (NOx). These pollutants combine under the action of sunlight to make a brown haze that irritates human eyes, noses, and throats. Smog is harmful to plant and animal life, and can damage plastics, paint, and rubber parts used in cars and other human-built structures. Los Angeles, with lots of vehicles, lots of sun, and a bowl-like layer of mountains surrounding the city that tends to hold pollutants in place, is a perfect place for smog.

Required by laws to deal with the problem, car companies were forced to contend with three kinds of vehicle emissions: Crankcase blow-by and oil fumes that would otherwise enter the air through crankcase breathers, exhaust gases that enter the atmosphere from a vehicle's tailpipe, and evaporative emissions from the vehicle's fuel system.

Crankcase emissions were one of the easier problems to fix. These combustible fumes could easily be sucked back into the engine along with the charge mixture and burned. Before 1963, crankcase emissions spewed directly into the atmosphere through a vent tube designed in such a way that an air draft produced by the vehicle's forward motion tended to suck or scavenge fumes from the tube into the atmosphere.

It was an easy thing to design a positive crankcase ventilation (PCV) system that uses engine vacuum to suck fumes from the valve cover into the intake manifold near the throttle plate. Open PCV systems use a PCV valve located in-line in the hose connecting the valve cover and intake manifold that contains a small orifice and a spring-loaded check valve that is sucked closed at idle or other times of high vacuum, only allowing suction through the small orifice. During heavier engine loading, spring action overcomes lower engine vacuum to allow more flow.

The oil filler cap contains a mesh filter that allows air to enter the valve cover from the atmosphere. At very low or zero vacuum, heavy blow-by could still escape into the air through the filler cap. In 1968, this system was modified to be closed so the oil filler cap was not vented to the air, but to a location inside the air cleaner, taking advantage of the fact that fumes could be sucked into the air cleaner at times of very low manifold vacuum when there is a lot of air being sucked into the engine through the air filter. At idle, or other high vacuum conditions, crankcase emissions would enter the engine downstream of the throttle plate via high manifold vacuum. This strategy of handling crankcase emissions was successful and took care of one-third of all engine emissions.

Handling exhaust emissions was a different story. The struggle to clean up exhaust emissions continues today. The problem is that when you improve one component of exhaust emissions, you tend to make other emissions components worse. Although perfect combustion produces heat energy plus harmless water and carbon dioxide, in reality, combustion is not perfect. Rich mixtures (lack of oxygen) produce CO_2, water, and carbon monoxide (CO, a deadly gas capable of asphyxiation by replacing oxygen in the blood of humans) plus unburned hydrocarbons (gasoline that made it through the combustion process without burning).

Carbon monoxide disappears as mixtures lean, but hydrocarbons rise again due to lean misfires when there isn't enough fuel. And the surplus of oxygen and the abnormal heat of lean burn mixtures around 16:1 also tends to produce peak oxides of nitrogen, while even stoichiometric air/fuel mixtures can have the same problem. A perfectly tuned engine with no emissions-control devices produces excellent (although not peak) power at stoichiometric mixtures, which implies high heat and high pressure in the cylinder. High cylinder pressure with heat tends to break down normally inert diatomic nitrogen gas and oxidize it to form NOx, a component of smog. HC and CO both fall off nicely at 14.7 parts air to fuel. But with mixtures leaner than 13:1 or so, excess oxygen becomes available to oxidize free atomic nitrogen. NOx emissions reach a peak at about 16:1 and then fall off with even leaner mixtures as combustion pressure falls off. Unfortunately, lean mixtures are exactly what you want for best economy, particularly at less than full throttle.

In 1966, the state of California applied standards to tailpipe exhaust emissions and toughened these again in 1973. Vehicles had to be certified by undergoing a test that simulated a 20-minute drive through downtown Los Angeles. Test equipment collected the exhaust gases, producing weighted average emissions figures for the three pollutants. In 1973, the U.S. government came up with the Federal Test Procedure (FTP) in which vehicles are tested on a chassis dyno that simulates a 23-minute, 7.5-mile drive after a 10-hour cold soak period to make sure the vehicle is undergoing a true cold start (the time when it produces the worst emissions). The chassis dyno is loaded in proportion to the actual vehicle's induced drag and weight. Exhaust gases are mixed with atmospheric gases in a specified proportion (while monitoring the composition of the atmospheric gases to correct for pollutants in the ambient air. Gases are collected in plastic bags.

For 1975 cars, the FTP was extended to include a rerun of the first 8.5 minutes after a shutdown to simulate a hot start. and averaged in with the 8.5 minutes after the initial cold start. More recently, California began to require emissions inspections of certain heavier vehicles, such as motorhomes and trucks, even if these were not required to pass emissions testing when new. California requires CARB Exemption Orders for these vehicles. Since there may be no emissions data from the vehicle model when new, manufacturers of add-on or modified parts for these vehicles will probably have to run comparison testing of the vehicle without the modified parts, and then again with the parts, and possibly a third test with the parts removed.

In summary, the problems automotive engineers had to deal with result from formation of CO in rich mixtures with too little oxygen to support full combustion. CO can never practically be eliminated completely, because even with a combustion chamber filled with an overall perfect air/fuel mixture, the gases are not perfectly distributed and local pockets of richness exist. Gasoline, consisting of numerous molecules containing hydrogen and carbon, releases heat energy when these are oxidized by burning in air. Hydrocarbon emissions in the exhaust gases are the products of gasoline that made it through the engine without burning completely. Eliminating CO and HC emissions is inconsistent with making maximum power, because to get best power you want to make sure sufficient fuel is present to make use of every bit of oxygen that you have been able to coax into the engine.

ONBOARD DIAGNOSTICS

The 1990 Clean Air Act specified emissions-related requirements for new cars in the 1990s and beyond as well as organizing requirements for aftermarket automotive equipment. The act extended the prohibition against removal or rendering inoperative the emissions-control devices to the consumer. It would now be a violation to manufacture, sell, or offer to sell parts where the principal effect is to defeat the emissions-control devices.

EPA memorandum 1A states that those who might be subject to claims of tampering as a result of installation of certain parts can be certain there will be no such claims if the manufacturer of the part(s) is able to certify that the vehicle with the parts in question installed could pass the Federal Test Procedure—or if they have received an Exemption Order (EO) from the California Air Resources Board. Any entrepreneur who wants to manufacture parts that might interfere with emissions-related systems on a vehicle (almost anything related to the combustion cycle of the engine) can obtain the "Procedures for Exemption of Add-on and Modified Parts" from the California Air Resource Board.

ENGINE SWAPS

According to the California Bureau of Automotive Repair (BAR), an "engine change" is the installation of an engine in an exhaust-controlled vehicle that is different from the one that was installed originally in the vehicle and does not qualify as a replacement engine. A replacement engine is defined as a new, rebuilt, remanufactured or used engine of the same make, size, and number of cylinders as the original engine with all original emissions controls reinstalled, or an engine that matches a configuration offered by the manufacturer for that year, make, and model of the vehicle with the appropriate emissions controls for the installed engine and chassis components present and connected. Licensed California smog check stations must refer vehicles with engine changes to the BAR for referee inspection. Once a vehicle has been inspected by the referee and a BAR label attached to the doorpost (listing the required emissions-control equipment), any licensed smog check station may thereafter perform an inspection. The vehicle must have all the emissions-control equipment listed on the BAR label.

Nonexhaust-control vehicles (cars more than 30 years old) may have any year engine installed and only require PCV (and appropriate retrofit devices, if required). These vehicles may be inspected at licensed smog check stations and do not need referee intervention. For newer vehicles, note the following explicit regulation: "An automotive repair dealer shall not make any motor vehicle engine change that degrades the effectiveness of a vehicle's emissions controls."

ONBOARD DIAGNOSTICS, LEVEL II

Onboard Diagnostics, Level II (OBD-II) is a series of regulations intended to reduce in-use vehicle emissions by requiring the OEM engine management system to continuously monitor the powertrain and its emissions-control systems for failures or deterioration. OBD-II includes provisions for standardization of diagnostic, repair, and other service-related information. The system was designed to reduce high in-use emissions caused by emissions-related malfunctions, to reduce time between occurrence of a malfunction and its detection and repair, and to assist in the diagnosis and repair of emission-related problems.

To be OBD-II compliant, an EMS must monitor virtually all emissions-control systems and components that can affect emissions. OBD-II requires a malfunction indicator light (MIL) visible to the vehicle operator that must be illuminated and a fault code set when either there is a failure of a monitored component/system *or* when any of the sensed parameters deteriorates to the extent that the vehicle's emissions would exceed the relevant standard by approximately 50 percent. In most cases OBD-II must detect malfunctions within two driving cycles.

OBD-II provides for strict enforcement to be implemented through increased penalties for noncompliance for both vehicle manufacturers and consumers. Prior to renewal of the vehicle's registration, a motor vehicle owner must provide proof that any required repairs have been made and that no fault codes are present.

OBD-II immediately required vehicle manufacturers to provide more robust warranty coverage for emissions-related engine and EMS systems. It required that the increased durability requirements (100,000/120,000) of the new, more stringent emissions standards would be accompanied by stricter reporting of component defects and failures along with lower triggering limits for recalls. The reading of fault codes during state emissions tests would now identify virtually all failures.

The Clean Air Act and Warranties

The federal Clean Air Act requires vehicle makers to provide a production emissions warranty as well as a performance emissions warranty. The production emissions warranty requires a vehicle maker to warrant that the vehicle has been engineered and manufactured so that it conforms with emissions requirements at the time of sale. The performance emissions warranty requires a vehicle maker to warrant that the vehicle will continue to comply with emissions requirements as tested in state vehicle emissions inspection programs for the warranty periods specified in the law. For model year 1995 and later vehicles, the warranty is two years/24,000 miles for all **emissions**-related parts and eight years/80,000 miles for the catalytic converter, electronic **emissions**-control unit, and onboard diagnostic device. The performance warranty is conditioned on the vehicle being properly maintained and operated.

Like the Magnuson-Moss Act, vehicle manufacturers may not refuse warranty repairs under the Clean Air Act's performance and defect warranties merely because aftermarket parts have been installed on the vehicle. The only circumstance under which the vehicle manufacturer can void the **emissions** warranties is if an aftermarket part is responsible for (or causes) the warranty claim.

OBD-II grew out of the more primitive OBD-I standard, which was adopted in 1985 for 1988 and later vehicles, and was designed to monitor ECU input devices and the EGR system. OBD-II was adopted in 1989 for all 1996 and later vehicles and some 1994 and 1995 vehicles. It was designed to address certain shortcomings in OBD-I. OBD-II expanded the scope of monitored to determine if a component or system had malfunctioned. And OBD-I provided substantial diagnostic information to help service technicians identify and repair problems.

Key OBD-II monitoring requirements include the following systems:

- Primary emissions control systems/components
- Catalyst
- Misfires
- Evaporative system
- Fuel system
- Oxygen sensor
- Exhaust gas recirculation system
- Secondary air injection system
- Heated catalyst system
- Comprehensive components

OBD-II required that any connectors through which the emissions-control diagnostic system would be accessed for inspection, diagnostic service, or repair be standard and uniform on all motor vehicles and motor vehicle engines, and that access to the emissions-control diagnostics system through such connectors should be unrestricted and should not require any access code or any device that is only available from a vehicle manufacturer. OBD-II required that the output of the data from the emissions-control diagnostics system through such connectors should be usable without the need for any unique decoding information or device.

OBD-I/OBD-II COMPARISON

OBD-I	OBD-II
Oxygen Sensor	Oxygen Sensor (Enhanced)
EGR System	EGR System (Enhanced)
Fuel System	Fuel System (Enhanced)
Electronic Input Components	Electronic Input Components
Diagnostic Information	Electronic Output Components
Fault Codes	Catalyst Efficiency
	Catalyst Heating
	Engine Misfire
	Evaporative System (leak check/function)
	Secondary Air System
	Diagnostic Information
	Fault Codes
	Engine Parameter Data
	Freeze-Frame Engine Parameters
	Standardization

BENEFITS OF OBD

OBD-II was designed to provide certain clear benefits. In addition to reducing air pollution, one of the most important was standardization. Standardized interfacing, diagnostics, and data-logging were of great concern to independent repair shops that might otherwise find it prohibitively expensive to acquire proprietary diagnostic tools for every make and model of vehicle serviced. OBD-II standardization reduced the cost and complexity of diagnosis by providing a known communications protocol for handshaking between the ECM and diagnostic devices.

OBD–II provided a physical diagnostic connector that would be the same on *all* 1996 and later vehicles (and a few 1994 and later vehicles) and—bottom line—allowed one diagnostic device to connect to any vehicle sold in the United States. Standardization basically enabled one scan tool to work on any vehicle. OBD-II provided a set of fault codes designed to help identify faulty components. OBD-II provided for real-time diagnostic information and, thus, continuously updated engine parameter data. It provided a facility to supply freeze-frame information, in which the system would store engine operating conditions upon detection of a malfunction.

OBD-II'S EFFECT ON VEHICLE DESIGN

OBD-II had a significant impact on the engineering of new OEM engine management systems. Compared to earlier factory engine management systems, powertrain control modules became much faster, with much more memory and more features, and cost and complexity increased significantly. OBD-II required additional components and systems for many of the monitors, including additional oxygen sensors, fuel tank, and EGR pressure sensors. Existing components/systems required substantial upgrades. Achieving higher standards required higher accuracy, durability, and reliability in addition to design changes that compensate for differences in the duty cycle or function of parts.

In addition to effectively revising existing functions, the global effect of OBD-II was to require that the base operating strategy, or algorithm, of the EMS be completely retooled to incorporate new features specific to OBD-II. Since the system required a substantial amount of monitoring to be performed while driving, a considerable effort was required to make sure monitoring was completely transparent to the driver. Older OBD-I diagnostic routines had to be significantly revised to support repairing malfunctions identified by the gamut of new monitors. Standardization of terms and equipment helped to some degree, but although the increased complexity of the overall system vastly improved diagnostics, it could actually result in more difficult *repairs*.

SMOG CHECKS AND OBD-II

OBD-II was designed for full integration into the smog check procedures required in many states and cities. At the time of the smog check, technicians would verify that the MIL was not illuminated (light and computer indication), that the I/M readiness code was fully set, that no fault codes were stored, and that proper calibration was installed.

The Specialty Equipment Marketing Association (SEMA)—the organization of aftermarket automotive equipment suppliers—has noted that various government agencies have been highly interested in implementing countermeasures to stop people from tampering with OBD-II engine management systems. Such agencies, says SEMA, "want to prevent passing of vehicles that have had their fault codes erased (or readiness code set) illegally by 'hackers.' The potential for such activity is purely speculative; there has been no proof it is occurring."

SEMA notes that the same agencies have had a strong interest in preventing consumers or non-OEM technicians from any tinkering whatsoever (recalibration/tuning) that would make adjustments to the vehicle's computer—effectively locking the hood.

SEMA points out that the OBD-II standard made no provisions for non-OEM calibrations and limited the performance aftermarket's ability to develop and sell many current emissions-compliant products since such calibrations would automatically fail I/M. Limiting parts and service to dealerships, says SEMA, reduces consumer choice and could raise costs.

In the meantime, says SEMA, aftermarket calibrations/computer chips have proven they do not increase emissions and many have California Air Resources Board (CARB) exemption orders. In fact, SEMA notes, CARB has agreed that aftermarket recalibration should remain possible under VC 27156/38391, though the OBD-II standard does not explicitly allow this. "Bottom line," says SEMA, "electronic tampering has never been verified as a real problem. It is unlikely it ever will be since the required sales volume, equipment investment and knowledge are prohibitive for all but the aftermarket [industry itself, *not* individual consumers]. The latter are too responsible/visible to consider such risks." The above statement is fairly true.

TAMPERING

In fact OBD-II initially required that computer-coded engine operating parameters should not be changeable without the use of specialized tools and procedures, for example, soldered or potted computer components or sealed (or soldered) computer enclosures. However, subject to executive officer approval, manufacturers could exempt from this requirement those product lines unlikely to require protection. Criteria to be evaluated in making such an exemption included—but were not limited to—current availability of performance chips, high-performance capability of the vehicle, and sales volume.

Manufacturers using reprogrammable computer code systems (for example, EEPROM) were required to employ proven methods to deter unauthorized reprogramming, which could include

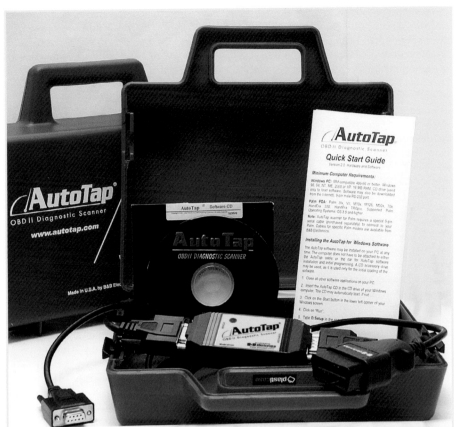

Auto Tap's OBD-II diagnostic scanner plugs directly into the 16-pin diagnostic port to tell you what the factory EMS thinks is going on in the engine and related sensor systems, as well as a vast array of specific malfunction codes. The purpose of OBD-II is to keep engine system components from breaking in ways that pollute the air without anyone knowing, but its usefulness goes way beyond emissions.

BARRIERS TO ENTRY: THE PERFORMANCE AFTER-MARKET AND OBD-II

OBD-II has definitely presented a challenge to aftermarket firms in the business of providing performance recalibrations on PROM-based OEM engine management systems. Using the new EEPROM technology, OEMs implemented write-protect features that removed the ability to perform aftermarket recalibrations required for compatibility/compliance for performance equipment packages.

OBD's sensitivity to changes and modifications increased the risk of incompatibility. The need to uniquely calibrate each engine family further added to the complexity. Extended warranties and increasingly higher vehicle useful life requirements have expanded emissions liability for both OEMs and the aftermarket, providing more chances for unexpected use, wear, adaptations, failures, and so on. To make street-legal products, aftermarket performance firms need a full understanding of the powertrain control/OBD system operation to minimize potential incompatibilities and false MILs over the vehicle's useful life. Given that OEMs continue developing OBD systems while they are being produced, OEMs have a nearly insurmountable advantage in having the information needed to maximize compatibility of aftermarket parts.

copyrightable executable routines or other methods. Beginning with the 1999 model year, manufacturers were called upon to implement enhanced tamper-protection strategies including data encryption with countermeasures to secure the encryption algorithm, as well as write-protect features requiring electronic access to off-site computers maintained by the manufacturer.

By 1997 it was clear that specialty products could still be used on pollution-controlled motor vehicles. SEMA was making a strong point of the distinction that SEMA members' products made provisions for emissions-control devices or were considered replacement parts; they were not defeat devices. It was clear by this time that the use of add-on or modified parts would not be considered a form of tampering if the product has been granted a CARB EO number (California and 49 states) *or* meets the requirements of the EPA's antitampering policy document, Memorandum No. IA (49-state only).

Hundreds of products of all types had been granted EO numbers for thousands of applications, with many specialty products sold by OEMs with EO numbers and many manufacturers representing their products as meeting Memorandum 1A. It was clear that such products did meet emissions and applicable criteria for an EO or Memo 1A, and that enforcement provisions/penalties designed to prevent fraud would be considered adequate to prevent fraud. In fact, elimination of components present in the originally certified configuration would be allowable under certain circumstances (for example, GM's Camaro package).

With the world going OBD-II in 1996, SEMA launched a lawsuit against the federal EPA, California's EPA, and the California Air Resources Board, basically to prevent these governmental organizations from enforcing OBD-II and the Clean Air Act of 1990 in ways that were economically ruinous and anti-competitive toward legitimate aftermarket equipment and performance suppliers capable of building parts and equipment that genuinely did not degrade vehicle emissions or even improved emissions performance. SEMA was looking to force OEMs to meet the following conditions with respect to the $200B specialty equipment aftermarket:

1. Provide initial calibration development training on applicable power trains.
2. Designate a liaison from each participating OEM.
3. SEMA acts as specialty aftermarket liaison.
4. Provide information to aftermarket for product development/recalibration.
5. Provide access to equipment/tools needed for development.
6. Provide ongoing training/communication as vehicles are updated.
7. Provide specifications, documentation, and other items needed to get an exemption order (information could be provided to J2008/central databases).

Aftermarket manufacturers would be liable for proper use of information and equipment provided, as well as security.

In spite of the above challenges, after some initial pessimism, aftermarket companies and individuals have succeeded in reverse engineering the EMS of some of the most important OEM performance vehicles to the point of making aftermarket recalibration feasible, so OBD-II has not been the lock on the hood that some people feared. Software like LS1edit readily hacks into the EEPROM of the millennium C5 Corvette and other LS1/6 vehicles. Highly sophisticated piggyback and interceptor computers are able to lie to OBD-II systems so effectively (and disobey them so cleverly in the pursuit of performance) that the government and OEMs may now understand that it is somewhat pointless to worry too much about providing insurmountable barriers to entry around the sacred OBD-II OEM calibration.

OBD-III

OBD-III is a proposal designed to minimize the delay between detection of an emissions malfunction by the OBD-II system and repair of the vehicle. In the service of this, the two basic elements are the ability to read stored OBD-II information from in-use vehicles and the authority to direct owners of vehicles with fault codes to make immediate repairs. There are three proposed ways to send/receive data: roadside reader, local station network, and satellite.

Proposed enforcement mechanisms include the following:

Incorporate into biennial I/M program (read fault code to screen for vehicles that need complete testing, pass or short test for vehicles with no fault code)

Out-of-cycle inspection (compile and screen data, mail notice to vehicle owner requiring out-of-cycle inspection within 10 days, require certificate of compliance on next registration/resale or within 30 to 60 days, issue citation for non-compliance, enforce citation via court or DMV penalty at next registration, and so on)

Roadside pullover (CHP flags down vehicles with fault codes, technician verifies problem by inspecting or testing vehicle, issuance of notice requiring out-of-cycle inspection, and so on)

The concept of OBD-III is rife with legal issues. OBD-III imposes sanctions based on "suspicionless mass surveillance" of private property. The concept includes random, possibly frequent, testing. There would be no advanced knowledge the vehicle will be tested. The results of testing would not be immediately available (unless roadside pullover follows). There might be no opportunity to confront or rebut, and there is the possibility of the system's use for other purposes (police pursuit/immobilization, tracking, cite speeders).

OBD-III raises Fourth Amendment search and seizure privacy issues: "The right of the people to be secure in their persons, houses, papers and effects, against unreasonable searches and seizures shall not be violated." From a legal perspective, the scope of the OBD-III concept is unprecedented; previous cases have looked at surveillance of individuals.

CARB has requested proposals for incorporation of radio transponders into vehicle onboard diagnostic systems, with the objective "to demonstrate the feasibility and cost effectiveness of replacing the current emissions-based periodic Inspection and Maintenance (I/M) program with automated inspections based on the OBD-II system and an on-vehicle radio transponder. The study will test, evaluate, and demonstrate the viability and cost of equipping new vehicles with various transponder technologies and assess how these technologies can be effectively used to improve the convenience, effectiveness, and cost-effectiveness of the I/M program."

The idea was to save money. Currently the entire vehicle fleet must be tested in order to identify the relatively small number of

Anatomy of an OBDII Message

The diagnostic scanner and the vehicle communicate with each other by exchanging small pieces of information, called diagnostic messages. The message transmitted from the scanner to the PCM is called a command. The message transmitted back from the PCM to the scanner is called a response.

Once the OBD11 scanner tool is connected, it will transmit a command message to the vehicle. The PCM microprocessor will receive the message and verify it. If the message is valid, the PCM will respond.

A typical scanner command looks as follows:

C4 10 FI 23 FF 06 8D

C4	Required Header Byte
10	Destination (PCM)
F1	Originator (Scanner)
23	Command ("Request data")
FF 06	Data ("Address FF06")
8D	Message checksum

In the above message, the scanner is requesting data from the PCM memory at address FF06.

vehicles that are likely to fail, and 10 million vehicles per year are required to undergo an I/M inspection that results in passing scores for 70 percent of the vehicles.

The idea was that transponder signal receivers would be capable of performing these functions storing a query and received data in a database format that would include the following:
- Date and time of current query
- Date and time of last query
- VIN
- Status (OK, Trouble, or No response)
- Stored codes
- Receiver station number

Besides emissions-related functions, authorities were very interested in *cooperative techniques* for police pursuit/mobilization. The California Air Resources Board noted that, "The cooperative techniques comprise devices that are installed on automobiles, which would receive a coded radio frequency signal that would produce a progressive speed reduction or shut down the automobile. The speed reduction and shut down could be incorporated in the Onboard Diagnostic (OBD) III system, which is planned to have a radio transponder for reading out automobile status including vehicle number and smog equipment fault codes."

The major obstacle to overcome, notes CARB, "is to get public acceptance of a device that they have to pay for and that can disable their automobile. One way to obtain public acceptance may be to offer incentives by including this device as part of a package that provides other benefits. These benefits could be an anti-theft device and/or a smog readout device. The smog readout device could eliminate the need for costly and time-consuming periodic inspections at smog stations. With the idea that the only time you would need to go to an inspection station is when the automobile exceeds smog-generating limits. In addition to public acceptance, this approach will require federal government, state government and car manufacturer cooperation. The use of a cooperative device has strong appeal because of effectiveness, safety and ease of use. Incorporating the overall system as part of a larger subsystem would reduce cost and make it more attractive."

EMISSIONS AND ONBOARD DIAGNOSTICS

CHAPTER 16
HOW TO HOT ROD YOUR CARBURETED SMALL-BLOCK CHEVY

Hear that little cough two blocks back under acceleration? Sounds a little lean somewhere—the kind of annoying little thing that's a nightmare to get rid of when you're shaking out a new carb, cam, and intake manifold.

Fortunately, this '55 Chevy does not have a carb. Instead of pulling the car over to the side and attacking the carb with a screwdriver, we reposition the cursor on the laptop PC resting on the bench seat of the 400-horsepower '55 Chevy, put that little arrow on the bar graph representing the exact rpm range and engine loading conditions, and hit a few function keys. As always, the computer is highlighting the current rpm range exactly when the engine coughed, so we know which graph to diddle. The laptop is connected by a thin black wire to the programmable electronic control unit driving the EFI-conversion 350 Chevy small-block V-8. Under the hood, four Weber look-alike throttle bodies with stacks sit hissing atop the Chevy mouse V-8. Exhaust rumbles out the back with just enough lope so you know this car has a good cam even though it works through a ratchet-shifter turbo 350 automatic. We lengthen the bar graph, hit the enter key, and immediately data is stored in electrically erasable programmable read-only memory in the Haltech ECU—and punch the gas pedal hard. The instant the hammer goes down, the eight Bosch injectors spritzing fuel directly at the intake valves will change their opening time by a tiny fraction of a millisecond and fatten up the mixture at the right rpm under the right load as measured by manifold absolute pressure.

EFI IN 1955

In the 1990s, Ivan Tull concentrated on building integrated fuel injection and EFI turbo systems for hot rod Chevys. A critical success factor was understanding that when people spend a lot of money on their car, the results had better look great. Tull injection systems display the exotic look of multicarb Weber induction systems, with great attention to detail. This kind of high-buck add-on engine system can't just look great. It has to perform noticeably better after than before.

Tull likes to call turbo-EFI conversions the last best speed secret for hot rod vintage musclecars and trucks—the difference between a cylinder or two filled *almost* right with fuel and air, and every cylinder filled *perfectly* with exactly the right amount of fuel as the air enters the combustion chamber. That difference spells horsepower—sometimes a lot of horsepower!

Tull induction systems use multiple throttle bodies and port fuel injection—from the simplest naturally aspirated powerplant utilizing double cross-ram throttle body pairs to a maxed-out full-bore motor with eight throttle bodies, sixteen injectors, twin turbos, and over a thousand horsepower. Tull designed one such configuration for use on a Donovan aluminum small-block Chevy V-8 in a custom-designed laminar-flow airframe built to take on P-51 Mustangs at the Reno Air Races—where the competition was hot rod World War II war birds running supercharged mega-cubic-inch Rolls Royce Merlin V-12 motors. The Tull Systems' sixteen-injector motors idled nicely on eight injectors sized reasonably for a smooth slow idle—while the other eight injectors were designed to kick in only for high-end power.

The great thing about programmable EFI is that when Tull was done messing around with a normally aspirated cross-ram Chevy V-8 with Haltech EMS, he simply plugged his IBM laptop computer into the Haltech ECU, fired up the graphical display of injection pulse width for various engine rpm and loading, and modified the bar graphs for loads beyond zero vacuum as reported by the two-bar MAP sensor.

Just a red '55 Chevy "Shoebox," still popular among older hot rodders, equipped with the original small-block Chevy V-8, of which the General would eventually manufacture and install about 60 million in trucks and cars into the 21st century with power ratings from about 100 SAE horsepower to 405 in the high-water mark early-1990s ZR1 'Vette. Its age made this car perfect street-legal hot rodding material that would forever be exempt from modern emissions requirements—and yet cleaner than it had ever run when carbureted.

A 5.7-liter custom cross-ram EFI system with programmable aftermarket injection and twin turbos seemed like a much better idea than the original V-8.

Building a beautiful and complex add-on turbo system to jack up the power to extreme levels was no trivial thing. The project involved installing new cast-iron exhaust manifolds designed to bolt directly to twin Garret T04 turbos. It required plumbing in high-pressure oil lines to lube the turbos and the larger low-pressure oil drain lines to return oil directly to the oil pan. It involved plumbing cooling water from the water pump to the center sections of the turbos to keep them running cool in south Texas' scorching summers. And it required building air boxes around the outboard side of the fuel injection throttle bodies and plumbing in the big 2-inch lines feeding boost air from the turbo compressors.

But once all that was accomplished, recalibrating the Haltech involved creating a superset of the existing calibration. The Haltech behaved exactly the same at lower loading, until you jumped on the throttle and the big turbos whistled up to speed, blowing in 7 or 8 psi more air than that Chevy 350 mouse motor could otherwise gulp down. At this point, the Haltech would obey the high-load graphs for air pressure higher than atmospheric, keeping the engine rich and happy. The engine would fire right up and idle perfectly. Later, he'd review the calibration under actual road-test conditions, or possibly on a dyno to get the mixture exactly where he wanted it. All this would be absolutely legal on a '55 Chevy, because America had no emissions laws for cars this old.

A customer looking for an EFI conversion could just bolt on junkyard TPI parts to a small-block Chevy V-8, but if he wanted an exotic-looking system with multiple TWM DCOE Weber-type throttle bodies on a cross-ram Weber-bolt-pattern manifold, this type of system is hard to beat. This system—available with either down-draft or side-draft throttle bodies—offered exotic looks and tremendous flow capacity but would not be cheap. Throttle bodies of 45 millimeters and above will flow plenty of air for any small-block-Chevy motor, while eight individual runners flow a massive 1,800 cfm with quad 45-millimeter TWM throttle bodies with two butterflies per throttle body.

The '55 Chevy pictured here uses a medium-compression 350 Chevy small block. Tull modified the fuel tank's sending unit

to accept a fuel return and installed a fuel return line. A high-pressure fuel pump replaced the old mechanical fuel pump, and a block-off plate sealed up the block where it had been. TWM Weber DCOE bolt-pattern throttle bodies bolted to the cross-ram manifold without modifications, and Tull constructed a two-piece fuel rail from D-section aluminum, ready to accept O-ringed injectors. The pressure regulator connected to the outlet end of one side of the rail, the pressurized inlet fuel line to the other, with a crossover hose connecting the two rails.

After calibrating the fuel curve for the first normally aspirated iteration of the hot rod, recalibration for turbocharging basically meant changing the flat segment of the loading bars at higher than atmospheric pressure to the correct pulse widths for fueling the higher turbocharged volumetric efficiency. Tull would remove the K&N air cleaners from the throttle bodies and fabricate plenums from intercooler end-tanks to pressurize the throttle bodies. The turbos were a simple bolt-on to the cast Chevy small-block turbocharger exhaust manifolds. Tthat's the great thing about small-block Chevys—whatever you want, somebody makes it! The rest was muffler shop-type stuff, piecing together constant radius tubing to feed air from the two compressors to the throttle body plenums.

The great thing about such a car is that it is radical but very streetable, no hangar queen but rather a daily driver. And a legal one at that.

The "Last, Best Speed Secret"—high-compression V-8 power, programmable EFI, and turbocharging—turned this old dog into a raging power freak that could idle smoothly.

CHAPTER 17
SUPERCHARGED
JAG-ROLET

It's no secret Yankees have been V-8-ing Jaguars for years with Chevy or Ford pushrod powerplants and using them to humiliate factory Jaguars in both straight-line acceleration and handling. Of course, Jaguar is now *owned* by Ford.

It's not news that Jaguar's spectacular post-1960s fall from performance grace led to legions of hybrid 1970s and 1980s Jags in the United States with a V-8 heart transplant. But consider this: It could've been worse. After all, Jaguar never produced a *diesel* sports car or a Jaguar station wagon, and at least Jaguar never diversified into construction equipment or lawnmowers or weed-whackers. And it was downright *decent* of Jaguar to faithfully make sure there was always sufficient space in the big cat's engine compartment to install an American V-8 without too much fuss. Jaguar also faithfully made sure there'd be a weight reduction as well as a performance advantage to be gained by junking your expensive aluminum-head DOHC six or V-12 and installing a high-performance dirt-cheap cast-iron pushrod American V-8.

A LITTLE HISTORY

Jaguar. Here was a company with the vision and cojones to build a multi-carb double-over-head-cam crossflow six in the *1940s* with more deep-breathing going on than an obscene phone call—at a time when most automobiles were still powered by wheezing flathead powerplants with all the high-technology and specific power of a prewar farm tractor. The XK engine powered a legacy of performance sports cars and sport sedans in the glory days that included the XK-120, the D-type, the Mark II sedan, and the legendary E-type—a car so gorgeous there are plenty of car people whose first sight of a new E in full flight will surely pass before their eyes in their final moments.

With such a legacy, one truly has to marvel that Jaguar managed to find the executive "talent" that replaced the E-type

with the XJS, introduced some of the more gutless cars to ever carry the "Jaguar" marque in the 1970s and 1980s, and formulated the brilliant strategy of milking a good name rather than wasting time and money producing cars with genuine thoroughbred performance like Porsche or Ferrari or BMW.

Ah yes, but Jaguar management did all of the above, and for a time, the supply of dirt-cheap gutless Jag sedans in the United States and the even larger supply of Chevrolet small-block V-8 engines made the mating of the two a marriage made in heaven. There are a lot of these hybrids still out there driving around in the United States.

It is only now that Ford-owned Jaguar has lashed back viciously against its modern detractors by introducing a scratch-built four-cam V-8 that it becomes a legitimate question whether tuners armed with a humble hot rod pushrod American V-8 as a weapon are still up to the Jaguar factory's challenge. Jaguar's own V-8-powered sedan—supercharged, no less!—has 370 crankshaft horsepower as delivered in the XJR and XKR, which might make you wonder how even the best Chevy-Jag V-8 conversion could possible measure up to the new supercharged V-8 sedan (when it finally arrives in the United States).

Dallas supertuner "Maximum Bob" Norwood was the perfect candidate to respond to the latest figurative shots Jaguar had fired across the bow of every Jaguar tuner with the newly introduced factory-blown V-8s in the XJ Sedan and XK8. Norwood responded by completing a state-of-the-art custom V-8 conversion of an electronic-injected 1990 Jaguar sedan. He weaned the beast from its original 3.6-liter Jag-computer-controlled in-line six in favor of a 1991 pushrod "Corvette Challenge" L98 electronic-injected Chevrolet 350-ci Tuned Port Injection V-8, with Motec programmable engine management and a Vortech high-efficiency S-Trim centrifugal supercharger—along with a GM 4L80E programmable four-speed overdrive

Bob Norwood had a fuel-injected late-model Jag sedan with the 4.0-liter inline six—and a love for supercharged Chevy V-8 power. Converting to a hot rod L98 Chevy small block while maintaining all the original engine and chassis electronic functionality would be a massive challenge.

electronic transmission. It there's any Yankee-engined hot rod Jag that could live up to the evil reputation of past hybrid Jags, this would surely be it.

FACTORY FIREPOWER

Norwood chose a 1989 Jaguar XJ6 3.6, which weighs in at more than 4,000 pounds stock, and when he pulled out the 3.6-liter naturally aspirated Jaguar six, he had to cut a claw and a half from each of the front springs to bring down the ride height, due to the nearly 200-pound weight advantage of the 1991 aluminum head, aluminum manifold, iron-block Corvette small-block V-8 found in certain 1991 Corvettes. In fact, the new Chevy engine that arrived in Dallas straight from GM in a crate was something special: The Corvette Challenge Chevy crate motor is a special version of the L98 TPI V-8 with 11:1 compression, rated by General Motors at 375 horsepower, and designed especially for Corvette Challenge Series factory racers. Although a later-model Corvette LT1 engine with low-lash hydraulic cam would have been an attractive option for the swap,

As always, wiring is the nightmare when combining two EFI engine management systems with functions that don't always overlap. Norwood techs built a custom harness using a Motec "starter" harness (an ECU connector with a huge bunch of multi-colored wires hanging out).

Norwood needed a motor with a distributor drive to power a load-leveling Jaguar hydraulic pump, hence the L98.

Mind you, 375 horsepower was before Norwood went through the engine to maximize it with improved large-runner Accel intake manifold with huge plenum and tuned runners, camshaft, and improved Motec engine management, not to mention the GM 4L80E electronic automatic transmission—the only unit Norwood found to be capable of handling the super-duty torque he planned to direct through the unit's torque converter. And this was *way* before Norwood went so far as to custom-fit an S-Trim Vortech centrifugal supercharger to the Chevy V-8. Curb weight for the Norwood-Jaguar is 3,824 pounds, roughly a 200-pound advantage compared to the stock supercharged XJR sedan.

All this looks good on paper, but when you're looking to establish big power for fast-car bragging rights, these days you head straight to the Dynojet chassis dynamometer to measure horsepower and torque as delivered at the wheels. When it comes measuring actual acceleration and speed, in Dallas you head for the Ennis Motor Speedway—a quarter-mile drag strip—to check out acceleration. With reliable data from the Dynojet and the drag strip, we could now compare the supercharged Norwood Jag-rolet to the new factory blown V-8.

In fact, the 303.4 rear-wheel horsepower found by the Dynojet on the factory XJR represents a rather typical drivetrain and tire power loss of 18 percent under Jaguar's advertised SAE horsepower. Magazine tests show the factory car can be flogged to 0 to 60 in 5.6 seconds, with the standing quarter-mile requiring 14.2 seconds and achieving a trap speed of 101 miles per hour, which is clearly about what you'd expect.

NORWOOD FIREPOWER

Testing on a Dynojet inertial chassis dyno confirms that the Norwood car produced 432.1 rear wheel horsepower at 5,000, and 484.7 ft-lb torque at 4,000 at the rear wheels, which is a calculated 540 horsepower at the crankshaft. At the strip, the car does 0–60 in as little as 3.8–3.9 and goes through the quarter-mile in the high 12s or low 13s depending on the driver and track conditions.

These are the bare-bones statistics. But what is the *character* of this Norwood-built tuner car, and how and why did it come to be born?

There have always been tuners who produced enhanced Jaguars built for racing or performance street use—including hybrid cars like the famous postwar Lister-Jaguar—that mostly used Jaguar powerplants, but also a few Chevy V-8s. Anyone who remembers glory-days Jaguar performance of the past will not be disappointed by the Norwood-Jaguar package, which was designed with a goal of nothing less than producing the fastest, most powerful, streetable Jaguar sedan ever built, while maintaining all of Jaguar's refinement and luxury.

1. THE RAW MATERIAL: 1990 JAGUAR XJ6/THE CHALLENGE: SEAMLESSLY MATING BRITISH AND U.S. ELECTRONICS

"The Jag sedan was just a sweetheart of a car—a nice-running bone-stock '89 four-valve 3.6-liter Sedan with a perfect body I bought from a customer," says Bob Norwood. Norwood and his crew of technicians pulled the Jaguar 3.6 engine and promptly set to work building a viciously powerful supercharged Chevy V-8 conversion for the '89 Jaguar, envisioning a genuine

Norwood Autocraft, which specializes in building and tuning complex and powerful EMS systems on exotic cars, made the TPI engine fit in the Jag with the sanitation of a factory installation.

2. THE CHEVROLET ENGINE

Norwood acquired a Corvette Challenge 350 V-8 as raw material for the supercharged powerplant. The Corvette Challenge engine is a special engine that was used at General Motors R&D for development work. The L98 engine is the last of the traditional-design small-block Chevy pushrod V-8s. It was produced in 1985 to 1991 Corvettes, Z-28 Camaros, Trans-Am Firebirds, Chevrolet Impalas, and some Cadillac vehicles, and has roots and backward-compatibility going all the way to the first overhead-valve (OHV) Chevrolet 283-ci V-8 engines built in 1955.

Beginning with the Challenge motor, Norwood Autocraft installed a performance-oriented Cam Motion low-lash mechanical roller blower camshaft, and a special Accel port fuel injection intake manifold, designed with large plenum and much-bigger-than-stock volume intake runners that connect the lower manifold to the floor of the plenum and can be equipped with bolt-on port-matched extension tubes of tunable-length that protrude into the plenum a varying distance according to the application and rpm range of the engine.

The cam profile was essentially designed to move a greater quantity of air through the supercharged engine, with special emphasis on additional lift and duration on the exhaust side. Along the same lines, the engine was equipped with Norwood Autocraft custom stainless steel headers of 1.75-inch diameter tubing welded to 3/8-inch-thick flanges sealed with copper-seal exhaust header gaskets, each collector equipped with a Bosch wideband O_2 sensor for highly accurate Motec air/fuel closed-loop mixture control and programming under full supercharger boost conditions.

The header collectors feed into twin 2.5-inch exhaust and twin Flowmaster mufflers, each with two outlets—one that feeds into the stock mufflers and tailpipes, the other connected to an exhaust-pressure-activated dump-valve that is normally closed to provide quiet running during ordinary driving, but opens for maximum performance at higher levels of supercharger boost under control of a modified turbocharger wastegate actuator diaphragm.

3. THE SUPERCHARGER SYSTEM

The Jag-rolet uses a Vortech S-trim centrifugal supercharger designed to produce 10 pounds of boost at 6,000 rpm (10 ribbed belt). The system uses a custom Norwood Autocraft billet-aluminum crank pulley and 10-ribbed serpentine belt system. Centrifugal superchargers provide the highest thermal efficiency on any supercharger design (meaning they heat the air the least amount over the minimum theoretical adiabatic heating of

supercar (and also, with its purely stock Jaguar exterior, something of a stealth supercar).

From the beginning, Norwood designed the conversion around a Motec engine management system. This aftermarket programmable onboard computer system has the power and flexibility that allows *all* stock '89 Jaguar systems to function seamlessly with a new engine, transmission, and engine management system—a far more complex task than the typical V-8 conversions done so frequently on earlier Jaguar cars due to complex new Jaguar electronics introduced in the 1990s for electronic fuel injection, computer-controlled ignition, complex OBD-II emissions controls, electronic transmission controls, self-leveling hydraulics, and computer-controlled anti-lock brakes. Even with the formidable Motec computer system onboard, this was not going to be easy.

In fact, the commitment of time, technical, and financial resources involved in the complex R&D required to analyze, research, or reverse engineer all Jaguar electronic and hydraulic systems—and then design the special custom mechanical and electronic equipment required for an EFI V-8 engine conversion that maintains 100 percent functionality of all systems on both the Jag chassis *and* the Chevrolet 350 V-8 engine—had already bested some well-known Jaguar conversion experts. In fact, the difficulty of a late 1980s-plus conversion is perhaps best illustrated by the brand-new but unused 1989 Jaguar sedan at a certain well-known Jaguar V-8 conversion facility and conversion kit supplier in Texas, which has been moldering silently, frozen in place, partially disassembled, for almost a decade—with the V-8 conversion effort abandoned in action.

Jag-rolet Main Tables
Fuel (% IJPU)

Eff\RPM	0	150	300	450	600	750	900	1050	1200	1350	1500	1650	1800	1950	2100	2250	2500	2750	3000	3250
200	27.00	27.50	27.50	27.50	27.50	27.50	27.50	27.50	27.50	27.50	28.00	28.00	28.00	28.50	29.00	29.50	30.50	31.00	33.00	35.00
180	28.50	28.50	28.50	29.00	29.00	29.50	29.50	29.50	29.50	29.50	30.00	30.50	31.00	31.50	32.00	32.50	34.00	35.50	37.00	39.00
160	30.50	30.50	30.50	30.00	30.00	30.50	31.00	31.50	31.50	31.50	31.50	31.50	32.50	33.00	34.50	36.50	37.50	39.50	41.50	43.50
140	31.50	31.50	31.50	31.50	32.00	32.50	33.00	33.00	32.50	32.50	32.50	33.00	34.00	35.50	36.50	39.00	41.00	44.00	45.50	46.50
120	19.00	21.00	21.50	22.50	24.00	25.00	25.50	26.00	27.50	30.00	32.00	33.00	34.50	36.50	38.00	40.50	42.50	44.50	46.00	47.00
100	16.50	20.00	21.00	23.00	24.00	25.00	26.50	27.00	28.50	30.00	31.00	32.00	33.00	34.00	36.50	38.50	40.00	41.00	41.50	42.50
80	15.00	17.00	18.50	20.50	22.00	23.50	25.50	27.00	28.50	30.00	30.50	31.50	32.00	33.00	33.50	34.00	35.00	35.50	35.50	
60	14.00	15.50	17.00	18.00	19.50	21.50	23.00	25.00	25.50	27.00	28.00	29.50	30.00	30.50	31.00	31.50	32.50	32.50	35.50	
40	15.50	17.50	19.00	20.00	21.00	21.50	22.50	22.00	24.00	24.50	25.00	26.00	26.50	26.50	26.50	27.00	27.00	27.00	27.50	28.00
20	15.50	16.50	17.00	18.00	18.00	18.00	18.50	18.50	18.00	17.50	16.50	17.00	17.00	17.00	17.50	18.00	18.50	19.50	20.00	21.00
0	13.50	13.50	13.50	13.50	14.00	14.50	15.00	15.00	15.00	15.00	15.50	15.50	15.00	15.50	16.00	16.50	17.00	19.00	19.50	19.50

Ignition (Degrees BTDC)

Eff\RPM	0	150	300	450	600	750	900	1050	1200	1350	1500	1650	1800	1950	2100	2250	2500	2750	3000	3250
200.00	22.00	22.00	22.00	22.00	22.00	22.00	23.00	23.50	23.50	23.50	23.50	23.50	23.50	23.50	23.50	23.50	23.50	23.50	23.50	23.50
180.00	22.00	21.50	22.00	22.00	22.00	22.00	22.00	22.00	22.00	23.00	23.00	23.00	23.00	23.00	23.00	23.00	23.00	23.00	23.00	23.00
160.00	21.00	22.00	22.50	22.50	22.50	22.50	22.50	22.50	22.50	22.50	22.50	22.50	22.50	23.00	24.00	24.00	24.00	24.00	24.00	24.00
140.00	10.00	21.50	22.50	23.00	23.00	23.00	23.00	23.00	23.00	23.00	23.00	23.00	23.00	23.00	23.00	24.00	25.00	25.00	25.00	25.00
120.00	7.00	16.00	20.50	22.50	24.50	25.50	27.00	27.50	28.00	28.00	28.00	28.00	28.00	28.00	28.00	28.00	28.00	28.00	28.00	28.00
100.00	5.00	10.50	16.00	19.50	22.50	24.50	25.50	26.50	28.00	29.00	30.50	31.50	33.50	34.50	35.00	35.00	35.00	35.00	35.00	35.00
80.00	5.00	8.00	13.00	18.00	21.00	23.00	24.00	26.00	27.00	28.50	30.00	32.00	34.00	35.50	37.00	38.00	38.50	38.50	38.50	38.50
60.00	5.00	8.00	11.00	18.00	20.00	21.00	23.00	24.50	26.00	28.50	30.00	32.00	34.00	36.00	39.50	43.00	43.50	43.50	43.50	43.50
40.00	5.00	6.00	9.50	14.50	17.00	19.00	20.00	21.50	23.50	25.00	27.00	30.00	32.00	34.00	40.50	44.00	44.50	45.00	45.00	45.00
20.00	5.00	6.00	8.50	10.50	13.00	15.00	17.50	20.50	22.50	25.50	29.00	32.00	36.50	41.00	45.00	45.00	45.00	45.00	45.00	45.00
0.00	5.00	6.00	7.50	9.50	11.50	13.00	16.50	19.00	22.00	26.00	29.00	31.50	35.00	40.00	43.00	45.00	46.00	47.00	47.00	47.00

compression predicted by thermal dynamics equations). The S-Trim blower provides 72 percent maximum efficiency and as much as 800 cfm maximum airflow without overspeeding the compressor wheel and gears. Like any compressor, pumping efficiencies are increased by intercooling. (Jaguar uses air-water intercoolers on the Eaton blower. Norwood, limiting boost to 10 psi, gets by without intercooling by using Motec programmability that provides additional fuel for combustion cooling, and retards ignition under boost as an anti-knock countermeasure.) Sure, an intercooler would make more power, but 10 psi on this engine gets the job done just fine without intercooling.

THE NORWOOD JAGUAR V-8 CONVERSION
Auxiliary Engine Equipment
A 16x27-inch three-row custom radiator and Mr. Gasket high-performance electric fan provide cooling for the most extreme levels of power on hot Texas summer days with the air-conditioning system on full-blast, simultaneously providing sufficient clearance for the supercharger system.

The air conditioning system on the Norwood-Jaguar conversion is basically stock, with the exception of a stock '97 Chevy truck compressor, brackets, and tensioner.

Transmission
The super-duty 350 V-8 was mated to a 4L80E GM super-duty four-speed electronic automatic transmission, normally used by GM on 454 truck motors or 502-ci performance engines. Normally controlled by a GM Powertrain Control Module (PCM) that manages both engine and transmission, the 4L80E transmission electronics on the Norwood-Jaguar are controlled by an Accel/DFI programmable transmission computer (independent of the Motec engine management system) that permits programming customized shift points and line pressures via laptop personal computer to control speed and harshness/speed of gear changes.

Needless to say, a very robust transmission was required to survive under the power and torque of the supercharged engine,

and the 4L80E has a large torque converter. According to Norwood, the biggest problem in adapting the engine and transmission was getting the huge torque converter to work with the 350 engine's relatively small flywheel. A specially machined flex-plate (bell housing access from the bottom) and reduced-diameter starter gear provide sufficient header clearance.

Norwood adapted a GM performance aluminum driveshaft to the Jag's 3.58 rear end, designed for a six-cylinder engine.

Hydraulics and Accessories
The Jaguar hydraulic system produces 1,100 psi pressurized mineral oil (Rolls-Royce "Green") to operate the self-leveling suspension and the ABS brakes, and all 1989 and later Jag sedans have it. With the L98 Chevrolet V-8's ignition provided by Mitsubishi direct-fire ignition coils, Norwood Autocraft was free to eliminate the stock Chevy distributor and mate the Jaguar three-cylinder hydraulic pump to a custom-built billet distributor shaft running directly off the cam (simultaneously driving the oil pump), which also houses a Hall Effect sync signal alerting the Motec computer when top dead center compression stroke number one cylinder is approaching.

The Jaguar alternator was discarded in favor of an 1989 GM high-capacity unit. The power-steering pump is also 1989 GM, while the water pump is an aftermarket Edelbrock unit. The stock Jaguar three-cylinder hydraulic pump (relocated onto a custom Norwood drive mechanism driven off the cam like a distributor) provides energy for anti-lock braking as well as ordinary braking power boost, with backup working hydraulic pressure in case of pump failure provided by hydraulic fluid kept under pressure by the stock Jaguar 1,200-psi nitrogen gas cylinder/accumulator.

Instrumentation, Electrics, and Computer Control
With the Jaguar computer removed from the car, Norwood programmed the Motec to provide a Jag-compatible warning system. The Motec M-48 computer has logic that provides a strategy to run the dashboard interface and warning system, that

Jag-rolet Main Tables
Fuel (% IJPU)

3500	3750	4000	4250	4500	4750	5000	5250	5500	5750	6000	6250	6500	6750	7000	7250	7500	7750	8000	8250	Eff\RPM
36.50	38.00	38.50	39.50	40.00	40.50	41.00	41.00	41.50	41.50	41.50	41.50	41.50	41.50	41.50	41.50	41.50	41.50	41.50	41.50	200
40.50	41.50	41.50	42.00	42.50	42.50	43.50	43.50	43.50	43.50	43.50	43.50	43.50	43.50	43.50	43.50	43.50	43.50	43.50	43.50	180
44.50	44.50	45.00	45.00	45.50	45.50	45.50	46.00	46.50	46.50	46.50	46.50	46.50	46.50	46.50	46.50	46.50	46.50	46.50	46.50	160
46.50	46.50	47.00	47.00	47.50	48.00	48.00	48.50	49.00	49.50	49.50	49.50	49.50	49.50	49.50	49.50	49.50	50.00	50.00	50.00	140
48.00	48.50	49.50	50.00	50.00	50.50	51.00	51.50	51.50	51.50	51.50	51.50	51.50	51.50	51.50	51.50	51.50	51.50	51.50	51.50	120
43.50	44.00	44.50	44.50	44.50	44.50	44.50	44.00	43.50	42.50	42.50	42.00	41.00	40.50	39.50	38.50	37.50	36.00	35.00	33.50	100
36.00	36.00	36.00	36.50	37.00	37.00	37.00	36.50	36.50	35.50	35.00	34.00	33.00	33.00	32.50	32.00	32.00	30.50	30.50	28.50	80
32.50	33.00	33.50	34.00	34.00	33.50	34.50	34.50	35.00	34.50	33.50	33.00	32.00	31.50	30.50	30.00	29.00	28.50	28.00	27.00	60
28.50	28.50	29.00	29.00	29.00	29.50	30.00	30.00	30.00	30.00	29.50	29.50	29.00	28.50	27.50	27.00	26.00	25.00	24.50	24.00	40
22.00	24.50	25.50	26.00	27.00	27.50	28.00	28.00	27.50	27.00	27.00	26.00	25.50	25.00	24.50	23.50	22.50	21.50	21.00	19.50	20
19.50	20.00	20.00	20.00	20.00	19.50	19.50	19.50	19.00	18.00	18.00	17.50	17.00	16.00	16.00	15.00	15.00	14.50	14.50	14.50	0

Ignition (Degrees BTDC)

3500	3750	4000	4250	4500	4750	5000	5250	5500	5750	6000	6250	6500	6750	7000	7250	7500	7750	8000	8250	Eff\RPM
23.50	23.50	23.50	23.50	23.50	23.50	23.50	23.50	23.50	23.50	23.50	23.50	23.50	23.50	23.50	23.50	23.50	23.50	23.50	23.50	200.00
23.00	23.00	23.00	23.00	23.00	23.00	23.00	23.00	23.00	23.00	23.00	23.00	23.00	23.00	23.00	23.00	23.00	23.00	23.00	23.00	180.00
24.00	24.00	24.00	24.00	24.00	24.00	24.00	24.00	24.00	24.00	24.00	24.00	24.00	24.00	24.00	24.00	24.00	24.00	24.00	24.00	160.00
25.00	25.00	25.00	25.00	25.00	25.00	25.00	25.00	25.00	25.00	25.00	25.00	25.00	25.00	25.00	25.00	25.00	25.00	25.00	25.00	140.00
28.00	28.00	28.00	28.00	28.00	28.00	28.00	28.00	28.00	28.00	28.00	28.00	28.00	28.00	28.00	28.00	28.00	28.00	28.00	28.50	120.00
35.00	35.00	35.00	35.00	35.00	35.00	35.00	35.00	35.00	35.00	35.00	35.00	35.00	35.00	35.00	35.00	35.00	35.00	35.00	35.00	100.00
38.50	38.50	38.50	38.50	38.50	38.50	38.50	38.50	38.50	38.50	38.50	38.50	38.50	38.50	38.50	38.50	38.50	38.50	38.50	38.50	80.00
43.50	43.50	43.50	43.50	43.50	43.50	43.50	43.50	43.50	43.50	43.50	43.50	43.50	43.50	43.50	43.50	43.50	43.50	43.50	43.50	60.00
45.00	45.00	45.00	45.00	45.00	45.00	45.00	44.50	44.50	45.00	45.00	45.00	45.00	45.00	45.00	45.00	45.00	45.00	45.00	45.00	40.00
45.00	45.00	45.00	45.00	45.00	45.00	45.00	45.00	45.00	45.00	45.00	45.00	45.00	45.00	45.00	45.00	45.00	45.00	45.00	45.00	20.00
47.00	47.00	47.00	47.00	47.00	47.00	47.00	47.00	47.00	47.00	47.00	47.00	47.00	47.00	47.00	47.00	47.00	47.00	47.00	47.50	0.00

is, the M4-8 checks for correct fluid levels and provides error messages. Fan control strategies are based on programmable voltage readings that can be customized to the particular sensors available and the appropriate temperatures/sensor voltages on which to activate or deactivate electric fans.

The original AJ-6 engine and computer used low-current earth-line switching in which the switch only carries signals. The stock Jaguar electronic control module provides all diagnostics and warnings. "We decided to run the Jag microprocessor just for diagnostics, but now the Jag gets diagnostic information from the Motec," says Norwood.

The correct signal needed to drive the stock Jaguar tachometer (designed to register voltages from a six-cylinder engine) is a matter of customizing the pulse width–modulated (PWM) signal supplied from a Motec M4-8 auxiliary output, but mating the Lucas tach to a foreign electronic control unit was not trivial and involved reverse engineering the Jaguar tach signal. "Making the dash work exactly as stock was a substantial amount of work," says Norwood.

Hot Rodding

The Vortech S-Trim compressor runs a 10-rib serpentine belt system. A stock six-ribbed belt runs all other accessories. The blower initially achieves 8–11-psi boost, depending on the drive ratio of the pulleys. Norwood Autocraft performed minor head porting, opening the combustion chambers and smoothing the surface. The stock pop-up pistons and small-cc chambers provide a compact combustion chamber, and with the Wedge head configuration providing significant squish and quench, the Motec ECU is able to provide fueling and spark timing that enable the engine to run at 10-psi boost on 93 octane pump gas without detonation.

Why not turbocharging? "I actually built a turbo Jag a long time ago," says Norwood, "and if I'd kept the four-valve AJ-6, I'd probably have built a bomber four-valve in-line turbo six. But with the Chevy V-8, clearances are a problem.

You've got a V-shaped engine compartment with no room at the rear for turbo manifolds. We've got the blower mounted out front, on the left side, with rubber mounts attaching it to the frame, and anti-torque struts to make sure the engine doesn't lift. We adapted the Vortech plumbing for a remote (cold air) air cleaner."

The stock oil cooler is located in front of the radiator. Norwood moved it downward and to the left side of the radiator. The trans cooler is now on one side of the radiator, the P/S cooler on the other, with the advantage that the radiator cools the fluids when they're hot, but warms them when they're ice cold.

DRIVING IMPRESSIONS

"I drive it every day to work," says Norwood, who clearly enjoys the Jag's comfortable yet precise ride over the mean streets of Dallas. He also appreciated the performance capabilities of the car as equipped with a supercharged hot rod Chevy V-8. The 1990 Jaguar is designed with a relatively low 0.37 coefficient of drag, which makes high-speed cruising all that much easier. With 550 horsepower on tap, the engine is clearly rpm-limited rather than power-limited at top speed.

The Jag-rolet is unlike any other sedan you're likely to encounter on the streets of Texas, combining stock Jag luxury and handling with the brutal hot rod characteristics of a competition NASCAR-type blown-V-8 powerplant, which is tamed to the relative docility required for a reasonable auto-transmission idle with the powerful engine management logic of a Motec M4-8 onboard computer. Clearly, this thing can hold its own in acceleration with not only the stock supercharged Jag XJR (both in-line six and V-8), but just about any European tuner sedan as well. The great thing about driving this car is that a stab of the throttle subdues virtually anything you might encounter on the streets of Dallas. Unless you come up against one of Norwood's other creations, like a twin-turbo 1,008-rear-wheel-horsepower Ferrari Testarossa.

SUPERCHARGED JAG-ROLET

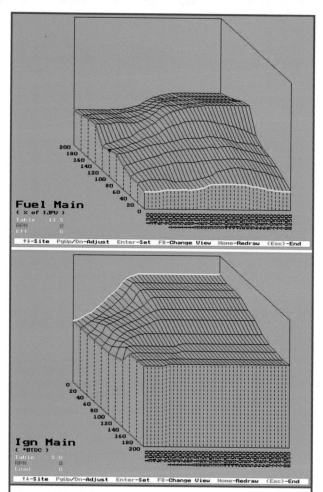

Fuel Main
< % of IJPU >
Table   13.5
RPM   0
Err   0

↑↓-Site   PgUp/Dn-Adjust   Enter-Set   F8-Change View   Home-Redraw   <Esc>-End

Ign Main
< °BTDC >
Table   5.0
RPM   0
Load   0

↑↓-Site   PgUp/Dn-Adjust   Enter-Set   F8-Change View   Home-Redraw   <Esc>-End

Motec M48 main fuel and ignition graphs. On the fuel graph, load increases into the positive Z-axis, and injection pulse width increases. On the ignition chart, load increases in the minus-Z direction, with timing coming out as supercharged load increases. Timing increases steeply with rpm at lower engine loading.

Specification

Engine: 1991 350 Corvette Challenge crate motor, Norwood Autocraft custom supercharging and other VE improvers
Crankshaft horsepower: 375 (nonsupercharged Corvette Challenge engine)
Wheel horsepower: 500 (as tested on Dynojet's 248C chassis dynamometer)
Intake manifold: Accel
Compression ratio: 11:1 (static)
Supercharger: Vortech S-Trim, 800 cfm capacity (uses adapted Chevy truck mounting kit for serpentine belt system and custom Norwood bottom pulley)
Engine management system: Motec M48 (Jaguar computer removed)
Camshaft: Cam Motion mechanical roller cam with solid roller tappets, 113 lobe separation
Intake lift: 0.483 (net of 1.5 ratio roller rockers)
Exhaust lift: 0.496 (net of 1.5 ratio roller rockers)
Intake duration at 0.050: 221.3 crankshaft degrees
Exhaust duration at 0.050: 231.1 crankshaft degrees
Headers: Norwood custom stainless-steel tube headers with ceramic thermal coating, copper-seal exhaust header gaskets, Bosch four-wire wideband O_2 sensor mounted in each, EGT sensor probes in each runner for Motec engine management
Cooling system: Mr. Gasket high-performance electric fan, custom alloy radiator (16x27, three rows)
Air conditioning: Stock 1990 Jaguar, excepting Chevy Vortec pickup truck compressor, brackets, and tensioner. Custom hoses.
Exhaust system: Twin 2.5-inch pipes to twin Flowmaster mufflers with vacuum-operated dump valves
Fuel system: 1990 Jaguar, with Walbro staged high-volume, high-pressure fuel pump, adjustable fuel pressure regulator
Transmission: GM 4L80E with Accel/DFI Tranmap programmable electronic transmission controller. Adjustable shift points, line pressure. Huge 4L80E converter adapted by Norwood to fit a small-block flywheel. Starter gear notched for small bolt, small starter gear ('Vette lightweight) to clear headers.

CHAPTER 18
1970 DODGE
CHALLENGER B-BLOCK

What did it take to change a 1970 Dodge Challenger B-block from a dinosaur into a creature of the new millennium? Perhaps good modern tires and wheels, some suspension gimmicks, hardened valve seats in the heads, things like that. And an upgrade to electronic fuel injection.

The application of electronic fuel injection to American V-8s and turbocharged fours and sixes in the late 1970s and 1980s spawned a performance renaissance that could not otherwise have happened. Fuel injection directly led to greatly improved emissions and economy, while maintaining or improving the power output of modern V-8s and smaller turbo engines. Injection improved drivability by increasing and broadening engines' torque curves with tuned dry manifolds and improved fuel distribution and more accurate air/fuel mixtures.

Electronic fuel injection improved efficiency by allowing higher compression ratios on modern pump gas, and improved power by eliminating the need for certain emissions-control strategies such as retarded ignition timing and low compression. Since electronic fuel injection requires a computer for calculating injection pulse width, the existence of an onboard computer on an injected vehicle made it not only feasible to compute and time sequential injection pulses, but to apply computer control to spark advance and to dynamically coordinate injection control with ignition and emissions-control devices such as EGR, making all these systems work together on a millisecond-to-millisecond basis—yielding clean power with good economy. A win-win situation. This is what you get when you buy a modern performance car like a millennium-vintage V-8 musclecar like a Mustang, Camaro, or Corvette.

Although the Mopar V-8s of the 1960s and early 1970s are famous for high-performance, high-output Mopar V-8 cars departed the scene in the mid-1970s along with *big-block* Mopar V-8s in anything but trucks. However, it is very possible to retrofit electronic fuel injection and computer-controlled spark advance—even emissions controls—to older carbureted Mopars. I had a 1970 Dodge Challenger with the big-block 383 V-8, and electronic fuel injection seemed like a great idea.

1970 Dodge Challenger B-block 383 four-barrel with sidepipes: the retro look.

PHASE 1. THROTTLE-BODY INJECTION

Holley's original TBI Pro-Jection system was designed as an affordable self-contained injection system that is simple to understand, install, and tune. It was designed to replace the four-barrel carb on older V-8s, and it is still selling for this purpose nearly 15 years after its introduction (though the control unit now uses a digital microprocessor instead of analog circuitry and logic). And you still tune the TBI Pro-Jection by adjusting five pots on the side of the control unit with a screwdriver. Holley started with a single-point-injection throttle body with two injectors that essentially had been developed as an OEM replacement part for certain late-model vehicles and made it part of a kit that also included an MSD-designed and manufactured control unit, wiring, a sensor or two, and some miscellaneous mounting adapters. The Pro-Jection was simple and cheap.

Naturally, there were tradeoffs.

An installer paying the half a grand or so for a Pro-Jection 2 system (now superseded by Holley's 670 CFM Two-Barrel Pro-Jection Digital system) could expect the following advantages compared to a carb:

• There is no float bowl, so hard acceleration and cornering or harsh terrain cannot degrade the operation of the fuel system as it can with a carb.

• Fuel delivery is precise and infinitely variable because the Pro-Jection provides fuel in multiple bursts or pulses that can be timed in increments of less than a thousandth of a second. Since the Pro-Jection system does not use separate and overlapping fluid/air systems as a carb does for idle, midrange, top-end, transient enrichment (accelerator pump) and cold-start operations, a properly tuned system has the potential to conform more perfectly to an ideal air/fuel curve, yielding more power and economy compared to a carb.

• The Pro-Jection allows adjustable enrichment for cold starting and includes a fast idle solenoid.

• The Pro-Jection system can be tuned with a set of screwdriver-adjustable dashpots while the vehicle moves down the street or rolls on a dyno—obviously an advantage compared to the repeated disassembly that may be required when making jet changes on a carb.

• The Pro-Jection, being a pulsed fuel injection system, is not subject to mixture problems resulting from reversion pulses flowing backward through a carb—particularly in a wild cammed engine with a single plane manifold—resulting in the carb over-richening the mixture by adding fuel to the air twice as it reverses direction.

• Since the Pro-Jection instantly cuts fuel delivery when you turn off the key, dieseling cannot occur as it might with a carb.

• Vapor lock is much less likely under the 12–20 psi Pro-Jection fuel pressure, compared to the reduced pressure of a mechanical vacuum pump sucking fuel forward to the engine.

• Hot air percolation, in which gasoline boils and overflows the float bowl(s) in a carb in hot weather—additionally pressurized by the action of vapor pressure in fuel lines—is not possible with the Pro-Jection.

183

On the other hand:

• One of the tradeoffs of the simple and affordable Pro-Jection is that fuel is injected from only two injectors at one place—at the two-barrel throttle body. Compared to supplying fuel from individual injectors at every intake port, throttle-body injection's ability to deliver equal fuel and air to every cylinder is subject to the same limitations and design flaws of any intake manifold that must handle a wet mixture. Gasoline tends to fall out of suspension in the air of the intake manifold during cold starting or sudden throttle openings that rapidly change manifold pressure—and then tear off the surface of the manifold unpredictably in sheets of liquid in a way that can interfere with equal distribution to individual cylinders. There is simply no way to guarantee equal distribution with single-point fueling. In addition, high air velocity is required to keep fuel suspended in the air at low rpm, which means that the Pro-Jection cannot use the radical dry manifolds of port fuel injection that provide great high-end breathing while relying on port injection to put fuel directly into the swirling dry air at the intake valve at low speed for equal distribution and consequent good idle.

• The Pro-Jection 2 did not control ignition spark advance or other emissions-control devices.

• The Pro-Jection estimates air mass entering the engine based on throttle position and rpm, with no provision for measuring actual mass airflow (MAF) or manifold absolute pressure (MAP), which means it was not able to compensate for VE changes to air entering the engine at equal rpm and throttle position based on, say, the action of a blow-through turbocharger. There is also no provision to raise fuel pressure (a relatively low 15 psi) to compensate for air pressure changes.

• Obviously, the Pro-Jection cannot time fuel injection for individual cylinders as in sequential fuel injection (a tactic used to improve emissions and fuel economy).

When I first began to work on this '70 Challenger, it was running an ancient Carter AFB carb on a single-plane Edelbrock Streetmaster 383 aluminum intake manifold. The engine was still running the stock compression ratio (9:1) but now ran a fairly hot auto transmission Competition Cams camshaft. The L code on the car told me that the vehicle had originally been built as a 290-horsepower two-barrel 383. The engine was now equipped with a new Carter AFB four-barrel carburetor.

With fresh paint and interior on the Challenger plus a freshly rebuilt engine, new air conditioning compressor and a host of other details like a new wiper switch and door handles, I later set about to install the Pro-Jection. Although the Pro-Jection 2 is a two-barrel throttle body, it will *not* fit on a stock two-barrel manifold. The Pro-Jection bolted to an adapter plate made to bolt on a *four-barrel* manifold and is workable on most small- and medium-displacement American V-8s, flowing up to 650 cfm.

It is very easy to install this type of Holley EFI system. You begin by designing a way to return fuel to the gas tank. Like virtually all injection systems, the Holley system must provide constant pressure to the injectors, regardless of the engine fuel requirements. This is done by providing a fuel loop that pumps pressurized gasoline from the tank to a pressure regulator in the engine compartment and then back to the tank. The regulator,

This car started out life as a 290-horse 383-ci two-barrel big block. With four-barrel Carter AFB carburetion and some other mild hotrod parts (Edelbrock "Streetmaster 383" single-plane manifold, better cam, headers, and so on), we guessed the "Commando" V-8 was up to the standard of the factory 335-horse "Magnum" four-barrel. Everything worked, including the air conditioning.

integral to the throttle body in the Pro-Jection, pinches off return flow to the degree required to maintain a preset pressure in the fuel supply line. The amount of excess fuel returning to the tank varies depending on engine fuel consumption. A lot of fuel is returning at idle, very little if the engine fuel consumption approaches the capacity of the fuel pump under high-rpm heavy loading.

I removed the fuel filler neck pipe, cleaned it carefully, and welded on a fitting to attach the returning fuel line. An alternative return might also have been possible by removing the sending unit for attachment of a return fitting. Obviously, you can't simply drill a hole in the tank. If it were necessary to remove the tank for drilling or welding, you should have the work done by a fuel tank specialist. An empty fuel tank, inevitably loaded with fuel vapor, is incredibly explosive!

Mount the fuel pump in a safe low spot near the fuel tank, preferably where it will be primed by gravity rather than the reverse. Holley suggests mounting the fuel pump no higher than the top of the tank, and suggests locating the pump at the bottom of a loop of fuel line so the pump will always have liquid fuel to pump when it first starts. When you turn on the key, the computer will momentarily activate the pump to pressurize the injectors. If fuel is not immediately available to the pump, the engine may be very hard to start. Naturally, on an old dog like a '70 Challenger, you'll want to remove the mechanical fuel pump from the engine, and blank off the hole in the block with a plate (available at most speed shops).

I next removed the old carb, and mounted the Holley adapter plate and throttle body, connecting the fuel supply line and a return line I installed and connected to the return fitting on the filler neck. Chrysler automatic transmission cars, using a mechanical kick-down linkage, may require an adapter to attach the linkage to the Pro-Jection. Holley builds a Mopar adapter, but doesn't include it in the standard kit. I got one at a local speed shop. The linkage on the Challenger, having already been doctored in the conversion to the Carter AFB four-barrel, required some fudging to get it perfect.

Aftermarket fuel injection computers, other than incredibly expensive mil-spec racing units like EFI Technology's Competition unit, tend to work better in a friendly environment, away from lots of heat and electrical interference, which means inside the car or in the trunk. I mounted the Pro-Jection computer on the console ahead of the shifter, and fed the wiring harness through an existing hole in the firewall to connect to the Pro-Jection throttle body, the new EFI water temperature sending unit (which I'd mounted in a 1/8-inch fitting in the water pump), and the fuel pump. That left several wires that needed to be connected to constant positive 12-volt, switched +12 voltage, and negative 12-volt coil or tach-drive voltage.

Start up time.

Since an EFI fuel pump usually only runs a second or two on initial energizing before start, it's a good idea to jumper the fuel pump to +12 voltage on the battery in order to fill the fuel system with gas, and then to listen carefully when you turn on the key to make sure that it is operating full time (if you wire it wrong, some ignition switch connections specifically do *not* provide voltage while cranking).

With the fuel pump working, I verified that the dashpots on the computer were at the correct position specified by Holley for initial start up, then cranked and watched the injectors with the air cleaner removed from the throttle body. It is easy to see the big throttle-body injectors spraying fuel, and in a few seconds the engine came to life. I immediately went to work on the idle mixture dashpot, hunting for the position that would give the smoothest idle. Disconcertingly, this was as lean as it would go.

Consulting the manual, I decided that perhaps the throttle position switch (TPS)—a linear potentiometer mounted on the throttle shaft on the side of the throttle body—might not be calibrated correctly. Since the computer determines engine loading directly from throttle position and rpm, it is essential that the switch read the right voltage. Using a digital resistance meter and some jumpers, I was able to watch the resistance of the switch change as I opened the throttle. Bingo. The switch was reading too high with the throttle closed. Loosening the TPS, I was able to rotate it until the meter showed the correct voltage for a closed throttle. Starting the engine, the idle mixture dashpot now yielded the best mixture farther up the scale as it should.

Holley's installation manual lists procedures for tuning the midrange and high-end power, which basically involves tuning for best mixture at 3,000 rpm unloaded in neutral, then using a stopwatch to time acceleration runs at wide-open throttle (WOT) until the Pro-Jection is set up for best power as indicated by the stopwatch. Pro-Jection 4 units have one extra pot for tuning high-end power. Of course, the best procedure is to attach a wideband air/fuel-ratio meter and tune for 12.5–13.5:1 mixtures.

Out on the highway, I noticed an interesting thing: Although the car felt—seat of the pants—as if it had better low-end torque, what was missing was the sudden burst of power under hard acceleration you get with a four-barrel as the secondaries open up under WOT.

The Pro-Jection system was a vast improvement over the carburetor on the Challenger. It started quickly in all weather, provided fast idle when cold, and was free of the hot weather starting problems encountered by the AFB in the heat of a south Texas summer. Throttle response was good under all conditions. The system had no way of compensating for the air conditioning compressor coming on line, but had it really mattered, I might have rigged a relay to trigger the cold start solenoid.

PHASE 2. PORT FUEL INJECTION

In the meantime, I got access to a Haltech ECU capable of controlling multiple port injectors on a V-8. I decided to convert to the Haltech. It turned out this was extremely easy, given that I had already made the conversions for the Holley TBI. I already had a fuel return line. I already had a mounting port for a coolant-temperature sensor. It turned out to be possible to use the Holley throttle body to meter air. The throttle position switch on the Holley throttle body is compatible with the Haltech. The injector/regulator/fuel fittings assembly is a one-piece unit that bolts to the main section of the Holley throttle body with three Allen screws. I removed it and fabricated an aluminum plate to cover the hole.

The single plane Edelbrock manifold on the Challenger uses runners that are perfectly in line with each other. I milled holes for injector bosses in the roofs of the runners working through the runners from the underside of the manifold, and welded the injector bosses in place from the bottom, grinding the excess boss material away so as not to interfere with airflow. I installed O-ringed Lucas disc fuel injectors in the bosses that were fitted for hoses on the fuel input side, constructing a D-section fuel rail with hose barbs to supply fuel.

The biggest problem had to do with the fact that 1970 Chryslers used a mechanical kick-down linkage that had originally connected to the carb linkage. Later, with adaptations, I had attached it to the Holley throttle body. Unfortunately, the injector assembly and fuel rail were now exactly where a linkage support had bolted to the intake manifold. Fortunately, I discovered it was possible to move the kick-down linkage outboard of the left bank of injectors by cutting the linkage support in two pieces and welding it back together with the addition of some mild steel bracketing. I cut the linkage itself and inserted a 3/4-inch inward offset so it could properly connect to the Holley throttle body.

It was an easy matter to connect the fuel lines to the new fuel rail (using 900-psi hose at $0.33 an inch!). I used a 39-psi pressure-referenced regulator from an L-Jetronic injection system, and I robbed the Bosch fuel pump I'd used in the original XKE turbo-EFI system featured in another chapter of this book, installing it in-line in the return side of the Challenger's fuel rail. The fuel pump was necessary since the Pro-Jection pump is not designed to provide the high pressure required to provide stable fuel pressure to eight batch-fire port injectors.

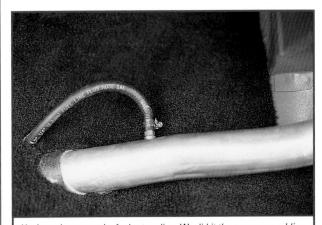

You're going to need a fuel return line. We did it the easy way, adding a hose barb to the fuel filler neck so unneeded fuel that bleeds from the fuel rail can return into the fuel tank. We also installed a coolant temp sensor, 15-psi electric fuel pump, fuel return line, and a few other tricks.

Wiring the Haltech ECU was simple, and programming is discussed in other sections of this book. Haltech provided a startup map designed for a modified V-8 that provides a nice rich set of base fuel maps as a good starting point for a car like the Challenger. Although this big-block Mopar 383 engine is not particularly radical, given the hotter-than-stock cam, the headers, and the large-port intake manifold, the modified V-8 map seemed like a good place to start, since it is always preferable to begin rich and work lean when calibrating an engine.

I installed an air temperature sensor for the Haltech ECU in the air cleaner inlet. This is much better than milling a port hole in an intake manifold runner, which is much more likely to heat-soak the sensor and lean out the mixture as the ECU tries to correct for unrealistically hot intake air.

Finally, I had to build a connector to adapt the Holley TPS connector to the foreign Haltech wiring harness. I located the Haltech ECU inside the car on the side of the transmission hump near where the Pro-Jection system had been. I removed the Holley wiring harness and equipment and set it aside. It would be easy enough in the future to replace the Holley throttle body with something else and use the whole Holley system on some other project.

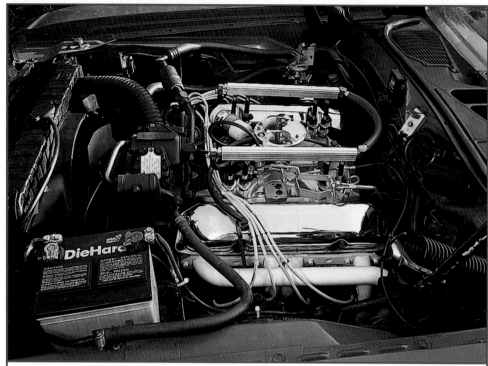

Starting with the Pro-Jection TBI system made the port-injection conversion a little easier. We removed the two big injectors from the throttle body, drilled and welded-in injector bosses, constructed a fuel supply rail, and installed a high-pressure inline fuel pump. The Haltech EMS generic harness plugged into the injectors and sensors straightforwardly. We plugged in a laptop, fired up the car with a library calibration, and tuned the car for drivability with the Haltech mixture trim module as best we could on some Texas back roads. Later on we found some free power with a Horiba wideband AFR meter.

There is a wonderful crispness to multi-port injection that is properly set up, given that the injectors are spraying fuel at high pressure directly at intake valves only an inch or two away on a setup like that of the Challenger. The Haltech ECU offered excellent resolution, basically providing a 512 element matrix of speed-load breakpoints that define the engine fuel map. Obviously, this provides greater flexibility that the Pro-Jection system that defines the base fuel map with three control knobs.

Is multi-port injection worth the additional expense over a programmable throttle body system? The point about the Holley system is that it was easy, affordable, and had clear advantages over the AFB four-barrel carb it replaced. A properly tuned Pro-Jection system will always outperform a poorly tuned software-programmable port-injected system, and a system like the Pro-Jection is much easier to tune. Gasoline-fueled engines can be reasonably forgiving about mixture, with a fairly wide margin in air/fuel ratio separating lean best torque and rich best torque.

The fairest thing to say about whether it's worth it depends on what you are asking the simpler Pro-Jection system to do. How good is the manifold you're using for wet mixtures? How smooth and consistent is the volumetric efficiency of the engine as power and airflow increase; how peaky is the engine in its torque curve; how much does torque vary from a truly flat curve? How high is the compression? How important is fuel consumption? Do you have the equipment and patience required to really take advantage of the programmable port EFI system's flexibility? It you have a radical manifold on a peaky engine with a lot of compression, it is very important to have the flexibility to build an oddly shaped fuel curve to match torque, and then you'd clearly want a programmable multi-port engine management system.

But on many engines, considering only peak power at wide-open throttle, the Pro-Jection is right up there with the best of them. Of course, it will clearly not have the wonderful low- to midrange broad torque range of a tuned multi-port injection system with long runners. But that's a different story. And when you've got 383 cubic inches under the hood, you've always got torque.

CHAPTER 19
FOCUS TURBO

"**P**eople like the Focus because it's roomy, quiet, and smooth," says Brice Yingling, owner of Alamo Autosports, a Dallas-area tuner that specialized in sport compact performance long before it was cool. "When they find the car can make horsepower, they're even more excited." Alamo zeroed in on the Focus very early on with the intention to develop a broad line of engine performance, handling, appearance, and custom-electronics equipment for the car.

"We wanted to do something new and cutting-edge," says Yingling. "Something no one else was doing. We wanted to take on a new form of car." Alamo immediately began building a hot rod ZX3 Focus show car to showcase pretty much every part available from anyone that could be made to fit on a Focus.

The key ingredients in building a streetable 400-plus horsepower Turbo Focus that can compete with modern factory performance cars and survive are a big turbo, a massive intercooler, significant cylinder head airflow improvements, a fully programmable aftermarket engine management system, and the right combination of super-duty internal engine parts. The stock Focus motor *will* break at power levels over 250 front-wheel horsepower at high rpm and high boost, and Alamo's SHO Focus build up is designed to nullify that problem nicely. However, Alamo's package of SHO Focus longevity upgrades is not for the timid because it involves installing serious internal engine parts: billet chrome-moly-steel X-beam connecting rods, reduced-compression (down from 9.6:1) forged pistons, multi-layer steel head gasket, billet main caps, high-strength head and main studs, cryogenic metallurgy parts treatment, increased oil pressure, and more.

WHERE FOCUSES BREAK

Designing a streetable 2.0-liter four-cylinder Focus long block with the strength to live forever at 400-plus horsepower would be trivial for Ford Engineering. But nothing is free; there'd be a cost. A super-duty 400-horsepower-capable short-block is going to weigh more. The weight is going to add cost at the gas pump every time you fill up. What's more, the added weight of super-duty block, crank, rods, and pistons takes a direct toll on acceleration performance and hurts the engine's transitional response and free-revving capabilities. A heavy engine is purely performance-negative on a 130-horsepower variant like the stock, naturally aspirated Focus economy car.

On the other hand, designing a high-output four-cylinder powerplant with exotic alloys and metallurgy can provide strength and longevity without a weight penalty, but clearly drives up the cost. Piston oil-squirters are a wonderful thing to keep pistons cool on a turbo-supercharged version of an engine, but they simply add cost on a low-output, high-volume production engine like the stock 130-horsepower Focus. Big valves and head ports capable of flowing 400 horsepower worth of intake charge and exhaust gases at moderate turbo boost are a ball and chain on the around-town drivability of a stock NA 130-horsepower Focus because you want high charge-gas velocity for really good low-end and launch torque on a four-cylinder engine. All of which is why the 2.0 Focus powerplant is most definitely *not* 400-horsepower capable as it arrives at the Ford dealership, and why making one that is 400-horse capable is an impressive trick for tuners like Alamo.

The stock Focus engine system is reliable and strong—at stock levels of power. The stock redline is 6,700 rpm, and Ford's rev-limiter is very soft, incrementally pulling out power quite early to the point that you almost don't notice the limit. It's just as if the engine simply runs out of breath and absolutely will not exceed 7,050 rpm. Which is clearly a good thing if you're running stock rods and rod fasteners.

From a performance point of view, the stock Focus rods are, well, *cute* is a kind word for what they are. They *will* fail catastrophically if you run them at high boost and high rpm and will probably take the block with them to that big junkyard in the sky. "We sell 7-psi street-type Focus turbo kits," says Yingling. "Nine psi and above is not good for the stock rods." In fact, Alamo managed to break all four rods at once while racing a Z06 Corvette at high rpm and more than 150 miles per hour in a Focus turbo at 9-psi boost.

Stock Focus pistons are lightweight castings with fairly high compression (9.6:1) designed with thermal expansion characteristics that eliminate the scary-sounding slapping noise higher-priced forgings make until they warm up. Cast pistons will not survive spark knock, and they are problematic at high rpm and high boost.

The stock composition head gasket will blow if you make much power.

The stock main caps, according to Alamo, will simply not hold anything over 350 crankshaft horsepower; they'll walk around and then cause crank bearing problems.

The crankshaft itself is subject to harmonic problems above 7,000 rpm if you remove a lot of mass, says Yingling. Alamo balances, but will never lighten or knife-edge, a stock Focus crank. "We were highly advised by people at Ford who know not to knife-edge or lighten a Focus crank due to potential harmonic and torsional problems," says Yingling, who received performance consulting on the Focus from the staff personnel at Ford Motorsports. A stock unmodified Focus crank with excellent balance, according to Yingling, is good to over 8,000 rpm.

The stock Focus engine oiling system is, obviously, designed for the stock engine, that is, to provide the thick-film hydrodynamic lubrication that uses pressurized oil to prevent all contact between moving metal surfaces under normal loading. The

A Focus Turbo in action.

increased loading that results from much higher boosted cylinder pressures (including the nonnegligible possibility of pressure spikes from spark knock on a performance R&D engine) or the exponentially higher G-forces on crank journals and wrist pins from increased redline could potentially cause the thick film to fail.

Focus cylinder head ports and valves are tiny for efficient production of low-rpm torque with high charge velocities at low speeds on engines designed to make peak power of 130 horsepower. Unfortunately, the intake manifold, being of plastic construction, cannot be ported or extrude honed.

The stock plastic intake manifold (O-ring-sealed to the head and built to match the tiny head intake ports), is designed for stock-type drivability. Being of plastic construction, it is impossible to modify. The stock intake is good for a maximum of 350 to 360 horsepower. It has almost no plenum; the throttle body flange area diverges directly into the four intake runners. Not a problem on a stock Focus, but definitely suboptimal on a heavy-breathing high-flow turbo mutant.

The stock Focus ignition—identical to that used on Euro Formula Fords—is quite good, fully up to the job of firing 10,000 rpm on turbo motors. Good thing too; if you try to overdrive Focus coils by increasing voltage with an ignition enhancer, the processor components in factory direct-fire coils turn themselves off at 14 volts, due to the excessive voltage and heat, according to Alamo, which has tried this trick. There is little to gain by doing anything to the ignition.

The stock Focus fuel supply is good for maybe 300 front-wheel-horsepower worth of fuel. The U.S.-market Focus has no standard injector rail fuel loop but rather a small loop plus pressure regulator at the fuel tank, with a deadhead line plumbed off the loop to deliver controlled-pressure fuel to the injector rail. The fuel pump flow rate is dynamically controlled by the stock Focus computer, with a fuel pressure sensor in the injector rail. The computer monitors pressure at the injectors and adds or subtracts voltage to the pump to maintain target pressures across the range of operating conditions.

The Focus OBD-II engine management system doesn't exactly break when you hot rod a Focus, but emissions, longevity, fuel pressure, and traction-control logic can be a serious pain. Maximum rpm is absolutely limited to 7,050. The computer simply does not like excessive rates of acceleration in lower gears and will initiate traction control countermeasures that stop acceleration dead in its tracks.

The stock Focus throttle body with its unique throttle butterfly (referred to affectionately at Alamo as the "potato chip" due to a pale color and geometric profile that looks something like a Pringle) would constitute a definite bottleneck on high-output "all-motor" Focus hot rod engines. However, it is amazing how much air you can force-feed a 130-horsepower throttle body with a serious turbocharger like the Turbonetics T04.

ALAMO FOCUS SHO PERFORMANCE TACTICS

A stock Focus ZX3 2.0-liter powerplant weighs in with 130 brake horsepower at 5,300 rpm, and 135 ft-lbs at 4,500. Alamo's Focus SHO project was designed to achieve more than 350 horsepower at the driving wheels on the Dynojet chassis dyno, which translates into well over 400 horsepower at the crankshaft. Such a radical powerplant requires improvements in parts strength and volumetric efficiency, a high-pressure intercooled turbo system, plenty of fuel and engine lubrication, and superior engine management strategies. Alamo used the following tactics to pump up the volume of the silver checkerboard Focus to antisocial levels.

Prauter Machine billet 4330 chrome-moly-steel X-beam design rods provide increased strength and improve aerodynam-

ics as the rod whips through the crankcase air at high rpm. They also reduce crankcase windage compared to the stock H-beam rod design. Other longevity upgrades include forged 8:1 dish-top pistons, moly rings, World-Rally-type multi-layer steel head gasket, cryogenic treated block and crank (and other parts), precision crank balancing, billet main caps, increased oil pressure stock oil pump, and fully synthetic oil.

Focus SHO VE improvements start with bigger Esslinger mild street camshafts, designed to provide maximum possible lobe lift with stock cam buckets. The stock cam buckets are so small that truly radical cam lobes begin to run off the edge of the buckets, quickly destroying both cam and bucket. "We could machine the head for bigger buckets, but since we're running a big turbo on the Focus SHO, we don't actually need that much bigger a camshaft," says Yingling.

Alamo installed adjustable cam-timing sprockets for optimizing cam phasing on the dyno and installed 2-millimeters-larger-than-stock intake and exhaust valves with requisite valve seat changes. An Alamo ported and polished cylinder head increases gas flow "roughly 30 percent." The Focus SHO package provides an upgrade from the factory cast-iron exhaust manifold or Alamo Stage 1 tube header with a switch to a higher-flow JBA header. Identical header/ exhaust manifold geometry allows Alamo's FocusParts turbo and wastegate assembly with inlet adapter to bolt directly to the stock cast exhaust manifold or either performance header. According to Alamo, the SHO's JBA header provides reduced backpressure through the turbo, higher maximum horsepower, and slightly faster spool-up.

The 430-horse Alamo Focus SHO turbo system is based on a large Turbonetics T04E/T3 turbo (an upgrade from the rewheeled T3 used in the standard high-output Alamo Focus turbo kit). Alamo selected this turbo for quick spool and drivability on the standard turbo kit, while the Focus SHO system required a larger turbo for power on the top end and the ability to rev freely. According to Yingling, Alamo was less concerned with rapid spooling on the SHO because the engine has more top-end rpm capability. A huge front-mounted air-air intercooler on the SHO Focus keeps charge air temperatures safe and dense, while a special SHO mandrel-bend exhaust system helps keep turbine-out exhaust gas backpressure low.

The JBA directional-flow header adds a little responsiveness to turbo spool-up by directing exhaust pulses smoothly into the turbo so they are not affected by interference. While the standard FocusParts.com turbo kit makes 238 front-wheel horsepower, the Phase II car will achieve more than 300 front-wheel horsepower at 15 psi and has achieved 366 horsepower at the wheels on 18-psi boost. Assuming a drivetrain/rolling loss of 15 percent, this translates to 430–460 at the flywheel.

The Focus SHO engine requires an upgrade to Haltech E6K engine management from Alamo's reworked OEM Focus engine management system with O_2 simulator used on lesser Alamo Focus turbo kits. The Haltech E6K incorporates special new firmware to interface with the stock 2.0-liter Ford crank-position multi-toothed wheel and Hall Effect pickup. Alamo's strategy is to install a Haltech E6K engine management system and piggyback it on the stock ECU, which continues to control basic functions unrelated to performance, such as air conditioning, idle, dash instruments, and the stock fuel pump (augmented at high boost with a Kenne-Bell Boost-a-Pump).

"The stock fuel supply limitations probably slowed this project more than any other single thing," says Yingling. "To properly convert to a traditional fuel rail with pressure regulator and fuel tank return loop, you need the European Ford ECU, the Euro

A Focus Turbo powerplant with all the good internals to push 500 rear-wheel horsepower and a Turbonetics TO4E-50 compressor.

fuel lines, the engine fuel rail, and the fuel tank from the European Focus. Some people's attitude was, just drill a hole in the tank or fuel-filler tube, and install a second fuel line for returning excess fuel." This is OK for a drag car, says Yingling, but unacceptable for a street car due to the additional fuel vapor it generates, which can increase fuel pressure and cause vapor lock and consequent drivability problems.

Alamo uses a Kenne-Bell Boost-a-Pump to jack-up the stock fuel pump's maximum fuel supply at high boost, which also easily provides fueling for a 40-shot fogger nitrous injection available for turbo spool-up purposes when required on the Phase II Turbo Focus.

"The SHO turbo system is very responsive," says Yingling. "We installed the nitrous system prior to turbocharging the car, and it's currently less a performance necessity than an interesting conversation piece. At this point, the bottlenecks currently limiting Focus SHO power in the range of 350–400 front-wheel horsepower are the intake manifold, the 'potato-chip' throttle body, maybe the cams, and the fuel system."

Alamo equipped the SHO Focus with a Quaiffe limited slip differential to keep the engine from routinely boiling the tires in lower gears when you hammer the gas pedal. At the same time, Alamo disassembles the transmission gear stack to strengthen it cryogenically. Alamo recommends a similar cryogenic process as useful for strengthening axles, CV joints, and the aluminum transmission case itself. Alamo is confident that the SHO drivetrain is up to the job of living a long time in street-performance driving conditions.

DYNO-TUNING AND ROAD TESTING THE ALAMO FOCUS SHO

Once Alamo had interfaced the Haltech E6K's wiring loom to the stock Focus electrical system and engine management sensors and actuators (including the cam trigger), the main challenge was to keep an eye on fuel supply while super-tuning the E6K on the Alamo Dynojet chassis dynamometer. The E6K is a programmable speed-density computer system that does not directly read mass airflow from a hot-wire or film sensor, but instead determines ignition timing and fuel injection pulse width based on extensive internal tables of numbers that approximate the volumetric efficiency of the engine. These VE tables must be developed based on exhaustive dyno and road-test experimentation at all achievable combinations of speed and manifold pressure.

The process of scratch-building Haltech E6K ignition and fuel speed-density maps begins when you select a startup fueling map from the Haltech library and supply a set of engine parameters to Haltech's auto-mapping software, which builds conservative best-guess ignition maps based on user-supplied engine parameters. These fuel and timing maps enable the engine to *start*, at which point a tuner must manually optimize best torque (LBT) across a matrix of breakpoints of engine speed and manifold pressure. This is done on the dyno with a laptop interface program connected to the onboard computer and the help of a sensitive wideband air/fuel-ratio meter or gas analyzer. Following this, the tuner must then verify drivability on the street and track. Once tuning is close, the car typically becomes the daily-driver for the chief tuner for a while, with a laptop on the shotgun seat for minor drivability adjustments, particularly during cold-start and warm-up.

This sort of tuning is routine for dyno shops like Alamo, but in the case of the Focus SHO, the main trick was to keep a close eye on fuel pressure to make sure that (with the help of a Boost-A-Pump when the turbo is working) the Focus' partially disconnected stock computer was kept happy enough to continue commanding a reasonable fuel supply from the variable voltage fuel pump under both naturally aspirated and boost conditions without going into a coma while the E6K took over fueling and timing tasks.

"Tuning just takes time," says Brice Yingling. "Not a problem."

TRICK ELECTRONICS

The 430-horsepower Focus show car was equipped with a Kicker stereo system that includes an Eclipse voice-activated head unit. Alamo interfaced the head unit's speech-recognition capabilities to the Haltech E6K so the driver can open the nitrous bottle and arm the nitrous system by *voice command*. This system does not require voice training by a particular operator, but it has a limited vocabulary, and the operator must simply speak in a normal tone of voice. Screaming "Turn on the nitrous!" at the top of your lungs is not going to work; "Auxiliary one on" is more like it.

DRIVING THE FOCUS SHO

Let's face it, the Focus SHO is a rolling dessert cart of Focus parts designed to showcase what's available, so no one was really looking hard at the cost-benefit of every last trick part. The total package, however, had to be fun and safe to drive at the hands of hardcore types who like to get behind the wheel and go mental.

"Turn on the nitrous," I command as I adjust the seat on the Focus SHO, sitting next to Yingling, ready for a shakeout mission. I'm only kidding. On a super-high-output turbo car like this, the nitrous injection is not necessary for street-type launches, and I want to see what it can do all-turbo.

What we mainly want to know is, what can this bad boy do when you whip it good? Moving 3,200 pounds of inertial dyno rollers, beginning in fourth gear at 15–20 miles per hour and wide-open throttle, the car takes roughly 10 seconds to reach peak power near 75 miles per hour. With the turbo working at full howl, the Focus SHO explodes from 4,000–7,000 rpm in fourth in four seconds. We're confident that a generic Ford Focus is entirely viable as a performance car with the right aftermarket parts. And that the Focus SHO from Alamo Autosports is a true highway thoroughbred capable of winning respect at the drags on a Saturday night. This SHO may be a show car, but it is capable of cleaning the clock of a Corvette Z06.

CHAPTER 20
REAL-WORLD TURBO
CRX Si

Overall, the philosophy of this project was simple: This would not be a drag-race car, not a VTEC engine swap, not a midengine conversion, not a tube-frame superfreak—just a 1.6-liter turbo Honda designed to blow away plenty of Mustangs and Porsches. So, the plan was to hot rod for *torque* at the wheels. To hot rod for a powerband with plenty of area under the curve, that is, high *average* horsepower.

Gen II stock Honda CRXs are perfect practical hot rodding raw material: brilliantly engineered, trouble-free, fun, and, by now, dirt cheap. If you've never driven a CRX Si before, you'll probably be pleasantly surprised at the responsiveness of Honda's 1.6-liter 16-valve single overhead cam (SOHC) engine in the lightweight 2,100-pound car.

When I first twisted the key on this project car, the fuel pump hummed and the engine fired immediately, settling into a crisp smooth idle. In fact, tests at Alamo Autosports would later reveal that the 1.6 Honda engine was indeed healthy, with excellent cylinder leakdown and before driving-wheel power at 96.9 horsepower, which is as good as it gets on a stock U.S.-market CRX.

REAL-WORLD CRX POWER GOALS
In contemplating the parameters for this project, I had to consider what it would take to truly spank a stock Mustang-Camaro-Porsche. How fast and powerful was the competition? What figures did we need to beat to get respect? A 215-horse 1990s-vintage Mustang GT weighed in at roughly 3,350 pounds and had perhaps 175–180 wheel horsepower on a good day. A late-1990s GT did 0 to 60 in 6.8 seconds. A Porsche Boxster weighed roughly 3,000 pounds and made an estimated 171 horses at the wheels, with a 0 to 60 of 6.9 seconds. A 252-horsepower Boxster S is probably good for 200–215 horsepower at the wheels.

So power-to-weight of the Mustang was 18.3 pounds per horsepower, the Boxster, 17.5—which we'll correct upward to 18.5, assuming the standard aviation weight of 170 pounds for a driver. To equal an 18.5 power-to-weight ratio, the 2,100-pound CRX (2,270 pounds with driver) would need 122 driving-wheel horsepower. By the way, a 1998 305-horse Mustang Cobra, or a 296-horse Porsche 911 with 3,170 pounds and 250 rear-wheel horsepower carries in the range of 12.5–13.5 pounds per wheel horsepower.

What would it take to jack the power of the project CRX by 55 percent at the wheels, from 97 horsepower at 6,200 to 150? Well, a 50-horse nitrous kit would get pretty close. To calculate potential *turbo* power, a ballpark assumption is 8 percent more torque per psi of additional boost. Starting with the car's peak Dynojet driving-wheel power of 97 at 6,200 rpm based on Alamo's Dynojet chart, we discovered torque at 6,200 rpm was down from the peak 92.4 to about 82 ft-lbs. Five-psi boost added 40 percent more torque, for a total of 115 at 6,200. Converting torque to horsepower with the formula HP = (Torque x rpm) ÷ 5250, we find the car's power would now be 131 at 6,200. But assuming turbo boost moves peak power upward from 6,200 to the redline at 6,500 with turbocharging—and no other hot rodding tricks—calculated maximum power would now actually be 142 rear-wheel horsepower at 6,500. Alamo Autosports reported that although the stock CRX-Si redline was 6,500, they had

1989 Honda CRX, with 17-inch wheels and paint. The wheels looked good, but they actually made the car slower due to higher final gear ratios and added weight.

pushed stock-type Honda turbo motors to 7,200 many times without a failure.

OK, but 142 is not quite enough. However, with a *6-psi boost*, torque would be up approximately 48 percent at 6,500 to 121.36. Converting this to horsepower, we get 150.25. OK, we're there. A 10-psi boost makes 148 ft-lbs torque at 6,500, or 183 rear-wheel horsepower. Bottom line is that 12.4 pounds per horsepower is surely enough to ambush a 911 Carrera or Mustang Cobra on a good day.

Other considerations: The Honda's effective compression ratio increases from the static CR of 9.0:1 to maybe 11:1 at 7-psi boost, nearly 12:1 at 10 pounds. A four-valve pentroof engine like the Honda, with intercooling, fuel enrichment, and timing-retard anti-knock countermeasures can be made to handle 9 or 10-psi boost on 93 octane fuel, if the pistons, rods, and block are upgraded for the job, or maybe if you feel lucky. Without intercooling, 6–7-psi boost is certainly achievable without detonation.

CRX 1.6 STRENGTHS AND WEAKNESSES
Effective engine management is critical to the longevity of a CRX.

The defining characteristics of a hot rod CRX Si are front-wheel-drive, 2,100-pound curb weight, and a lightweight 1.6-liter SOHC aluminum Honda engine with 9:1 compression. Honda is known for building a jewel of an engine that is lightweight and powerful for its size. The CRX 1.6-liter engine is a 16-valve design, with a single overhead cam (SOHC) aluminum head bolted to an aluminum block. The cylinders are aluminum, with nonsleeved silica-aluminum (hardened) bores. Pentroof four-valve heads breathe extremely well for the valve lift, and this combustion chamber is an excellent shape for resisting spark knock, always a concern with turbocharged engines. Pentroof chambers provide a very short path from the centrally located spark plug to the farthest end gases in the combustion chamber—improving the chances normal combustion will progressively and smoothly burn

With stock Honda EMS, the first thing we tried was a wet nitrous kit, which operates independently of the Honda EMS. It can be tricky getting enrichment gasoline introduced just ahead of the throttle body to distribute well in a dry EFI manifold. On about the third dynorun, we crisped two exhaust valves.

the air/fuel mixture before any of it has a chance to explode in the increasing heat and pressure of combustion.

Honda's SOHC 1.6-liter aluminum block is lightweight but, unfortunately for hot rodders, less rigid than a production iron block. The problem is made worse by Honda's decision not to sleeve the block. In fact, Honda cylinders are barely thick enough for traditional sleeving, requiring welding around the cylinder bases to provide sufficient strength for installation of steel or cast-iron sleeves on competition blocks. The 1.6 uses a full-floating design, which means there is no continuous flat deck surface that presses against the head surface. The four cylinders are attached together, though each floats semi-independently like an island in the water jacket and presses individually against the cylinder head.

Full-floating cylinders are cheaper to cast, but combustion pressure can shift the cylinders, blow the head gasket, and in extremely high output engines like those found in drag cars, actually crack and blast pieces of the cylinders into the coolant jacket like shrapnel from a grenade. Push the 1.6 hard enough and eventually the cylinders will disconnect from the block. Some of the world's-fastest drag cars do not run coolant through the block at all. The water jacket is filled with cement and epoxy to stiffen the cylinders and protect the block from being destroyed.

Aluminum cylinders in an alloy block tend to distort as you torque down the head, and they distort even more under high piston speed, heat, and pressure. It is more difficult for rings to seal on a cylinder that squirms under really heavy combustion and reciprocating forces. In fact, production alloy-block engines typically make less power than equivalent iron blocks when hot rodded, and they tend to have less margin for handling power upgrades. Severely hot-rodded Honda engines (of the type that can drag-race the quarter-mile in the 10s and 11s) may actually be strengthened with a CNC-billet aluminum support structure like

the Nuformz Block Guard, that installs between the cylinders and block like a piece in a jigsaw puzzle to keep the tops of the cylinders from moving around at extreme power levels.

The CRX 1.6 engine's biggest problem is the cast pistons and lightweight rods, which are designed for a lightweight, high-revving engine making less than 100 driving-wheel horsepower. Stock Honda rods are toys compared with the stout rods we're accustomed to on larger performance engines. According to Alamo Performance, if you can prevent the 1.6 Honda from detonating, then you'll bend the rods first with too much power. Otherwise knock will kill the pistons first. Cast pistons are desirable in production engines because they are quieter and less susceptible to piston-slap due to reduced piston-block clearances (the coefficient of thermal expansion is lower than that of a forging).

Unfortunately, cast pistons are more susceptible to damage from the heat and shock of spark knock. Knock should not happen on a well-designed engine, but when you are exploring the strange new worlds of turbo or supercharger boost and significantly higher specific output (power per displacement)—and tuning fuel and spark systems for boost conditions—knock can happen. If it happens much, especially on a Honda, something will break. According to Alamo, head gaskets may blow, the piston ring lands can break, and as a worst case, cast pistons can actually melt with too much boost.

The stock CRX has almost none of the factory countermeasures found on factory turbo cars, such as overboost warnings, overboost fuel-cut, boost ignition retard, alternate fuel and spark maps, and knock sensor.

The stock Honda exhaust system is a big bottleneck to begin with, even before installation of a turbo. The stock cat will handle perhaps 165 horsepower at the wheels, but on this vehicle you get more power with the cat gone.

The Honda PFI engine management system estimates base fuel requirements (injector open time, also referred to as pulse width) using a speed-density strategy based on engine rpm and manifold absolute pressure (MAP). The MAP pressure is limited to slightly above 1 bar (atmospheric). Boost fuel must be supplied by altering the Honda computer's calibration, by dynamic fuel pressure changes, with additional injector equipment, or with an aftermarket programmable computer system.

According to Alamo Autosports, the Honda CRX fuel delivery system—centrifugal high-pressure fuel pump, regulator, fuel lines, and so on—is designed to supply 100 horsepower worth of fuel plus a small margin, and will run out of capacity quickly on a hot rod turbo conversion. The stock fuel injectors are out of capacity by 6-psi turbo boost, which we figure is about as low boost as it's worth bothering with on a turbo or supercharger conversion. The stock CRX fuel pump has a 60-psi bypass valve, meaning that the fuel pressure cannot be raised enough to support a full 6-psi boost without a replacement or additional high-pressure fuel pump. The stock pump's volume has very little capacity beyond stock horsepower, and essentially all turbo conversions—whatever the fueling strategy—require a fuel pump upgrade. Fuel lines do not require upgrading for high-output street turbo cars.

The bottom line is that it is fairly easy to hurt a Honda engine with serious hot rodding, because it is easier to make power with turbo boost than it is to strengthen the engine, which requires disassembly.

WET NITROUS DISASTER

Nitrous injection has a reputation for leaping tall buildings in a single bound, as the fastest, cheapest path to unmanning bigger engines in faster cars. It also has a reputation for being a loose cannon.

Basically, nitrous is a safer source of oxygen than pure oxygen, and you use it to enrich the O_2 content of ordinary air, which is slightly more than 20 percent pure oxygen. You put nitrous in the intake, add extra fuel, and presto—instant, kick-in-the-pants torque and power. Oh sure, I was going to write about how you've got to be careful, how too much nitrous or too little gasoline can toast your engine in a matter of 2 or 3 seconds and how nitrous can explode the intake or crankcase.

I had a nitrous fogger kit, we know people who've used just such a wet kit on the 1.6 Honda with great results, and the whole thing bolts on in a few hours. This is a binary sort of kit in which a fixed amount of nitrous and enrichment fuel are injected at a single point from one fogger nozzle just upstream of the throttle body. On a CRX, a 50-shot nitrous kit adds 50 percent more power and blasts torque to the stratosphere at low rpm where nitrous can instantly double stock twisting force.

I made a single awesome nitrous run with the fogger kit—pulled the CRX onto the Dynojet, and made a benchmark run at stock power. I brought the engine back up to 2,500, put the hammer down hard to 100 percent, and as the engine revved through 3,000, flicked a dash-mounted arming switch to bring the nitrous on-line. The CRX's engine let out a low-pitched growl like Frankenstein's monster clearing his throat, gained super powers, and spun up those two giant 1,200-pound Dynojet rollers in a nitrous-enhanced dash to the redline, then reverted instantly back to its mild-mannered secret identity as a Honda CRX economy car when I dumped the throttle. We graciously accepted the applause of the onlookers and drove (fast) to the local saloon for beers, where pretty young barmaids draped flowers around our necks and kissed us on the lips.

Unfortunately, that last sentence never happened. As the CRX's roar fell back to an idle on the dyno, it was instantly clear something was very wrong. The car was stumbling and missing like a zombie, hitting on maybe half the cylinders. Following exactly one nitrous run (which actually made pretty good power before nitrous fried the engine), the CRX was left with approximately enough guts to labor off the dyno under its own power and limp across the lot into an empty parking spot where it immediately stalled into a deep coma. The crowd of on-lookers stood around in the sudden stillness, letting out half-hearted rebel yells and making farting noises with their underarms.

As we expected, testing would reveal the 1.6 engine's two center holes had *zero* compression. I hoped the problem was something easy—for example, a blown head gasket. But forensic analysis by yours truly on a marathon Saturday night cylinder-head R&R session would reveal that three of the Honda's eight exhaust valves appeared to have been attacked with an oxyacetylene cutting torch. In the space of a few Saturday night hours at Alamo Autosports, I removed, disassembled, and resurfaced the head, replaced and lapped the bad valves, made sure everything else (pistons, etc.) appeared to be in good shape, then cleaned and block-sanded the deck surface of the block, reinstalled the head, and buttoned up the engine.

Following Saturday night's emergency surgery to the head and valves after the nitrous disaster, we fired up the newly repaired engine at about 2 A.M. Sunday morning. None of us looked too good by this time, but I must have been living right. The engine fired up immediately and ran like a top. The Dynojet proved the 1.6 had almost exactly as much power as before I'd flame-torched three valves. Things could've been much worse—for example, melted or cracked pistons, broken ring lands, squeezed or broken rings, or flame-torched valves.

Analysis: Clearly, we had a relative lack of gasoline in the center cylinders. Possibly this had to do with an interaction between freezing-cold boiling nitrous and chilled gasoline vapors that subsequently condensed into drops of liquid while dodging past the throttle blade or turning corners in a dry intake manifold and departed the air stream, spuriously resulting in a lack of enrichment gasoline in cylinders two and three. Clearly, the Honda's two center holes had received a surplus of nitrous that oxidized the hell out of all the available gasoline and then, with nothing better to do, burned up three exhaust valves where they'd been preheated from previous combustion cycles.

They no longer sell fogger systems for naturally aspirated EFI cars like our Honda. Again, such systems often worked well, but can nuke an engine under the right conditions. When fogger systems didn't work right, the problem was virtually always enrichment fuel starvation at one or more cylinders. The fluid dynamics of freezing-cold vaporizing nitrous and raw gasoline entering through a fogger nozzle in the air intake and then blasting around the throttle plate and finding a way into the howling winds of a long-runner EFI manifold can obviously be somewhat unpredictable.

Keep in mind that old-style fogger systems mix nitrous and fuel in one nozzle near the throttle and spew this fog mixture into the intake charge. You have to hope the air, fuel, and nitrous stay nicely mixed all the way past the throttle and through the dry intake manifold plenum and runners, so that each cylinder gets the proper dose of nitrous, gasoline, and air in one well-mixed homogenous cloud of vapor. The trouble is, liquid gasoline and gaseous nitrous have different fluid dynamics, and they don't corner equally well in a twisting, turning manifold.

Carburetor wet manifolds were specifically designed with special dams, ribs, cross-sections, and gentle turns to keep gasoline from dropping out of the air and condensing in the manifold. Except on race engines, carburetor manifolds are always heated to improve vaporization and keep gasoline suspended in air (which hurts volumetric efficiency, particularly at lower engine speeds).

Modern dry-port EFI manifolds like our CRX's are designed to use inertial and resonance tuning effects to cram the maximum amount of air in the cylinder—with no consideration given to keeping gasoline suspended in air like wet carb or throttle-body-injection manifolds. Modern EFI manifolds are designed to get the maximum equal air charge to all cylinders, but are not required to deal with distributing gasoline evenly to the cylinders (the port fuel injectors do that).

A fogger system would've probably worked fine on a turbocharged CRX engine. Dumping gasoline and nitrous near a turbocharger blowing out hot air at maybe 100,000 rpm is something like dumping a load of manure through the prop of an old B-29 bomber on a maximum-performance engine run-up. Nothing bigger than a quark is going to survive, and you know the stuff's gonna get spread around. Actually, according to NOS and Alamo Autosports, a lot of people have run fogger systems on EFI Hondas with good results, sometimes with 80-shot or bigger nitrous jets. Our engine coulda-shoulda-mighta lived with the fogger nozzle in the *perfect* place.

Bottom line? Nitrous: Be afraid. Be very afraid. People who are afraid of smoking their engines will do everything right and their nitrous engines will make supernatural power and live a long time.

EFI NITROUS: THE SEQUEL

After the first nitrous disaster, we installed a state-of-the-art NOS 5122 50-horse EFI kit on the CRX. This was very much a fail-safe type of kit. You'd have to be really clever to hurt your car with one of these.

This NOS EFI kit injects nitrous liquid near the throttle like a fogger system, in a steady stream, independent of engine speed. But the EFI kit provides enrichment fuel not by wetting the manifold dumping in enrichment fuel with the nitrous but *by jacking up fuel pressure in the injector rail* and artificially raising the manifold pressure reference to the fuel pressure regulator diaphragm.

When you arm an EFI nitrous system and put the pedal to the metal, a throttle-activated microswitch triggers a special nitrous solenoid that delivers pressurized nitrous gas to the fuel pressure regulator via a T-fitting inserted in the reference line (with a check valve to prevent nitrous from entering the manifold through the reference hose). This action substantially raises fuel pressure from the normal 36 psi maximum, to between 60 and 70 psi, so that each fuel injector squirt provides enough additional gasoline to adequately fuel the additional oxygen supplied from the nitrous. Once fuel rail pressure reaches at least 50 psi, an NOS pressure switch installed in the fuel rail or supply line opens the second regulator. Located downstream of the first solenoid, the second regulator actually delivers nitrous liquid to an injection nozzle in the intake tract that must be located 4–12 inches upstream of the throttle body.

Always install the fail-safe fuel pressure switch. If fuel pressure ever drops below a threshold of 50 psi, the NOS pressure switch will immediately cut off nitrous flow, preventing lean-mixture engine damage. So what conditions could contribute to rail pressure drop? Start with the engine's normal fuel requirements when high rpm and loading result in a rapid-fire series of injections of long pulse width. Add to this air-density increases due to cold weather or lower altitude that add even more to required pulse width, stressing fuel requirements to the maximum stock level. Combine this with an armed and activated EFI nitrous system immediately resulting in a sudden requirement for much higher fuel pressure. Fuel pumps can lose efficiency due to aging, and this could be the month that pumping losses reach a critical threshold. Or there could be a minor—almost invisible—fuel leak that's the final factor that causes rail pressure to drop or surge under the right set of circumstances. So install the fuel pressure switch. A lean nitrous mixture can kill an engine so fast you won't believe it.

The NOS EFI nitrous kit only puts nitrous in the intake manifold for oxygen enrichment, not fuel. Injected nitrous liquid boils almost immediately into a gas (simultaneously chilling the intake air and making additional "free" power available via a denser air charge). Gaseous nitrous behaves exactly like air in the manifold, simply enriching the oxygen content of the intake charge by a few percent and hitching a ride with charge air into the cylinders. By the time the nitrous reaches the intake valve, it has warmed up enough that it has little detrimental effect on gasoline vaporization. When nitrous is heated to 565 degrees Fahrenheit in the combustion chamber, the oxygen and nitrous molecules break apart and the oxygen becomes available for combustion.

How much additional power can your CRX-vintage street Honda engine stand before destroying itself? Experts like David Vizard used to say you shouldn't add more than 30 to 50 percent more power to an engine unless you upgrade the engine's internals (particularly the pistons and rods). According to NOS, 50 horsepower is the maximum reliable increase you can use on a stock 1.6 Honda, though NOS knows there are people out there driving time-bomb stock Hondas with 60- or 80-horse port-injection NOS systems that are still living. These systems are logically similar to our 50 system, but with larger nitrous jetting, recalibrated nitrous pressure to the Honda fuel pressure regulator for increased fueling, and individual nitrous lines running to each intake runner with separate nozzles installed for each cylinder.

The NOS EFI nitrous system was advertised to deliver 50 crankshaft horsepower. Testing on the Alamo Autosport Dynojet chassis dyno revealed a gain in peak power of 38 *rear-wheel* horsepower—increasing power from 96 to 134 horsepower with the nitrous active. This is a substantial increase, considering that a new 1989 CRX Si had 108 advertised crankshaft horsepower.

The really great thing about nitrous injection is that the entire 50 horsepower (and additional torque) is all available instantly—and all at once—at 2,500 rpm (the lowest activation rpm recommended by NOS), and linearly provides this continued boost all the way to redline. Actually, when the nitrous hit, we were somewhat worried about clutch slip with the car strapped hard to the Dynojet where tire slip is virtually impossible no matter what the torque boost.

BOOSTED-ENGINE ANTI-KNOCK COUNTER-MEASURES

When Alamo installed the EFI NOS system on the CRX, they added two devices to increase the safety margin of the system. One was a 550-horse inline electric fuel pump from Nitrous Oxide Systems, and the other was a J&S Safeguard individual-cylinder knock-retard computer/coil-driver. The NOS in-line fuel pump can be installed anywhere upstream of the fuel rail, but Alamo preferred to install it downstream of the fuel filter. Alamo installed it in the Honda's engine compartment and wired it to an automotive relay so that it only ran when the nitrous system was armed (by turning on the dash-mounted arming switch). Of course, Alamo verified that the auxiliary fuel pump did not interfere with the pumping of the stock EFI fuel pump when the auxiliary pump was off.

We wired the Safeguard into the stock Honda ignition, and in this configuration the Safeguard could remove up to 20 degrees of timing in 2-degree increments from any or all individual cylinders found to be knocking via a crystal microphone (knock sensor) screwed into the block or intake manifold. NOS specifies that the 50-horsepower Honda EFI nitrous system will work with stock timing and premium street fuel. In case of additional heat and pressure from nitrous combustion or bad gasoline-induced detonation, the Safeguard is designed to kill knock before it can damage the engine. A dash-mounted LED meter lights one or more indicators according to how much timing is removed from one or more cylinders. An adjustment screw permits adjustment of the sensitivity of the Safeguard so that normal engine sounds do not trigger spark retard. The J&S indicator module can also display air/fuel mixtures if you install an auxiliary O_2 sensor in the exhaust system in or near the manifold.

The knock threshold for a cylinder depends on a lot of factors; rarely would all cylinders begin knocking at once, and the great thing about the J&S is that it listens for knock, correlates the knock to a particular cylinder or cylinders, and only retards ignition timing on the slackers causing trouble rather than retarding all cylinders due to one or two problem areas. Individual-cylinder knock retard maintains higher power and lower coolant-jacket temperatures, while protecting the engine from knock-induced damage. Knock is knock, meaning that the Safeguard can protect from boost-induced knock just as well as nitrous-induced knock.

The J&S includes an auxiliary wiring harness, processor module, indicator module, and the knock sensor, as well as power and ground leads. Tapping into the low-voltage ignition wiring located inside the CRX's integrated distributor/coil module as Alamo did is the best method to supply power to the module. Ideally, the knock sensor should be mounted at the top center of the block, but we didn't find a convenient threaded hole and instead mounted it on the driver's side of the block, near the end of the block.

NITROUS FUEL THROUGH THE INJECTORS: NITROUS TIME VERSUS INJECTION PULSE WIDTH VERSUS RPM

Since nitrous injection occurs in a steady stream per time, the amount of nitrous arriving in a given cylinder varies in a way that is directly related to engine speed and the time available to make a full four-cycle power stroke. During this time, nitrous gas is flowing constantly, constantly enriching the air with oxygen. Ideally, total enrichment fuel per power stroke should similarly be automatically modified according to time available for nitrous delivery each power stroke, meaning according to engine speed.

By contrast, electronic fuel injection delivers one (or two) injector squirts of a certain computed length of time per power stroke, correlated according to the engine's torque curve and actual engine loading. The amount of fuel injected is *not* directly affected by time available per power stroke (unless the injectors reach 100 percent duty cycle, which you normally never want to happen). Of course, since engine volumetric efficiency (cylinder filling) is affected by engine speed and throttle angle, time has an indirect effect on injection pulse width. But since nitrous is normally only injected at wide-open throttle, engine loading does not effectively change injection pulse width and is not a factor that must be considered in calculating ideal enrichment fuel delivery.

However, since the delivery of nitrous *is* steady over time, increasing rpm will slice the amount of nitrous actually delivered to a cylinder into smaller and smaller amounts as engine speed increases.

Actual wide-open-throttle (WOT) fuel-injection pulse width as calculated by a Honda computer will increase with rpm as the engine approaches peak-torque rpm, then decline somewhat as the torque curve noses over a bit as rpm approaches peak power.

Computer-controlled nitrous fuel enrichment logic must take into account the time available for nitrous injection in computing the required delta in injection pulse width needed to fuel the nitrous. When this is not done, the air/fuel mixture becomes increasingly rich as engine speed increases due to the falloff in time available for nitrous (oxygen) induction per power stroke.

Of course, EFI nitrous systems would cost a lot more if they all required programmable fuel injection. Fortunately, the spread between air/fuel ratios producing rich and lean best torque for gasoline is fairly wide. And the acceptable air/fuel ratio range gets even wider on the rich end if you're willing to tolerate suboptimal power increases, and excessively rich mixtures are a kind of cheap insurance, if you want to look at it that way. Most EFI nitrous kits live with increasingly suboptimal over-enrichment as rpm increases.

For practical purposes, as a nitrous run proceeds, you want the air/fuel mixture to get increasingly rich for safety's sake, and if it gets a little richer than ideal, that's OK given the cost advantages of a simple mechanical system like the NOS EFI nitrous setup.

In the real world of street nitrous systems for street cars with stock engines and stock redlines in a very price-elastic marketplace, increasingly rich air/fuel ratios are a fact of life on systems where the priority is to deliver a bunch of dirt-cheap power rather than wring out every last horsepower.

NITROUS POST-INSTALLATION CHECKLIST

• Verify the nitrous and/or fuel jetting sizes.
• Change the fuel filter.
• Purge the nitrous line of any debris.

• Verify the nitrous-off fuel-rail pressure (36 psi with vacuum reference hose disconnected).
• Pressure Make sure the pressure gauge is in place to verify fuel pressure rise during nitrous injection.
• New Install new spark plugs installed one to two heat ranges colder.
• Ensure that the Nnitrous tank is oriented per instructions so G-forces don't interrupt the flow of liquid nitrous on hard acceleration.
• Retard Static static spark timing retarded 2 degrees
• Consider optional low-rpm limit switch to prevents nitrous engagement at too-low rpm (some Hondas may have trouble providing tach signal).
• Perform a compression test/leakdown on engine (before and after).
• Flow-test fuel injectors for uniformity and high-flow rate.
• Use unleaded race gas only in fuel tank (at least initially, then 92-plus only on street).
• Inspect the stock ignition system (install new plugs, wires, distributor cap, rotor).
• Upgrade coil/, plugs/, and wires (always a good idea on power-modified engines).
• What happens if the tank vents from over-pressure? Will the car fill with nitrous-plus and possibly suffocate you, cause you to crash, or make you nauseated from sulfur dioxide inhalation?
• Are anti-knock countermeasures in place? (J&S Safeguard knock sensor or other boosted-conditions timing retard device)?
• Does the fuel pump provide sufficient volume?
• Is nitrous tank pressure OK?
• Is a fire extinguisher on hand?
• Is the nitrous system armed?
• Is there any chance nitrous has been released into the intake manifold during shutdown (can cause huge explosion on startup)?
• Is the engine above 2,500 rpm before full-throttle nitrous run begins?
• Use the smallest nitrous-shot as a starting place
• Is the clutch up to the job (depending on tires)?
• Is the nitrous nozzle perpendicular to throttle and centered in air intake or throttle body?
• Is an extra car available, just in case you smoke the engine?
• Is a change of underwear available, in case you don't?

THE TRUTH ABOUT NITROUS OXIDE INJECTION

Simply put, nitrous oxide is not a high-powered fuel, but rather a source of additional oxygen for an engine, enabling more fuel to be burned. Nitrous does not make power. Additional fuel burned by nitrous-supplied oxygen produces more power. Nitrous oxide consists of two atoms of nitrogen and one of oxygen. Air is, for practical purposes, 80 percent nitrogen, which is an inert gas that is just along for the ride. Nitrogen does not participate in combustion but only exerts pressure against a piston due to expansion as it is heated by the burning of fuel and oxygen. Oxygen—and only oxygen—is what burns fuel.

Most people consider oxygen too dangerous to inject into an engine in its pure form (although this has been tried) due to its combustible nature. The reason for not injecting pure oxygen begins with the fact that too high a concentration of oxygen produces dangerously hot and fast combustion that can kill an engine in a matter of two or three seconds. But only about a third of the weight of pure oxygen as nitrous oxide is necessary to achieve an equal oxygen concentration in upgraded air, making pure oxygen injection more sensitive to precise metering—meaning small tuning mistakes could be catastrophic. What's more, liquid oxygen

The next trick we tried was switching to an NOS "dry" nitrous kit, which introduces fuel during nitrous injection by jumping up the fuel pressure to the stock injectors. If fuel pressure ever drops off to the reference port on the blue fitting, it immediately chokes off the nitrous flow, saving the engine. No problem.

requires cryogenic handling techniques, while bottled oxygen *gas* lacks the intercooling advantages of working with bottled liquid nitrous. No one uses pure oxygen injection.

Nitrous oxide injection is safer because each oxygen atom in nitrous is strongly bonded to two nitrogen atoms, and the oxygen is normally unavailable for combustion. Not only is the oxygen unavailable at ambient temperatures, but when the nitrous molecule becomes dissociated, the freed oxygen is heavily diluted with molecular nitrogen. It is not until initial heat of combustion heats nitrous oxide above 565 degrees Fahrenheit that the oxygen molecule dissociates from the nitrogen and becomes available for oxidizing fuel. This delay produces a buffering effect on violent combustion and an added margin of safety, though the nitrous dissociation itself produces heat and the additional concentration of oxygen in a nitrous-air mix immediately induces faster, hotter combustion in nitrous engines.

Nitrous oxide is a liquid at standard pressure between -128 degrees and 97.7 degrees Fahrenheit, at which point it vaporizes into a gas. At temperatures below 97.7 degrees, it is possible to extract either liquid or gaseous nitrous from a nitrous tank, depending on the location of the pickup in the tank. Liquid nitrous exiting from a tank will expand rapidly due to the pressure drop from 760 to 14.7 psi (atmospheric pressure at sea level), and will consequently undergo substantial cooling, becoming super-chilled. This chilling effect intercools inlet air (improving volumetric efficiency), as well lowering combustion temperatures—helping to offset nitrous' tendency to increase the possibility of detonation.

Nitrous oxide is 36 percent oxygen by weight. The chemically correct nitrous-gasoline ratio is 9.6:1, and fuel management for nitrous injection is often handled as an independent variable from the normal fueling system. For example, an entirely separate fuel delivery system such as a spray bar under a four-barrel carb, additional fuel injectors, and an independent calculation and addition to injection pulse width for nitrous fueling is necessary. At the very least, an independent increase in system fuel pressure designed purely to provide additional fuel for nitrous to combust should be installed.

However, nitrous and fuel for nitrous are not actually entirely independent of air and normal fuel intake, because nitrous (particularly gaseous nitrous) and nitrous enrichment fuel will displace air in the intake tract and combustion chamber and affect the total oxygen-fuel mixture ratio. In addition, the intercooling effect of liquid nitrous injection can cool the intake air enough to signifi-

cantly affect the density of the intake air charge. These factors must be considered in computing the amount of fuel to provide for the nitrous-injection conditions.

In addition to basic combustion fuel requirements for nitrous-enhanced air, the total fuel requirement (nitrous enrichment and normal fuel) almost always includes an additional excess fuel factor beyond the chemically ideal mixture for enrichment and normal fueling—to help keep combustion temperatures down. Excess fuel, which cannot be burned due to a lack of oxygen, soaks up a certain amount of combustion heat. Similarly, lean mixtures can be disastrous in a nitrous motor. With no excess fuel to keep combustion temperatures down, hot spots can appear in the combustion chamber, which in turn can cause the onset of pre-ignition and detonation, further raising temperatures to the point where combustion chamber metal, valves, and piston surfaces can become red hot. Excess oxygen will attack the red-hot surfaces and burn them away exactly like an oxidizing acetylene torch.

Lean nitrous-fuel ratios must be avoided at all costs. With gasoline, the maximum horsepower in a normally aspirated engine occurs at air/fuel ratios 15 to 20 percent rich of stoichiometric. Under these conditions, additional power is gained by the intercooling effect of gasoline's heat of vaporization (which improves engine volumetric efficiency) and by improved combustion characteristics in which the excess gasoline improves the odds that all oxygen molecules will find fuel to combust. Nitrous motors burn a gasoline-fuel mixture with enhanced oxygen content, but less nitrogen, which means there is less inert nitrogen to dampen combustion temperatures, automatically causes combustion temperatures and flame speeds to increase.

There is no such thing as factory nitrous injection. Nitrous Oxide Systems (NOS) almost owns the market for add-on nitrous injection, to the extent that NOS is nearly a generic term for any nitrous injection. Low-power nitrous kits (up to 150 additional horsepower) are simple and cheap. It has been nearly 10 years since Detroit has built a carbureted engine, but there are still many V-8 hot rods running four-barrel carburetors with fogger or spray-bar nitrous systems. Modern EFI vehicles with port fuel injection use intake manifolds with long tuned port intake runners and a large plenum.

A dry manifold setup, which not designed to handle wet (air/fuel) mixtures, will probably cause fuel distribution problems with a spray-bar type system introducing fuel into the manifold. Dry manifold nitrous systems inject nitrous at the throttle body or into individual ports, using a fuel pressure regulator to massively raise the fuel pressure in the injector rail so that even though injection pulse width (open time) remains the same, each squirt injects much more fuel. Typically, nitrous kits cost $500 to $1,000, and include a nitrous bottle and fittings, braided-steel/Teflon hose, and additional items. A black box, custom computer chip, or other means of retarding ignition timing under nitrous combustion conditions is necessary. This is true because combustion proceeds much faster after oxygen dissociates from the nitrous oxide.

The most advanced nitrous kits use programmable engine management systems that activate and deactivate nitrous based on various engine parameters such as throttle angle, manifold pressure, and rpm. Some kits also factor nitrous activation into spark timing and provide all fuel management from the primary injectors using fuel algorithms, which make fuel injection synchronous with rpm and nitrous injection asynchronous with rpm. Some computers can activate several stages of nitrous or even cycle the nitrous solenoid rapidly on and off to regulate the amount of nitrous reaching the engine, which is useful in drag racing to prevent smoking hell out of the tires for faster times.

REAL-WORLD TURBO CRX Si

Ignoring complex high-end racing systems, nitrous is cheap and easy to install on nearly any vehicle, and easily makes 10 or 20 real additional horsepower per cylinder. On the other hand, nitrous is only useful for a limited number of killer acceleration runs, after which the nitrous tank must be refilled. Nitrous can damage engines when pushed to super-high output levels. Since most nitrous systems are 100 percent off or 100 percent on, nitrous can cause uncontrollable wheel spin in drag racing situations. Nitrous systems use fuel pressure changes to increase the fuel output of an EFI system are relatively limited in total amount of power to approximately 50 percent torque increase; beyond that, addition fuel must be provided by more complicated additional injectors or computer reprogramming, and so on.

Nitrous injection can be combined with turbocharging and supercharging very effectively. It is particularly well suited to turbocharging because it provides its maximum percentage increase in torque at low end, which is the opposite of turbocharging. Many turbo setups use a jolt of nitrous to get the car moving and generate a large blast of hot exhaust gases to spool up the turbo(s) more quickly, with the nitrous then deactivated at 3–8 psi turbo boost.

Most people would consider a nitrous kit a relatively easy installation for the average performance enthusiast. Ordinary tools will do the job. But a potential downside of installing nitrous is the possibility of making mistakes. Since you're modifying the vehicle's fuel system and radically increasing power, there could be a serious fire or major engine damage. Other than that, nitrous is a breeze (so to speak).

TURBO BOOST: 148.2 HORSEPOWER OF LOW-BUCK, LOW-BOOST, NON-INTERCOOLED TURBO POWER

With the NOS kit temporarily disconnected, using a nonintercooled Alamo Autosports/RD Racing turbo conversion system, we made roughly 150 horsepower at 7-psi boost using a Turbonetics Super-60 turbo. Based on nonoptimized project CRX dyno results using a Turbonetics Super-60 ceramic ball-bearing turbo, we were sure that with equal modifications in place, the ball bearing unit would make 160 wheel horsepower—an estimated 185 horsepower at the flywheel!

Turbo systems work not so much by coaxing in additional air, but by jamming it down the engine's throat under pressure, and then (in the simpler turbo kits, such as this one) increasing fuel pressure as a multiple of increasing boost pressure to match the increased air charge with additional fuel.

This CRX turbo kit retained the stock Honda computer EFI and stock engine management system and used a variable-rate-of-gain fuel pressure regulator to provide fuel enrichment under boost conditions. The VRG regulator installed in-line downstream of the stock regulator and increased fuel pressure as a multiple of boost pressure (producing more fuel per injector squirt). You really need an add-on electronic ignition timing-retard device to fight detonation under boost. The alternative was static timing distributor retard, but this would definitely hurt performance in the normally aspirated and low-boost ranges.

We already had a J&S Safeguard left over from EFI-nitrous experiments on the CRX, so no problem there. We used a boost gauge mounted on the A-pillar to provide the driver instant feedback on manifold pressure. Our turbo kit was designed to allow retention of the stock catalyst, muffler, and exhaust system—though there was plenty to be gained with exhaust-flow upgrades. The turbo system initially used a Bell Engineering fuel management unit (VRG regulator) for boost-conditions fueling and a J&S Safeguard to provide boost ignition retard on an individual-cylinder basis when

required. Our plan was to test the CRX's power with the most basic system, nonintercooled at 6 psi, then crank up the volume later with intercooling, programmable EFI, and so forth.

A Honda CRX is equipped with computer-controlled port EFI of the speed-density variety that uses a 1-bar Honda manifold absolute pressure sensor. Rather than directly measuring engine airflow with a mass airflow (MAF) sensor, the Honda's computer estimates airflow based on intake air temperature, manifold pressure, throttle angle (which correspond to engine loading), known engine displacement, and engine speed. These values are used to index into an internal table of numbers corresponding to the correct injection pulse width for a list of possible engine speeds and manifold air density. According to Hondata, the stock Honda MAP sensor is actually capable of detecting an 11 psi (0.75 bar) boost.

The CRX 1.6 engine is equipped with four-valve pentroof combustion chambers, which are very effective at breathing deeply and resisting detonation. The CRX Si's static compression ratio is over 9:1, which is enough to make the torque you'll need to have reasonable low-rpm grunt at engine speeds below the threshold of sufficient exhaust energy to generate any turbo boost—before the turbo spools up and boost arrives, where you're still effectively naturally aspirated. On the other hand, this compression ratio is not so much as to constitute a serious pro-knock problem at higher levels of boost. The Honda engine block is not particularly robust for super-high-output (actually, it's quite robust at stock levels of power, but definitely not suitable for power levels vastly above stock).

Most experts agree that you can run 7-psi boost on a stock 1.6 CRX Si all day without problems by using a well-designed basic turbo system, which will not require an intercooler (though an intercooler will increase performance). A more sophisticated turbo conversion that never permits knock can probably survive 9 psi, maybe 10, using a very good intercooler and great engine management air/fuel and spark calibrations. A really good individual-cylinder-retard knock sensor like the J&S Safeguard is excellent engine insurance, and we've used it to great effect with the NOS nitrous kit and now with low-boost turbocharging. Of course, using race gas during testing or competition adds additional engine life insurance or permits increased levels of power.

The consensus seems to be that no matter what you do, with a stock-block Honda 1.6, you're living on borrowed time at boost levels above 9 or 10 psi. However, some people will point out that some Honda engines just seem more robust than others: Some blocks seem to live forever at 10-psi boost (assuming *rigorous* detonation control), while others seem to quickly go away at 8 or 9. The problem is, CRX rods, pistons, and block structures are *toys* compared to the super-duty components typically found on factory turbo engines, and they absolutely *require* upgrading or strengthening for more than 7-psi boost.

REAL LIFE HIGH-BOOST MAYHEM

Here's what typically happens when you turbocharge your "toy" Honda. Six or 7 psi makes at least an extra 55 driving-wheel horsepower, that is, more than 60 percent more power than the stock CRX Si's 90–96. And it feels great. At 2,150 pounds and almost 150 horsepower (14 pounds per horsepower), your CRX is now quicker than plenty of cars on the road, and driving is suddenly a whole lot more fun.

For a while.

Then, one fine day, your quick turbo CRX gets blown into the weeds by some big-dog V-8 pony car. Even before your adrenaline-drenched brain and heart-rate get anywhere close to normal, you're already thinking, with a little bit more boost, I could have had him. The next morning, even before the sun is fully awake, you're out in the driveway tightening down the wastegate spring, installing an air bleed in the wastegate manifold pressure reference line, or installing an electronic boost controller or some other neat trick to ramp up maximum boost.

On the initial test runs, you're careful at first, listening for detonation, gradually tweaking the boost to 7, or 8, maybe even 9 psi. But that extra boost feels good when you put the pedal down.

And you *will* put the pedal down.

Here's how it happens: You're cruising, perfectly mellow, and suddenly your heart stops. There's that breathed-on V-8 pony car that you hate. And this time things are just the same. Only different. This time your girlfriend is riding in the passenger seat. This time, as the 'Stang starts to pull away, you nail the gas pedal and simultaneously reach down to the electronic boost controller display and pump up the volume (boost). As the boost goes nuts, you can feel the J&S taking out ignition timing by the truckload to kill knock, the engine trying to die with up to 20 degrees of timing retard—so you reach down and turn the sensitivity on the J&S all the way down.

Instantly, the CRX growls like a weightlifter on crank and steroids trying to snatch too big a load, and you're thinking, I can't *hear* any knock (forgetting that high-rpm, high-boost spark knock is often impossible to hear). The next moment, with a mighty lunge, the little Honda pulls even with the pony car—Mustang/Z28/Whateveritis—and then lunges ahead into cool, clear air. (Pretend it's a Mustang: As you go by, you turn and stare at the driver, mouthing the words, "Ford: Fix or Repair Daily.") Maybe the guy figures you said another word starting with "F" or doesn't like your face or something, because you notice he's shaking his fist at you.

You smile, and—just for fun—as you pull farther ahead, you hold up your fist, middle finger erect. You've now arrived at a plotpoint in life: You're the King of the World, girlfriend beside you gazing adoringly up at you, a gleam in her eyes that means . . . well never mind. Actually, you're not sure what it means, but you figure it has to mean *something*. Anyway, there you are, the two of you flying through clear air, arms metaphorically outstretched like Leonardo Di Caprio and Kate Winslet on the bow of the Titanic, whooping it up, thanking the god that created turbochargers.

And then, like a beautiful dream that dissolves gradually into a nightmare, something changes deep in the Honda's heart, and you can feel something wrong in the engine's song. For an instant you're not certain, but now you can definitely feel a weariness taking your Honda's engine like a fatal disease. You can feel the CRX starting to lose it, power leaking out from under your right foot like the CRX just got figuratively kicked in the groin, hard. If you had even one brain cell left that was not drunk on testosterone, you'd back off the pedal instantly and pull over. But that last brain cell is screaming, "No fear!" or maybe "Go for it!" or maybe even "Kill!" and your foot's still glued to the floor.

Unfortunately, by now the Honda 1.6 is fully involved in an unscheduled work stoppage. The Mustang flashes by, and the guy riding shotgun is flipping you off. As the Mustang takes the corner ahead, you catch sight of the guy in the back seat—hairy butt pressed hard against the rear window. You glance over and notice that your girlfriend's expression is no longer adoring (but to hell with that, something's wrong with your baby—the Honda, that is). You pull over and jump out for a look and a listen: Damn! The engine is missing and shaking, and it's spitting water out the exhaust. And is that steam, or maybe smoke drifting out the 3-inch chrome tip exhaust? But you're still smiling, thinking, I whupped the son of a bitch, and it was worth it! Yeah, it was worth it. It was sure worth it. . . . Worth it. . . .

Your testosterone-enhanced reverie is interrupted by the sound of the passenger door slamming. Your girlfriend is walking toward you. Smiling. Cool. But wait . . . What's this? She's smiling past you. What the . . . ? You turn, and there's the Mustang rolling to a stop. Some guy is holding open the door. "Baby, wait," you whisper. But it's too late, she's lost to the heavy metal snarl of a Mustang five-point-oh V-8, and a moment later you're eating tire smoke.

That's how it'll happen.

"I blew a head gasket!" you tell yourself. You got that right, Sport.

And maybe your cylinder head gasket really is leaking, but probably not because the gasket blew out. Unfortunately, further analysis will probably reveal that one or more of the Honda's floating cylinders has shifted downward a few thousandths of an inch from the head surface, allowing coolant and combustion gases to escape the water jacket and combustion chambers. That's not all: You'll probably find that the main journal support structures—which, by the way, help keep the cylinders in place—have also migrated. You may discover broken rings or ring lands if the Honda was knocking severely as well as overloaded mechanically.

OK, maybe you were just cruising when the day turned bad, maybe you were actually being good for once. Maybe you're sure the engine was *loafing* when the problem seemed to come from nowhere, the engine just suddenly missing, stumbling, spitting water for no apparent reason. That intense and rare feeling of virtue that arises from driving legally and sanely—for once—has overwhelmed the memory of scores of past-tense heavy-duty full-boost kamikaze-style banzai acceleration runs with no apparent consequences. This has started to give you the idea that your overboosted CRX truly *is* immortal and that a 10- or 12-psi boost actually is a "divine wind.") Either way—guilty now, guilty then—to use a technical term, you're screwed.

Honda engines do *not* take well to boost abuse, and once cylinders and main support webbing have moved, your Honda engine will never be the same. The technical term in this case is *totaled*. You may burn up money trying to machine things back to righteousness in the block, but fix one thing and you'll find something else moving out of whack, and after you fix that, that something else shifts out of place. And so on. Full-floating Honda cylinders move kind of like a landslide moves: Once it starts, it's not gonna quit until you've got a pile of rubble in hell.

Read my lips: None of this needs to happen if you keep the CRX 1.6L's boost to the 7-psi range and don't tempt yourself with a cockpit-adjustable boost controller. When you're ready for the big power, dismantle the engine and strengthen it internally. A 7-psi boost on a 1.6 CRX makes a dramatic difference in power, a dramatic difference in the car's power-to-fun ratio.

With a turbo kit on your CRX, you've essentially got a new car. We tested several different turbo system configurations, including Turbonetics ball bearing and conventional T03 turbos, a hot engine-compartment air versus isolated cold-air intake system, and a full-cat muffler and stock exhaust versus complete exhaust dump. In general, turbo boost comes in early, builds very quickly to 7 psi, produces greatly improved average torque, and makes peak power and torque with the conventional Super-60 turbo of nearly 150 wheel horsepower at 6,400, and 128 ft-lbs wheel torque at 5,000 rpm (and there'll be another 10 wheel horses with the Turbonetics ball bearing turbo installed).

The turbo CRX was much stronger from just above idle all the way until the redline and rev-limiter hit at 6,500. Compared to the stock CRX's 92–96 horsepower, this is a gain of more than 60 percent power—on "free" air! This is more than we expected from a nonintercooled turbo kit and almost double what we obtained from the nitrous system (though that could change in a hurry by swapping a jet or two on the NOS kit). The thing is, turbos and nitrous make a different kind of power. Nitrous power comes in all at once, instantly, with the greatest impact at lower rpm, while the turbo power surges into action over a larger rpm range, continuously increasing engine power and torque as exhaust and intake gases feed back and boost increases with rpm to the max.

THE TURBO CONVERSION

Fuel Enrichment: Engines converted to turbocharging *always* need additional fuel under boost conditions merely to maintain stock wide-open-throttle (WOT) air/fuel ratios, and they virtually always need richer than stock air/fuel ratios to realize full power potential and fight detonation.

The Alamo turbo conversion initially provided fuel enrichment during boost using a Bell Engineering fuel management unit (FMU) installed downstream of the stock Honda fuel pressure regulator in the fuel return line. The FMU is an in-line variable-rate-of-gain fuel pressure regulator. The FMU is normally adjusted to have no effect during naturally aspirated conditions at idle or light cruise. Under boost conditions, the FMU increases fuel rail pressure as an adjustable multiple of manifold pressure in a range configurable from 4 to 7 pounds increase in fuel pressure per pound of boost. Higher fuel rail pressure causes injectors to inject more fuel per timed squirt, maintaining or richening the air/fuel ratio.

An FMU-enriched turbo system's maximum boost is limited by the inability of fuel injectors and driver circuitry to function reliably at levels of fuel pressure much over 80–100 psi (essentially limiting maximum boost to 5–10 psi). Another constraint is the fact that all electric centrifugal high-pressure fuel pumps deliver decreased maximum fuel volume with increasing rail pressure. Some fuel pumps are equipped with a high-pressure bypass valve that firmly limits maximum pressure. In many cases, even a low-boost turbo conversion will require an upgraded or auxiliary fuel pump.

Careful testing on the Alamo Dynojet revealed that the stock CRX fuel pump reaches its volumetric limit (at the supernaturally high fuel pressure of the FMU) at roughly 135 driving-wheel horsepower. As discussed elsewhere in this book, the method for determining unknown fuel pump capacity—and remember, aging fuel pumps may no longer flow according to design specs—essentially involves energizing the fuel pump with the engine stopped and artificially raising fuel rail pressure at the regulator with a hand-operated air pump. Measuring regulator fuel output to the return line per time reveals the maximum amount of fuel that can be removed from the rail by injectors (or the regulator) before rail pressure falls (which can be a piston-destroying disaster).

An FMU-fueled turbo conversion is *not* compatible with the NOS EFI kit's method of using nitrous pressure to drive up fuel pressure at the regulator. (Of course, the stock engine block itself almost certainly can't handle both nitrous and turbo power at once.)

Alamo had previously installed a 550-horse NOS inline auxiliary fuel pump along with the EFI nitrous system, so that it ran only when the nitrous system was armed. For the turbo conversion (and possible future reintroduction of nitrous injection), we rewired the auxiliary pump so that a relay activates the nitrous oxide system's pump either when the nitrous system was armed or when manifold pressure exceeded 2 psi.

The plan was to shortly install a Gen 6 DFI engine management system, and we figured that once the project CRX's DFI programmable computer was installed, we'd remove the FMU and experiment with using transient nitrous at wide-open throttle to make instant torque and spank up the turbo quickly toward full boost. We'll cut out the nitrous at 3–5-psi boost to keep extreme

combustion pressures from hurting the engine (a side benefit of this method of nitrous use is that a tankful of N_2O lasts forever when used in this limited way). However, combine nitrous and full boost at peak power and boost will kill your stock Honda 1.6 engine! Wait until you upgrade the engine with internal modifications.

Turbocharger: The Alamo Autosports turbo system on our project CRX was designed to use a Turbonetics T03-type turbocharger—specifically using a Turbonetics Super-60 compressor. (For those of you old enough to remember the Mustang SVO 2.3-liter Turbo, the Super-60 compressor wheel is a proprietary Turbonetics design a little larger than the standard Garret 60-trim wheel found on the SVO.) The turbine section uses a 72 trim turbine wheel (standard T03 trim) and a 0.48 A/R turbine housing, which is quite small, for improved low-end response on a 1.6-liter motor that's on the small end of the Super-60 envelope. The Super-60 is available from Turbonetics in both ceramic ball bearing and standard bushed designs. Turbonetics has repeatedly demonstrated the ball bearing turbo's free-spinning advantages in increasing engine performance via faster spool-up and higher boost pressure per engine rpm. The company has also proven the ball bearing unit's ability to survive extremely high side-loading forces in race-type conditions, withstanding as much as 50 times greater loading without failing compared to the conventional bushed turbo design.

The Turbonetics Super-60 is designed to efficiently deliver 220–485 cfm and is usable on 150- to 335-horsepower engines. When everything is perfect—using an extremely efficient intercooler—the Super-60 compressor is capable of making as much as 350 horsepower, or roughly 500 cfm (which does, however, clearly put the turbo on the hairy edge of over-speed and compressor choke).

We made Dynojet runs with both types of turbo. Our experience was that the Super-60 BB is always the equivalent of a few hundred engine rpm ahead of the conventional Super-60, with the corresponding dyno power graphs almost parallel and the ball-bearing turbo's power and torque constantly ahead by several hundred rpm worth of power and torque all the way up to redline.

The CRX was now unquestionably more fun to drive with the Super-60BB, accelerating harder and sooner, with the car's gas pedal much more responsive. Peak power near redline was up by 10 driving-wheel horsepower with the Turbonetics ceramic ball-bearing unit. We ran the ball-bearing turbo first, then switched to the standard unit, saw the 10-horse loss (from 135 down to 124.6), then experimented with intake, exhaust, fuel supply, and tuning mods that brought the standard turbo power up by almost 25, to 148.2 wheel horsepower.

Time constraints prevented our switching back to the ball-bearing unit, but we estimated maximum power would have maintained or increased the 10-horse advantage of the ball-bearing unit, thereby producing almost 160 front-wheel horsepower on race gas at 7-psi boost with no intercooling. Performance on the street with 93-octane premium was subjectively equal to that with 100 octane, except on the few occasions when the J&S Safeguard took out timing to kill detonation in 100-plus degree weather in south Texas.

Wastegate: The Alamo turbo system uses an external Turbonetics Deltagate to limit boost to 7 psi (which, again, I consider the highest safe level of nonintercooled boost-pressure for the relatively fragile Honda CRX Si engine). The Deltagate is a special dual-ported wastegate. In conventional mode, manifold pressure is routed to an exhaust-side diaphragm port that can be used to push open the wastegate poppet valve to control boost by diverting exhaust gases around the turbine section entirely, effectively reducing the power of the turbine and, therefore, the maximum cfm capacity of the compressor. The secondary Deltagate port, located

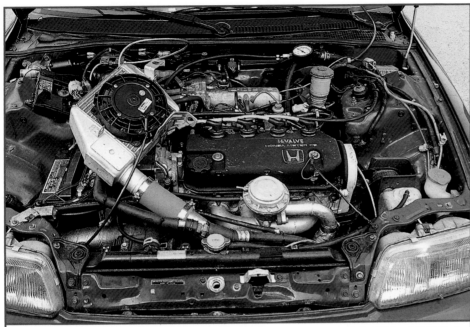

When we pumped up the boost to higher levels and added intercooling, we installed a DFI Gen 6 EMS. Back on the dyno, we tested with a fan-driven air-air-intercooler.

turbo compressor to the Honda's throttle body, mounting an HKS pop-off valve to prevent compressor surge on sudden throttle closing. We wanted to try nonintercooled operation first.

The Honda computer will freak out if the stock MAP sensor reports higher-than-atmospheric manifold pressure (and it can measure at least 0.6 bar of positive MAP pressure!), turning on the check engine light and switching to limp-home mode. To prevent this, Alamo Autosports used several check valves to provide accurate manifold pressure to the MAP sensor under vacuum conditions, while preventing boost pressure from entering the MAP reference hose when the turbo is working. How it works is this: When positive boost pressure moves toward the MAF sensor, check valves that are T-ed into the reference line to the MAP sensor open, bleeding off pressure in the reference line to prevent boost from reaching the sensor. Under vacuum conditions, the check valves close and the reference line provides accurate vacuum to the sensor.

The Bell Engineering FMU is mounted downstream of the stock fuel pressure regulator, inline in the fuel return hose. Rate of gain is set with a bleed valve, while onset pressure (normally zero vacuum) is adjusted with a bolt and lock-nut diaphragm controlling internal spring pressure. In general, tightening the bleed increases the rate of gain. Always start rich for safety, and work lean for power.

All turbo conversion engines should use very cold spark plugs to fight detonation. Use the coldest plugs you can find that won't foul at idle.

Make sure ignition static timing is set at stock specs if you're using a J&S Safeguard individual cylinder knock retard device to control knock (as we are) or follow instructions supplied with other boost timing retard devices. Do not attempt to run a turbo without some electronic boost-timing retard device. To prevent knock at peak boost, you'll be forced to retard static timing enough to severely reduce drivability under naturally aspirated conditions.

You'll need an air cleaner—the bigger, the better. Do not make the common mistake of mounting a K&N air filter in the engine compartment where it ingests heated air. We found that an isolated cold-air intake that could breathe air only from outside the engine compartment made five additional wheel horsepower on race gas, and even more relative power on street gas by supplying air so much colder that the J&S was no longer forced to retard spark as much (or at all) on boost.

We have found the J&S Safeguard effective and reliable, greatly increasing drivability and power by permitting maximum possible ignition timing advance. The J&S is sensitive, with adjustable knock-detection threshold. Keep in mind, however, that it is always harder to stop knock once started than it is to prevent it with timing retard. If the Safeguard is having to work a *lot*, you might do better to install an auxiliary boost-retard device that

on the opposite side of the diaphragm, can be used to exert an opposing force that tends to help the internal spring keep the wastegate closed. Standard practice in Deltagate mode is to feed full-boost pressure from the compressor outlet to the primary Deltagate port, simultaneously routing manifold boost pressure (regulated by various means to an adjustable lower value) to the Deltagate port, thereby increasing the wastegate maximum allowable boost by an adjustable amount.

The T-3 turbine housing is designed to allow installation of an internal wastegate. Although internal-type wastegates essentially eliminate wastegate plumbing fabrication work, the integrated turbo-wastegate concept effectively increases the turbo's size. Where space is a problem as it is on the CRX, an external unit offers more options in packaging the turbo/wastegate system.

We integrated a Deltagate mounting flange on the exhaust header so the Deltagate bolted directly to the top of the RD turbo manifold. Unfortunately, hood clearance and space under the Deltagate are now so tight that Alamo was forced to reclock the Deltagate actuator to relocate the primary diaphragm port, and, further, to remove the Deltagate actuator diaphragm spring tensioner bolt so the hood could close without contacting the bolt. Radiator and air conditioning compressor clearance problems forced us to remove the secondary radiator fan that Honda installed on air-conditioning-equipped CRXs. Alamo was forced to grind some metal from the air conditioning compressor itself for clearance with turbo plumbing.

Many lower-quality Honda turbo systems simply dump wastegate overboost exhaust gases directly into the engine compartment, unmuffled. Not only is this setup noisy at full boost, but the engine quickly becomes covered with the black soot of dead-rich full-boost air/fuel mixtures. Alamo fitted the Deltagate with a bolt-on outlet flange and pipe, which slides into a return snout in the main turbine-out exhaust downpipe. We had a quiet, clean engine compartment with a wastegate dump pipe that is easily removed.

Miscellaneous: With the main turbo components in place, Alamo reworked some RD intercooler tubing to route air from the

routinely removes a configurable amount of timing per-psi boost. You could leave the Safeguard in place as insurance that permits minimal routine boost retard, and still protecting the engine under rarely encountered worst-case pro-knock conditions.

We always begin tuning with race gasoline, in this case a mixture of 108 unleaded and 93 octane street gasoline. When you are looking for maximum power, start with lots of octane, find the limits, and then experiment with lower octane fuel combined with more aggressive timing retard and richer air/fuel ratios.

We decided to experiment first with nonintercooled boost, which greatly reduces the cost of an entry-level system. I was surprised that we could make so much power without an intercooler on the Honda.

Always work with a good, fast, accurate (expensive) air/fuel-ratio meter like the Horiba. Unless you've got dual O_2 sensor ports, you'll have to remove the stock O_2 sensor and screw in the (expensive) Horiba wideband sensor. Combined with a Dynojet chassis dyno, a good air/fuel-ratio meter is invaluable in tuning both peak power at higher rpm, and peak efficiency at idle and light cruise.

DYNO-TESTING AND ANALYSIS

With stock cat and exhaust still in place, with the Turbonetics Super-60 ball-bearing turbo sucking hot air coming through the radiator, the Alamo/RD system made 135 wheel horsepower with the Super-60 turbo. This is almost 40 more wheel horses than stock and slightly more than the NOS nitrous kit. I drove the car around Texas for a few weeks, then came back to Alamo for more fun and turbo system testing with the conventional turbocharger.

With no other changes besides switching to the nonball-bearing turbo, the conventional Super-60 turbo immediately made 124.6 peak wheel horsepower off the shelf at about 6,300 rpm. Installing a cold-air intake added 5.1 peak horsepower, and roughly 5 horsepower from 4,5000 to more than 5,000 (and marginal added power at lower rpm).

Next we disconnected the cat and exhaust and tried some dyno runs, increasing power to 134.3. Further tuning of the VRG and activation of the auxiliary NOS fuel pump increased power to 148.2. The ball-bearing turbo with this setup would surely have added at least 10 wheel horsepower, for an estimated total of nearly 160 wheel horsepower at 7-psi nonintercooled boost.

Off the dyno, back on the streets with 93-octane fuel, the Honda 1.6 CRX Si pulled hard to full boost and redline, with the J&S Safeguard hardly working (it lights one or more LEDs that indicate ignition retard in progress in 2-degree increments).

Even with just one radiator puller fan in 100-degree Texas weather, the air conditioning worked fine almost all the time.

Anyway you look at it, these were impressive results.

MAXIMUM STOCK BLOCK TURBO HORSEPOWER

The goal now was to extract every bit of power from our 1989 project CRX Turbo's 1.6-liter SOHC stock block without killing the engine. We made nearly 200 driving-wheel horsepower this time, doubling the stock 96 wheel horsepower at just 7.5-psi turbo boost, without nitrous on-line. What is it like to drive a 2,150-pound car featuring a long-stroke, 9:1 compression powerplant with fast-response ball-bearing turbo where each crank horsepower hauls around just 8.6 pounds? In Texas, we have a technical term: Yeeehaaa!

For maximum stock block power and torque, our strategy included adding the following major components: a programmable engine management computer, intercooling, intermittent nitrous injection, electronic boost control, upgraded fuel supply, and a higher-flowing exhaust system—plus reinstallation of the ceramic ball-bearing turbo.

Accel's Gen 6 programmable engine management system was designed to control engine fuel and ignition events by reading GM-type engine sensors, consulting internal tables of calibration data stored during tuning, and initiating engine actions by electronically directing the operation of GM-type fuel injectors, ignition module, and other actuators. Although now often superseded by the Gen 7 system, the Gen 6 was equipped with sophisticated logic to control idle and light-cruise air/fuel mixtures in closed-loop mode via exhaust gas oxygen sensor feedback. The DFI could further improve idle quality with feedback-based micro changes in timing.

The DFI was able to control idle speed via a GM IAC throttle air-bypass stepper motor, and implement stall-saver strategies to prevent stalling under highly dynamic throttle conditions. The DFI provided a large variety of enrichment and enleanment offsets to base fuel and ignition computations to improve starting, warm-up, throttle transition, and more. The DFI was equipped with advanced nitrous control logic, and can direct up to three stages of nitrous delivery and matching fuel enrichment with an added safety margin.

Tuners calibrate a DFI system using a DOS or DOS-Windows laptop computer running DFI's CalMap graphical interface software, connected to the DFI computer with an RS-232-type interface cable. DFI internal data structures include a 16¥16 matrix to store injection pulse width for 256 combinations of engine loading and speed, with a smaller matrix storing ignition advance in the same manner. DFI identifies engine speed and loading via engine position and manifold absolute pressure sensors, and uses these values to look up injection time and spark advance in the internal tables on an ongoing basis.

When it came to installing the DFI System, we decided to leave the Honda computer and most of its wiring and sensors in place and to install the DFI in parallel for easy return to stock controls if ever required.

DFI's Gen 6 engine management kit for the Honda CRX includes an electronic control unit (ECU) equipped for reading inductive engine-position pickups, a generic wiring harness adaptable to most vehicles, and a basic set of GM-type engine sensors (with additional sensors available optionally). You'll also need high-pressure fuel supply and return lines, a high-pressure fuel pump and regulator, injectors correctly sized for expected engine horsepower, a throttle body, and so forth. All of this was already on the turbo EFI CRX, though we upgraded some components for higher-horsepower turbocharged operation.

The DFI ECU is not designed for the harsh environment of an engine compartment. Alamo Autosports installed it below the dash, routing the wiring harness through the firewall into the engine compartment, where the DFI injector harness plugged seamlessly into a set of Honda-compatible 310-cc/min RC Engineering fuel injectors that upgrade fuel delivery potential 50 percent over the stock 220-cc injectors.

Next, Alamo installed the engine sensors, drilling and tapping threaded bosses into the compressor-out air tube and cooling system for installing DFI GM-type coolant and intake air temperature sensors, allowing the stock (and DFI-incompatible) Honda sensors to remain in place and connected to the stock Honda computer wiring harness. Alamo installed a 2-bar GM MAP sensor (with sensing range accurate from high vacuum to 15 psi pressure) to measure intake manifold pressure.

Manifold pressure correlates well to engine loading for a given camshaft profile when engine displacement, speed, and air temperatures are factored in, giving the DFI an excellent estimate of engine airflow for use in air/fuel computations. The MAP sensor mounted neatly on the firewall so the sensor's pressure port could be referenced to the intake manifold with a short length of vacuum hose. The Honda throttle position sensor (TPS) is a 0–5 volt linear potentiometer fully compatible with GM's TPS, which conveniently allowed Alamo to discard the GM TPS and replace the DFI TPS harness connector with a Honda-style connector for connection to the stock TPS.

The final sensor related to DFI fueling was an optional three-wire GM-type heated exhaust gas oxygen sensor, used by the DFI computer to deduce current intake air/fuel ratios based on residual oxygen content in the exhaust gases for on-the-fly air/fuel trim. This sensor screwed like a spark plug into a K&N sensor boss Alamo had previously welded into the exhaust just downstream of the turbo.

At this point, Alamo connected the DFI's fused continuous 12-volt power line directly to the Honda battery, and wired in the switched +12 volts line to the ignition circuitry so the DFI always sees 12 volts on this wire during both starting and running modes. Finally, Alamo connected a special DFI interface cable into a pigtail interface connector near the ECU, and connected the other end to a battery-powered laptop PC running DOS or DOS/Windows and DFI's Calmap graphical calibration software. The laptop enabled us to first define global parameters in the computer (such as number of cylinders), and then to build fuel injection pulse width and ignition timing tables.

We did not make use of the DFI's optional IAC idle speed control at this time. And though the DFI can control an optional GM knock-sensor module, we decided to retain the J&S Safeguard—with its individual-cylinder retard and sensitivity-adjustment capabilities—as a last line of defense anti-knock countermeasure.

The DFI computer's ability to precisely control ignition timing from idle all the way to high horsepower turbo boost is essential in optimizing engine performance. In order for the DFI to control ignition functions, crank or cam trigger engine position information is mandatory in the form of one 180-degree inductive pickup/trigger. In fact, there are three inductive pickup/flying magnet triggers on the Honda distributor, indicating top dead center compression-stroke number one cylinder, 180-degree increments of crank rotation, and 120-degree increments of crank rotation.

All are essential to keep the Honda computer happy and functioning, which we would have preferred, since the Honda computer controls air conditioning, alternator charging, fast-idle, the tachometer gauge, and other nice drivability features. Unfortunately—as is typical—the Honda inductive pickup circuitry was not powerful enough to drive the input sensor circuitry of two computers at once. Choices were to add amplification circuitry to the 180-degree trigger, to add another 180-degree trigger to the distributor or cam (four flying magnetic bolts required) or the crankshaft (two flying magnetic bolts required), or to effectively make do without the Honda computer online at all—which was what we did. This would require minor rewiring to the alternator and air conditioning control circuitry.

Following initial testing with the Honda computer handling timing and the DFI handling fueling, Alamo disconnected the Honda computer from the distributor and wired in the DFI so it could read the Honda's 180-degree cam-speed magnetic inductive pickup for determining engine position. In this mode, the DFI provides timed low-voltage ignition pulses defining spark timing to an OEM ignition module or aftermarket coil-driver box (Accel 300+). For simplicity's sake, we continued to use the J&S Safeguard module's coil-driver capabilities to drive the stock distributor-integral Honda coil.

For compatibility reasons associated with GM EST home-mode ignition-timing requirements, without correction, DFI-based timing would be retarded 10 degrees on a CRX. This requires that the distributor be rotated until the DFI's internal timing tables correspond correctly to actual engine position relative to the current compression-stroke cylinder location at the timed spark event. The distributor rotor must then be reclocked so the rotor geometry corresponds precisely at firing time with the correct plug-wire electrode in the distributor cap. Alamo modified the plastic rotor to this purpose, though it is also possible to heat the Honda distributor shaft with a torch until the pressed-on rotor-orienting stub shaft can be turned 10 degrees.

In order to extract maximum performance and economy from the 1.6 Si engine in our CRX, the Honda alternator charging rate is partially controlled by the computer, which can lower or even zero charging rates at full throttle to reduce parasitic accessory horsepower losses. The computer can also maximize charging during deceleration. Without the DFI computer online, the alternator will not charge at all. Alamo jumpered two leads together in the alternator-controller connector plug to provide conventional (non-computer-based) charging.

Similarly, the Honda computer controls air conditioning compressor clutch operation in order to disengage the air conditioning clutch at wide-open throttle to further reduce accessory drag. The solution is minor rewiring of air conditioning control circuitry to eliminate the Honda ECU from the circuit and allow direct compressor control from the dash controls via a 30-amp relay. One could even wire in an NOS full-throttle microswitch to disable the compressor at full throttle, just like the computer.

Upgraded fuel supply is essential to any CRX engine making more than about 140–150 horsepower. The stock Honda in-tank pump and an auxiliary NOS 550-horse inline unit provided more than enough fuel, and we continued using the stock fuel supply and return lines. The DFI ECU controls a fuel pump relay for pressurizing the fuel rail for a few seconds at initial key-on time and during starting and running. For safety reasons the computer also kills the fuel pump quickly if the engine stalls. We had 700-horse worth of fuel pump capacity on tap at 40 psi and roughly 300 horsepower worth of injector capacity at stock fuel pressure.

Generally speaking, colder plugs are the right way to go on a turbo conversion powerplant. Stock spark plugs may work OK with very low boost, and hotter plugs are clearly better at resisting fouling from rich idle mixtures. But when you begin pushing the limits of cylinder sealing, ring land integrity, and block and head structural strength, the goal is to avoid detonation at all costs. Two or more heat ranges colder than stock is a good place to start. Just remember that such plugs can easily foul if grossly rich mixtures accidentally occur while calibrating custom fuel injection.

The car begins running terribly, and the last thing you'll suspect is the plugs—because just a few minutes ago they were brand new. *Suspect the plugs.* Keep an extra set on hand. And, by the way, turbo engines require top-quality plug wires. It's normal for tuners who know what they're doing to set the plug gap at 0.029 instead of 0.039 or whatever it says in the manual for a stock NA powerplant. Plan on upgrading the plug wires, coil, and coil driver when you begin to pump up turbo boost. Even if the engine still sounds good, upgraded ignition can add 20 or 30 horsepower on a turbo-four.

Alamo knew from earlier experiments with CRX turbocharging that the stock Honda exhaust is a bad joke when you start to crank up the boost. In fact, we were amazed that the 1.6 engine made as much as 135 wheel horsepower with such a restrictive exhaust and cat installed. We were not a bit surprised to make another 15 horsepower by disconnecting the exhaust and retuning. This time, for reasons of expediency, Alamo fabricated a custom 2.5-inch shorty high-flow exhaust system that bolted to the RD Racing turbo downpipe and included a 2.5-inch resonator with straight-down dump. It was illegal and embarrassingly loud, but the plan was to install a high-flow cat and a quiet—but high flow—muffler.

The HKS EVC IV electronic boost controller is an ingenious electromechanical device that lies to a turbo wastegate about manifold pressure in order to raise average and peak boost. The EVC uses fuzzy logic (a sort of artificial intelligence) to observe how fast the turbo can build boost under maximum loading, how fast the wastegate opens, and how effective it is once open. Once it understands the turbo system's capabilities, the EVC can keep the wastegate fully closed until an instant before full boost is achieved. It can then slam open the wastegate quickly and to the degree required to prevent over-boost, greatly increasing midrange torque compared to traditional mechanical wastegates, which begin to bleed off a certain amount of boost long before maximum boost has actually arrived, degrading potential midrange torque.

The HKS device includes a dash-mounted controller with a few wires you'll connect to constant and switched 12-volt power and ground, plus a harness that connects through the firewall to the HKS electro-pneumatic actuator unit. The controller connects via vacuum hoses to the intake manifold, compressor discharge (highest available boost pressure), and the standard wastegate port.

Once installed and educated, the EVC can be adjusted while you drive from the comfort of the cockpit. This can be deadly if you like to street race and have no judgment about what pressure will overboost and kill your engine. Some people should not mount this device in the cockpit—you know who you are. We used the EVC to crank up boost from the Turbonetics Deltagate's nominal wastegate spring pressure of 5.5 psi to a higher 7.5 psi maximum boost, and to increase midrange boost and torque by 36 percent. As you'd expect, 36 percent more boost makes an amazing difference in performance.

With the DFI properly calibrated on Alamo's Dynojet, the ball-bearing turbo back in position, the HKS boost controller online, we tested two different intercoolers to simultaneously lower combustion temperatures (to control detonation) and stuff the engine with denser cooler air. Neither cooler was designed to fit the Honda engine compartment, but we kludged the coolers into place with temporary plumbing for Dynojet testing with the hood up. Later on, we removed the hood for street runs. The air-air unit was designed for a serious 2.2 Mitsubishi engine, while the air-water unit is a custom Norwood Autocraft design.

The air-water cooler was the more interesting, because it can easily achieve thermal efficiencies beyond 100 percent (meaning the compressed intake air from the turbo is cooled *below* ambient air temperatures). We ran the air-water cooler on the project CRX on the Dynojet using cold tap water from a hose to get excellent, consistent results. For street driving, you need an efficient heat exchanger (radiator) with fan and bilge-pump-type circulation system to keep the intercooler nice and chilly (though a little water will cool a *huge* amount of air). For drag racing, you can circulate ice water from an onboard tank through the intercooler for even denser, colder intake air.

Other advantages of an air-water system include smaller intercooler size per BTUs of heat-removal capacity and the fact that separating the air-cooler unit from the water-cooling heat exchanger can add layout flexibility that makes custom installations much easier. On the other hand, air-water systems require electrical circulation pumps (with moving parts and electrical controls) and are more complex and prone to failure than simple rugged one-piece air-air intercoolers.

Our intercooler allowed us to safely run more boost (7.5 psi versus 5.5), and the denser, colder, higher-boost inlet air clearly contributed to the 45–55 additional wheel horsepower we made in final Dynojet testing with the DFI in control. With 500-plus-wheel horses worth of intercooler cfm capability, we knew there'd be zero pressure drop at current power levels on the Honda. We also knew the cooler could keep intake air cold and safe and dense.

A nice feature of the new computer-controlled modified-NOS nitrous system is its ability to control nitrous solenoid activation, then provide precise fuel enrichment through the port injectors by increasing base injection pulse width, with perfect fuel distribution to all cylinders guaranteed. The DFI also automatically provides spark retard during nitrous injection by subtracting a programmable nitrous offset timing retard value from base spark timing (which will also contain an independent boost-based timing retard offset).

Since our turbo system is clearly capable of making as much power as we have the guts to make at this time, we elected to use intermittent nitrous to spank up the turbo quickly with a sudden blast of exhaust pressure and heat on full-throttle nitrous activation, at the same time radically and instantly increasing torque to full-boost-type levels. Then, as boost goes through 4–5 psi on our CRX, an adjustable NOS-supplied pressure switch disarms the nitrous system as boost is still rising.

The DFI system must be configured to define rules for nitrous activation (arming switch on, minimum engine rpm, 100 percent throttle position, and so on). You then define timing retard on nitrous activation and fuel enrichment parameters (which must be properly synchronized with the numbered nitrous jet installed in the injection nozzle). You may select a variable delay to nitrous delivery after fuel enrichment begins in order to make sure nitrous never arrives in cylinders before fuel, avoiding incredibly destructive nitrous lean-outs.

Alamo started with a 50-shot nitrous charge, and the stock clutch let go and the engine surged for the redline. A 30-shot yielded similar results. A 20-shot produced three nitrous-boosted dyno runs before the stock clutch gave out entirely. Since we wanted dyno results in a hurry, we installed a Stage IV Clutchmasters performance metal clutch system, capable of handling 375–400 wheel horsepower on the Honda with a reworked 1,400-pound clamping force pressure plate (double stock) and a spring-loaded, damped street-strip metal-puck clutch disc. This clutch was a mite touchier engaging from a full stop than a Stage III Clutchmasters system with the carbon-Kevlar disc, but the carbon clutch required 500 miles of break in, and we couldn't wait. Clutchmasters changed the geometry of the clutch fork to keep pedal engagement pressure down to near stock or even below stock level. This clutch system is extremely streetable.

Programmable ECU calibration and tuning should begin with race gasoline or at least race-gas-fortified 93 octane—in our case, a mixture of 108 unleaded and 93 octane street gasoline. When you are looking for maximum power, always start with lots of octane, find the limits of your system, and then experiment with lower-octane fuel combined with more aggressive timing retard and richer air/fuel ratios.

Always work with a good, fast, accurate air/fuel-ratio meter like the Horiba. Unless you've got dual O_2 sensor ports, you'll have to remove the stock O_2 sensor and screw in the Horiba wideband sensor. Combined with the Dynojet chassis dyno, a good air/fuel-ratio meter is invaluable in tuning both peak power at higher rpm and peak efficiency at idle and light cruise. If at all possible, always work on a chassis dyno, where you can immediately see the posi-

RESULTS AND ANALYSIS

The CRX project produced some of the nicest dyno charts I've ever seen: smooth, linear, and predictable. With all the good stuff in place, we produced multiple runs that were very repeatable. Peak power with air-water intercooling and a nitrous shot between 2,500 and 3,500 rpm was in the 190s at the wheels just above 6,000 rpm, with power still on the way up when Alamo stopped the run near redline. Torque was nearly flat at 170 ft-lbs from 2,500 to redline (nitrous armed), or from 3,500 to redline without nitrous.

It is interesting to compare the results of the turbo CRX, running non-intercooled at 5.5-psi boost, to an Eaton-supercharged Honda running 7-psi boost. The CRX made 10.7 more maximum horsepower and 12.0 ft-lbs more peak torque, *on less boost*. The supercharger made more power and torque below 3,000—unless you armed the CRX's intermittent nitrous—in which case the CRX made a ton more torque anywhere above 2,000, and about 3 ft-lbs less below 2,000. The graphs were otherwise very similar until 4,500, at which point the turbo CRX began to pull ahead significantly. On less boost. At roughly equal levels of boost (7-psi blown, 7.5-psi turbo boost), the stock-block turbo CRX made 70–80 more peak horsepower! Which car do you think is faster? There is something to be said for efficient compressors and good intercooling.

There's nothing like air-water intercooling on the dyno, because you can easily achieve more than 100 percent cooling efficiency cooling air with tap water or ice. With a rigged air-water cooler and DFI engine controls, we more than doubled the stock 96 wheel horsepower to roughly 200.

tive or negative effects of changing tuning parameters without the risks inherent in street tuning.

The initial process of making certain all engine sensors are functioning correctly and in sync with the global parameters of the DFI requires patience and exhaustive checking, but making sure everything is right before startup can save a lot of grief and time later.

Actual calibration (mapping) of an engine management computer requires establishing a start-up map good enough to keep the engine running until it's warm, then tuning the engine at zero load across all speeds, setting injection pulse width at light load, then gradually loading the engine at various speeds while observing air/fuel ratios, exhaust gas temperatures, and dyno power and torque to get the perfect injection pulse width for lean best torque (richer if fuel cooling is required) at heavier loads. All elements of the DFI's 16¥16 air/fuel matrix had to be set with the laptop PC, while avoiding dangerously lean mixtures at all cost. Optimized and safe ignition timing is vital but generally a little more predictable in advance.

Using extensive experience with hot-rod 1.6 Hondas, including various turbo CRXs, Alamo's Steve Webb got a headstart building a custom air/fuel map project CRX Si by working with a map from a reasonably similar turbo CRX. You absolutely cannot optimize tuning and power on a programmable EFI car without good diagnostic equipment. Period. Even with good equipment and experience, it typically takes one to four days to get a perfect map under all conditions.

I found the J&S Safeguard to be very effective and reliable in controlling knock with boosted and nitrous conditions, greatly increasing drivability and power by permitting stock-type ignition timing advance under most conditions. The J&S is sensitive, with adjustable knock-detection threshold. Keep in mind, however, that it is always harder to stop knock once started than it is to prevent it with timing retard. This is the reason we programmed automatic turbo boost and nitrous timing retard into the DFI ignition maps that routinely remove a specified amount of timing per-psi boost (also modified according to other engine and atmospheric conditions). We left the Safeguard in place as a coil-driver, and to provide insurance on a street car, allowing DFI calibration for minimal routine boost retard, while still protecting the engine under rarely encountered worst-case pro-knock conditions (such as a tank of lower-octane fuel).

Minor problems included nonoperative tach (controlled by stock computer only), too-loud exhaust, and lack of a cat, which is illegal and unnecessary. Good cats do not have to hurt performance on a balanced, truly streetable car. There are better ways to get power (like another pound or two of boost). A well-tuned DFI can meet emissions standards on an '89 Honda, though it may not technically be legal in some states.

REAL-WORLD TURBO CRX Si

CHAPTER 21
HONDA DEL SOL Si TURBO

The world changes. . . . When designing the earlier CRX turbo conversion project for this book, the choices were to retain the stock Honda engine control system and then add on electronic and mechanical aids to provide boost fuel enrichment and retard spark timing or to just junk the Honda EMS in favor of an aftermarket ECU. We did both.

By the time I was working on a somewhat more modern '95 del Sol Si project, there were several new EMS solutions on the market that enabled an enthusiast to retain the stock Honda system but to convert it to user-programmability for reconfiguration and recalibration. Another possibility was the plug-and-play aftermarket EMS from AEM.

I decided to contact Hondata about using their software-hardware programmability conversion solutions for the del Sol project's stock Honda ECU, and I also contacted AEM to obtain a replacement aftermarket ECU.

THE HONDATA TURBO DEL SOL

After reading several Hondata installation and user manuals, we pulled back the carpet on the passenger-side kick-panel, unbolted and unplugged the del Sol's ECU, and shipped the 1.6-liter D-16Z SOHC VTEC engine's electronic controller off to Hondata in California. Hondata socketed the main circuit board to permit installation of an aftermarket PROM chip, added some resistor-jumpers to the board that enable the stock microprocessor to read from the new PROM, and provided a socket on the board suitable for connecting a communications cable to a Hondata user interface box.

Modern high-output multivalve engines like Honda's SOHC VTEC 1.6 with port EFI are intentionally designed from the factory with very flat torque curves, and such engines now typically use tricks like variable-length induction systems and variable valve timing (and lift, in the case of Honda's VTEC) to keep torque from falling off too fast in the upper rpm range in a quest for highly linear torque and high specific power that hangs on at high rpm. Flat torque curves make for simple, straightforward EMS calibrations (though VTEC engine management has been forced to deal with a transient dip in torque that tends to occur where the engine changes to the high-speed cam and charge velocity momentarily dips).

Flat torque curves make for relatively simple EMS calibrations, but turbo engines—particularly ones with large, high-slow (and lazy) compressors—are more challenging. Since it would not be very interesting to reprogram the stock ECU for a stock engine (it's already optimized for the job; there's nowhere to go), to make things more challenging and fun, I designed, built, and installed a custom turbo system while the ECU was at Hondata.

Turbo engine management *is* interesting, and, done right, capable of producing extremely high levels of specific power (hot rod–class streetable 1.8-liter turbo Honda drag engines are now making around 1,000 horsepower on methanol). Turbo systems tend to have trickier torque curves because turbo boost is never available immediately or in the lowest rpm range, since the turbo must spool up to 40,000–80,000 rpm to make much boost, and the exhaust energy is simply not available at all times to drive the compressor full-tilt.

Small engines with big turbos lack low-end torque before the boost kicks in, but then torque increases greatly in a huge whoop

1995 Honda del Sol SOHC VTEC on the Alamo Autosports Dynojet for Hondata recalibration of the stock Honda EMS.

of boost as the compressor comes up to speed, and turbo powerplants thus have extremely wide dynamic power range and certainly represent a challenge to calibrate. Not only does volumetric efficiency move around less predictably on turbo engines (and you may be pushing the envelope by turning up the maximum boost along with everything else when you calibrate the engine), but turbo engines are far more prone to detonation than NA engines and usually need at least some fuel-based combustion cooling. This means that finding the optimal air/fuel ratios is critical but also less predictable. Turbo engines make much more use of advanced features and high-resolution engine management system tables to keep air/fuel ratios and timing optimal and safe under all conditions and to retain excellent drivability while providing high specific power.

With a Majestic T3-T04 turbo on an otherwise stock Honda SOHC VTEC 1.6-liter engine as the challenge, Hondata's programmer environment made available high-speed and low-speed cam fuel and timing tables in the Honda ECU with at least 20 predetermined rpm ranges and 13 load ranges, of which four represented 0.6, 3.9, 7.9, and 11.0-psi boost, respectively.

The Hondata ECU arrived back in Texas with the new socketing and jumpers on the circuit board, ready for installation of an aftermarket Hondata-format PROM to calibrate the Honda EMS. Unfortunately, although I built the turbo system in central Texas, we planned to calibrate the Hondata-upgraded EMS at Alamo Autosports near Dallas, which had the package of additional Hondata components awaiting installation, including the new PROM, interface box, cabling, and so on. So we had to figure out a way to get the car to Dallas without an auxiliary PROM installed.

Fortunately, if you interrupt one of the new Hondata jumpers on the main Honda circuit board, the original Hondata PROM calibration data will still run the car from the stock Honda PROM. At this point, the problem was that a turbo conversion with stock EMS calibration could be expected to produce poor drivability and

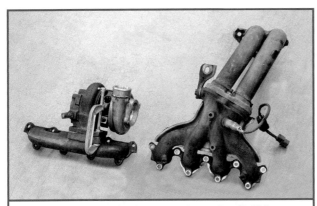

To make the del Sol recalibration project interesting, the highest priority was to get a turbocharger on the car. A special cast turbo exhaust manifold replaces the stock cast-iron header and downpipe and tucks a T28 turbo close to the engine behind the radiator.

dangerous detonation without fuel enrichment under boost and perhaps a little timing retard under boost. In the end, we cut the J1 jumper and then road tested the car to get baseline street and dyno results—first naturally aspirated with the turbo system pulled running the stock calibration, then with the turbo system reinstalled and fuel enrichment provided by a Bell Engineering Group fuel management unit.

BELL ENGINEERING'S FMU

As described in the CRX project chapter in this book, the Bell FMU works by jacking up fuel pressure during boosted conditions as a *multiple* of the manifold pressure when the engine is under load and the turbo is making positive pressure. The FMU does not replace the stock fuel pressure regulator but installs downstream of the stock regulator to provide fuel enrichment under boosted conditions.

Under NA conditions, the stock Honda fuel pressure regulator delivers normal fuel pressure, which is over 40 psi at atmospheric manifold pressure (wide-open throttle, no boost), with a manifold pressure reference signal allowing the stock regulator to maintain fuel pressure as a *fixed amount* above manifold pressure so fuel pressure is actually *reduced* at idle when vacuum is high and the fuel injectors are injecting into reduced pressure in the manifold. (The stock fuel pressure regulator would also similarly *increase* fuel pressure with manifold pressure in the presence of positive boost—except the FMU located in the fuel return line downstream of the stock fuel pressure regulator renders the stock regulator ineffective by increasing the pressure *even more*). We adjusted the FMU to kick in at 1-psi boost and provide about 5- to 6-psi gain in fuel pressure for each pound of boost for a total of 70 to 80 at full boost of 6- to 7-psi.

We spent some time testing the turbo system on the twisting roads near Bell Engineering in the Hill Country northwest of San Antonio under boosted conditions with fuel pressure and boost gauges duct-taped to the windshield, listening for knock and summoning up good karma as an anti-knock countermeasure. The Bell FMU is adjustable for onset manifold pressure of the rising rate of fuel pressure gain, as well as for the rate of gain itself. The initial turbocharger was a hybrid Majestic T3-T04 that was quite large for the engine size, which made it quite efficient at higher rpm, but unavailable much below about 3,500 rpm. We discovered the car would knock under heavy boost at higher rpm as boost built to approximately 6 to 7 psi approaching redline.

INSTALLING AND ACTIVATING HONDATA

At this point I drove the car without beating it to death to Alamo Autosports near Dallas, where Alamo techs installed the rest of the Hondata system by wiring in the small blue Hondata add-on auxiliary processor box to the main circuit board in the Honda ECU. The blue box contains circuitry and a 9-pin delta connector enabling attachment of a laptop computer to the Honda ECU, which can then be used to datalog the Honda processor when the car is running. The Hondata box also contains an LED that can glow steadily or blink to signify various internal events.

The stock Honda MAP sensor is capable of measuring slightly more than 11 psi (0.75 bar) positive boost pressure, even though the stock Honda ECU will freak and report an error at anything more than about 1-psi boost. The stock calibration provides fuel delivery and ignition timing advance for *only* the expected range of manifold pressure for the stock normally aspirated SOHC 1.6-liter Honda engine, with the stock Honda cam.

In order to provide a turbo calibration, Hondata provides a calibration that describes fuel delivery all the way to 11-psi boost (or up to 29-psi boost if you elect to take advantage of Hondata's ability to enable the stock Honda ECU to support a GM 3-bar MAP sensor).

Hondata provides a selection of turbo calibrations for the D-16Z powerplant; Alamo loaded a likely candidate into a rewritable PROM and installed it in the new socket installed by Hondata, and then it was onto the Dynojet.

DYNO-TUNING THE HONDATA-IZED HONDA EMS

Once the OEM Honda ECU has been socketed for an aftermarket (Hondata) PROM and the blue interface box installed, there are two methods of tuning:

1. Remove the Hondata calibration PROM and install an emulator in its place, plugging the Hondata PROM into the emulator (the stock Honda PROM is permanently resident in the Honda ECU; it's just not doing anything unless the Hondata J1 jumper/resister is cut).

2. If you do not have an emulator, the alternate is to use Hondata's ROMeditor software to make changes to the calibration offline on a laptop (probably connected to the Hondata blue box so you can datalog dynoruns) following evaluation and analysis of the results of the previous dynorun and associated Hondata datalog or dynamometer wideband AFR meter logged data). You can then burn a new PROM and install it into the Hondata Honda ECU socket for testing. There are two kinds of PROMs: those that can be written once and more expensive PROMs that can be rewritten many times.

An emulator is a fast microprocessor device with a lightning-fast memory buffer containing an image of the PROM that will be tuned (the emulator sucks in the data from the PROM and stores it in a buffer). A short cable (short because even at the speed of light, long cables can negatively affect the timing of memory access) plugs into the PROM socket in the Honda ECU, implementing the PROM multi-pin protocols needed to access PROM memory. The PROM image in the emulator can be changed on the fly with a Windows laptop PC while the Honda ECU is running the engine (Hondata's ROMeditor has commands to read and write calibration data from the emulator), which makes an emulator extremely valuable for tuning and debugging.

When the Honda ECU executes a read operation to PROM memory, the emulator—as you'd expect—faithfully emulates the electronic handshake signals of the Hondata PROM and supplies PROM-image data in place of the data that would be supplied by the real PROM. When you are done modifying the PROM image

A custom-turbocharged 1.6-liter Honda waiting for the next new Hondata PROM. Hondata mods make the stock Honda computer programmable and capable of datalogging. Hondata's "Hondalogger" box allows a tuner to datalog the complement of stock Honda and modified sensors during test runs, including, optionally, wideband O_2 data.

to recalibrate the engine management system, the PROM-image calibration can be offloaded from the laptop into a PROM burner (in our case, Alamo used a PocketProgrammer) and a new PROM can be blown (ROMeditor provides a burn PROM command).

Some emulators are fast enough that the PROM image calibration data can be changed while an engine is running without causing problems. More economical units might hiccup during a change, disturbing smooth operation. Even if you've got a fast emulator, dynamic changes are never a good idea during high-load operation. The best alternative is to make dynoruns on a chassis dynamometer while running Hondata's Hondalogger datalogging software (or logging the air/fuel ratio with a wideband AFR sensor connected to the dyno), and then make change to the emulator between dynoruns.

Tuning the del Sol at Alamo, we worked without an emulator (Alamo's was out for repair). We installed a rewritable PROM in a special carrier designed for easy R&R in a PROM socket without damaging the delicate pins of the PROM itself.

One thing to keep in mind when tuning an engine using a wideband O_2 sensor is that the sensor is *not* located in the combustion chamber but is farther downstream in the exhaust header or exhaust system. There will always be at least a tiny delay between the parameters that generate a particular combustion event and the arrival of the exhaust gases at the wideband sensor where the exhaust gas oxygen can be converted into a voltage for interpretation, say, as evidence of a particular air/fuel ratio in the intake.

If the engine is operating at steady state, at a particular speed and loading, there is no problem, and, in any case, exhaust gases travel *very* fast in an engine under load at higher rpm. But it is still useful to keep in mind that wideband O_2 data is actually *past history* on an engine accelerating very hard at high rpm. Keep in mind also that engine conditions will have changed at least a little (and possibly more than a little on a really fast-revving engine) by the time you log exhaust gas oxygen data. Alamo owner Brice Yingling likes to assume that Dynojet wideband data on Dynojet is 100 rpm or so old (some dynamometer systems enable you to program out delay in correlating wideband data with rpm or miles per hour).

In fact, our initial Hondata calibration had one particular spot on the dynorun at 6,500 rpm where the air/fuel ratio plunged deep into rich territory to such an extent that it was off the rich

end of the map at 10:1. Keeping in mind the past-history nature of logged wideband data was useful as we worked to smooth out the air/fuel ratio at wide-open throttle on the Dynojet.

You must also always keep in mind that ECUs virtually always provide fuel and timing based not only on the exact speed and loading cell that coincides best with current operation, but with an average of the best, closest cell *plus a weighted average with surrounding cells.*

We made a succession of about 12 dynoruns at Alamo, working in between to smooth out the worst peaks and valleys in the Dynojet's air/fuel ratio graph by using ROMeditor and the laptop's mouse to select fuel cells in the immediate vicinity of the worst peaks and valleys, and also highlighting (selecting) much larger areas of the graph to raise or lower using the percentage change function of the Hondata ROMeditor. Since the stock Honda O_2 sensor was uninstalled in favor of the Dynojet's wideband, we turned off closed-loop mode during the tuning process (although closed-loop was, in any case, configured only to operate below 790 millibars, about two-thirds of the way between high vacuum and zero vacuum).

Naturally, the default Hondata turbo map was extremely rich under most of the loading range, so it was mainly a matter of removing fuel, though occasionally we added a bit back in selected areas. With no knock sensor capability on the car, we kept a sharp ear out for detonation (I find this is easier if you hold your hands over your ears to block out the main roar of the engine and exhaust). On the occasion or two where we heard a bit at high rpm near redline, we immediately added fuel or pulled timing or both in that immediate vicinity of the calibration on the next dynorun.

Reviewing the dynoruns, it is clear that the default Hondata calibration was right enough that we didn't add a whole lot of power, but we certainly increased fuel efficiency and emissions by cleaning up the calibration.

In the end, the Hondata programming solution was easy. You need electronic technician's skills to socket the Honda ECU and install the jumpers, but once that's out of the way, a laptop and PROM burner—or maybe also an emulator—makes turbocharging a Honda straightforward and rewarding. This is absolutely one of the best solutions for a street Honda turbo vehicle, and Hondata has some super-high-output race vehicles running on modified Honda factory ECUs.

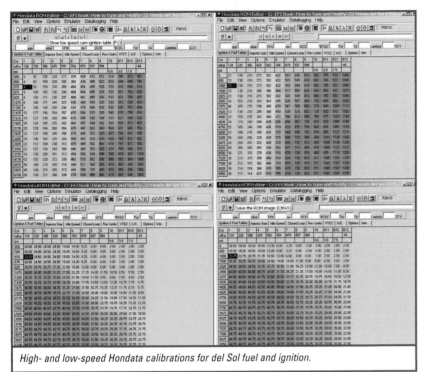

High- and low-speed Hondata calibrations for del Sol fuel and ignition.

ROMEDITOR

Hondata's ROMeditor provides a tuner with, first of all, the ability to modify the main high- and low-speed fuel and timing tables, each of which has 20 hard-coded rpm ranges and 13 loading ranges for a total of 260 speed-density breakpoints in each fuel table and 520 total. You can view fuel tables in normal mode as a spreadsheet-type matrix of integer fuel value numbers that are not pulse width but give an idea relative to each other of how much fuel the EMS will deliver to the engine for the various possible combinations of speed and loading.

ROMeditor can provide background shading for the table that bleeds from turquoise through green to red as fuel delivery increases in various segments of the table. You can also elect to view the fuel table as a set of injection *duty-cycle* (percentage on-time) numbers rather than the fuel values (anything over 100 percent indicates the injectors will be open constantly, which can spell trouble). A tuner can select the option of viewing or not viewing the four boost-table loading ranges starting at 0.6 psi and to view or not view the extended rpm above 9,000.

A tuner can also view the fuel tables as a 2D graph, where the 13 load ranges are plotted as a line graph, each a function of rpm and fuel value, or as a single 3-D graph. You can also view a lambda (air/fuel ratio) table, a lambda differential table, and the target lambda table.

The high-speed and low-speed cam ignition tables are very similar to the fuel tables, with the exception that they display degrees of advance before top dead center (BTDC), without, of course, any of the lambda options.

ROMeditor also has a set of configuration parameters that include the ability to:

Scale injector size for larger injectors (you specify the injector multiplier as a percentage of the stock injector flow rate, so 0.50 would automatically scale injection pulse width to 50 percent of normal to compensate for injectors with twice the fuel flow). ROMeditor also allows special injector scaling for idle and acceleration conditions.

Set idle target rpm by increasing or decreasing the idle air control (IAC) duty cycle.

Choose to enable or disable closed-loop feedback mode (the Honda EMS targets the values in the target AFR table), and to set the threshold loading above which the Honda EMS will switch to open-loop mode and disregard the O_2 sensor.

Set a rev-limiter rpm and a boost fuel cut pressure in psi.

Disable or enable VTEC cam-switching, and specify the rpm and loading above which the Honda EMS will command a switch to the high-speed cam(s).

Select the rpm and loading above which the EMS will turn off the air conditioning compressor.

Select 3-bar MAP sensor mode, and disable or enable the knock sensor, ELD, injector error, barometric pressure sensing, and the O_2 sensor heater.

There is a rich set of editing commands for the tables, including undo, redo, undo history, cut, copy, paste, select all, increase selection, decrease selection, adjust selection, interpolate selection, and create boost tables.

Emulator commands include download table or entire ROM image to the emulator, go to current cell, increase or decrease the current cell, provide real time update, and reboot the ECU.

File commands include new, open, reopen, save, save as, protect ROM, close, print setup, print, import or export tables, convert ROM version, burn EPROM, settings, and exit.

Datalogging commands include connect, record, ECU information, beginning, rewind, play, fast forward, end, load recording, and save recording.

There are also several help commands.

Some of the most important pull-down commands are also available as buttons on one of two toolbars.

BUILDING THE DEL SOL TURBO CONVERSION

I had an HKS cast-iron turbo exhaust manifold lying around, designed to mount a tiny T-25 or T-28 turbo directly in front of the engine block just behind the air conditioning condenser on a 1992 to 1995 D-16Z SOHC 1.6-liter VTEC powerplant of the type found on various versions of the Civic and del Sol. However, we were keen to experiment with a T-3/T04 hybrid turbo, which is a significantly larger turbocharger on both the compressor and the turbine side.

Majestic Turbo supplied a low-inertia hybrid with a small 0.48 A/R nozzle. I was hoping it would spool up fast enough to provide decent response, yet provide the high thermal efficiency required to provide relatively cool compressed air at low-to-medium boost, while offering the possibility of producing crossover conditions at high boost in which intake manifold pressure is actually higher than exhaust manifold pressure—producing *great* power—should we ever get around to strengthening the engine enough to withstand crazy levels of power.

In the meantime, it was summer in south Texas when we were fabricating the turbo system, with temperatures hovering near 100 degrees, and maintaining the full efficiency of the air conditioning system and cooling systems seemed like a right-on idea. Meaning we did not want to hack up the fan shroud or

HONDA DEL SOL Si TURBO

207

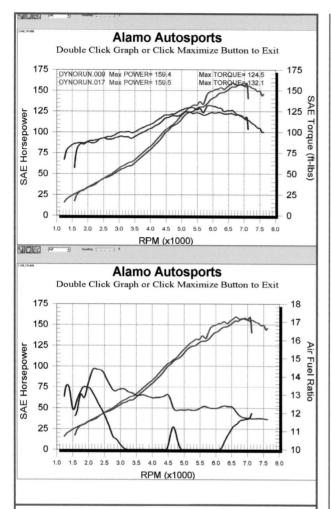

Alamo Autosports
Double Click Graph or Click Maximize Button to Exit

DYNORUN.009 Max POWER= 159.4   Max TORQUE= 124.5
DYNORUN.017 Max POWER= 159.5   Max TORQUE= 132.1

Optimizing the Hondata turbo calibration on the Alamo Dynojet. The "sensitivity" of gasoline is fairly low; there is a fairly wide range of air/fuel ratios that produce peak or nearly peak power. Note how the air/fuel ratio in the bottom graph jumps richer simultaneously with the VTEC switch to the high-speed cam at 4,500 rpm.

The large turbo proved capable of about an 8-psi boost on the 1.6-liter engine near redline without the wastegate coming into play, which was, in any case, all the engine could safely tolerate without internal modifications. Boost was nearly linear, something like what you'd expect with a Vortech-type centrifugal supercharger.

INSTALLING AEM'S PLUG-AND-PLAY SYSTEM

When it was time to try out the AEM system, I discovered installation was about as simple as installing a new printer on your Windows PC.

You open the passenger door of the del Sol and peel back the carpet that covers the stock Honda ECM located in the outside kick panel next to a passenger's right foot. You get out a 10 millimeter socket and loosen/remove the three bolts and nut that hold down the ECM. You pry off the sheet-metal cover that protects the connectors on the ECM from being damaged if your passenger kicks the kick panel. You'll see three connectors on the bottom of the ECU, each secured in place by a locking plastic tab. One at a time, squeezing on the plastic tab and connector (*not* the bundle of wires entering each connector), pull (not force) each of the three connectors out of the ECU. The three connectors are not interchangeable.

Remove the stock ECU. In some applications, you may need to cut one or two stock wires entering the ECU connectors, but normally you are immediately ready to install the AEM system. It may be easier to plug the AEM laptop serial cable into the ECU before you connect and secure the ECU. The AEM unit actually has four connector sockets, but only three are needed for the Honda wiring (unless you decide to use some of the myriad of fancy programmable AEM features, such as nitrous control, boost control, and so forth—which is one huge advantage of the AEM over the Hondata solution). AEM provides a forth plastic connector and pins that you can crimp to auxiliary wiring to control fancy performance equipment, but that is optional. Using some sturdy cable ties, secure the AEM system to the kick panel and route the serial cable under the carpeting to a convenient location to connect the laptop. Push the carpeting back into place. Get out your laptop and plug in the AEM ECU to the serial port (you *do* have a serial port, right?).

Done (well, except for installing and configuring the SEM software). If you have not already done it, insert the AEM CD-ROM in your drive and follow the instruction for installing the EMS software and documentation (manuals).

If you are running a stock Honda engine, you'll need to select the correct start-up map and download it into the AEM ECU from the laptop, and you should be able to load the calibration, fire up the car, and drive away. If you are running a turbocharger like we were, there will be a turbo calibration you can install, but *do not count on the calibration being right.* You will have to tune the car using the tricks described in the EMS Tuning 101 chapter in this book. You will definitely need a wideband AFR meter (or AEM's auxiliary wideband controller and UEGO sensor) to do this right. If you do not know what you're doing, it could be a great idea to start up the car and make sure everything seems to be running fine, and then have the car towed to an AEM tuning expert—preferably one with a chassis dyno.

In our case, the Hondata was running the car so smoothly and well that there was no opportunity for the AEM system to do any more than prove that it had equally good drivability and power. The AEM installation is as simple to install as you can possibly image. But an AEM system has a huge repertoire of tricks that can be used to handle add-ons like nitrous (four—count them—stages), and so forth.

remove any electric fans or anything, and we preferred to keep the hot turbine housing far from the radiator and condenser. On the del Sol, the radiator and air conditioning condenser are actually located side-by-side, with separate electric fans, which is nice because it keeps either system from interfering with the other's full thermal efficiency.

We discovered that there was room for a remote turbocharger in the vicinity of the stock air-cleaner box—if you removed the airbox and installed an inline K&N filter in the cold-air-intake port inside the passenger-side fender. So we built an exhaust tube connecting the T-25 manifold to the big Majestic T3-TO4 turbo mounted on a special bracket atop the transmission. We then built a downpipe to connect to the stock exhaust system below the engine, and connected the compressor to the air cleaner and throttle body with custom mandrel-bent steel tubing. It was a simple matter to remove the oil-pressure sending unit below the intake manifold at the center-rear of the engine and fit a T to allow pressurized oil to exit through a braided-steel hose to lubricate the turbo. I pulled the oil pan and welded on a large fitting for oil drainback, above the level of the sump.

CHAPTER 22
TURBO-EFI JAGUAR XKE 4.2

The last 4.2-liter six-cylinder XKEs were built more than 30 years ago. The E-type's triple-SU carb system had its own kind of beauty, but by modern standards these old carb systems are inefficient, unreliable, and dirty. Even though many surviving E-types have now been relegated to collector-car status, people with street-driven Es know they can be a lot of fun to drive rather than just to have and appreciate for their timeless beauty.

For those who want the reliability, power, and low exhaust emissions of modern engine management, it is surprisingly easy to convert these old cars to programmable fuel injection and spark controls. What's more, for people who appreciate the XKE's race-bred performance capabilities—and might even want to enhance the car's performance to 1990s standards—electronic engine management is fully compatible with headers, reground cams, high-compression pistons, ported and polished heads, turbochargers, superchargers, port nitrous injection, and all other performance equipment.

In fact, compared to tuning tricks of the 1960s and 1970s, programming a modern engine management computer with a laptop PC interface while driving or on the dyno is much easier, and the results are immediately apparent, particularly on the dyno. This is about how we converted one old E-type to computer controls and turbocharging, and then optimized tuning and power on a Superflow engine dyno. We ran the engine on the road and dyno with three different engine management systems: a vintage Haltech F3 fuel-only EFI system, a sequential EMS from EFI Technology, and later with a Haltech E6, a flexible batch-fire EMS with spark control and staged-injector capability.

This 4.2-liter Jaguar XKE was born in 1969, fighting for breath through twin Stromberg emissions carbs. It was the slowest of the entire XKE line, having lost its lovely triple SU carb induction system to the battle against smog in North American cities.

Jaguar no longer advertised horsepower in 1969, but clearly this car had less than the E-type's previously advertised 265, which was known to be optimistic. Although Jaguar advertised 265 horsepower in the XKE, experts say a really "on" early three-carb engine might measure 220 horsepower at the crankshaft, while the smogged-out engines used in later two-carb six-cylinder XKEs and the carbed or injected Jaguar sedans that continued as an

application for the 4.2-liter motor through 1987 had as little as 150–200 horsepower.

By the early 1980s, V-8 conversion kits, particularly using the compact 5.0-liter Ford small block, were more popular than ever for hot rodders who wanted to give an XKE a little more edge. Lots of people were attacking Jaguars with hacksaws and hammers.

I was willing to give this car some help, but it had to stay *Jaguar*, engine and all.

The 4.2 XK Jaguar engine was always a machine with great potential and was only eclipsed by four-valve pentroof technology in the late 1980s. Like the Offenhauser four that dominated Indy 500 racing for many years, the twin-cam Jaguar six engine had decades of development behind it, with roots going back before World War II. The block was produced in 2.4, 3.4, 3.8, and 4.2 versions (and even an experimental 12-cylinder version that was basically two siamesed flat-sixes, which predated the 5.3-liter V-12 engine used in the final Series III XKE from 1971 to 1975). The 4.2 XK engine, which continued in production through 1987, used Weslake hemispherical combustion chambers in an aluminum head, twin overhead cams, large ports, multiple carbs, and a tough seven-bearing crankshaft with castle-nut rods above a 9-quart cast-aluminum oil pan.

With tremendous low-end torque off idle, the car seemed to beg for turbocharging: the limited slip rear end would handle up to 600 horsepower; the semigradual buildup of turbo boost is reasonably forgiving on the clutch; the XK crankshaft and bearings are very strong and can easily handle large power increases; and the hemispherical combustion chambers resist detonation very effectively.

This project actually began in northern California in the early 1980s and continued into the late 1990s. The world didn't know as much about turbocharging in the early days; even factory turbos were not always very successful. Researching the project, I'd heard rumors of a turbocharged XKE or two, but that was it—rumors. The project required extensive research and experimentation, from reading SAE papers on turbocharging to fabricating parts and tuning and road testing. Over the years, the car had six different carburetion and fueling systems, various fuel pump and fuel pressure regulator combinations, two turbo man-

A 1969 Jaguar E-type in full flight. The car weighs in at 2,550 pounds, at least 600 of it engine and transmission.

EFI Turbo 4.2 Jaguar, Power and Torque				
Speed	Torque (1)	Torque (2)	Power (1)	Power (2)
1750	-	322	-	107
2000	285	311	108	118
2250	262	297	112	127
2500	268	303	128	144
2750	281	337	147	177
3000	289	418	165	239
3250	290	436	180	270
3500	313	440	209	293
3750	326	442	232	315
4000	352	438	268	334
4250	365	433	296	350
4500	380	415	326	356
4750	381	400	345	362
5000	238	369	227	351
5250	254	336	254	336

(1) 4.2L Six, 1 Rajay 300 series turbo, .6 A/R, 6 237cc/min Lucas disc fuel injectors, 1 Jaguar throttlebody, Turbonetics 5 psi Deltagate (actually allowed 12 psi boost!)
(2) 4.2L Six, 3 Airesearch T-2 turbos, 12 218cc/min Lucas disc fuel injectors, 6 TWM throttlebodies, no operative wastegates (12 psi boost)

To provide boost retard, the turbo conversion used a primitive vacuum-advance, boost-retard distributor canister to pull timing under boost as an anti-detonation countermeasure. MSD

ifolds, and several turbine housings and wastegates. There were some failures involving cracked pistons and broken rings, but ultimately success.

When converting an old Jaguar to EFI, there are two main ways to go:

Bolt on a Weber-type intake manifold in place of the stock SU or Stromberg manifold, and then, rather than installing the three dual-throat Weber DCOE carbs typical of this conversion, install three TWM Weber-type throttle bodies, identical in overall size and bolt pattern on both ends, with each throat equipped to install one or more Bosch-type electronic injectors downstream of each butterfly in the dual-throat throttle body. You must also install a high-pressure fuel supply and excess-fuel return line and then install an engine management computer, wiring harness, and sensors from an aftermarket firm like Haltech, Accel/DFI, Electromotive, Motec, EFI Technologies, Zytek, and others. You program the onboard computer by connecting it to a laptop PC running a graphical online interface and calibration software.

Adapt a 1978 or later Bosch L-Jetronic EFI manifold from an EFI 4.2 Jag sedan to the XKE engine, and either use the stock Jag sedan L-Jet computer, sensors, injectors, wiring harness, and fuel pump, or install and program an aftermarket computer as described above, perhaps using larger fuel injectors for higher-out-

put 4.2 engines. The L-Jet manifold bolts seamlessly to any 3.8 or 4.2 engine, but in the E-type engine compartment, the L-Jet manifold will require a 1-inch spacer-plate to move it outboard from the head plus minor modifications to fit within the E-type's front space frame.

There are many other potential E-type EFI conversion methods. You could potentially mount three one-barrel throttle bodies directly on the triple SU intake manifold, and then drill and weld six aluminum injector bosses into the six runners of the Jaguar SU manifold. You could fabricate an aluminum "sheet-metal" manifold from scratch, using a single Holley four-barrel-type throttle body. There are many possible conversion methods, but the L-Jet and Weber-type methods mentioned above are simple and make use of meticulously optimized existing intake manifold technology.

The project began with a complete strip down to bare metal, a new midnight blue lacquer paint job, a new interior, and a complete engine rebuild. That was when I tore off the emissions carbs and the bizarre OEM wraparound intake manifold that runs part-throttle intake air through an extra couple feet of heated runners on its way into the engine and bolted on an triple carb setup from a Series I XKE.

I equipped the engine with 8:1 compression cast aluminum pistons designed for 4.2-liter engines in Jaguar four-door sedans. (In those days, prudent people considered stock 9:1 XKE pistons to have too much compression for high-power turbocharging; 1978 Porsche 930 Turbos had 7:1 compression from the factory!) I painted the motor Ford-red from top to bottom like a Detroit diesel truck motor I had admired on a 12,000-pound super-modified pulling tractor. I siamesed two cast-iron Jag headers into a turbo flange, bolted on a 0.6 A/R Rajay 300E turbocharger, pressurized the carburetor float bowls, and rejetted the carbs. A Turbonetics Deltagate and miniregulator would enable cockpit-adjustable maximum boost anywhere from 5 psi to blowup (or maximum-achievable boost of the Rajay compressor).

More complicated was how to route the boost air from the turbo's compressor around the crossflow head to the carbs and where to route the exhaust from the turbine, all in a very tight engine compartment.

And what to do with all the extra heat.

As a first step toward improved Jaguar 4.2-liter performance, I had replaced the twin Strombergs with triple SUs. These are constant-velocity carburetors based on the following principle: As engine speed increases, liquid fuel flows more readily than air, meaning that, uncorrected, a standard carburetor would run increasingly rich at higher engine speeds. Constant-velocity carbs effectively change the size of the venturi with a sliding airfoil that enlarges the venturi as manifold pressure decreases, keeping air flowing through the venturi at a constant speed, which means only minor air correction is needed. This is easily handled with a tapered needle that gradually withdraws from the main jet as the constant velocity slide opens.

The trouble is, turbocharging at medium-to-high levels of boost requires highly enriched fuel mixtures to produce maximum power and prevent detonation. Installing jet needles with a richer taper helped but did not cure the problem, and given the fact that SUs were not designed for forced induction, the selection of jet needles was extremely limited. And it was hardly feasible to modify the jet needles with the ease that you might drill main jets to richen up the mixture.

Once the constant velocity slides were open wide, the air velocity would tend to increase—which would tend to richen up the mixture—but this was not predictable and not always repeat-

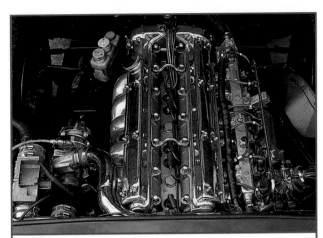

No XKE Jaguar ever had electronic fuel injection, but the 4.2 engine outlived the XKE and acquired EFI in the late 1970s. The initial EFI Jaguar 4.2 used in this project was a fuel-only Bosch L-Jetronic system from a 1979 XJ-6 Jag sedan with the distributor-points ignition. The L-Jet system adapted nicely to the XKE engine, with the VAF air meter relocated across the engine to the exhaust side, upstream of the turbo's compressor inlet. Fuel enrichment came from the boost pressure reference to the stock L-Jet fuel pressure regulator—and twin auxiliary fuel injectors located about 9 inches upstream of the throttle body for better mixing under control of an AIC from Miller-Woods.

able, given the wide variety of air pressures possible on either side of the throttles, combined with various possible throttle positions.

Trying to keep the SU running well under low-to-medium throttle and yet achieve high performance under boost was nearly impossible. To add insult to injury, the turbocharger on the E-type, blowing through the triple SU carbs, had a tendency to bang open the constant velocity slides under heavy boost with great force, sometimes causing them to stick. The moral is, don't ask a carburetor to do a fuel injector's job. By modern standards, carbureted turbo systems just don't work.

Meanwhile, companies like BAE and Spearco were achieving excellent results with turbocharging fuel-injected motors like the Datsun Z-car with its Bosch L-Jetronic injection or the Volkswagen water-cooled fours with K-Jetronic. Fuel injection seemed to offer possibilities. What's more, Jaguar had switched to Bosch L-Jet injection in 1978 on the same basic 4.2-liter engine used in the E-type (though this engine in reality probably didn't make more than 150–180 horsepower). But perhaps it was possible to convert over to EFI and make an L-Jet system work with turbocharging.

As luck would have it, I was able to buy a nearly complete 1979 Bosch L-Jet fuel injection system from a California company that converted Jaguar sedans to American V-8 power.

What was missing was the fuel injection electrical harness, a self-contained wiring loom that connects all the fuel injection components to the computer and has only a handful of connections that must integrate into the car's electrical system: tach/coil, continuous power, switched power, starter voltage, and so on.

Building a nice custom harness was appealing, but I was unsure at the time whether custom-building a wiring loom (I had the Bosch schematic!) would cause problems if the resistance of wires was not precisely correct, given that the analog computer controlled injection pulse width via sensor voltages. I bought a complete Jag wiring harness from a Jaguar salvage yard, which cost nearly as much as the rest of the injection system!

What was the fueling capacity of the L-Jet system in terms of horsepower? It turned out the stock L-Jet Jaguar injectors used a Bosch pintle design that flowed 214 ccs per minute. To convert this to horsepower potential (see the chapter on actuators), you multiply by 0.1902044, which means the stock injectors could deliver fuel for at least 250 horsepower at standard pressure (2.5 bar), perhaps a little more. But 250 is probably a 70–100 horsepower gain over a stock normally aspirated Jaguar 4.2.

I removed the old Jaguar intake manifold and carb system and tried out the new EFI manifold. Several problems: The space frame that surrounded the E-type engine was in the way of the EFI plenum. The plenum, if located outward somehow to miss a longitudinal space-frame rail, would then interfere with the space frame another place. The plenum interfered slightly with the firewall above the passenger foot well.

I solved these problems by building a 1-inch thick spacer that located the manifold directly outward, so the EFI runners passed over the XKE frame rail and located the plenum just outside the rail. I then modified the passenger-side outboard diagonal frame rail slightly so that the plenum would miss *it*. Finally, I reworked the firewall sheet metal slightly to give it clearance above the passenger footwell.

The manifold now fit in place, but another problem appeared. The L-Jetronic Jaguar 4.2 manifold is built to fasten to a 4.2-liter engine with studs and nuts. The geometry of the downward-curving intake runners and the fabricated spacer thickness would not allow bolts of sufficient length to make it all the way through to be inserted through the manifold flange holes on the bottom side. But the use of studs uncovered a different problem: although the manifold now fit in place, it did not have the clearance to *slide* straight onto the long studs needed to secure the manifold and spacer-plate.

The answer was to slide the studs through the spacer and manifold flange from the engine side, lower the whole assembly into place from above, then screw in the studs (by hand, with Vise-Grips, with double jam nuts, whatever worked), then assemble the nuts and tighten. Some modern studs are built with a provision for tightening the stud in place with a recessed Allen head in the outer tip, which would have made the operation a little simpler, but I found no such thing easily available. This was not an operation you'd want to do very often (though through the early life of the project, I got pretty good at it).

With the Jaguar L-Jet intake manifold mounted, it was a simple matter to route the turbocharger compressor outlet plumbing around behind the head from the driver-side turbo into the EFI throttle body on the passenger side of the crossflow head.

I relocated the L-Jet vane airflow meter (normally near the throttle body) to the driver side of the car to be upstream of the turbocharger (a velocity air meter must be upstream of the turbo to avoid damage from the intermittent surging action of the turbocharger's output. A mass airflow (MAF) sensor should be *downstream* of the turbo and the intercooler to correctly measure air density after compression heats the air. Locating the 1979 Bosch L-Jet system upstream of the turbo initially seemed to be impossible, given the space constraints imposed by all the intake, exhaust, and lubrication-system plumbing in the vicinity of the turbocharger.

But eventually I was able to design inlet-air plumbing between the air meter and the turbo compressor inlet that allowed the cockpit heater-blower unit to remain on the car (initially removal had seemed essential), while permitting reasonably efficient airflow into the VAF meter without turbulence that might have adversely affected sensor readings. Then, by carefully routing the unmodified stock L-Jetronic wiring harness, it was possible to make the wiring connection to the air meter while allowing the harness to mate with other L-Jet EFI engine sensors and injectors on the right side of the block.

The original L-Jet system was connected to an electronic distributor, but the coil-negative terminal turned out to provide a fine engine-speed trigger signal for the L-Jet computer.

The stock L-Jet ECU is located in the trunk of '79 Jag sedans (a good place that is out of harsh weather and far from EMI radio interference in the engine compartment). The harness was easily long enough to reach the XKE trunk, only requiring cutting fairly large holes in the firewall and rear trunk bulkhead with a hole-saw mounted on an electric drill.

I augmented the stock XKE in-tank electric fuel pump (low pressure) with a high-pressure inline Bosch L-Jet unit, connecting the new pump directly to the L-Jet ECU's pump electrical connection. The relay-activated L-Jet fuel pump circuit is designed to activate the pump for a few seconds upon initial energization of the ECU, run the pump during cranking and while the engine is running, and kill the pump in case of accident if the engine stops with the ignition still energized or—if the engine is not dead—if the multi-G crash-activated inertia switch gets triggered.

Since an L-Jet system (like all pulsed electronic EFI systems built before the mid-1990s) requires a fuel return line, I had to fabricate a second fuel line so that excess fuel exiting the fuel pressure regulator on the fuel rail would have a way to return to the fuel tank. XKEs have a bolt-on metal access plate with a gasket on top of the fuel tank that contains the fuel tank gauge-sender and fuel pickup, and it was a simple matter to remove the plate and drill and tap it for a fuel return dump.

Had I wanted to get fancy, I could have constructed a small tank or bucket within the main fuel tank so returning fuel would enter the main tank through the auxiliary tank, from which fuel would also be drawn for the engine. In this way, violent multi-G maneuvers and steep hills will not cause the pump to starve when fuel level is very low, and returning fuel can immediately be suctioned up for return to the engine when fuel is nearly exhausted. However, since the XKE already incorporates a deep auxiliary well in the bottom of the tank for the stock returnless fuel system's pickup, I concluded this was not necessary. However, anyone building a return line should make sure a fabricated return line is not spraying fuel on the fuel level sending unit. For an EFI fuel filter, I used an inline unit from a Datsun Z-car with stock L-Jet injection, locating the filter below the fuel rail in the engine compartment.

I did not initially use the stock O_2 sensor that came with the L-Jet EFI system, since the exhaust manifold did not have a threaded boss for one. Very early Jag L-Jet systems did not use an O_2 sensor, and later systems will work fine without one (assuming you don't ground the O_2 sensor wire). The advantage of an O_2 sensor is that it permits the ECU to trim air/fuel mixtures at idle and light cruise for lowest emissions and good cruise fuel economy by using closed-loop feedback algorithms in which the ECU correlates the results of recent injection pulse width trim to residual exhaust gas oxygen, making adjustments to pulse width to target the chemically perfect 14.7:1 air/fuel mixture.

I wired the EFI harness's constant ECU power connection to the battery and switched power to the key behind the dashboard (plenty of harness length).

With everything connected, it was time to start the vehicle.

The engine cranked, it made a little stumble, then nothing. I began checking electrical connections everywhere. In the end, I finally pulled all the early EFI-style hose-fed electronic injectors

out of the manifold, plumbed everything back together and cranked the engine.

Pfffuuuufff! All injectors batch-fired a *huge* burst of fuel twice per power stroke. The injection was working, but something was wrong.

Eventually I discovered that a single tiny connector in the six-wire plug to the air meter was pushed out of place, causing the sensor to report maximum airflow and triggering a huge injection pulse width, immediately flooding the engine. Once this was this fixed, the engine fired right up and ran great.

That is, it ran great until I *drove* it, whereupon there was a big, ugly flat spot during warm-up, after which it ran fine at normal operating temperatures. It turned out the XKE engine had the wrong thermostat for EFI (too low temperature to ever get the engine management all the way out of warm-up enrichment mode. It also had a defective coolant temperature sensor (keep in mind that sensor problems, vacuum leaks, and so on, are the biggest cause of EFI problems and failures).

When it came time for a wide-open throttle run, it quickly became clear that the car was leaning out during boost above 5 psi. The initial solution was to wire the L-Jet cold-start injector to kick in above 2-psi boost. An L-Jet cold start injector is a binary, nonpulsed device (either off or constantly on) designed to squirt a continuous stream of pure fuel directly toward the throttle plate when the engine is cranking at cold temperatures. The cold-start injector is controlled by a thermo-time switch and a cranking-voltage sensor. I discovered it was feasible to wire it up with relays so it could still perform its cold-start function yet also provide enrichment fuel under boost. This was a very crude method of fueling. The cold-start-type enrichment helped make power and stifle knock, but I realized there were one or two lean cylinders when boost began to get serious, and that, as you might expect, enrichment fuel was not making it evenly to all cylinders.

Eventually I acquired an additional injector controller (AIC), a computer box that could pulse one or two auxiliary injectors under boost to provide accurate enrichment in response to engine speed (via negative coil connection) and manifold pressure (via reference hose connected from a hose barb on the compressor-out plumbing and running through the firewall to the AIC's internal MAP sensor). Installation instructions called for silver-soldering or MIG-welding the two injector bosses to the inlet air plumbing 9 to 12 inches upstream of the throttle body. It was a simple matter to supply fuel for the additional high-flow injectors by tapping into the high-pressure fuel supply line anywhere upstream of the main fuel rail. It was a similarly simple matter to tap into the negative side of the coil to acquire engine speed (exactly like the L-Jet ECU).

Under boost conditions, the AIC kicked in to pulse the two additional injectors at least three times per engine revolution to provide the 12.5:1 air/fuel ratio required to fight detonation and provide a big power boost.

In order to check the efficiency of the fuel pump at higher levels of fuel flow and pressure, I plumbed a mechanical gauge into the fuel line where you could see it while driving. Fuel pressure actually topped out around 45 psi, indicating that the stock L-Jet pump was at its limit, and that the engine was not actually receiving the full benefit of the AIC at levels of boost above 7 or 8 psi. I installed a more powerful in-line fuel pump and verified that fuel pressure climbed in a one-to-one correspondence with boost pressure from 38 psi at zero vacuum, to as much as 53 psi at 1-bar (15-psi) boost.

There was one more required modification to help fight detonation. In the early days of this project, I acquired a GM HEI vacuum advance-boost retard canister, modifying it to fit the

Engine dynos have a selection of common bell housing/flywheel/clutch adapters (i.e., Chevy small block, Ford Windsor, and Chevy BB)—but not Jaguar 4.2. I built this dyno adapter by cutting down an auto-trans Jag bell housing and welding on a rear plate designed to mate to a Chevy SB transmission.

Jaguar points distributor and running a manifold pressure reference line from the intake plenum. Adjustable stop-screws limited total advance or retard.

There was a *dramatic* increase in performance with the extra injectors, enabling boost to increase without detonation from an intermittent 8 or 9 psi to as much as 15 psi (which illustrates the importance of a good rich mixture under higher boost). Out on the highway, power arrived on the turbo Jag with a bike-like rush. All was well in the world. For a while. But power is addictive.

When it came time to consider additional modifications, I formulated the following goals:
1. Standalone programmable ECU, to eliminate the kludge aspect of running two computers to do what could be done by one.
2. Eliminate any lingering fuel distribution problems by providing *port* fuel enrichment under boosted conditions rather than enrichment at a single point with additional injectors.
3. Huge fuel pump to make sure the injectors always had enough fuel pressure under the most extreme conditions.
4. More power.
5. Completely eliminate turbo lag.
6. An exotic look.
7. More efficient engine compartment layout that eliminated certain space considerations of the earlier system that made it very hard to service certain parts of the engine.
8. Intercooling.
9. Programmable electronic ignition advance as opposed to the earlier-generation vacuum advance-boost retard distributor that protected the engine but was not an ideal ignition advance curve.
10. Clean emissions.
11. Reliability. This had to be a street-driven car.
12. Something different from anything else around.
13. Eliminate intake restrictions from the L-Jet air meter and the complex plumbing that enabled the air meter to fit in place upstream of the turbo while retaining the stock heater.
14. Electronic overboost control instead of the mechanical dial-a-boost used in the earlier-generation car.

15. Eliminate the water injection system if at all possible.
These were all reasonable goals.
The initial idea was to install a Haltech fuel-only EFI system on the car—basically just as it was. This would have met many of the above goals, but not all. It would have been simple to remove the Jaguar L-Jet ECU along with its wiring, air meter, and other sensors and install the Haltech on the L-Jet Jaguar injection manifold with its single turbo. With a larger fuel pump and perhaps a regulator change, goals 1, 2, 3, 6, 9, 10, 12, and 14 would be accomplished.

Eliminating turbo lag could be done by adding nitrous oxide injection to turbocharging, which is an excellent combination of power-adders in which the nitrous provides great low-end torque and power while the turbo provides great mid- and high-rpm power. Perhaps a second set of staged injectors would work out well to provide additional fuel to burn the nitrous.

However, staging injectors meant that there had to be a way to mount two injectors per cylinder.

TWM was making DCOE Weber-type throttle bodies with bosses to accommodate two injectors per intake port. Of course, that meant that you needed a Weber-type manifold. TWM, it turned out, could provide that too.

From this point on, the design goal became take it to the limit: twin turbos with liquid cooling. No, more than two turbos—try three! Custom-fabricated turbo exhaust manifold . . . twelve injectors . . . one throttle per cylinder . . . port nitrous . . . speed-density injection with computer-controlled nitrous . . . crank-triggered ignition.

I acquired three tiny Airesearch T-2 turbochargers, the kind once used on the old 1.5-liter Nissan 200SX Turbo or the Buick Skyhawk Turbo—a small, efficient turbocharger with integral wastegate and actuator. The challenge was to build a good turbo exhaust manifold that could mount three turbos. After building two 1/2-inch-thick header flanges (front three and rear three cylinders, just like the stock dual headers) and turbine inlet and outlet flanges, I removed the original Rajay 300 turbo exhaust manifold and spent some time playing with the three turbos, searching for the right design, the right geometry, the right materials. You couldn't build robust headers with standard thin-wall header tubing. In any case, there didn't seem to be any way to get excellent exhaust flow. Log manifold? Not!

Finally, using tight-radius black steel pipe, mounting the turbos in a transverse position next to the in-line engine, it turned out there *was* a layout to make it all work. Working with a band saw and a TIG welder to build turbo headers, when it was finished—all surfaces ground smooth and ready for HPC coating, I took a look and began laughing. The design was beautiful, the flow smooth, poetry in black steel. But somehow the thing had the look of three babies' butts mooning you. So we dubbed it the Heiney Manifold, and proceeded to build a downpipe and exhaust. The final result was a triple exhaust system that bolted to the three turbine-out flanges.

On the intake side, a fabricated three-into-one compressor-out pipe crossed over to the right side of the engine to plumb into the Weber manifold. The original idea was to have an air-water intercooler on the intake side of the engine connecting to all three turbos, but at this point forced-induction expert Corky Bell designed a way to construct a custom intercooler that mounted in front of the Jag radiator. Bell notched the top of the Jag radiator, lowered it about an inch, and constructed a gigantic intercooler from Thermal Transfer Systems (Dallas–Fort Worth area) intercooler cores. Even though the intercooler blocked direct airflow into the radiator, the intercooler was set at such an angle that there was room for air to rush around the sides to provide reasonably good radiator efficiency. Bell fabricated an intake plenum outboard of the TWM throttle bodies and plumbed intercooler-out air to it from the front of the car. The final touch was a heat shield that kept exhaust manifold heat away from the compressor-out piping.

In the meantime, I installed 12 288-cc/min Lucas disc injectors in the TWM throttle bodies and moved over to the ignition side of engine management, where the first thing we needed to run was engine position information.

I removed the distributor and modified another Jaguar electronic distributor to include the flying magnet cam trigger (sync) you need in order to run sequential injection or direct ignition.

PREPPING FOR SEQUENTIAL INJECTION AND DIRECT IGNITION

We obtained a sequential injection EMS from EFI Technology, and set about to prepare the Jaguar 4.2 for one of the most sophisticated aftermarket engine management technologies available. This was a top-of-the-line race system typically used on expensive tunercars and cost-is-no-object Indy-500-type race cars, competition offshore boats, and other all-out vehicles in which flexibility and reliability are vital. The EFI Tech ECU was designed with mil-spec wiring and enclosures to be absolutely bulletproof.

EFI Technology has added logic to activate nitrous, calculate nitrous fueling via port injectors, and to pulse a nitrous solenoid rapidly like an injector to provide *proportional* nitrous flow. Proportional nitrous can come on very gradually or in a scorching sudden blast, and the nitrous flow can be regulated using pulse width–modulating techniques. The ECU provides spark timing, using a cam trigger set up to provide sync information 45 degrees before TDC as the required reference for compression stroke that the number one cylinder needs for sequential injection. For a six-cylinder engine, six flying magnets provide crank position information to a crank ref trigger six times per revolution. I removed the crank damper to drill and tap holes that could be used to mount the six small flying magnetic bolts next to the pulley groove in the damper.

When using distributor and crank triggers to provide spark timing and sequential injection, it is necessary to coordinate the position of the distributor body (which may or may not be adjustable), the position of the sync trigger in the distributor body, and the clocking of the positions of the distributor rotor, distributor flying magnet, crank trigger, and flying magnets on the crank.

It was a simple matter to offset the number one cylinder crank magnet 15 degrees before TDC, separated from the other flying magnets in 60-degree intervals. This must be done by a competent machinist to high standards of precision to avoid spark scatter and other EMS problems. For example, really sophisticated engine management systems might use exact timing of crank reference pulses to detect micro-changes in crankshaft acceleration and deceleration that indicate misfires.

Next, we positioned the crank pulley timing mark exactly over the timing pointer at TDC to index the distributor. The trick here with the distributor was that I had built a slot that held the cam position sensor and allowed 45 degrees of possible sensor movement. With EMS-controlled spark using crank reference triggering on the crankshaft, rotating the distributor will no longer affect ignition timing, but the rotor must be more-or-less centered in the vicinity of the number cylinder, so that when the spark occurs (always between, say, 0 degrees TDC and 40 degrees BTDC) so that all degrees of timing advance and retard provide

a narrow gap for the spark to jump from the rotor electrode to the high-voltage distributor cap plug terminals when the ECU, coil driver, and coil generate a precisely timed spark.

Accel/DFI recommends the following procedure for indexing the distributor using a strobe light and a clear distributor cap :

Compute the mid position between least spark advance and most spark advance.

Put the engine at this position on compression stroke number one cylinder with a degree wheel, using degree markings already on the flywheel or damper, or by measuring the distance from TDC to some marked BTDC position (for example, suppose 10 degrees BTDC is one inch from TDC, and you are trying to locate 21 degrees BTDC. 21 ÷ 10 x 1 (inch) = 2.1 inches from TDC. Or suppose 10 degrees is 3/4 inch from TDC. 21 ÷ 10 x 0.75 = 1.575 inches.

It is not critical to get the cam trigger perfect on systems with crank-triggered engine position sensors. The cam trigger is a reference indicating compression stroke, not a marker of engine position, but it should be within 10 degrees of 45 degrees BTDC, according to EFI Technologies. However, it is truly critical to get cam triggering perfect if you are not using a crank trigger for ref position.

With the engine at this position, point the rotor in a convenient direction for organizing plug wires and such (remove the distributor, if necessary, to index the distributor drive gear). The Jaguar distributor can only point in two directions since it engages a slot in a driven gear, not the crankshaft or camshaft itself (though the distributor shaft drive gear can be set in various positions on the crankshaft drive gear from the bottom of the engine with the oil pan off!).

Install the distributor cap, place the number one plug wire into an appropriate hole in the cap, and center it over the rotor.

Now move the engine to 45 degrees BTDC, observe the position of the flying magnet, and adjust the trigger so that it is exactly above the magnet. That's it.

We had a valid and optimized Haltech fuel map available from a Jaguar 4.2 engine in another vehicle with a similar TWM induction system, and I decided to use it to start the XKE with a Haltech F3 fuel-only ECU, get things sorted out and working correctly, and then move the Haltech to another project vehicle and switch the Jaguar over to the EFI Tech EMS. This Jag 4.2 had higher-flow injectors, but the map could be rescaled for the bigger injectors using the Haltech Delta feature, which changes entire map ranges or segments of ranges, or even all ranges by a percentage. Naturally, we'd have to calibrate the speed-loading breakpoints above 1 bar with tune-and-test operations on the Dynojet or on the street, and we'd have to calibrate the staged injectors. With the recalibrated Jag map in the ECU, I turned the key on the XKE; it fired up immediately and idled nicely.

After getting the vehicle running nicely with one EMS like the Haltech, it should be a relatively straightforward matter to switch over to another system like the EFI Technology unit. This is slightly less true when switching from a batch-fire system like the F3 to a sequential system like the EFI Tech system, because the sequential system only fires each injector once per power stroke, while a batch system fires all injectors at once, once per engine revolution. The translation goes further to hell if you are switching from an EMS like the Haltech—which runs six primary injectors all the time and stages the second set of six injectors only during heavier engine loading—to a system like the EFI Tech EMS which fires two injectors per cylinder all the time, unstaged. Assuming both ECUs use peak and hold or equivalent saturated circuitry to open injectors at the same rate, the translation can be a simple matter of some easy calculations to adapt the program-

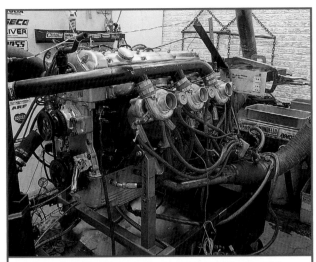

On the Kim Barr Superflow engine dyno, we tested the Jag 4.2 in single-turbo, L-Jet intake configuration, then installed the triple-turbo, individual-throttle intake. The engine made more than 300 ft-lb torque at 1,750 rpm, over 460 at 4,750.

4.2 XKE EFI CONVERSIONS ON THE ENGINE DYNO

In the end, we tuned and tested both Jag XKE EFI conversion packages on a Superflow engine dynamometer, and for good measure built up separate single- and triple-turbo forced-induction manifolds and plumbing to increase the power massively. The key to making everything work right was tuning and dyno-testing our new engine systems on a Superflow engine horsepower dynamometer. Our turbo 4.2 Jag was such an oddball that we had to meticulously develop our own air/fuel and spark timing maps. For most engines (including stock 4.2 Jags), excellent maps have been already been developed by experts and many are available from your aftermarket engine management system dealer; you may still want to fine-tune the 4.2 engine on a dyno to truly optimize the power and torque. Here's how.

The Superflow engine dyno is very flexible. It can be used to perform automated power pulls, which last a matter of seconds and provide a readout of power and torque across the entire rpm range from as low as 1,750 rpm to redline. This quick and simple measurement reveals your engine's capabilities and produces the chart of horsepower and torque versus rpm we're all familiar with from countless magazine articles. But an engine dyno is also extremely useful for tuning and computer calibration.

In manual mode, eddy-current or water-brake dynos like the Superflow can hold the engine with a specific load while you work to calibrate tuning of the engine's onboard computer. By varying the dyno load and engine speed and moving exhaustively through all the available breakpoints of the computer's internal air/fuel and spark timing tables, you can manually build raw fuel and spark tables, cell by cell. Each defined breakpoint in the fuel or spark tables is optimized for best torque or best economy, or lowest emissions, and the engine management system provides fuel and spark for speed and loading that falls between breakpoints by computing a moving average over several of the closest defined breakpoints.

Most programmable engine management systems provide automated tools that can speed up this process by quickly setting certain classes of elements in the fuel and spark tables at once to the same value, by linear smoothing of the fuel delivery or spark timing for a range of elements between two specified breakpoints, or by modifying the values of a specified range of elements by a percentage. However, particularly where perfect tuning is important—especially on competition or emissions-controlled engines—tuners typically spend many hours with engine dynos, working to squeeze every last pony, or every last mile-per-gallon, or the cleanest exhaust from the engine—fine-tuning every breakpoint.

There are other advantages to using an engine dyno for tuning EFI conversions or heavily modified EFI engines. An engine dyno can work an engine harder than it might ever be worked on the streets, holding extremely high fixed levels of power over long periods when necessary. By carefully monitoring and datalogging exhaust gas temperatures, coolant temperature, and intercooler efficiency on the engine dyno, it is possible to develop an engine and EMS combination that is far more robust than what is possible on an inertial chassis dyno or safely achievable on the street or track. The engine dyno can load the powerplant to levels of torque and airflow that might be impossible to achieve on the street other than momentarily, due to tire spin or other factors—for perfect tuning of even what are otherwise the most transitory engine events.

On the dyno, you can scientifically test various engine modifications and get the real story about what works, when it works,

ming results of one ECU to the other. In any case, a known-good, optimized calibration is a great place to start.

It was straightforward to install the EFI Technology EMS. Our GM TPS was a little different from the shaft-mounted TPS used by vintage Haltech systems, though certainly not a problem on the TWM throttle body (one of which has a special boss for the TPS). The NOS system had recently gained the capability to run in closed-loop mode with an O$_2$ sensor, so we had the option to weld an EGO sensor boss somewhere in the exhaust system, possibly using a heated sensor if the location was downstream of the heat-robbing turbos.

All EFI Technology wiring harness connections are unique, so it is impossible to confuse which connector connects to which sensor or actuator. One nice thing about the EFI Tech Competition ECU was that it was set up to provide diagnostics with a check-engine-type light located somewhere within the driver's vision. If there is an ECU-detected failure, for example, of a sensor, the light will flash out a trouble code, enabling you to know what the system thinks is wrong, even if there's no laptop handy.

The EFI Tech unit was designed for racing use, so it did not do frivolous things like stabilize idle via IAC, or control knock sensor (in racing, the engine is expecting good race fuel, not 87 octane junk). However, the EFI Tech EMS did have some really nice features, such as the ability the change the granularity of the fueling map in a nonlinear fashion, for example allowing rpm-based breakpoints to occur more frequently at some parts of the scale than others. This is a great feature for engines in which a linear interpolation between certain standard breakpoints of loading and rpm do not approximate an ideal fuel curve. You can change the granularity until the linear interpolations *do* sufficiently approximate the perfect fuel curves. Our EFI Tech ECU could control proportional nitrous, and the ECU could be programmed to control electronic wastegates. In fact, the ECU was set up to potentially control at least 15 classes of sensors and actuators!

An EFI Tech EMS of this vintage represents various segments of the fueling map as line graphs, in which pulse width is plotted as a function of rpm and loading. If you change the map, the change first appears as a dotted line, then changes permanently to a solid line when you save the changes to the ECU.

and how well it works. The guesswork is gone. A driver can't accurately detect 8 horsepower, for example, by seat-of-the-pants testing, but the dyno can. Not only can one develop an exceptionally high state of tune, but any problems or weak points can be revealed in the strenuous environment of the dyno and fixed before the engine goes in the car.

THE DYNO CELL

In the dyno room itself, the engine sits on a stand with its clutch and bell housing connected to the water brake section of the dyno, with the engine mounted on a stand. The dyno environment provides a fuel supply, coolant to the engine's water pump, a flexible exhaust system routing gases to the outside atmosphere, and an oil cooler to keep oil temperature under control.

At Kim Barr's Superflow dyno in Dallas, where we tested and tuned our conversion Turbo EFI 4.2 Jag engine, a huge aircraft oil-cooler attaches to the engine's lubrication system, and the cooler itself is submerged in a barrel of water; total-loss cold water continuously flows through the barrel at a designated rate and into the dyno building's plumbing system. Enormous fans suck outside air through the dyno room to keep things cool, and portable fans blow directly on the intercooler, exhaust, and any other components that require special cooling. With the fans off even momentarily, a large turbo motor like our 4.2-six radiates massive heat. The headers will rapidly glow red in a darkened room if the engine is running even briefly at high loading with the fans off. With two 2,000-gallon backup water storage tanks filled with water that will be violently churned and heated as it passes through a 1,200-horsepower water brake, Kim Barr's dyno has been upgraded to allow it to measure up to 2,000 horsepower without turning the water into a boiling cauldron. Just to be on the safe side, we ran 107-octane Sunoco race fuel through the turbo Jag 4.2 continuously and never saw a hint of detonation.

CHASSIS DYNOS

Once an engine is off the Superflow and back in its vehicle, inertial chassis dynos like the Dynojet are very useful. On the Dynojet, you maneuver the driven-wheels of a vehicle onto two large 1,600-pound single knurled rollers and strap the vehicle in place for testing. The Dynojet's computer and sensitive accelerometers measure the rate at which the engine is able to accelerate the drivetrain and rollers, and then the computer transforms this data into charts of torque and horsepower numbers and graphs of horsepower and torque throughout the acceleration range. Systems like the Dynojet are extremely useful, although these inertial dynos are usually used for quick 2,000-to-redline runs, similar to the automated Superflow pulls described above, rather than the steady state tuning of the Superflow. Good tuners routinely use Dynojet logs to very effectively optimize tuning; just usually not in real time.

THE ENGINE

We dyno tested a freshly rebuilt 1982 4.2-liter six-cylinder EFI Jaguar engine on the Superflow. The long block had 0.030-over 8:1 compression forged pistons, chrome rings, an O-ringed block, and an annealed copper head gasket, and was otherwise internally stock. The engine, which had been extensively garage-tested out of the car at zero-load before arriving at the dyno, was to be the powerplant for a 1969 Jag XKE convertible that I was using as an R&D project vehicle.

Externally, the engine was most definitely *not* stock. The stock Stromberg carbs, induction system, and exhaust headers were gone. In their place we had constructed two complete EFI/intake systems and two complete turbo systems for the engine, which were standing by, ready for testing.

The engine controller was a programmable Haltech E6 engine management system with the ability to control up to 12 port injectors, providing batch-fire injection (two squirts per power stroke) simultaneously to all six primary injectors, with secondary injectors staged into operation at a preconfigured airflow or manifold pressure.

The plan was to initially install and test our stock-type 1979 4.2 Jag sedan-type L-Jetronic manifold with six replacement high-flow injectors and a stock single throttle body, good for 300-plus horsepower worth of fuel. We would then test a TWM triple Weber-type intake manifold equipped with three DCOE-type dual-throat TWM throttle bodies with one throttle per cylinder and twin port-injectors firing directly at each intake valve. This system had the ability to deliver more than 500-horsepower worth of fuel and was equipped with a massive air-air intercooler for the turbocharger.

On the exhaust side, we would first test a single-turbo system based on stock XKE headers fed into a custom two-into-one Y-pipe plus a large Rajay-type single turbine flange bolted to one large turbo and a single exhaust system with an external Turbonetics Deltagate. We then planned to test a triple-turbo power-adder system using three custom two-into-one headers built from tight-radius black-steel pipe bends welded to twin stock-type thick header flanges. Each individual two-port header supported a tiny Airesearch T-2 turbocharger mounted on its own separate turbine-in flange. Each of the three turbos was equipped with an internal wastegate, and each turbo had its own individual exhaust pipe.

We planned to calibrate our Haltech E-6 programmable engine management computer under controlled conditions where we could keep an eye on exhaust gas temperature (EGT) and air/fuel ratios as we loaded the engine to increasing levels of power.

In summary, our EFI-Turbo 4.2 Jag engine system consisted of a combination of 1970-vintage, 1980-vintage, and brand new parts plus newly fabricated equipment—all working together. Fitted with multiple turbos, 12-injector EFI, closed-loop O_2 sensor self-calibration, and automatic idle speed stabilization, this is a fairly complex engine. We wanted to get the engine running really right and did not want to burn it down on the way. We also wanted to test certain performance modifications and find out what really worked at medium to high levels of boost. We wanted to be scientific. This performance tuning process can be scary in the case of an oddball turbo engine like the 4.2 Jag, which no one has ever seen before with high-boost turbos and electronic injection, and which probably no one knows the limits of. We wanted to get past the guesswork and get the system working perfectly before we turned the key with the engine in the car.

We worked with Bob Norwood of Norwood Autocraft in Dallas. Haltech engineer Shane Scott, who wrote internal software that operates the Haltech E6 engine management computer, was also on the scene. Kim Barr, drag racer and dyno expert, owns Kim Barr Racing Engines in Garland, Texas, where the Superflow dyno is located. Kim's shop had built the Jaguar 4.2 short block.

COOL RUNNING

Working with the 1979 L-Jet manifold and single-turbo exhaust, we got to work. Before initial startup, we installed theoretical calibration data in the E6 targeted at the Jag 4.2 six. Based on about seven parameters, Haltech's QuickMap facility downloads computed fuel pulse width data and a best-guess timing map from a library of thousands stored on laptop floppy in the Haltech programmer directory.

Our engine started immediately and ran decently and warmed up while we dealt with some minor leaks and such. For a while we

This tri-turbo XKE engine is back in the car, complete with proportional port nitrous injection under EFI Technology EMS control.

engine in its six-injector, one-turbo configuration.

Following calibration, Kim Barr ran several 1,750–5,500 rpm dyno runs in auto mode. We achieved more than 381 ft-lbs of torque at 4,750, and as much as 345 horsepower at the same speed.

At this point, we let the engine cool and then converted to the 12-injector manifold with its three, dual-throat throttle bodies, each throat containing twin injectors of 218 cc/min capacity of the type used on Chevy 5.7-liter TPI motors. We also installed the second turbo system that supports three Turbo Power internal-wastegate Airesearch turbos. This configuration also included an intercooler. Each injector was capable of providing fuel for about 41 horsepower, for a total of roughly 500 horsepower with all 12 injectors in play. The E6 required an add-on driver box that allows it to

fought high-voltage spark problems. A minor problem showed up in the form of oil leaks from the cam covers and dipstick. In the meantime, once the engine reached operating temperature, it was clear our base fuel map was rich. Norwood worked on the idle maps with a laptop PC, lowering bar graphs with arrow keys to shorten injection pulse width at light loading manifold pressure.

With a good clean idle, Norwood now began calibrating the engine on the dyno in full manual mode. The process was:
1. Establish a speed and loading via throttle and dyno load control, work the Haltech mixture-trim pot to hunt for the right mixture, watching the UEGO and reading the dyno torque meter to find lean best torque (LBT).
2. Center the mixture trim dial and permanently change the appropriate element of the fuel map.
3. Move to another level of loading by raising the throttle and adding dyno load (an arrow on the laptop will point at the loading bar graph currently fueling the engine).
4. Adjust fuel for best torque, and move again—until all speed-loading breakpoints for this rpm range are adjusted.
5. Raise engine speed 500 rpm to match the next defined Haltech rpm range.
6. Adjust the mixture for various engine loading bars in this rpm range.

This process is especially tough with turbos. Compressors make boost, feeding back horsepower, which supplies increased exhaust gases, which add boost, which changes speed, which changes load, and so on. The process requires a deft hand, and plenty of experience. Tuning carbs, you'd change idle mixture, main jets, emulsion tubes, and power valves, and so on, but with EFI, you play with arrow keys on a graph and a computer. Spark timing is adjusted similarly to fueling.

Norwood eventually completed calibrating the 4.2 Jaguar

drive the additional six staged injectors.

With new injector sizes and changed volumetric efficiency, Bob Norwood began the meticulous, manual-mode recalibration of air-fuelair/fuel ratios and spark timing, coordinating dyno load, throttle, fuel, and timing.

In this configuration, the motor made 322 ft-lbs of torque at 1,750 rpm, peaked at 442 ft-lbs of torque at 3,750 rpm, and made 362 horsepower at 4,750 rpm, a 90 percent increase over stock! The big 4.2 six made it look and sound easy. Compared to the wild big-block Chevys that produce such radical noise outside the dyno building that Kim prefers to run tests at night so nearby businesses in a light-industrial area don't complain about the shattering roar of a wide-open V-8, the un-muffled turbo Jag had an eerily quiet, even-fire in-line-six purr you might mistake even at full throttle for the purr of a big turbo-diesel generator. Turbos help to muffle the exhaust note to the degree that, installed in the car, the engine does not even require mufflers for street use. The three turbos maxed out at about 12-psi boost with the wastegates closed, indicating that there is probably a lot of backpressure at higher rpm and boost, limiting the total amount of horsepower at higher rpm.

CONCLUSIONS

What did we learn?

We learned that a 4.2 Jag motor, stretching back to near-World War II vintage, will live on a dyno for long periods, making huge amounts of torque and impressive horsepower.

With the engine back in the car, we learned that a three-turbo, injected XKE with 441 ft/lbs ft-lbs of torque at 3,750 rpm is awesome to drive.

On the streets we learned that well-tuned EFI always starts, doesn't stall, operates smoothly, and supports extreme levels of power without detonation.

TURBO-EFI JAGUAR XKE 4.2

MATING A JAG 4.2 WITH THE SUPERFLOW ENGINE DYNO

Most dynos are designed to mate easily with a few common performance engines. Our Superflow had adapters for the small-block Chevy V-8, the big-block Chevy V-8, and certain large Ford V-8s. An engine dyno typically provides a simulated transmission input shaft designed to mate to one of the above engine's clutch/pilot bushings, plus a plate that bolts to the rear of the equivalent bell housing and connects via a semi-flexible coupling to the dyno's water-brake assembly.

We built a bell housing that was "Jaguar" on the engine side and "small-block Chevy" on the transmission side. We did this by band-sawing off the rear of a damaged Jag bell housing, and then welded on a thick aluminum plate machined to match the rear pattern of a Chevy housing—using a $25 junk Chevy three-speed tranny as a jig. By incredibly good luck, the 4.2 Jag's clutch disc used exactly the same ten-spline pattern as the Chevy. If it hadn't, we would have used a Chevy disc and pressure plate or riveted a Chevy disc center section to a Jag disc outer section—or something. A Chevy SB pilot bushing is larger than what's on a Jaguar, but it was a simple matter to fabricate a bushing consisting of a Jaguar 4.2's outside diameter and a Chevy transmission input shaft's inside diameter.

Certain late-model 4.2 Jag sedan oil filter assemblies are equipped with connections for an external oil cooler, and we lucked out, so it was simple to divert engine oil to the engine dyno's external aircraft oil cooler—and it is definitely important to keep oil at safe temperatures below 220 on the long dyno sessions that may be needed for a from-scratch EFI calibration.

We had previously drilled and tapped our cast-iron Jaguar turbo exhaust headers for six brass ⅛-inch pipe to ⅛-inch compression fittings. These fittings thread into the runners of a header system and securely hold EGT probes in place, preventing any exhaust leaks on a high-exhaust-pressure turbo motor, while providing dynamic combustion temperatures—which is an extremely accurate indication of combustion conditions and a good advance warning of impending trouble. In addition, we equipped each turbo with an exhaust gas oxygen sensor boss. Besides allowing installation of ordinary O_2 sensors, these sensor bosses allowed us to connect Norwood's Horiba UEGO fast wideband air-fuelair/fuel meter probe.

The dyno is set up to accommodate many different sizes and lengths of engines with adjustable motor mount connections. Thus, we only required a set of steel brackets that bolted to the Jag engine flanges at one end and the dyno brackets at the other. In fact, Kim Barr had a huge bucket of brackets, and we found two that worked without modification. Bob Norwood fabricated mandrel-bent tubing to interface both the single and triple turbo exhausts to the 6-inch flex pipes the dyno uses to evacuate exhaust gases to the outside atmosphere.

We connected the aviation oil cooler via braided steel hose to the fitting on our Jag oil filter housing, connected the dyno cooling system to the Jag engine thermostat housing and water pump, and added oil. We installed the Haltech E6 engine management wiring harness to the engine sensors and actuators, connected the ECU power to the dyno's continuous and switched 12-volt DC power sources, and then connected the fuel supply system to the engine and computer.

We installed the Superflow air-metering hat on the turbo inlet to yield standard cubic feet per minute air flowairflow. We connected the fuel supply through the Superflow's fuel-flow measurement equipment.

Finally, we connected the dyno's gauge probes to the engine (oil pressure, EGT, manifold pressure, etcand so on.), connected the ignition and E6 to dyno electrical power, and connected the dyno's throttle and starter actuators to the engine. We also ran data-link connections to the dyno control room from the E6 to a laptop PC and a Haltech mixture trim module, good for immediate dynamic plus-or-minus 30 percent changes to the air-fuelair/fuel ratio. We located the UEGO air-fuelair/fuel meter in the dyno room window, facing the control room for easy air-fuelair/fuel ratio viewing, and then got ready to fire up the 4.2 Jag engine for calibration and horsepower testing.

For dyno testing this type of engine, you'll probably need as a minimum the following test equipment:

1. Injector tester
2. Hall Effect tester
3. Ignition test equipment, including strobe timing light
4. A top dead center finding tool or degree wheel
5. A fuel pressure gauge
6. A laptop personal computer

WHEN JAG 4.2 TURBO ENGINES BREAK

The stock Jaguar 4.2-liter engine was a high-torque motor, with long stroke and very long rods, which is one reason it redlines at only 5,500 rpm. High-rpm 4.2 six racing Jaguar engines require very strong aftermarket rods (Carillos), special vibration dampers with the right balancing, and really good oil pressure. Turbo boost puts far less stress on rods than the momentum of the reciprocating piston/rod mass accelerating and decelerating. (The worst case is when the piston is being pulled downward on the intake stroke by the crankshaft and the rod bolts and cap are absorbing all the stress of stretching; the higher effective compression ratio of turbo boost actually helps slow the piston on the compression stroke, reducing the stress on the rods.) By keeping the redline stock on a turbo motor, we were not worried about killing rods. Really, it's rpm that breaks Jaguar engines, not boost.

Well, actually, boost will rapidly break a Jag 4.2 without serious anti-knock countermeasures. Knock under serious turbo boost will rapidly crack pistons, break rings, and melt or smash ring lands. Stock Jaguar pistons are cast, of 8:1 or 9:1 compression, either of which could be made to work with turbocharging and modern EFI engine controls with effective boost retarding timing strategies based on engine loading, or very effective and fast knock sensor-based timing retard algorithms—plus plenty of fuel. Some people would say any turbo motor should have forged pistons, but a number of very-high-output Japanese factory turbo cars like the Turbo Supra and Turbo MR2 have used pressure-cast pistons without problems. On the other hand, these cars do have very sophisticated anti-knock computer fail-safe measures, including low-boost fuel and spark tables, and computer-controlled wastegates with which the computer decreases turbo boost when the engine is hot or stressed.

Actually, given the radically long head studs on 4.2 Jags, cylinder head lifting under higher boost-enhanced combustion pressures can potentially result in loss of coolant or oil leaks, and can also result in blown head gaskets. We began dyno-testing the Turbo EFI 4.2 Jag engine with wire O-ringing installed in grooves machined in the cast-iron block, using a stock composition head gasket. After blowing the stock gasket, we fabricated a custom copper gasket, annealed and installed it, and had no further problems.

CHAPTER 23
TURBO-NITROUS LOTUS EUROPA 1.6-LITER ENGINE SWAP

The DFI/Accel's bread and butter has always been managing small- and big-block Chevy V-8 engines. The Accel engine management product line ranges from injection parts to turnkey EFI systems complete with everything from injectors to "special" GM V-8 induction systems. However, when it came time to work with a DFI system, what I had available was a Lotus Europa. Four cylinders, fiberglass body, 1,300 pounds, midengine, and British. Quick and nimble, even without much power.

The point was, what interesting things could a DFI Gen 6 engine management system do for an engine, and what is it like to program?

The original Lotus Europas were built in England in the late 1960s, but used a *French* motor. This was a Renault 1.6-liter four of wedge-head design that output 78 horsepower. A special version of the Renault 1.6 was built for the Europa with 10.5 compression. But it still used a wedge head and Solex carburetor on an old-fashioned integrated intake/exhaust manifold on the left side of the block. In their day, people developed plenty of hot rod parts for the Renault engine, with some really pumped Weber-ized motors outputting 125 or so horsepower.

Some later Europas were available with a high-output Ford twin-cam four of 1.5 liters, but these are very scarce, and very expensive, and therefore not suitable to swap into lower-class Europas. The twin-cam made probably 115 horsepower in stock form. Some Europa hot rodders have used Ford Cortina engines as raw material in ordinary Europas. These are much more common than the twin-cam (and have the same transmission bolt pattern), but these are not very interesting engines. In any case, the last Europas rolled out of the factory in 1974, now over a quarter-century ago.

The engine on this Europa smoked badly and looked ugly, right down to its aftermarket Weber carb, which still worked but

Lotus Europa: Longitudinal, mid-engine, fiberglass body, 1,250 pounds, tiny 1.6-liter wedge-head Renault engine. But that was then.

appeared to be in critical condition—like everything else. Rebuild time. Unfortunately, Lotus parts are ridiculously expensive. A set of the unique Europa-Renault high-compression pistons is something like $800. Other parts are nearly as extreme.

However, aluminum-block Renault engines similar to those used in the Lotus had continued in production into the 1980s in other vehicles, gaining such amenities as hemispherical heads, overhead cams, fuel injection, and turbocharging. Renault spent years evolving the little wedge-four motor into something better. Renault, which has done a lot of high-tech racing, has built some real terror turbo motors. The 1.6 engine was virtually modular. For example, 1980s-vintage Turbo Fuego motors used pushrod 1.6-liter hemi-head motors with Bosch L-Jet EFI and Airesearch T-3 turbos that made 125 horsepower or so at 10–12 psi of boost. These have bell housing bolt patterns identical to the OEM Renault Europa powerplant and conveniently have bosses for Lotus-style motor mounts.

I bought a Turbo Fuego motor, complete, at a junkyard for around $300 and set about figuring out how to fit it in the Lotus. And how to rebuild and modify the engine for some serious pumping. The plan was to install the injected Fuego motor, upgrade it and its turbo/injection system where required to make a lot more boost and power (from 10 to 15–20 psi), while adding a very mild port nitrous kit that would receive fueling via the fuel injectors rather than gasoline ports or spray bars. About the least nitrous you can port-inject is 10 horsepower per cylinder, so we decided to start with 40-horse nitrous jetting. The DFI Gen 6 ECU, harness and sensors had all the features needed to handle sophisticated raw fueling, idle speed control, nitrous fueling/enrichment, spark timing, and so on.

The rebuild began with a complete tear-down. If you're building a turbo motor, it's nice to start with a block designed to stand up to turbocharging. The rebuild involved installing new moly rings, new gaskets, bearings, and so on, as well as standard machining procedures. Besides a performance valve job, we converted the head to use a rear water pump instead of the front water pump used on Fuegos. This sort of thing might normally be a nightmare except the head contained vestiges of both front and rear pump mounting bosses and water passages. The conversion necessitated welding closed a passage on one end of the aluminum block, grinding open a passage on the other end, building and fitting a blank-off plate on the front of the engine, and fitting a rear water pump to the back.

Another tricky part came when it became clear that we needed the larger clutch assembly used on the Turbo Fuego motor instead of the wimpy unit used on the Lotus. The Fuego flywheel has a set of optical teeth behind the ring-gear starter teeth for use by the stock electronic ignition as a crank trigger. The teeth interfered with the Lotus bell housing and had to be machined off. We built a custom throw-out bearing combining parts from the Lotus and Fuego units, plus some light machining.

The stock Lotus did not have any crank damper or pulley on the front of the block, instead driving the water pump and distributor using a special cam that protruded from the rear of the block with an overdrive pulley. The Lotus cam will fit in the Fuego block, but the geometry is all wrong for the hemi-head valvetrain. A cam exists that will work, but did not have the combination of high lift and good, high rpm duration, but very little overlap that is the right way to go for a turbo. We ended up amputating the pulley-driving tail shaft from the Lotus cam, machining it and the rear of the Fuego cam to fit together in a press fit, and finally pinning the two pieces together. The Renault does not require a crank damper for engine balance, but we were able to machine a very small front pulley that cleared the rear Lotus firewall and the shifter mechanism (which is the real bear), and could be used to drive a possible future air conditioning compressor, and made a great place to mount flying magnets for a crank-triggered ignition setup via the DFI ECU.

With the body off for restoration and painting, I installed the finished motor in the Lotus frame for testing. I decided it made sense to bench test the motor in the frame by starting and running it with a carburetor to make sure everything was working right without introducing the complexity of a ground-up calibration effort of all tuning parameters. I found a carb manifold that would work with the hemi-head motor, and I already had an almost new Weber DGAV carb with the right jetting for the wedge-head motor. Since the Fuego distributor had no provision for centrifugal or vacuum advance, this being provided electronically with an optical crank trigger and electronic ignition controller, I borrowed an older but compatible points distributor that worked great.

The engine started immediately and ran well with excellent oil pressure. I eliminated an oil leak at the sending unit, set the timing with a strobe light, and ran the motor for 10 minutes at varying speeds to make sure everything was OK.

Time for fuel injection.

FUEL INJECTION LOTUS EUROPA

The next thing was to mock-up installation of all injection and turbocharging components while access was easy with the body off the car. This was mainly a plumbing issue—routing heater hose connections (including to the center section of the turbo where I'd replaced the original noncooled Fuego T-3 center section with a liquid-cooled T-3 center section that would help the turbo run cooler), routing turbo oiling lines, routing the injection wiring harness and sensors, locating the intercooler in the new engine compartment and plumbing it to the turbo and injection manifold with 2.25-inch steel tubing and silicone hoses.

The Europa engine compartment is large, with a fiberglass luggage compartment box located behind the engine, over the transaxle. However, because the front of the car has some space for luggage, and because this thing was a hot rod, I decided to eliminate the rear luggage box, instead locating the Fuego intercooler in its place just under the Fiberglas engine cover, fabricating a hole through which the stock intercooler fan could suck air from below the intercooler. Engine interference with the body turned out to be more of a problem than I'd first thought when examining the generous engine compartment space. The EFI intake manifold and turbocharger exhaust manifold projected outward from the block much farther than carb and standard exhaust manifolds, to the point where it became clear that the body would not lower into place on the frame with these items in place. Fine, they'll have to be removed to R&R the body. But how often are you going to remove the body? Maybe never again.

With a few modifications (for the cam-driven alternator, for example), this hemi-head Renault Turbo Fuego EFI engine bolts right in a Europa to make 125 stock horsepower at something like 10-psi boost from the intercooled Airesearch T3 turbo.

It appeared that Renault had provided very restrictive plumbing from the compressor to the intercooler, a liability I eliminated in this configuration. With the less restrictive turbo plumbing and exhaust system, and with the ability to reprogram the fueling and ignition, it would be possible to increase the turbo boost by tightening the threaded actuator rod on the wastegate to increase spring pressure in the wastegate diaphragm, keeping the wastegate closed to higher levels of boost.

The turbocharger oil return had to be rerouted to clear the Lotus frame, while the oil supply line, originally a metal pipe with short rubber hose connections at either end, was replaced with a flexible braided steel line. The stock plumbing that routed compressed air to the intercooler and into the engine was all wrong in the engine's new home, so we constructed all-new turbo plumbing out of mandrel-bent and straight mild-steel tubing using a silicon/bronze TIG welding rod. At the same time, we built a straight-through exhaust pipe dumping from the turbocharger straight out the rear of the car with no muffler. We also welded an EGO sensor boss to the exhaust pipe immediately behind the turbo so that the DFI fuel injection system would be able to operate happily in closed-loop mode, setting its own air/fuel mixture at idle and light to medium cruise conditions.

The stock Europa gas tank was a mess inside, and after attempting to clean and coat it with motorcycle tank treatment chemicals, I finally cut one side of the tank open and completely sandblasted the tank, inside and out, welded the tank closed (after welding in new fittings for fuel supply, return and vapor emissions purging), and then coated the inside of the tank.

Now it was time to replace the body and fuel tank around the engine, which was easy with the manifolds off the engine and with the help of some friends to handle the heavy lifting.

EMS INSTALLATION

The DFI Gen 6 injection unit is a batch-fire nonsequential system, and with Calmap 6.0+ firmware, included the ability to provide spark timing advance. It also has some nice nitrous-injection capabilities. I set the engine at 6 degrees BTDC number one cylinder and adjusted the crank trigger so that it was centered on the magnet. The Gen 6 injection system (now superseded by the newer Gen 7 system, but still found running thousands of engines around the world) batch-fired all injectors together twice per combustion cycle, that is, once every engine revolution.

The DFI wiring harness was actually a modified V-8 harness, with two injector connections amputated from each of the two sides of the V-harness. The injector harness unplugs from the main section of the harness, so it was an easy matter to remove the injector harness cover and do some minor rewiring to clean up the injector harness for appearance and ergonomics in working in what would now be a fairly crowded engine compartment. I wanted the air temperature sender downstream of the intercooler for accurate air temperature, but it was wired closely with the coolant sensor that on this engine is now on the front water pump block-off plate. More rewiring, but not difficult. Normally, DFI would have supplied a custom harness for a weird application like this, but due to time constraints, I had decided to use the GM V-8 harness, which cleaned up fine with minor rewiring for a sanitary installation.

Some EMS manufacturers void the warranty on the harness if you modify it; however, anyone who knows how to use a good crimping tool or a soldering iron can modify a harness with no trouble to shorten or lengthen wires. Make good hot solder joints or use crimp-on connectors *with the right crimping tools*, and cover everything properly with heat-shrink tubing, tape, or flexible cable conduit. Cable ties made of plastic that pull through themselves to tighten are great for fastening a wiring harness to almost anything.

DFI provided a GM-type idle air controller stepper motor and adapter plate that allows air to bypass the throttle body for faster idle in cold conditions and to stabilize idle at a particular speed after warm-up. DFI's Calmap software allows you to specify speeds for IAC control activation, as well as specifying how fast and to what degree the ECU and IAC will react to changes in idle speed, a useful feature that allows you to deal with stalling under sudden rapid throttle closing.

I plumbed the IAC into the pipe from the air cleaner to the compressor inlet, which would provide clean, cool bypass air. I located an air cleaner on the Lotus at the front passenger side of the engine, behind what amounts to a Fiberglas body gusset, next to the battery behind the passenger seat. This was a diesel Mercedes in-line-type unit with ducting connections on both the inlet and outlet sides, so that you can duct 100 percent cool ambient air into it (in this case from the front of the engine cover), and then duct clean air to the turbocharger compressor inlet.

Since this turbo plumbing was located near the hole through which all wiring passes through the front firewall separating the engine from the passenger compartment, the wiring harness connection for the IAC motor was already in easy proximity to the IAC. I routed the DFI wiring harness (two main cables) in a hole cut through the fiberglass with a hole saw close to the stock wiring harness.

With the manifolds and turbo in place, oil fitting connected, turbo coolant lines in place, and new gaskets locating the EFI intake, the end was in sight.

I located the high-pressure fuel pump below the fuel tank where it will always be self-priming, grounded it to the frame, and connected it to the fuel pump relay wire from the DFI ECU. I installed a pressure gauge in the stock EFI fuel rail so it would be clear if the fuel pressure was correct. I connected the fuel return line from the pressure regulator to the tank connection and jumpered 12-volt positive current to the large lead on the Bosch fuel pump. The pump came to life and began pumping fuel through the rail and back to the tank. With no idle reference vacuum to lower fuel pressure, the gauge showed 39 psi, exactly what it should be.

With the battery disconnected, I connected the crank trigger connector, the three-wire connector for the heated O_2 sensor,

the temperature sensors, and the MAP sensor, which I held in place near the intake manifold with cable wraps, using a short hose to reference pressure to it from the manifold.

The TPS sensor turned out to be a challenge. The three-wire Bosch TPS, designed for an air-metered EFI L-Jet system, seemed to supply continuity only on idle or full throttle, and since the DFI ECU likes to see continuously changing voltage as the throttle opens, it didn't appear possible to use the Bosch sensor. DFI normally supplies GM TPSs, which are triggered by a crank arm on the side of the sensor. It would have been very difficult and time consuming to construct an elegant way to interface this TPS to the stock Weber throttle body on the Fuego engine. On the other hand, it was easy to use the sensor Haltech supplies with their EFI system, which is allegedly a Ford OEM sensor from a Ranger.

With the sensor bolted to a fabricated plate on the side of the throttle body, with the ECU powered up and a laptop PC showing a log of engine sensors, the new sensor's range was 0 percent to 70-plus percent, or about 19 percent to 99 percent, which seemed reasonable. I adjusted it to the 19–99 percent range. Clearly this is not the whole range, but the ECU only looks at the delta for acceleration enrichment, and the throttle position is programmable for the position that will allow activation of nitrous, it seemed reasonable that this sensor would work OK. I'd find out.

It was now time to power up the DFI ECU and do some reconfiguration with a laptop DOS-type PC. First power up the PC, then load the Calmap software according to the DFI manual. A series of menu screens will route you through programming steps. In this case it was necessary to change certain engine parameters before even attempting to start the vehicle. These include manifold pressure range of the MAP sensor (1, 2, or 3 bar), number of cylinders, trigger rpm for closed-loop operation, parameters for torque converter lock up (if applicable), rev limit, and fan on temperature—that sort of thing. You must also specify the Firmware ID of your ECU (which allows the one version of Calmap PC software to configure/tune several different DFI ECUs).

Reviewing the base fuel map, I decided that it appeared plausible for initial startup, and indeed, the engine fired up, and after a few initial coughs, ran well—too well, in fact. The engine seemed to want to run too fast, which clearly indicated some kind of manifold air leak. Well, actually, it turned out there was a hose fitting from the manifold open to the atmosphere in a location that could not be seen or felt until, in desperation, I removed the intake manifold for troubleshooting.

Since the manifold was off, this seemed like a good time to check the condition of the injector bottom-side O-rings. I replaced the O-rings as a matter of course. The engine now ran great—with an apparent tendency to run hotter than you'd expect. Eventually it became clear that providing turbo coolant by tapping into the heater inlet hose and dumping coolant back by tapping into the heater-core outlet hose was not the hot ticket—well, that's the thing, it *was* the hot ticket. The cool ticket was to route the hot coolant from the turbo directly back to the hot side of the radiator for immediate cooling.

Looking at the stock injector capacity at 2.5-bar (40-psi) fuel pressure, Bosch 0280150125 injector rated at 188 cc/min, each injector would max out at roughly 36 horsepower per cylinder (188 cc/min x 0.1902044), which translates to roughly 143 horsepower in a four cylinder. If the goal was to make 200 horsepower, we would have to raise injector capacity to 262 cc/min, which would mean raising system pressure to approximately 75 psi. Given that we knew how much horsepower we wanted to make,

With more boost, free-flow intake plumbing, computer-controlled 40-horse port nitrous injection, and a DFI Gen 6 EMS, you've got a 200-horsepower engine hauling around 1,250 pounds of backbone frame and fiberglass. Can you say, "Scary"?

TURBO-NITROUS LOTUS EUROPA 1.6-LITER ENGINE SWAP

we could calculate the new fuel pressure required to raise our injector flow rate to the required level using the following formula from the fuel injector section of the chapter on actuators:

New Pressure = [(New Static Flow ÷ Old Static Flow²)] x Old Pressure

A fuel pressure of 75 psi would make 200 horsepower through the OEM Turbo Fuego injectors, which will provide fueling for 40 nitrous horsepower on top of 160 turbocharged horsepower.

The fuel pump, however, was going to need help.

Tuning the DFI Gen 6 system, you change break points by running in edit mode and typing new numbers into the fuel table. In fact, the discrete numbers in the main fuel table are not actually pulse width, but rather pulse width divided by 0.0627 (perhaps to avoid floating-point math in the ECU). This tuning task is massively easier with a gas analyzer and mixture trim control (which is a DFI option, and which has separate trim knobs for idle and wide-open throttle mixture).

You may find it useful to use the CalMap screen display in closed-loop mode to show what the mixture *would have been* if not for closed-loop mixture trim. You will be setting cranking pulse width, after-start enrichment, warm-up enrichment, possibly idle speed versus coolant temperature, idle speed control response versus idle speed error, idle control limits versus throttle position, air temperature corrections, ignition timing versus engine speed and load, additional pulse width versus acceleration, acceleration temperature correction, closed-loop speed ranges, nitrous oxide control (ignition retard versus speed, nitrous activation versus throttle angle, nitrous delay time, nitrous speed threshold, percent nitrous fuel enrichment, and percent nitrous correction versus engine speed), and a few other parameters. This might sound intimidating. In fact, DFI was giving the tuner tremendous flexibility in tailoring the ECU to individual needs. However, there is no reason, for example, why a user would ordinarily have to change the default air

temperature correction table—and the newer Gen 7 Accel/DFI system only provides this type of capability in a pro version of the EMS.

The Gen 6 DFI system has robust nitrous capabilities, with provision for providing multiple stages of nitrous injection. DFI points out that providing dry nitrous and supplemental gasoline enrichment via injectors avoids wall-wetting and distribution problems associated with fogger methods of nitrous fueling. Fuel injectors provide fuel particle size in the range of 150 microns. The nitrous itself can be considered a gas, therefore this nitrous fueling method maintains the dry flow characteristics of the intake manifold. DFI suggests that by using an enrichment factor that is multiplicative (percentage), a better air/fuel ratio is maintained when the nitrous is introduced.

DFI's literature makes the additional point discussed above that there is a limit to fuel flow through injectors at a given pressure, which occurs when the injectors are constantly open. At this point you lose control of fueling and if the mixture goes lean under turbocharging and nitrous, you can quickly turn the engine into a paperweight. An injector should never run static for the above reason as well as the fact that it can overheat and quit or permanently change flow rate if it's run static. If you decide to change injectors or raise the fuel pressure on an already-programmed system, DFI software allows you to shift the entire fuel curve up or down by that percentage.

The way to tune a complex engine system like the one on the Lotus, is to get the low-load base fuel map right, then acceleration enrichment, then high loading, high rpm running, then cold running enrichment, and finally nitrous enrichment. Remember, always start conservative and work toward hairy edge tuning, which means work from rich to lean, from ignition retarded toward advanced, and so on.

How do you compute an initial nitrous enrichment factor for an engine? You have to know how much power your nitrous is set up to make and then make some assumptions about what percentage this will be of the power the engine was making without nitrous when the nitrous comes in. Assume you know how much power the engine makes at maximum torque, as well as the maximum torque rpm. What percentage of this will the nitrous be? In our case, with the modified Turbo Fuego powerplant, assume the engine could make 100 horsepower at 3,200 rpm (this is *not* peak power rpm), and that this was maximum *torque*. Since nitrous provides 40 additional horsepower, total nitrous fueling at this rpm and power would be 40 percent. DFI can provide up to 32 percent fuel enrichment plus an additional correction versus engine speed of up to 10 percent. Placing a 40-horsepower nitrous kit on this engine is getting near the limit.

DFI's ECU has the ability to stage nitrous using two different size nitrous solenoids, so you bring in the smaller nitrous valve, switch to the larger valve (killing the smaller valve), finally bringing in both, something like a three-way light bulb with two filaments. I didn't try using staged nitrous, figuring this is something you'd want on a drag race car, which the Lotus is not.

We did not configure the DFI ECU to use a knock sensor; given the fact that it would only be operated by drivers familiar with the precautions you take with a very high output racing-type engine and would always have high-octane gasoline, we saw no urgent need for the protection of a knock sensor.

Bottom line, EFI provided the Lotus with excellent drivability under normal conditions, safe operation via excellent distribution and programmable fuel enrichment and ignition retard under high levels of boost, and additional retard and fuel enrichment for nitrous operation.

CHAPTER 24
MR2 TURBO

This is the story of a great performance adventure: The MR2 Turbo project used a variety of performance parts and engine management strategies to upgrade a 1991 Toyota MR2 Turbo from 200 to 600 horsepower.

We began the project with a 1991 2.0-liter MR2 Turbo advertised to have 200 horsepower, with 65,000 miles, in excellent condition. The car had always used synthetic oil. Running port fuel injection, unleaded 93-octane unleaded fuel, and synthetic oil, the engine could be expected to have negligible wear.

Actually, early dyno tests surprisingly revealed that with an HKS air cleaner, BEGI custom mandrel-bent exhaust, and larger intercooler, *this* car made exactly 139.5 rear-wheel horsepower on the Dynojet at the stock 7-psi boost, which translates to about 172–175 horsepower at the crankshaft. Since Toyota advertises 200 flywheel horsepower at 6,000 rpm and 200 ft-lbs torque at 3,200, figuring drivetrain losses of 15–20 percent, we had expected to see at least 160–170 horsepower at the rear wheels. So, in a way, we started at Less Than Zero.

But this was supposedly a *strong* little powerplant with a lot of upside. The MR2 2.0 engine is a 16-valve unit, a twin-cam aluminum head bolted to a cast-iron block. At low rpm, Toyota's Variable Induction System (TVIS) closes a set of throttle butterflies to block off the secondary intake runners for improved cylinder filling via super-high intake velocities. At higher rpm, above 3,250 rpm, the TVIS opens under electronic control and the pentroof four-valve heads breathe very well. And the combustion chamber is an excellent shape for resisting spark knock, always a concern of turbocharged engines.

There is a very short path from the centrally located spark plug to the farthest end gases in the combustion chamber—improving the chances normal combustion will progressively and smoothly burn the air/fuel mixture before any of it has a chance to explode in the increasing heat and pressure of combustion. An iron block is heavier but more rigid than a typical aluminum block. Iron cylinders in an iron block distort less as you torque down the head and distort less under high piston speed, heat, and pressure. And it's easier for the rings to seal on a truly round cylinder. In fact, iron-block engines typically make more power, and they also have more margin for handling power upgrades.

We would see.

Perhaps we're all guilty on occasion of discussing a performance engine as if it behaves like a precise computer program or a mathematical equation: A+B=C. Add this part to that engine, out comes perfectly predictable horsepower. A hot rod piston engine, however, is a chemical reactor—a relatively sloppy mess of chemicals and temperatures and parts that are dynamic, that may perform variably based on conditions, that may fail or partially fail, that are not entirely predictable until you actually try something. But the end-game R&D on our super-duty 2.19-liter stroker 3S-GTE turbo motor was more like solving a mystery or perhaps a game entitled "Where in the World is the Newest Bottleneck?" or maybe "What Might Blow Up Next?" or even "How Much Guts Do We Have?"

This project started out leaning on a stock 1991 Toyota MR2 Turbo sports car with the goal of doubling the stock 200 advertised SAE flywheel horsepower. We eventually achieved this goal with external bolt-on parts, without even opening the engine. And so the project continued. Down the road, following a lean-mixture dyno accident resulting from inadequate fuel supply, we eventually decided to rebuild and strengthen the engine and push on to a new goal of 500 crankshaft horsepower—which we also more than achieved.

As you'd expect, the last 50 horsepower were a lot tougher to come by than the first 50. In fact, completing the project was more difficult, expensive, and time-consuming than anyone expected, but that's the way it is with a science project: You learn things.

We discovered Murphy's Law is still alive and well ("If something can go wrong, it will!"), though we only had one minor disaster and a few lesser failures (blown head gaskets of various types, and so on), but most stages of the project were relatively trouble-free. We verified that making a very high specific output engine live requires both brute-force design and some subtle tricks of assembly. This is particularly true with head-sealing at extreme pressures, which is vital if the engine is going to live a long time with maximum reliability.

The stock MR2 Turbo EMS is a PROM-based system with distributor ignition, four fuel injectors, a fifth cold-start injector, VAF meter, MAP sensor, TPS, coolant and air temp sensors, and a trunk-mounted ECU that also controls the electric power steering pump, the air conditioning compressor, and the cruise control computer. From an engine management point of view, this

Toyota 3S-GTE EMS block diagram. This is standard stuff except for the TVIS (variable intake runner switching) and turbo vacuum-switching for normal and low-boost mode (when the engine's too hot). The stock EMS allows 7- to 9-psi boost with the wastegate, with a 12-psi fuel cut, and clamped-off 14- and 16-psi boost-fuel curves. With a few tricks, the EMS will run almost 16-psi boost and live.

project initially used the stock Toyota EMS, progressed with the addition of several auxiliary electronic and mechanical devices designed to modify the way the Toyota system actually delivers fuel and spark, and ended up with the installation of a sophisticated Motec M48 standalone EMS to take over all engine control tasks (though the stock ECU and several other functional factory computers remained on the car and alive). In this project, we did almost everything that you *can* do in terms of modifying and upgrading engine management and EFI systems.

The following represents our experience hot rodding the MR2 Turbo and its engine management system(s)—which are inextricably entwined. At every stage we measured horsepower and torque results at the rear wheels with a Dynojet chassis dyno, and we calculated that drivetrain and tire resistance consumed as much as 20 percent of crankshaft power. We optimized air/fuel mixtures with a fast Horiba wideband air/fuel-ratio meter. Once the Motec was on-board, we data-logged all events for review on a Windows laptop.

MYSTERY OF THE MISSING HORSES

Testing on the Dynojet chassis dynamometer quickly revealed that our project MR2 had a deficit of about 24 horsepower at the wheels, which is more like 29 or 30 horsepower at the crankshaft—and this was with custom performance air cleaner/intake, custom performance exhaust, and a larger-than-stock custom performance intercooler.

Unfortunately, there can be *many* possible causes in this kind of situation, from EMS- and tuning-related problems to inferior modified parts to a sick engine.

We speculated initially that perhaps the car's custom intercooler (not equipped with a custom electric fan like the stock intercooler) had become heat-soaked due to lack of cooling air during testing and that we were losing horsepower from air density losses along with consequent over-rich fueling from our Cartech (BEGI) fuel management unit that expected cool, intercooled, dense air at high levels of boost and had no way of compensating for air density changes.

To test this hypothesis we installed temperature probes just upstream of the throttle body and just upstream of the intercooler, which would give us a good idea how well the intercooler was working. We also removed the stock Toyota O$_2$ sensor and installed in its place the UEGO wide-ratio, fast sensor from Bob Norwood's Horiba air/fuel-ratio lambda-checker meter.

We now strapped the MR2 onto our test Dynojet chassis dyno, this time with an enormous fan blowing directly into the intercooler vent behind the MR2's right door—and a smaller fan blowing in the opposite-side inlet for the air cleaner. Adjusting boost via an air-bleed in the manifold pressure line to the wastegate actuator and lying to the wastegate about maximum boost to keep the wastegate closed longer, we made dynoruns at 7-psi boost as well as just below 12-psi boost. Even with the installed HKS Fuel Cut Defender, which is designed to truncate airflow sensor data as seen at the ECU in order to prevent a fuel cut, this was apparently the maximum we could achieve without the stock fuel cut kicking in.

Watching the temperature probe output, we discovered that the throttle body probe would heat-soak to about 110 to 120 degrees between dyno runs. When the car started and began to roll lightly through gears one to three with zero boost, temperature would work down to about 85 degrees as measured at the throttle body probe. But when we hit the Dynojet control to begin a full-throttle run, the temperature would quickly climb to 105 to 110 at 7-psi boost during the 5 to 10 second run, slightly higher at 12 psi. Meanwhile, at the intercooler probe, we meas-

Installing the HKS vane pressure converter and Fuel Cut Defender. The VPC adapter plugs in between the Toyota ECU and main wiring harness and replaces the stock VAF meter with its own MAP sensor and IAT. The FCD is basically a voltage clamp that keeps the ECU from seeing more than 12-psi worth of airflow.

ured a maximum of 170 degrees at 12-psi boost. Clearly the intercooler was working fine, at least so far. Unfortunately, the dyno results showed that we gained exactly four horsepower at 7-psi boost. Unfortunately, we were only making a tiny bit of additional horsepower—144 versus our last results of 139.5.

Meanwhile, the air/fuel ratio was fine across the rpm range until the car got into the high-rpm, heavy-loading range near 6,000 rpm—at which point the air/fuel ratio went dead-rich, in the 0.6–0.7 lambda range (lambda is 14.7:1 air/fuel ratio). That is, at high rpm and high loading, the air/fuel ratio suddenly dropped to below 10:1, which is extremely rich.

We removed the Cartech fuel management unit, made another dynorun, and discovered that the air/fuel ratio was still *plenty* rich. It appeared that we had not needed the fuel management unit to maintain fuel under heightened boost. It appears that the factory onboard computer has a strategy of dumping in fuel and taking out timing just beyond the 6,000 rpm stock peak horsepower point, which is clearly designed to protect the engine.

Looking at the dyno graphs, the uneven graph of the torque and power graphs from the Dynojet made us wonder if the spark might be a little weak, though the car seemed to be running smoothly under all conditions.

We decided to quit for the night; power was still lower than it should be, but compression and leakdown were fine. First thing in the morning, we decided to install some new HKS performance equipment.

First, we installed a thing called a vane pressure converter (VPC), which eliminates the vane airflow (VAF) meter in favor of a small add-on microprocessor box with its own auxiliary MAP and air temp sensors and a wiring harness that connects the VPC to the Toyota ECU—all of which work together to *simulate* the vane-pressure (airflow) data of the original airflow meter by substituting MAP data fudged according to inlet temperature and translated according to internal VPC data tables.

One advantage of the VPC is that you eliminate the restrictive spring-loaded air-door of the original vane pressure meter, needed to measure the velocity of air entering the engine as an estimate of air mass. A VPC has the additional advantage that it allows adjustment of air/fuel ratios in 2 percent increments for idle, across-the-board gain, and during transient (acceleration) operations (like an accelerator pump).

Stock Toyota MR2 VAF meter (left, with HKS air cleaner installed) and VPC replacement tube (also with HKS air filter). No more airflow bottleneck from the VAF air door, and the VPC allows adjustment of the idle and maximum airflow voltages from the cockpit with trim pots to tip and trim the entire MAP-based fuel curve. Acceleration enrichment is also adjustable.

The HKS vane pressure converter includes a complex wiring harness with a maze of wires and connectors that interpose between the ECU in the MR2's trunk and the stock Toyota ECU wiring harness connectors. The VPC wiring also contains a long pigtail connector that must be routed all the way forward from the trunk through the engine and both rear and front engine firewalls into the passenger compartment, where the VPC's adjustable-pot control unit mounts to allow dynamic adjustment while driving. Two additional wiring pigtails connect the VPC's temperature and pressure sensors in the engine compartment to the control unit. We installed the new sensors straightforwardly by removing a diagnostic temperature probe near the throttle body in favor of the HKS temperature probe, and by mounting the VPC's B-MAP absolute pressure sensor on a bracket located on the rear engine firewall close to an unused intake manifold reference port on the right rear of the engine.

At the same time, we removed the car's Fuel Cut Defenser (FCD), which we'd now concluded was inoperative since the fuel still cut above 12 psi. We reconnected the stock VAF-ECU wiring that had been cut for the original FCD installation, and then installed a new generation HKS FCD. We did not tamper with the HKS factory calibration of the FCD, which is visible through a window in the FCD case.

Next, we installed an HKS electronic valve controller (EVC), which would enable us to overboost the car with cockpit adjustability.

The EVC installs by routing a special wiring harness from the valve's pulse width–modulated (PWM) actuator unit that installs in the engine compartment through the front bulkhead into the passenger compartment where you then locate the EVC controller/display. The EVC is a fairly complex computer-controlled electro-pneumatic device that's essentially designed to *lie* to the stock wastegate actuator by preventing it from seeing true manifold pressure, thus enabling overboost by keeping the wastegate closed longer.

The EVC also enabled us to remove the far less sophisticated air-bleed device we'd been using previously to deceive the wastegate actuator but which could not be adjusted from the cockpit. This generation of HKS EVC includes a fuzzy-logic scramble boost mode, in which the controller watches the behavior of the turbo and wastegate in making stock boost as you accel-

erate in third gear three times from 1,000 rpm to high boost under direction of liquid crystal display commands on the EVC display. After it graduates from learning mode, the controller is able to provide very high-performance wastegate control. The EVC limits boost to specified levels of overboost, while providing very fast turbo spool-up by preventing premature gradual wastegate opening. The benefit is really good midrange torque.

Meanwhile, a customer showed up at the dyno in a red 1993 MR2 Turbo. We seized the opportunity to dyno-test the red car, and were chagrined to discover that this bone-stock MR2 made 163 horsepower at the rear wheels at stock boost, which was as much as our car had made overboosted to 12 psi and with performance intake, exhaust, and intercooler.

Something was clearly wrong with our project car.

We decided to look exhaustively for problems before making any additional dyno runs. We resistance-tested the project MR2's plug wires, which were around an acceptable 11,000 ohms. We obtained new plugs and a distributor rotor, but it was getting late, and we were unable to locate a new distributor cap or plug wires. We installed the plugs, scraped some sludge out of the distributor cap, and buttoned up the ignition system.

Back on the dyno, this time at stock boost, we worked to adjust the air/fuel ratio using the VPC mixture controls. The car was running rich both at idle and across the board. Unfortunately, adjusting the VPC's gain for perfect max-torque midrange air/fuel ratios still produced an over-rich top-end. Since we were trying for maximum high-end power, we adjusted fuel until top-end test operations produced a 0.8 lambda, that is, 11.8 air/fuel, which is still too rich, but better than the 0.7 we had been seeing.

At this point we did a full-power dyno pull in fourth gear.

It instantly became clear that the engine was now producing better power at stock boost than it had been making the previous night with 12-psi boost, though we still weren't sure why. As we worked to adjust the air/fuel, power came up even further. Now we began to raise the boost still further on the dyno, using the EVC controls from the passenger seat while Bob Norwood watched the Horiba. Now we began to see some real power. At 0.9 bar (13.2-psi boost), we were now making nearly 200 wheel horsepower at 6,000 rpm, which amounts to 230 to 240 horsepower at the crankshaft.

Now we cranked up the boost to 15 psi and at this point discovered a new bottleneck: The car began to misfire at mid-to-high rpm under heavy boost, backfiring through the intake manifold. At first, I wondered whether we were also hearing some knock, but the Toyota onboard computer really works to keep things safe, with extremely aggressive knock sensor-based ignition retard strategies and rich mixtures at high-load, high rpm conditions. You can actually see dips or "holes" in the dyno power graphs when the car knocks and the knock sensor takes out timing. Clearly, the stock ignition—although vastly improved with new plugs and rotor—could not fire consistently under the dense combustion chamber pressures of 1.0 bar boost.

Fortunately, we had an HKS Twin Power ignition system, designed to provide capacitive ignition at lower ranges and improved power spark at all rpm ranges. This is a powerful coil-driver box similar to an MSD 6-series, though not capacitive-discharge, just high-power inductive coil-driving.

We quit for the night, but quickly installed the Twin Power unit early the next morning at Norwood's. Driving the car on the dyno, the engine immediately began to miss. Although idle and very light cruise were smooth, the MR2 was losing a cylinder as soon as we loaded the engine virtually at all. I consulted with

Eff/RPM Fuel Main (% of IJPU)	0	200	400	600	800	1000	1200	1400	1600	1800	2000	2200	2400	2600	2800	3000	3200	3400	3600	3800
									Toyota 3S-GTE Motec M48											
333	39.5	40.0	40.5	41.5	42.5	43.0	43.5	45.0	46.0	46.5	47.0	48.0	49.0	49.0	49.5	50.0	50.5	51.0	51.5	51.5
300	40.5	41.5	42.0	43.0	44.0	44.5	45.0	45.5	46.5	47.5	48.0	49.0	49.5	50.0	49.5	50.0	50.0	50.0	50.0	50.0
267	41.5	42.0	42.0	42.5	43.5	44.5	45.0	45.5	46.0	46.5	47.5	48.0	48.5	48.5	48.5	48.5	48.5	48.0	47.5	47.5
233	42.5	44.0	43.5	44.0	44.5	45.5	46.0	46.5	47.0	47.0	47.0	46.5	46.5	46.0	46.0	46.0	44.5	44.0	44.0	44.0
200	41.5	42.5	43.0	44.0	44.5	45.0	45.5	46.0	46.0	45.5	45.5	45.5	45.0	45.0	45.0	45.0	44.5	44.5	43.5	43.0
167	43.5	43.5	43.0	43.0	43.0	43.0	43.0	43.0	43.0	42.0	42.0	41.5	41.5	41.0	41.0	41.0	40.0	39.0	39.5	39.5
133	42.0	41.5	41.5	40.5	39.5	39.0	38.5	38.0	37.0	36.5	36.5	36.5	37.0	35.5	36.0	36.0	36.0	35.5	36.0	36.5
100	34.1	33.5	33.5	33.0	33.0	33.0	33.0	32.5	32.5	32.0	31.5	30.5	30.0	30.0	30.0	29.0	30.0	30.0	30.5	30.5
67	29.0	29.0	29.0	29.5	30.0	30.5	30.5	30.0	30.0	29.5	28.5	28.5	28.5	28.0	28.0	27.5	28.0	28.5	28.5	29.0
33	18.0	17.5	17.5	17.5	18.0	18.0	18.0	19.5	20.0	21.0	21.0	20.5	20.0	20.0	20.5	20.5	20.5	20.0	19.5	19.5
0	14.5	14.5	15.5	16.0	16.0	16.5	16.5	17.0	17.5	18.0	18.0	18.0	18.0	18.0	18.0	18.0	18.0	18.0	17.5	18.0
Ign Main (degrees BTDC)																				
333	22.0	20.5	21.5	22.0	22.0	22.0	22.0	22.0	22.0	22.0	22.0	22.0	22.0	22.0	22.0	22.0	22.0	22.0	22.0	22.0
300	18.0	18.5	19.5	20.0	20.5	21.0	21.0	21.5	22.0	22.0	22.0	22.0	22.0	22.0	22.0	22.0	22.0	22.0	22.0	22.0
267	21.0	21.5	22.0	22.5	22.5	22.5	22.5	22.5	22.5	22.5	22.5	22.5	22.5	23.0	23.0	23.0	23.0	23.0	23.0	23.0
233	20.5	21.0	22.0	22.5	23.0	23.0	23.0	23.0	23.0	23.0	23.0	23.0	23.0	23.0	23.0	23.0	23.0	23.0	23.0	23.0
200	15.0	15.0	17.5	20.5	22.0	23.0	23.5	24.0	24.0	24.0	24.0	24.0	24.0	24.0	24.0	24.0	24.0	24.0	24.0	24.0
167	5.0	5.0	8.0	13.0	18.5	22.5	24.0	25.5	26.0	26.0	26.0	26.0	26.0	26.0	26.0	26.0	26.0	26.0	26.0	26.0
133	5.0	5.0	8.0	12.5	17.0	20.0	22.5	25.5	27.5	29.0	30.0	31.5	31.5	31.5	31.5	31.5	31.5	31.5	31.5	31.5
100	5.0	5.0	8.0	11.5	14.0	16.5	19.0	21.5	24.5	27.5	30.0	32.5	33.0	33.5	33.5	33.5	33.5	33.5	33.5	33.5
67	5.0	5.0	7.5	10.0	12.0	14.0	16.5	18.5	22.0	25.5	28.0	30.5	32.0	33.5	34.5	35.5	36.0	36.0	36.0	36.0
33	5.0	5.0	7.0	8.5	11.0	13.0	15.5	18.5	22.5	27.5	30.0	32.5	33.5	34.5	35.5	36.5	38.0	38.5	39.0	39.5
0	5.0	5.0	7.0	8.5	10.5	13.5	16.0	20.0	24.5	28.5	32.5	35.0	37.0	38.5	39.5	40.5	41.5	42.0	42.0	42.5

HKS to make sure the Twin Power unit was correctly installed. At first, we assumed the problem might be our failure to adjust the Twin Power to match impedance with the stock coil, but allowing the vehicle to cool off for correct matching, and making the adjustment did not help.

We made another dyno run, with the same results. We decided to pull the Twin Power unit, returning to stock ignition, but the problem persisted. At this point, Bob Norwood began pulling plugs, wires, rotor, and so on. It turned out a hole was burned through one of the plug wire towers. And that's the problem with high-power spark: It tries to leak out of the ignition system at the location of any flaw on the way to the plugs, just like high-pressure water finds holes in weak plumbing—and once a leak starts, it tends to get a lot worse. Norwood regapped the plugs to 0.030, and we installed new plug wires. We fired up the car and ran a series of pulls. Again, we watched the lambda sensor to adjust idle and gain air/fuel ratios with the HKS vane pressure converter.

Now it was time to celebrate. We were back in business with the most impressive power runs we had yet made with the MR2: 215 ft-lbs wheel torque at 3,500 rpm and 197 wheel horsepower at 5,000 rpm (significantly lower rpm than previous dynoruns making about the same power). The Twin Power unit was now working great; all it needed was great plug wires. Given the MR2 Turbo's stock peak torque and power are at 3,250 and 6,000, we were running out of *something* at 5,000 rpm.

In order to get the really big power, we need to be making peak power at 6,000 or more. By the way, since a stock MR2 made 163 rear wheel horsepower from 200 advertised flywheel horses, we could assume an 18 percent loss. Applying this formula to the project MR2, working backward from 197 rear wheel horses, we could assume that the car was making 240 horsepower at the flywheel at 0.9 bar. To this point in the project we had gained 57 horsepower from the low-boost previous configuration of the engine!

Project Journal: "For now, we'll be looking to understand what's happening at high rpm. We probably still need more spark

There's a ton of power available by cranking up the boost, but the stock Toyota computer stands in the way, with the calibration and configuration in the PROM that is soldered on the motherboard. Here Tadashi Nagata desolders the stock PROM in preparation for installing the Techtom ROMboard.

at higher revs, and we need to find a way to increase midrange fueling without overfueling on the high end. Yesterday, the car got new spark plug wires and distributor cap—and we'll make sure all service items in the 60,000-mile checkup have been properly performed. We have a 3-inch HKS exhaust, and we plan to install a new turbocharger with more high-end airflow. We'd like to crank up the boost first with the stock turbo into the 18 psi range (1.2 bar) before installing the high-flow turbo and evaluating it. We'll evaluate the new gear plus take another look at the HKS Twin Power ignition module's effect—back on the dyno."

At this point the project proceeded rapidly to the 300 rear-wheel horsepower level with the addition purely of more turbo

Toyota 3S-GTE Motec M48																				
4000	4200	4400	4600	4800	5000	5200	5400	5600	5800	6000	6200	6400	6600	6800	7000	7200	7400	7600	7700	Fuel Main (% of IJPU)
51.5	51.5	51.0	51.0	51.0	51.0	52.0	52.0	52.0	52.0	52.5	52.5	52.5	51.5	51.5	51.5	51.0	50.5	49.5	48.5	333
50.5	50.5	50.5	50.5	51.0	51.5	52.5	52.5	52.5	52.0	52.0	51.5	50.5	50.0	49.5	48.5	47.5	47.0	46.5	45.5	300
47.5	46.5	47.0	47.5	48.0	48.5	48.5	48.5	48.0	47.0	46.5	45.5	45.0	45.0	45.0	45.0	44.5	44.0	43.5	43.0	267
43.5	43.5	44.5	45.0	46.0	46.0	45.5	45.5	45.0	45.0	44.0	43.5	43.0	43.0	42.5	42.5	42.5	43.0	43.5	44.0	233
43.5	43.0	42.5	42.5	42.5	42.5	42.0	42.0	42.0	42.0	41.0	41.5	41.5	42.0	42.0	42.0	41.5	41.5	42.5	42.0	200
39.0	39.5	40.0	40.5	40.5	41.0	41.0	40.5	40.5	41.0	41.0	41.0	41.5	41.5	41.5	41.5	41.5	41.5	42.0	42.0	167
36.5	37.0	37.5	38.5	39.0	40.0	40.0	40.0	40.0	40.0	40.5	40.5	41.0	41.5	41.5	42.0	42.0	42.0	42.0	42.0	133
31.0	32.5	32.5	33.5	34.0	35.5	37.0	37.0	37.5	37.5	38.0	38.0	38.0	38.0	38.0	37.5	37.0	36.5	36.0	35.5	100
29.5	30.0	30.5	30.5	31.5	31.5	32.0	32.0	32.0	32.0	32.0	32.0	31.5	31.0	30.5	30.0	30.0	29.0	28.0	27.5	67
19.5	19.5	19.5	19.5	19.5	19.5	19.5	19.5	19.5	19.5	19.5	19.5	19.0	19.0	18.5	17.5	17.0	16.0	15.0	14.0	33
17.5	18.0	18.0	18.0	18.0	18.0	18.0	18.0	18.0	18.0	18.0	18.0	17.5	16.5	15.0	13.5	12.5	11.5	9.5	9.0	0
22.0	22.0	22.0	22.0	22.0	22.0	22.0	22.0	22.0	22.0	22.0	22.0	22.0	22.0	22.0	22.0	22.0	22.0	22.0	22.0	333
22.0	22.0	22.0	22.0	22.0	22.0	22.0	22.0	22.0	22.0	22.0	22.0	22.0	22.0	22.0	22.0	22.0	22.0	22.0	22.0	300
23.5	24.0	24.0	24.0	24.0	24.0	24.0	24.0	24.0	24.0	24.0	24.0	24.0	24.0	24.0	24.0	24.0	24.0	24.0	24.0	267
23.0	23.0	23.5	24.5	24.5	25.0	25.0	25.0	25.0	25.0	25.0	25.0	25.0	25.0	25.0	25.0	25.0	25.0	25.0	25.0	233
24.0	24.0	24.0	24.5	24.5	25.0	25.0	25.0	25.0	25.0	25.0	25.0	26.0	26.0	26.0	26.0	26.0	26.0	26.0	26.0	200
26.0	26.0	26.0	26.0	26.0	26.0	26.0	26.0	26.0	26.0	26.0	26.0	26.0	26.0	26.0	26.0	26.0	26.0	26.0	26.0	167
31.5	31.5	31.5	31.5	31.5	31.5	31.5	31.5	31.5	31.5	31.5	31.5	31.5	31.5	31.5	31.5	31.5	31.5	31.5	31.5	133
33.5	33.5	33.5	33.5	33.5	33.5	33.5	33.5	33.5	33.5	33.5	33.5	33.5	33.5	33.5	33.5	33.5	33.5	33.5	33.5	100
36.0	36.0	36.0	36.0	36.0	36.0	36.0	36.0	36.0	36.0	36.0	36.0	36.0	36.0	36.0	36.0	36.0	36.0	36.0	36.0	67
40.0	40.5	40.5	41.0	41.0	41.0	41.0	41.0	41.0	41.0	41.0	41.0	41.0	41.0	41.0	41.0	41.0	41.0	41.0	41.0	33
43.0	43.0	43.0	43.5	43.5	43.5	43.5	43.5	43.5	43.5	43.5	43.5	43.5	43.5	43.5	43.5	43.5	43.5	43.5	43.5	0

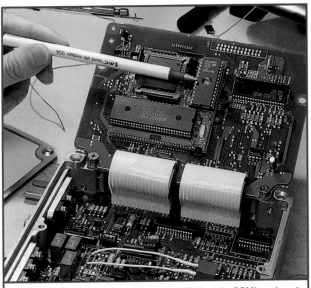

With the ROMboard in place, the stock PROM on the ROMboard, and an auxiliary processor redirecting memory reads to the new PROM calibration, the stock 3S-GTE EMS becomes programmable.

boost and a backup fuel pump. We switched over to an HKS electronic boost controller. We tested a larger HKS Sport Turbo, then an Aerodyne variable area turbine nozzle (VATN) turbocharger, and finally a big Turbonetics T04E-50 with custom Bell Engineering Group turbo plumbing. At 300 horsepower, we were approaching the limits of what's possible with external parts on a stock 3S-GTE powerplant, and making the final stock-block 35 rear-wheel horsepower would require some finesse.

FINDING THE LAST STOCK-BLOCK POWER

Temporarily stopped at 300 rear-wheel horsepower, this section tells how we found more power without even raising the turbo boost. The bottom line was an additional 35 horses that put us at 336 at the rear wheels—or a computed 422 at the flywheel! This put the project over the magic 400 figure we needed to meet our initial goal of doubling the stock 200 advertised horsepower.

By this time we had roughly 130 runs on the Dynojet chassis dyno and at 300 wheel horsepower had essentially hot rodded us a whole new breed of car. We were on our fourth turbo, a ceramic ball-bearing unit from Turbonetics thought to have the cojones to boost the 2.0 16-valve Toyota engine all the way to 30 psi or meltdown, whichever came first. The turbo's ball-bearing design also gives it 50 times the durability and greatly improved spool-up responsiveness. We had previously produced outstanding results from the stock turbo, from an HKS Sport Turbo, an Aerodyne VATN unit, and now from the Turbonetics piece.

At this point we were running an HKS cat-back mandrel-bent exhaust, an HKS boost controller, an HKS high-flow air inlet system, and an HKS Fuel Cut Defenser (which we had to recalibrate for the new Turbonetics turbo in order to nip in the bud a new level of fuel cut we had not seen before). We were still using an HKS vane pressure converter to convert from the stock MR2 airflow metering to manifold absolute pressure-based speed-density fuel controls—eliminating the bottleneck in airflow and providing air-measurement authority at cfm levels beyond which the stock airflow meter is wide open. We were still running a big Bell Engineering intercooler.

Having broke the 300 rear-wheel horsepower barrier, we took the project MR2 to Alamo Autosports in Arlington, Texas, where owner Brice Yingling and his techs methodically and scientifically proceeded to make more dyno horsepower than we'd yet seen.

Alamo began by installing a better clutch and limited slip differential from Toyota Racing Development. With the MR2 up in the air, Alamo also installed TRD lowering springs and SPAX adjustable strut inserts.

With the car back on the ground, we installed an RC Engineering big-bore throttle body, which not only has a *larger* butterfly but has an aerodynamically improved butterfly cross-section at wide-open throttle. A matching Alamo T-Body air inlet completed the picture on potential volumetric efficiency improvements. Alamo also installed an Alamo fuel rail kit, designed to eliminate fuel distribution problems at very high fuel flow by feeding fuel to both ends of the fuel rail and moving the regulator to the center. This is a beautifully engineered piece that uses rerouted braided-stainless hoses in place of factory-type rubber fuel hoses.

With the new pieces in place, we prepared the car for dyno testing and tuning. I cannot overstate how impressed I was with Alamo's approach to making power on the MR2. Using past experience with other MR2s, Alamo began by getting the Mister Two to a known state of tuning where they could achieve entirely repeatable consistent dyno runs on the Alamo Autosports Dynojet. We started off installing a new set of HKS Racing spark plugs two heat ranges colder than stock, gapped at 0.028. We then changed the oil to Amsoil 20W-50 fully synthetic performance oil. We changed all belts—including the timing belt—and replaced the timing belt idler pulley bearing when the setup sounded a little noisy.

Now we attached an engine analyzer to the MR2 and looked for trouble codes (verifying our guess that our O_2 sensor was bad and that there were no other check engine codes). The O_2 sensor had never been replaced, and a brief flirtation with high-octane leaded on a different Dynojet appeared to have recently sent it over the edge. Alamo installed a temperature probe downstream of the intercooler and attached an injection pulse width meter to watch how the stock computer-with-VPC affected injection time.

Now we were ready for the dyno. To be very safe, we fueled up the car with 116.5-octane Phillips 66 leaded race gas just before strapping the car onto the dyno. Although the goal was a state of tune that would work with street gas, ultra-high-octane fuel is good insurance when you're exploring strange new worlds on the dyno. Just make sure you don't get lead in your cat or O_2 sensor, since it will kill both (and it is just plain stupid to kill your cat, particularly since the stock cat is a very good flowing unit, you need it to pass emissions, and it's expensive).

The final step was to reinstall the Cartech fuel management unit (FMU), which we'd removed when we found it to be overkill at lower levels of power. The FMU is a variable rate-of-gain fuel pressure regulator that raises fuel pressure an adjustable 5–7 psi per pound of boost. Recently we had essentially moved to the ragged edge of the stock fuel system's capacity, running with injectors at 100 percent duty cycle (constantly wide open) at high-rpm maximum throttle and boost. The only way to get more fuel short of larger injectors and accompanying complex computer changes was to artificially jack-up the fuel pressure as a function of boost, increasing fuel pressure as a multiple of boost pressure increases.

With the FMU in place, we made some introductory dyno runs, checking for consistency before beginning the tuning process.

It took 24 dyno runs before Alamo had the project MR2 consistently making run after run above 330 horses with a smooth and excellent power curve. It is true that at this power level top-end fuel with our setup was on the ragged edge of lean. Tuning was a process of balancing the VPC's ability to set low- and midrange air/fuel mixtures as an across-the-board gain in plus-or-minus 2 percent increments from the cockpit—while incrementally increasing boost and calibrating top end air/fuel mixtures by calibrating the FMU in the engine compartment to optimize both the onset of rising-rate fuel pressure as well as the

rate of gain per-psi boost pressure. The idea was to tune idle and all-rpm-ranges fuel with the VPC, while retuning high-end, high-boost fuel with the FMU, effectively using the VPC for idle, low-, and midranges, and the FMU for top end. This was a relatively crude method of tuning compared to a programmable EFI computer, but it was what we had at this point.

Alamo began the tuning process by backing off static timing from 10 degrees BTDC to 2 degrees BTDC by rotating the distributor, and eventually working back to 8 degrees BTDC as tuning progressed. We made careful notes of every change to tuning and the dyno results and were careful to allow the car to reach equilibrium cool-down between dyno runs to keep the data crisp.

Shortly after installing the new Turbonetics T04E-50 turbocharger, we had again begun to see the factory computer instigating a fuel cut. We countered this by recalibrating the HKS FCD voltage clamp to 6, versus the previous 7 setting, which clamped the simulated air-meter voltage from the VPC a bit harder to make sure the Toyota factory computer never detected unacceptable overboost totaling over 12 psi worth of air-meter voltage.

One of the first new problems we encountered was a boost-creep situation. This problem occurs when the wastegate cannot divert enough exhaust gases around the turbine to effectively control maximum boost, even with the wastegate wide open, allowing boost pressure to creep somewhat higher before there's a new equilibrium. The solution is larger diameter wastegate exhaust plumbing and a possible change in wastegate geometry or size, but the result in the meantime was that with maximum boost set at 1.12 bar (16.5-psi boost), the boost would creep up to 1.35 bar (roughly 20-psi boost) at maximum power. Unfortunately, this meant that midrange torque was hurt, because Alamo was forced to limit nominal midrange boost to 1.12 in order to limit maximum boost at 1.35—after the creep—thereby ensuring that midrange operation saw only 16.5-psi boost worth of *torque*.

In the end, Alamo was certain there was more midrange torque to be found, but top-end was starting to go lean *even with the FMU in place*, probably indicating that the stock fuel pump was running out of capacity at the extreme levels of fuel pressure required for high-end fueling under FMU authority and that *rail pressure* was dropping. We had a super-duty NOS in-line pump ready to install, and had a Kenne Bell Boost-A-Pump (which sort-of supercharges the stock pump via raised voltage under boost conditions), but for now, 336 horsepower at the rear wheels seemed like a good place to stop.

We verified that the stock injectors where indeed going static somewhere around 5,000 rpm. With pulse width on the pulse width meter reading in the mid-20 milliseconds range in the high-4,000 rpm range, time available for injection decreases below 24 milliseconds as you cross 5,000 rpm, which mathematically indicates 100 percent duty cycle or static injectors (continuously held open). The thing you've got to know is that the faster an engine goes, the less time there is available for injection. At 7,000 rpm, the crank is turning roughly 116.7 times per second—meaning each two-rev time period available for injector squirt per power stroke whips by in slightly more than 17/1,000 of a second, or 17 milliseconds. Since the maximum injection pulse per power stroke always occurs at maximum torque (and maximum torque occurs at maximum volumetric efficiency, the point at which the engine breathes in the most air and requires the most fuel), and since you never want a situation where your injectors are too small to meet fueling requirements above maximum torque when time available for injection gets scarce, most aftermarket engine management systems limit maximum injection pulse width at all rpm ranges to 16 milliseconds. This is really a strategy to protect tuners from themselves by limiting fuel available at maximum torque to what will be available at redline (assuming redline occurs at 7,000 rpm).

By this time we were clearly having trouble getting sufficient cooling air through the BEGI intercooler and were now also making power in a range greater than the design limits of the intercooler. In fact, with 80-degree ambient air, the intercooler heat-soaked to 125–130 degrees temperature at idle, coming down to 110 degrees under light loading as the engine picked up speed to 3,500–4,000 rpm and the *intake air began to cool the intercooler*.

Unfortunately, at high boost and high rpm, the intercooler temperature skyrocketed to 170 at peak power. We rigged up a powerful squirrel-cage fan 6 inches from the MR2's passenger-side intercooler air inlet and built a cardboard shroud that we duct-taped to the MR2, which allowed you to feel air coming through the fan-less intercooler. With this arrangement in place, idle temp stayed about the same, light-load temp came down a few degrees, but mainly peak power temperatures were down in the 140-degree range (still way too high for best power). Twenty degrees above ambient is a good target for a really high-efficiency intercooler sized properly for maximum horsepower and airflow. With high air-temp at full boost, really high boost can start losing in density what it gains in pressure—a zero-sum game in which additional power is increasingly elusive.

Without even a hint of detonation, we completed four 330-plus horsepower runs at 2:30 A.M., and then we shut down. The next day we drove the project MR2 back to Dallas where Bob Norwood brought by a Horiba fast air/fuel-ratio meter to watch the mixture over a few quick runs on the Dynojet. We did several pulls, noting that the air/fuel ratio was a little fat (rich) in the midrange. The trouble was, the tuning tricks we had available to modify engine management were a little crude, and it was nearly impossible to optimize fuel delivery everywhere.

Clearly, before we went much further we were going to need bigger injectors and a programmable computer to control them. We also needed to handle the boost-creep problem. We needed to solve the intercooler temperature blowout problem at high boost, which might require an air-water system. We also wanted to design a way to provide tons of cold air on the intake side while muffling the sounds of the turbo sucking air and the compressor bypass valve burping on sudden throttle closure. We knew we'd need to provide more fuel pump capacity.

In general, we wanted to put some attention into heat management in terms of intake temp, exhaust heat, intercooler efficiency, and so on. We believed there was potential to gain some pumping efficiency by straightening out the turbo intake plumbing and drastically shortening the compressor-to-T-body path by locating the intercooler directly over the engine. And this guess was borne out by the 8-horsepower gain we saw on Harrington's dyno when we removed the entire air cleaner system and directly connected a pipe to the 5-inch 90-elbow on the compressor inlet of our T04 turbo. Alamo recommended porting the exhaust manifold and tapping it for pressure and EGT sensors. Meanwhile, Bob Norwood was predicting the possibility of significant added rear-wheel power with better engine management.

REMOVING THE BIG-POWER BOTTLENECKS

We decided *not* to quit while we were ahead. In a new quest for 500 crankshaft horsepower, we decided to hack the MR2's factory computer, reprogram it to a higher state of tune, and rework the calibration to control big Toyota 550-cc injectors. We'd need

---

Project MR2 Turbo Motec Configuration
(83 lb/hr Bosch injectors, 2.1L 3S-GTE)

Injector Scaling Value = 13 (used to insure that the fuel table has sufficient range and resolution). The base pulsewidth is calculated by multiplying the number from the fuel table (as a percentage) by the IJPU parameter, that is, if IJPU = 13 msec and the fuel table number = 50 percent then the base injector pulsewidth will be 6.5 msec for that RPM.

Crank Index		93.6 BTDC
IGN Trim		5%

Fuel Accel
Accel Sensitivity		-1
Accel Decay Rate		2.6

Cold Start (the larger the figure the more additional fuel)
Cold Warm Up Enrich		-1 (use Fuel Engine Temp Comp table)
Post Start Enrich		40
Cranking Enrich		20
Post Start Decay		20
Cold Accel Enrich		5

Sensor Defaults
Throttle Position	TP	70
Manifold Pressure	MAP	290
Air Temp	AT	36
Engine Temp	ET	80
Aux Temp	Aux T	0
Aux Voltage	Aux V	100
Barometric	BAP	100.0

Main Setup
Injector Scaling	IJPU	13
Injector Current	IJCU	3.0
Injector Battery Comp	IJbc	3 (2.5 ohms vs .5, 2.0, 2.2, 2.4, 3.0 4.5, 16)
Injector Operation	IJOP	0
Efficiency Calc Method	EFF	3 (Method used to calc efficiency point)
Load Calc Method	LOAd	3 (3 = MAP/BAP vs TP. MAP aux volt, MAP, MAF, etc)
Load Sites Selection	Ld S	330 (Selects the Load and EFF site range)
Number of Cylinders	CYLS	4
Ref Sensor Type	rESn	2 (1 = Hall Input; 2 = Mag Input/Mag Sensor etc)
Sync Sensor Type	SYSn	2
Ref / Sync Mode	rEF	14 (14 = multi-tooth)
Crank Ref Teeth	cr T	12 (Number teeth per crank rev)
Tooth Ratio	TOr	65
Crank Index Position	criP	93.6
Ignition Type	ign	1 (1 = fall trigger; 2 = rise; 100+ special types)
Number of Coils	COIL	1
Ignition Dwell Time	dELL	0.0 (Disable (sq. wave); -0.1 (User-defined table)
Ignition Delay Time	dLY	50 (delay time of the Ignition Syst (usec)-typical

Miscellaneous Setup
Boost Table Type		0
Injection Timing Table		1 (1 = RPM and EFF Point Dependent; 0 = RPM-only)
Injection Timing Pos		0 (0 = end of injection; 1 = start of injection)
Over Boost Cut		0
Ground Speed Limiting		0
Fuel Used Calc		0
Primary/Secondary Inj Ratio		0.00
Primary/Secondary Enrich		0
Primary/Secondary Enlean		0
Stopped Fuel		0
First Injection		40 (units percent of normal)
Telemetry Baud Rate		5
Radio Data Set		0
Internal Log Set		0
Internal Logging Rate		20
Advance Tuning		1
Password		0

to commensurately upgrade the stock fuel pump, air intake, exhaust manifold, compressor inlet, intercooler, and wastegate.

Pre-OBD-II Toyota onboard computers are fairly resistant to tampering. This surely gladdens the heart of the EPA, and it certainly makes life easier for Toyota executives looking to control the costs of bogus warranty claims arising from out-of-control hot rodders who might push their turbo Supras and MR2s and Celica All-Tracs over the edge with a dangerously high state of tune and then swap the stock chip back to score free warranty repairs for blown-up engines. But the PROM's inaccessibility definitely complicates achieving a seriously high state of tune.

What does an easily hackable factory engine management system look like? Pre-1994 GM computers store ignition timing curves, air/fuel mixtures tables, and other tuning calibration data in a standalone programmable read-only memory (PROM) chip that easily unplugs from the PROM socket on the main circuit board to change the calibration. An entirely separate Motorola microprocessor is soldered in place on the main circuit board (motherboard), with yet a third scratchpad random-access memory (RAM) device providing a kind of working storage for on-the-fly processor calculations. Tuners quickly reverse engineered the GM data structures back in the mid-1980s and happily began selling premium-fuel-only performance chips that offered at least minor increased horsepower and torque.

Toyota's pre-OBD-II engine management computer design for the MR2 was entirely different from GM's. While not user-programmable as stock, the architecture seems to have been an excellent design from the beginning, sufficiently robust to get the job done on a variety of Toyota cars and trucks for many years without a major redesign. In the Gen II MR2, the heart of the onboard computer is a slightly modified version of a standard Nippondenso chip that integrates the microprocessor, PROM, and RAM together on one large multipin chip soldered in place on one of two main circuit cards where it's essentially inaccessible for recalibration.

Not only does chip removal require electronics technician-level expertise, but if you remove the vehicle's PROM calibration, you're also removing the processor and RAM. The inaccessibility of calibration data that directs the operation of Toyota computers' fuel, timing, boost control, fuel cut, and rev-limiting calculations is especially frustrating for would-be tuners, since the extremely conservative factory calibration sacrifices as much as 150 horsepower on performance cars like the Gen II Turbo Supra.

We'd been getting ready to jettison the entire MR2 factory computer in favor of a programmable aftermarket engine management system when we had a conversation with Tadashi Nagata, a wizard computer programmer, car guy, and electrical engineer who founded G-Force Engineering near Los Angeles, and is now part of an outfit called Technosquare. Tadashi is the U.S. importer for Techtom, a Japanese automotive electronics firm that sells circuitry, diagnostic equipment, and laptop PC interfacing and tuning software that effectively converts Toyota and most other Japanese automotive computers to full programmability.

Techtom's mission includes providing U.S. tuning professionals with Techtom programming, tuning equipment, and expertise so that professional shops can hot rod Japanese car computers for their customers. Tadashi quickly discovered that marketing credibility required they demonstrate the dramatic power gains possible with the Techtom tools by providing not just the tools for hot rod tuning, but actual performance calibrations for specific cars.

You can send your Japanese onboard car computer and money to Tadashi Nagata, who will modify the computer hard-

Tadashi burns a PROM with MightyMap DOS PC software, removing the 3S-GTE's low-boost mode, removing the 12-psi fuel cut, turning on the 14- and 16-psi boost-fuel curves, removing a touch of timing at really high boost, and removing a bit of the excessive fuel the Toyota EMS throws at the engine at high-boost, high-rpm conditions.

ware, reprogram the tuning data to your specs (or a preprogrammed higher state of tune), and send it back to you. However, I wanted to photograph Tadashi in action, and I wanted him to customize the MR2 fuel map to support big 550 cc/min turbo Supra injectors on the project MR2—and shake-out the new calibration on a Dynojet chassis dyno while we watched the action with a fast air/fuel-ratio meter.

"People come in here all the time with MR2s running badly," says Tadashi. "Sometimes the check engine light is on, and they've been driving it that way ever since they first installed certain bolt-on hot-rodding equipment. The car's performance is typically very suboptimal, and the driver may not even know it. People often want to know why the MR2 Turbo seems to run so much better at certain times than others. In fact, they're not just imagining this." We discovered Tadashi had the expertise to solve many MR2 problems for hot rodders to find horsepower, and—based on our experience—a lot of hot rodders must leave his shop with a big smile on their faces. There's a lot going on behind the scenes that can limit maximum performance on the MR2, but we'd come to the right place. Tadashi removed the MR2 computer and placed it on the bench for rework.

MR2 COMPUTER REWORK

Tadashi began the G-Force MR2 computer upgrade by manually desoldering selected pins on the Toyota processor-memory device that are bent over to retain the device as the circuit board is stuffed prior to wave soldering. Tadashi then used a power desoldering device that's something like a cross between a soldering gun and a tiny electric drill, with a built-in miniature vacuum cleaner, and is commonly used for reverse-engineering purposes. Working from the underside of the circuit board, the tool first melts the solder on a selected pin, then rotates while vacuuming solder from around the pin. One by one, the pins are desoldered, at which point the processor-memory-PROM module falls from the board.

Following processor removal, Tadashi manually soldered a multipin socket into the mother board in place of the processor-memory module that allowed a special Techtom daughter board to be manually plugged into the socket without the need for soldering. The stock processor module is first soldered into the Techtom board, which also contains its own processor, as well as an auxiliary PROM socket. When fully stuffed, the Techtom daughter board uses the stock processor to manage the MR2 engine, but allows G-Force to substitute a new calibration for the original tuning data by loading the data into a removable PROM chip that is plugged into a second socket on the daughterboard. In order to direct the Toyota processor in managing the MR2 engine when you start the car, the Toyota processor is instructed to access the auxiliary PROM rather than its own PROM on board the processor module.

Tadashi Nagata can upload your original calibration into a Windows PC and display all data tables for review or modification—or simply work from a library of air/fuel and timing maps to build a custom PROM calibration. G-Force keeps on hand many performance calibrations that will produce increased power in both hot rod or stock Japanese cars. Once the calibration is customized for the individual application, Tadashi downloads the completed calibration into a PROM burner to program a new PROM chip, which is then carefully plugged into the Techtom daughterboard.

This custom calibration can be modified while driving or on the dyno by using an emulator. An emulator is a specialized microcomputer that plugs into a PROM socket and substitutes data maintained in random-access memory for the original PROM data. In our case, the drill was to remove the auxiliary Techtom PROM chip from the socket and plug it into a socket in the emulator. A ribbon-cable from the emulator plugs into the Techtom socket, whereupon the emulator reads the PROM calibration tables into RAM as a sort-of virtual PROM and emulates the PROM chip when the Toyota processor requests data—providing virtual PROM data from RAM.

Meanwhile, the emulator is connected by a second cable to a laptop PC running an interface program for display and modification. Under laptop control, the emulator can modify the RAM-based virtual calibration, which immediately affects tuning and engine management on the fly for observation or measurement on the dyno. When finished, the emulator calibration must be transferred to a PROM burner to generate the new PROM. The new calibration becomes permanent when you remove the emulator and plug the new PROM into the Techtom daughterboard socket.

After installing the daughterboard on the MR2 turbo computer, Tadashi relocated two capacitors on the second Toyota computer circuit board, providing clearance to button up the computer and reinstall it in the car. The project MR2 fired up immediately and ran flawlessly.

RECALIBRATING THE MR2

Tadashi Nagata points out that there are a number of important areas in which the MR2 Turbo computer can be modified for improved performance. Tadashi says the MR2 Turbo engine makes use of a Yamaha-designed four-valve pentroof head that achieved near perfection in design. The strong short block (forged pistons and stronger rods are used in racing 2.0s), excellent breathing, and superb combustion chamber design made the engine management task on the MR2 turbo much simpler than that of the late-model Turbo Supra, for example, where Toyota designed in multiprocessors and more complex control logic to safely achieve high power while protecting the engine.

Tadashi began by scrapping our very early MR2 calibration in favor of one intended for later 2.0-liter computers. Later computers have modified algorithms directing certain minor aspects of engine management, on the order of changes to the way idle stabilization is coordinated with air conditioning compressor-cycle operation. Our new calibration was actually based on calibration data designed for the heavier Celica All-Trac, the only other nonracing Toyota vehicle besides the MR2 that uses the 3S-GTE 2.0-liter Turbo engine. Tadashi reports the stock All-Trac calibration provides certain minor performance advantages over the MR2, and makes building a custom performance MR2 calibration slightly easier.

The first customization for our new MR2 calibration locked out the low-boost mode ignition timing and fueling maps. Toyota's MR2 calibration uses tables of timing or injection pulse width offset values that are organized from left to right in 16 columns that represent increasing rpm ranges from 0 to 9,000, and from top to bottom in 15 rows that represent increasing manifold pressure, from 25 to 30 inches vacuum, to 1.1 bar. Specific pulse width and timing for breakpoints of rpm or loading falling between specific points defined in the tables is defined via a weighted average of surrounding data points in the tables.

The critical thing for hot rodders to understand is that stock Toyota computer calibration includes entirely separate timing and fuel maps for low-boost mode and high-boost mode operation. The computer is designed to switch to a fail-safe low-boost mode if it decides the engine is running too hard and hot. In low-boost mode, the computer directs the wastegate to lower maximum boost from 9 to 7 psi. While maximum spark advance on either map is defined at 53 degrees BTDC (at high vacuum and rpm), full-boost ignition timing is retarded from 22.5 degrees advance to as low as 19 degrees when operating from the low-boost mode map.

Tadashi's new calibration restricted the computer to always remain in high-boost mode. According to Tadashi, the stock calibration is extremely conservative, and low-boost mode is not required on hot rod MR2s. In fact, even the stock high-boost timing and fuel calibrations are overly conservative for knowledgeable hot rodders who always maintain their cars, always use super premium fuel, and always keep an ear open for sounds of detonation—and back off instantly if they ever appear. The high-boost maps can be safely modified for increased responsiveness and torque. More on that in a minute.

Tadashi next modified the calibration to eliminate the high-boost fuel cut. An overboost fuel cut is actually defined for three separate rpm ranges, the first of which specifies maximum boost at up to 3,800 rpm, the second range to 5,000 rpm, and the third range to redline. Tadashi eliminated all fuel cuts by changing the three hexadecimal (base-16) numbers that specify maximum boost in the three ranges. *Our Fuel Cut Defenser would no longer be required.*

Tadashi now showed us that although stock boost is normally limited to 12-psi boost by the fuel cut, Toyota has actually provided two additional internal fuel and timing speed-density tables for 14- and 16-psi boost—though these last two ranges are normally locked out and unavailable for managing the MR2 engine—even if you are using a Fuel Cut Defenser device and overboosting above 12 psi. Although the high boost timing curves exist in PROM, with these clamped off, all injection pulse width and spark timing for boost above 12 psi are limited to fueling from the 12 psi numbers. Tadashi unlocked the 14 and 16 psi calibration numbers in our new calibration.

We decided to leave the rpm-limiter at the stock 7,250, although Tadashi says MR2 drag racers often want 9,000 capability. However, high-rpm operation is much harder on an engine than high boost, assuming you've got the correct tuning and fuel to prevent detonation at any cost. So our plan was to make all the power we wanted within the stock envelope. Extreme-performance Celica WRC (rally) powerplants vary from street MR2 turbo engines only in that they use direct-fire distributor-less ignition, better rods and pistons, and a bigger cam. Although such engines rev instantly like a superbike, it would appear they aren't that different from the race-bred street version. Given the 2.0's race-bred heritage, Tadashi speculated that exceeding the redline on a street MR2 mainly risks valvetrain damage.

The new calibration included some additional modifications to the high-boost timing map. Timing was advanced about 7 degrees at light load as well as at very heavy loads. Bell Engineering had recently installed a larger A/R turbine housing on our Turbonetics T04E-50 spec'd with the larger 0.63 turbine nozzle appropriate for all-out banzai max-horsepower dyno runs—at the cost of somewhat reduced boost at 2,500–3,500 rpm. Tadashi had predicted that added timing advance would help compensate for the A/R change by providing increased torque under naturally aspirated and low-boost conditions (which proved to be true on the street). Discussing detonation safety margins for the new timing, Tadashi said experience indicates that we could safely run another three degrees of timing without a problem. We decided to stick with the three degrees of margin.

As we pushed 335 horsepower, we had been running a Bell Engineering FMU to maintain fuel delivery under maximum power conditions in the rarefied atmosphere over 300 horsepower. Given that our plan was to compare torque and power for the stock and G-Force calibrations while retaining the stock injectors, we decided to remove the FMU and limit boost to 1.1 bar for a fair comparison. Under these conditions, Tadashi actually modified the stock calibration to take out some high-end fuel. Above 1.1 boost requires additional fuel at the high end, but we had plans to handle that via larger injectors.

According to Tadashi, the stock MR2 air/fuel map is inferior to the All-Trac's at the top end as well as at medium-rpm during hard acceleration when the small stock turbo kicks in and boost increases very rapidly. To avoid detonation under these conditions, the stock calibration dumps in excess fuel. Tadashi recommends that engines like this with bigger turbos should have leaner fuel curves in the midrange at the onset of boost, but increased fuel at the high end for really big boost numbers.

Testing on the R&D dyno found increases in torque and power with the new calibration. Following power runs on the R&D Dynojet, we decided to pull the fuel rail and install larger 550 cc/min injectors.

The plan was to provide the gross pulse width changes needed to compensate for the big injectors by reprogramming the vane pressure converter tables that translate manifold pressure and engine speed into a 0–5-volt simulated airflow voltage to the main computer. Tadashi Nagata planned to modify the VPC program, and then fine-tune the car with VPC idle and gain controls and—if necessary—additional calibration changes to the main computer via Techtom PROM-data modifications. The Dynojet and Tadashi's Horiba lambda air/fuel meter would tell us when we had it right.

Basically, he planned to arrange for the VPC to lie about measured airflow in such a way that the 25 percent larger injectors would provide correct fueling all the way from idle to high levels of overboost. This would be a little complex because the lowest vane pressure conversions had to translate MAP pressure in such a way as to underreport airflow by 25 percent via subnormal voltage. On the other hand, the translation would have

to do something different for the higher ranges, since the normal translation of VPC MAP numbers to simulated Bosch vane airflow numbers would convert MAP numbers around 1 bar boost into the full-scale VAF meter voltages approaching 5 volts.

By modifying the VPC program so it would require numbers above 16-psi boost manifold pressure before it generated the maximum 5-volt value that simulates a stock wide-open airflow meter, we could effectively extend the computer's airflow-based fueling authority from the stock high vacuum-through-16-psi domain to make it a system that effectively mapped airflow numbers from high vacuum all the way through 2 bar (30 psi) boost into the same 0- to 5- volt range. The engine-load granularity of the computer's timing and fuel tables would be reduced, but with the addition of the larger Supra injectors, we'd now have the potential to provide fuel and spark timing for maximum torque occurring at boost levels far above the 16-psi limitation of the stock computer!

For example let's say you remapped the VPC to assign a 1.5-bar MAP pressure to 5 volts. The ECU internal tables would assume that it was seeing the stock airflow meter maxed-out at 1.0-bar boost, and call for nearly full pulse width. But since the engine would have 25 percent larger injectors, the engine would actually receive 25 percent more fuel.

If we did it right, the new VPC calibration would simultaneously correct for the increased flow of the big 550-cc Supra injectors at both ends of the scale. Hopefully, the main fuel map in the Techtom PROM would require only very minor changes. The biggest trick would be to maintain idle quality in a situation where the effect of relatively minor changes to vacuum at idle would be multiplied by the VPC changes, and appear to the computer as larger changes to airflow—given the reduced granularity and effectively increased spread of the VPC map. However, given that the oxygen sensor is king for delivering correct air/fuel mixtures at idle and light cruise, we were depending on closed-loop feedback mode to maintain correct mixtures for excellent quality idle.

With the Toyota computer hacked and a new calibration in place, fuel supply changes were relatively simple. We removed the Cartech FMU from its installed position downstream of the stock fuel pressure regulator, and disconnected the FMU's boost and vacuum/pressure lines from the intake system. We used a double-ended hose barb to join the fuel return line back together. We removed the throttle body and injector rail, swapping in the new Supra injectors, which are externally identical to the stock MR2 injectors.

We found that a NOS high-volume, high-pressure fuel pump can be feasibly run in series with the stock in-tank fuel pump, mounted in-line upstream of the fuel rail. Alternately, a Kenne-Bell Boost-A-Pump control unit could be installed between the stock fuel pump and power supply. The Boost-A-Pump must be wired into switch power, with a specified thick-gauge constant power source attached directly to the battery. The boost sensor is installed in the intake system in order to trigger the Boost-A-Pump's control unit to supply increased voltage to the stock pump—or, for that matter, to an external in-line pump like the NOS pump. Admittedly, two pumps and a Boost-A-Pump are overkill, but we tried it just for grins.

With final stock fuel supply and injector benchmark dynoruns out of the way at R&D Dyno and the Techtom ROM-board system working perfectly, G-Force owner Tadashi Nagata and I removed the MR2's throttle body and air inlet, unbolted the fuel lines from the Alamo Autosports dual-inlet center regulator fuel rail, and removed the rail. Space is tight; to R&R the

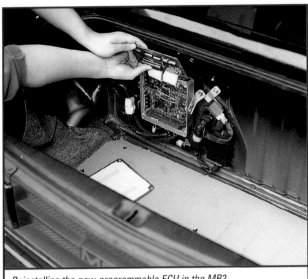

Reinstalling the now-programmable ECU in the MR2.

rail requires some patience and finesse. With the rail out, we unbolted the injector bracket and pulled the stock O-ring type injectors from the rail, replacing them with outwardly identical (other than the color) 550-cc/min Turbo Supra injectors available through Toyota dealers. The 3.0 turbo Supra six-cylinder motor has equivalent displacement per cylinder to the 2.0 four-cylinder MR2 turbo, and since the Turbo Supra obviously works fine with 550-cc injectors, we felt confident that the new injectors could be made to idle smoothly with a pulse width (open time) sufficiently lengthy for repeatable fueling.

Our plan was to use the HKS VPC's idle and gain mixture controls (which enrich or enlean fueling 2 percent per click up to a delta of plus or minus 12 percent) to get the MR2 started and running with the big injectors, allow the Toyota computer to recalibrate idle mixture using the oxygen sensor, and try to get a few trial dyno runs while carefully watching the Horiba lambda air/fuel meter.

With the VPC at maximum lean, the car did indeed start and settle into an increasingly smooth idle. Tadashi adjusted idle mixture, higher rpm air/fuel gain, and transient enrichment on the VPC while I drove the car on the Dynojet. We had verified that the fuel pump was maxed out at 1.35-bar boost with the stock injectors installed and a new G-Force calibration in place, so we went ahead and installed a large in-line NOS high-pressure fuel pump in series with the stock in-tank pump.

On its own, the NOS pump is good for 550 (crankshaft) horsepower, so we knew the two pumps together would provide more than enough fuel for any remotely achievable power we'd see on the dyno today. In fact, we were a little worried there'd be too much fuel, that the rail might be overpressured at idle even with the regulator wide open and both pumps going. Therefore, I set up the NOS pump with a manual switch, although, as it turned out, the fuel return line seemed to do the job just fine. After some random driving on the dyno, during which Tadashi tinkered with the calibration and watched a Horiba lambda meter, we decided to try a live horsepower pull.

We did not have any illusions that we were going to make 400 rear-wheel horsepower at this time. Increasing levels of boost beyond the design limits of the BEGI intercooler produced a 5-psi pressure drop through the intercooler as well as a loss of thermal efficiency, and we had plans in progress for a high-efficiency

air-water intercooler. However, it was not yet on the car, meaning the really big numbers were still somewhere in the future.

Nonetheless, I wanted to see what the car could do with a recalibration. With the EVC digital boost controller still set at 1.35 bar, I operated the car while R&D's Darren worked the Dynojet controls. On the maiden 550-cc injector run, the car made 334.2 peak horsepower and 281.3 ft-lbs torque—about where we'd been with the stock injectors and fuel pump. However, the Horiba now indicated the air/fuel ratio was on the rich side everywhere. At last we had plenty of fuel, and now a 1.5-bar run seemed feasible.

Now, it was getting on into the evening by this time, and we were all pretty tired (I'd been up until 4:30 A.M. the previous night). Tadashi still wanted to reprogram the HKS VPC to compensate for the big injectors and to enable the computers to correctly fuel a greater range of boost airflow—as would be required eventually to achieve 400 rear-wheel horses—but this was not necessarily going to be simple. At this point the plan became, do a 1.5-bar run, then drive the car on the streets with the lambda meter in place to watch the air/fuel ratios over a wide range of conditions, then build a new VPC map.

In the meantime, with the car on R&D Dyno's Dynojet and a 1.5 bar run imminent, Tadashi Nagata connected an emulator and went to work removing timing from the high-rpm, high-load section of the spark timing table of the Toyota computer—just to be on the safe side.

I shut down the MR2's engine to keep it cool during reprogramming, but left the key on to provide power to the Toyota computer. Tadashi did his thing on the calibration with the laptop, following which I restarted the car. At this point, a dyno technician watched the dyno screen, Tadashi watched the Horiba and the EVC display, and I watched the MR2's gauges.

I quickly brought the car up to 2,000, punched the record button on the Dynojet, and mashed the gas to the floor. Boost and rpm began building fast, and you could immediately tell the car felt much stronger at the new higher boost. The Dynojet would later reveal that torque had reached nearly 300 ft-lbs as low as 4,750 rpm and the car was accelerating the Dynojet violently—but suddenly the lambda meter went dead lean, and I heard Tadashi's voice shout a warning. The tach hit 6,000 as I dumped throttle and listened to the sound of the Dynojet's rollers coasting down.

Why the lean surge? The answer was quickly apparent: I'd forgotten to turn on the NOS fuel pump. However, the car never lost power, and initially idle seemed OK. I climbed out and switched on the auxiliary fuel pump. Back in the car, we began a new dyno pull at 2,000, and this time there was plenty of fuel all the way to redline.

Falling back to idle, though, the engine's note was suddenly lumpy, and smoke was now emanating intermittently from the exhaust. Something was clearly wrong. We checked for excessive breather blow-by, then shut down the engine, pulled and examined the plugs, and looked through the plug holes at the piston tops. The good news was that the plugs were not embedded with foreign matter and blow-by didn't seem any worse than usual. The bad news was that you could see visible—although perhaps not terribly severe—pitting on the tops of pistons. What's more, a compression test revealed that while the right-most cylinder read 190-psi cranking pressure, the other three were now in the 115 to 125 range. There was no question: One or both of the last two dyno runs had hurt the MR2's engine. And yet, the dyno chart showed that the car had smoothly reached nearly 350 horsepower at the wheels!

MOTEC PROGRAMMABLE EMS INSTALLATION

At this point, we clearly needed a new super-duty engine, and we needed a way to manage it. Bob Norwood lobbied for a Motec M48 programmable EMS. The new 3S-GTE powerplant shaped up quickly as the following:

• Paul's 5 millimeter welded-stroker crank with TRD/Toyota main and rod bearings
• JE custom stroker forged pistons with Performance Coatings ceramic thermal barrier coatings
• Ported and polished head with HKS performance cams and adjustable sprockets
• Norwood Autocraft aviation-polished Turbonetics ball-bearing T04E-50 turbo
• Norwood custom air-water intercooler
• Siamesed and ported intake manifold (four runners instead of eight) and eliminated TVIS intake butterflies
• Billet underdrive crank pulley

On the MR2, Bob Norwood was particularly concerned about achieving maximum power with large injectors, while retaining a streetable idle and low-end on an engine with extreme dynamic power range. Norwood considered it possible that even the 550 Supra injectors would not be sufficient for the reworked MR2 at full howl, maximum horsepower, that we'd want to manage even larger injectors for peak power. He also wanted the capability to tune out *individual cylinder* air/fuel and timing limitations, without hurting power on other cylinders.

We decided to use a Motec M48 system for maximum power results. (We *knew* there'd be a use one day for the port exhaust gas temperature [EGT] probe holes that Alamo Autosports had tapped into the exhaust manifold during the exhaust port job, and EGT is a great way to keep combustion safe while you push the envelope on a water-brake dyno.) Motec's M48 system is a powerful engine management environment with the ability to allow programming of dozens or even hundreds of parameters affecting engine operation that do not even exist on most factory and aftermarket engine management computers.

The Motec is extremely flexible regarding sensors and actuators. Our only concerns had to do with TPS and engine position sensor interfacing. As it turned out, the M48 will work fine with the stock Toyota throttle position sensor (normal three-wire TPS sensor with a fourth wire attached to a full-throttle microswitch). It will also work fine with the 3S-GTE engine's engine position sensors. The 3S-GTE has a two-in-one cam-speed trigger wheel in the distributor with 24 teeth in one row with a pickup for crank position (12 teeth per crank revolution). It also has a second pickup reading a single sync tooth for top dead center positioning needed for direct ignition and sequential injection. Motec systems have been used before on the 3S-GTE, and Motec was able to fax down an interface-wiring diagram illustrating connections for the Toyota four-wire distributor trigger wiring connector.

The Motec M4, M48, M800, and M880 engine controllers are serious EMS computers with serious software to control engines. They are designed to be extremely flexible in design, to give dedicated tuners control over every imaginable aspect of engine and vehicle performance. The Motec M48 ECU is built around sequential injection and distributor-less (direct coil) spark controls. The Motec also includes individual cylinder fuel and spark trim (timing and injection pulse width can vary between cylinders). Injector scaling is available, in which granularity of injection pulse width adjustment is scaled according to the available time for injection at redline and minimum and maximum injection times needed for a given engine and set of injectors.

The Motec provides extremely robust acceleration enrich-

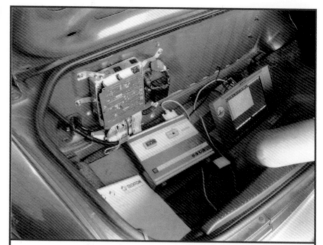

With Techtom Toyota ECU PROM modifications, Tadashi Nagata connects an emulator (center of photo, with PROM installed in left socket) to the Techtom daughterboard on the ECU (front firewall, upper left), then connects a laptop running MightyMap to the emulator, ready to optimize the calibration on the R&D Dynojet.

ment, including sensitivity and decay features: Accel Clamp, Accel Decay, Accel Sensitivity, Decel Clamp, Decel Decay, and Decel Sensitivity. By contrast, some less expensive engine management computers have offered only acceleration percent enrichment and sustain, sometimes a fuel cut on deceleration. The Motec provides a rich array of warning alarms: Max Manifold Pressure, Max Air Temp, Max Engine Temp, Min Battery Voltage, Max Aux Temp, Min Aux Temp, Max Aux Voltage, Min Aux Voltage, Max Speed 1, and Max Speed 2. The Motec system provides an extensive set of boost controls, including Boost Limits on Trim, Air Temp, Engine Temp, Aux Temp, and Aux Voltage. Boost tables can be based on rpm, rpm and throttle position, or rpm and gear. The Motec defines objects such as radio telemetry data sets and many other factors.

The Motec M48 requires twin MAP sensors for barometric and manifold absolute pressure, which are trivial to install. The Motec reads essentially any O_2, coolant temp, intake air temp, manifold temp sensors, although we decided to install a second set of sensors for these items that would allow the stock computer to remain alive for cruise control and so forth.

In our case, the project MR2 would be taking advantage of a Motec capability not directly related to programmability, namely the ability to control injectors precisely at extremely short pulse widths.

Normally, the Motec wiring harness must be custom-built for a particular engine installation. The harness consists of an ECU connector attached to a huge bundle of wiring, and various hardware for installation of sensor and actuator connectors on the wiring. Conduit or heat-shrink tubing must be installed following customization of the harness.

MOTEC START UP AND TESTING

The engine fired up immediately and ran well on the first try at Norwood Autocraft with a best guess custom Motec calibration. Bob Norwood's hands flew over the keys of his laptop as he calibrated fuel pulse width and ignition timing maps on the Motec by altering the complex shape of a 3-D graph while the car ran. Norwood also specified target air/fuel ratios for various speed-density cells, set up the Motec to control the stock fuel pump and to automatically stage in our 550-horse NOS in-line high-volume auxiliary fuel pump at 1 bar boost. With the calibration roughed in and looking good, we drove the MR2 onto the Alamo Autosports Dynojet chassis dyno for additional power testing and debug.

The new engine would prove to be rock-solid, with the initial exception of the head gasket and valve springs. We measured intake manifold and exhaust manifold backpressure, intercooler in and out temperatures, individual-port exhaust gas temperature, instantaneous wideband air/fuel ratios, and many other parameters. We would eventually experiment with looser turbine housings (Turbonetics 0.63 and then 0.82 A/R versus the original 0.48 housing) and larger 775 cc/min Bosch injectors (versus the 550 cc/min Toyota Supra side-flow injectors). As we would discover, when such an engine starts to rip, hang onto your shorts because things happen quickly and you better make sure your boost controller installation is up to the job.

Bob Norwood maps a Motec on the dyno and street with the help of a lifetime of high-output engine experience plus Motec's extensive datalogging facility. The engine is virtually always equipped with a wideband O_2 sensor, and a super-duty engine will normally have EGT probes. The wideband sensor is lightning-fast and designed to supply air/fuel lambda ratios across the range rather than only near 14.7 air/fuel like standard zirconium O_2 sensors. The EGT probes allow the Motec computer to log combustion temperatures, permitting tuning to the ragged edge for maximum power.

Calibrating a computer like the Motec on a super-duty turbo engine is an exacting process that involves gradually operating the engine at every achievable combination of engine speed and manifold pressure while watching timing, air/fuel ratios, transient enrichment, EGT, torque, power, and other factors.

The new MR2 stroker engine immediately made boost from 1,800 rpm on the Dynojet with a larger 0.63 A/R Turbonetics turbine housing installed. This was a legacy of the high-energy stroker exhaust, the ceramic thermal coated pistons and combustion chamber, and the ball-bearing turbocharger, and was significantly better at low-end than the stock-block was with the tighter 0.48 housing. It was even better than the long-ago-and-far-away stock twin-entry turbocharger, which had, so far, been the champ. In fact, the new setup was the first performance turbo setup we had ever seen with *improved* bottom-end torque versus stock—not to mention 100 more ft-lbs torque in midrange and above.

By this time we had more than 40 dyno runs on the new engine, and it was definitely broken in. We had now seen consistent power levels of 370 horsepower at the wheels at almost 7,000 rpm, and 387 ft-lbs torque at roughly 4,500, but that was in spite of the effect of boost-controller problems that hurt both low- and high-rpm performance, and with intake valve springs that were floating at high boost and getting worse each run—blown open and hyper-extended by extreme boost conditions. When it comes to controlling boost, it is essential that a fuzzy-logic boost controller (HKS EVC electronic wastegate controller) properly experience the engine and turbo boost capabilities while in learning mode. It is essential that the manifold pressure boost reference lines for the EVC and wastegate be independent from other vacuum hoses feeding fuel pressure regulators, computer MAP sensors, and so on.

The first big-power dyno session ended explosively with a blown stock head gasket and steam exiting the tailpipe. However, it had also become clear on the first dyno session that the 550 cc/min injectors were already at 82 percent duty cycle.

Since we figured there was plenty more power to be had, the temporary strategy to deal with this was reinstallation of our old

Bell Engineering FMU to raise injector capacity with increased rail pressure during boosted conditions. This resulted in Norwood remapping the Motec, and it produced a somewhat strange-looking air/fuel map as the Motec calibration had to deal with the triple variables of engine fuel requirements, injection pulse width, and rapidly rising boost fuel pressure.

With the new steel gasket in place, we soon encountered new problems but pushed on with dyno testing beyond the point where we knew something was wrong. Though 30-psi runs ended more than once with an explosive backfire, the engine still performed fairly well, though when it buzz-sawed hard into turbo boost against the dyno's inertia, there was an increasing popcorn aspect to combustion that we decided was definitely not detonation or pre-ignition. Forensic analysis later showed it to be an indication of snap-crackle-and-pop valve-float misfires and minor backfires. The explosive combat-like multiple backfires that ended maximum-boost runs in a volley of sounds that shook the dyno were clearly, in retrospect, an indication of intake valves blown open by high boost.

We found that when boost got too high for the stock valve springs (installed stock at 45- to 53-psi pressure, shimmed for our application to slightly higher pressure), the intake valves would be pushed open when the exhaust valve was already open. This shot fresh air and fuel into the combustion chamber and beyond into the exhaust and superheated turbine, igniting explosively into both intake and exhaust, producing a shuddering series of backfires, followed by the compressor surge-valve's own blast of compressed air into the atmosphere as the operator released the throttle.

We pulled the MR2 head for a third time, verified that nothing was broken other than the head gasket, and, once again, began chasing parts.

The HKS performance cams were made for stock valves and springs; we were unable to find any bolt-on spring kits designed for the MR2. Norwood Autocraft located some 100-pound springs that would fit and figured out a way to install them with pressure in the 80- to 90-pound range, which is about what we wanted. Obviously, more valve spring pressure than you need just increases power-robbing friction and wears cams out faster, but too little pressure is even worse. In the meantime, Alamo and Norwood arranged to have Clark Copper Head Gaskets, Inc., copy a stock gasket in pure annealed copper.

In the meantime, we knew certain things: Engine disassembly for head removal on two separate occasions had proven that the internal thermal coatings were staying on the valves, combustion chamber, piston tops, and all intake and exhaust runners. The coatings keep heat in the exhaust gases rather than the engine, and until we had head gasket problems, the coolant jacket stayed cooler than before, even at very high torque and power. Exhaust gas temperatures were similarly low.

We had simultaneously tested intake boost and exhaust back pressure, and at low and medium boost pressure we observed the highly desirable crossover phenomenon in which intake pressures are higher than the exhaust pressure driving the turbo. Even at 30-psi boost pressure, backpressure was never higher than 35 psi, which we consider excellent, given that turbo engines often run two to three times boost in backpressure. Certainly increased horsepower will push more gases through the turbo, but the new 0.82 turbine housing we're installing can only help this situation.

The air-water intercooler was outstanding for making dyno power, with only minimal amounts of tap water through the intercooler producing intake air temperatures that never exceeded 18 degrees Celsius at 30-psi boost. We'll know soon what our heat exchanger can achieve in keeping intercooler water cold, but for drag racing-type situations, an ice-water reservoir is easy to do if necessary.

The adjustable cam sprockets for the first time gave us the ability to modify overlap of the exhaust and intake cam. We started with stock cam timing until we blew the stock gasket, then tried more overlap in the next dyno sessions with the HKS steel head gasket, but we were busy troubleshooting and did not get to do back-to-back cam tuning for maximum power.

The MSD 6A ignition was producing an extremely hot spark on a distributor-equipped four-cylinder engine on which there was only so much time for coil recharge. In fact, this ignition will smoke the coil wire if you're not careful to have perfect connections everywhere.

Fuel supply problems were a thing of the past. With a gauge reporting fuel pressure at all times (and a person assigned to watch it), the Motec-staged NOS fuel pump and stock pump were producing all the fuel we could ever want. Injectors, however, were a different story. We datalogged the 550-cc Turbo Supra injectors above 82 percent duty cycle and decided to add the VRG regulator. We built a new fuel rail able to mount standard 80 pounds per hour top-mount pintle-type Bosch injectors in the engine.

Final results would require a strong Alamo Autosports/Norwood Autocraft ported 2.19-liter stroker engine with all the best internal parts (JE forged stroker pistons, ARP head studs, Norwood high-pressure valve springs, Norwood cylinder O-ringing and copper gasket, HKS performance camshafts) and thermal coatings everywhere. Final results required a super-duty T04E-60 Turbonetics ball-bearing turbo and Racegate, HKS fuzzy-logic boost controller, a Norwood air-water intercooler running 70-degree Fahrenheit cold water, 850 cc/min injectors and a Motec M4-8 computer with the circuitry that can make such huge fuel injectors idle well at very short pulse widths. The results also required Bob Norwood's carefully optimized Motec computer calibration, VP 108 race fuel, and a serious Clutchmasters clutch.

With the above parts in place, under Motec engine management, the super-duty stroked MR2 powerplant would survive and thrive on numerous runs approaching 30-psi boost—several to almost 9,000 rpm. We fought minor demons: At very high boost, the stock fuel-rail seals tried to blow out of the cylinder-head injector bores mounting the new Bosch-type fuel injectors clamped under our new-design Norwood fuel rail (the stock side-feed injectors are O-ringed into the stock rail and clamped in place with a cover, while rubber-sleeved nozzles in the *rail* mate directly with the head bores). Additional O-rings handled the problem.

The Motec computer balked about out-of-range sensor data when turbo boost exceeded the GM 3-bar MAP sensor's limits at 30 psi. Initially, running rich at maximum power levels, the fuel injectors began to hit 100 percent duty cycle at certain speed-density breakpoints (injectors open all the time, or static)—and, indeed, we initially regretted using 83-pound instead of 93-pound injectors. But as Bob Norwood refined the Motec air/fuel and spark calibrations (working rich to lean, data-logging air/fuel lambda via the four-wire O_2 sensor, searching for maximum-power air/fuel ratio of 13.5 at zero vacuum, 12.5 at 1 bar, as rich as 11.5 at 2 bar), the injector duty cycles came down to levels below 90 percent. With all demons temporarily under control, at roughly 25-psi boost we had achieved 424 rear-wheel horsepower at 6,250 rpm and 420 ft-lbs torque at 4,750, which translates to estimated flywheel power in the 533 range, crankshaft torque of 528 ft-lbs.

In the meantime, we had encountered another problem: With fuel right-on and power climbing, our Turbonetics T04E-50 ball-bearing turbo began to run out of breath. The T04E-50—originally specified to make 325 wheel horsepower or so on our free-breathing four-valve stroker motor—was clearly approaching its airflow limits, actually dropping boost pressure from 26 psi at the 4,750 torque peak, down to 24 psi at the 6,250 power peak. Interestingly, when we tried disconnecting our HKS muffler for more power, we were surprised to see power *drop* nearly 30 horsepower. And we could only regain about half this power with air/fuel tuning on the Motec, which illustrates the importance of dyno tuning and of questioning all assumptions when searching for maximum power.

With measured exhaust backpressure unexpectedly low through the Turbonetics T-3 0.82 A/R turbine section (exhaust backpressure was less than manifold pressure until the highest levels of boost, at which point backpressure hit 2.5 bar [nearly 38 psi] which is still excellent), we felt the T-3 turbine could handle more compressor. Turbonetics recommended reworking our T04E-50 turbo compressor to higher-flow T04E-60 compressor specs (fitting a new compressor wheel and CNC-machining our aluminum compressor housing to new specs), which would theoretically provide at least 10 to 15 percent more airflow. This would turn out to be a good prediction; we would make at least 50 more rear-wheel horsepower with the T04E-60 installed.

While the turbo was at Turbonetics for the upgrade, Norwood Autocraft completed the air-water intercooler system for street use by fitting a heat-exchanger and triple fans to the engine cover for cooling the intercooler water. We had been cooling the intercooler using cold tap water while on the Dynojet and figured on using a self-contained ice-water tank for drag-type maximum-performance acceleration on the street or strip.

With our all-new killer setup installed, once more we headed back to the Dynojet. Following additional Norwood Motec recalibration of certain rpm and loading points we hadn't previously encountered, Alamo Autosports began optimizing the HKS Stage 1 adjustable cam sprockets at low (9-psi) boost. By retarding the exhaust cam in small increments to a total of 5 degrees, we were able to improve power at all rpm—with peak power up by 21 horses compared to stock cam timing, power at 7,000 rpm up by 60 rear-wheel horsepower!

At this point, the importance of optimizing wastegate control became really apparent when we realized that we were losing huge amounts of midrange torque with the new T04E-60 compressor due to the wastegate being forced partially open at boost pressures substantially below our desired maximum boost. The electronic valve controller (EVC) could control boost to 2-bar or nearly 30-psi boost, and when we began exceeding this maximum, the EVC controller began allowing boost to fluctuate violently.

Now, for tremendous overboost, we were forced to switch to dial-a-boost air pressure regulator control of the Deltagate secondary wastegate port. Unfortunately, this strategy greatly reduced torque from 3,200 to 5,200 rpm—as much as 140 ft-lbs at the wheels at 4,750 rpm range—even though we hit new highs of peak power further up the scale. Boost was down correspondingly in the midrange, with a maximum delta of minus-14-psi boost at 4,750 (11 versus 25 psi), and 5 psi versus 10 available at 3,500 rpm!

We experimented with various standalone wastegate controllers before eventually producing maximum performance on the dyno by running the turbo at maximum performance without boost controls—that is, by *disconnecting* boost reference to

Magnum opus: super-duty 2.1-liter stroker 3S-GTE, almost ready to rock 'n' roll. Boost pressure moved directly from the compressor into the top-mounted air-water intercooler and directly out the other side into the big-bore RC throttle body. Efficient and responsive.

the primary port on the wastegate actuator and actually feeding full boost pressure to the secondary port (Deltagate) side of the wastegate actuator to *help hold the wastegate shut*, thereby preventing exhaust backpressure from possibly forcing open the Racegate at any manifold pressure all the way up to maximum compressor boost and airflow.

The final moment of truth for the project came when the 2.19-liter 3S-GTE MR2 powerplant hunkered down on the Dynojet and ripped off a run we aborted at 461 rear-wheel horsepower, followed quickly by a 471.1 horse maximum-power run at 6,000 rpm. This is nearly 600 horsepower at the crankshaft using the conversion factor we've used to account for drivetrain horsepower losses. By contrast, a stock MR2 turbo makes 160–165 at the wheels on the Dynojet. Best torque was 431.9 ft-lbs at 5,500. And we're still not sure whether these figures were achieved despite the adjustable compressor bypass possibly bleeding off air above 30-plus-psi boost!

Disconnecting from the dyno cold-water intercooling system, we took the car out for road testing—a somewhat risky maneuver at these power levels, given the stock five-speed transmission's known reliability problems above 400–450 crankshaft horsepower. But driving back to Norwood's, we downshifted on the freeway into third gear and mashed the gas, lifting the front wheels into the air!

MR2 COMPUTER TROUBLE CODE COUNTERMEASURES AND ANTI-HOT RODDING TRICKS

When the MR2's check engine indicator lit up on the way to G-Force Engineering in California the car simultaneously lost most of its power under boost—and we weren't immediately sure what the problem was. This was not the first check engine problem we'd seen. We'd previously noted O_2 sensor problems on the MR2

on the Alamo Performance dyno after a previous brief flirtation with highly leaded fuel apparently killed off the 65,000-miles original equipment O_2 sensor. We'd replaced this $135 item immediately before leaving on the trip so we'd have the fuel-saving advantage of the car's closed-loop on-the-go self-tuning as we motored down the highway in light cruise. With a bad O_2 sensor, there really weren't any symptoms other than the glowing check engine light.

This time, as soon as the turbo started to make any boost, you would feel a massive bog, as if all the tuning and power had gone south. As the boost came on, the car would actually loose power, and if you let off the gas, you'd get this odd, backward, momentary sensation of increased power as the car came out of boost. Our car's Turbonetics ceramic ball-bearing turbo has so much airflow capacity that if you kept your foot in it, you could force the engine through the sluggish zone into modest power at very high boost. But clearly the computer thought something was wrong and was fighting our attempts to go fast.

Now, with a shop manual and a few jumper wires and a little dexterity, you can operate on the Toyota computer diagnostic plug at the right rear of the engine compartment and flash out codes on the check engine light (and then look them up in the manual), but at 4:30 in the morning in the wasteland between Phoenix and L.A. with the engine running fine when unboosted, pushing on seemed like a better idea than playing with jumper wires along the road by moonlight. For 500 miles, from Tucson to L.A., we drove essentially without turbo boost, at light cruise.

At G-Force, the computer immediately yielded up its trouble codes to Tadashi, telling us the computer thought there was both an airflow meter problem and a knock sensor problem. The knock sensor, it turned out, had died, and with no knock sensor, the computer's strategy is to take all the timing out of the engine on boost, which effectively kills power, particularly when you're running the stock turbo. We bought a new $125 sensor at the local Toyota dealer, which Tadashi installed in five minutes with the car on a lift, curing the first problem.

The airflow meter trouble code was interesting, because we were no longer even running an air-flow meter on the car. The problem, it turned out, arose from the combined efforts of our high-boost Turbonetics T04 turbo, our HKS Fuel Cut Defenser and our HKS vane pressure converter—and the computer's continuing efforts to detect possible sensor problems.

The explanation goes something like this: The stock MR2 Turbo uses a Bosch-type airflow meter, a device that provides a 0–5-volt signal to the computer depending on the position of a spring-loaded flapper door hung out in the intake air stream upstream of the turbocharger, the position that corresponds to engine mass air consumption when corrected for air temperature. The Toyota computer is constantly comparing airflow data as reported by the meter with data from two other sensors, one reporting manifold pressure, the other atmospheric pressure, which together provide a credible estimate of engine airflow. If the engine and computer system are healthy, there is a predictable relationship between reported airflow and manifold pressure at a given rpm. This fact is the basis of all speed-density engine management systems (including virtually all programmable aftermarket and racing systems).

Speed-density systems forgo airflow meters due to the flow restriction of the meter itself and instead estimate airflow via sensors measuring engine speed and manifold air pressure. Toyota MR2s use airflow metering, but also use a pressure sensor as a backup for the airflow meter and an indication of possible trouble with the airflow meter. If the airflow meter signal is very high, for example, but the manifold pressure is suspiciously low, this might indicate that there is a large air leak between the meter and the throttle body. The computer will set a trouble code and turn on the check engine light.

Factory MR2 wastegate springs limit boost to 7- to 9-psi by diverting exhaust gases around the turbine when manifold pressure gets too high, and the MR2 computer can further lower boost via electro-pneumatic input to a dual-diaphragm wastegate actuator if coolant temperatures get critical. If all is well, the Toyota computer monitoring engine conditions should never detect more than a 10-psi boost worth of airflow. But if the manifold pressure reference vacuum hose became disconnected from the wastegate diaphragm—or, as is the case on our project car, an HKS electronic valve controller was tampering with the maximum airflow voltage as seen at the wastegate diaphragm in order to lie to the wastegate and keep it closed in order to bring about increased boost and power—the computer actuates a fail-safe measure at a 12-psi boost threshold, a fuel cut condition in which the fuel injectors are not fired until manifold pressure drops below atmospheric pressure.

Again, an FCD makes sure the computer never sees more than 12 psi worth of airflow by truncating the VPC's output voltage. However, as mentioned previously, the computer has its own manifold pressure sensors, which are unaffected by the FCD. And if manifold pressure gets too far out of whack in relation to reported airflow, the computer will set an airflow meter trouble code. Given the stock turbo's flow limitations and the computer's arsenal of countermeasures to foil boost-junkie hot rodders, such an imbalance is extremely unlikely. But possible. And if you've got the massive boost potential of a Turbonetics T04E-50, and you keep turning up the wick, it's just a matter of time.

At first glance, the FCD's fuel-cut defeat strategy might seem a misguided or even dangerous way to permit high-boost operation. How can the Toyota computer provide correct fueling at high boost if you chop off reported airflow voltage at 12 psi? Actually, as Tadashi explained—and as we had noticed on the Dynojet while watching high-boost air/fuel ratios—under high-rpm, high-boost conditions, the computer tends to throw all the fuel in the world into the engine and provide extremely conservative spark timing boost retard, which are both effective engine-saving strategies.

There's also the knock sensor, which can trigger truly huge computer-generated decreases in spark timing to kill detonation. In addition, the VPC's adjustability allows tuners to increase the across-the-board fuel delivery. (We actually made more power for a long time by leaning the high-rpm fuel.) This VPC adjustability can be used on the Dynojet to target max-power high-boost fuel for tuning perfection. In fact, there are other built-in engine protections: The Toyota computer can activate an additional fail-safe engine-saving mode under certain conditions in which, under boost, the computer begins providing greatly enriched injection pulse width and retarded spark timing values from an entirely separate backup set of fuel and spark timing tables—low-boost tables, a strategy normally activated simultaneously with low-boost mode at the turbocharger wastegate via the dual-diaphragm wastegate computer pneumatic controls. These countermeasures are anti-abuse tactics designed to keep warranty claims down on a turbo motor and to keep maniacs from killing their engines. Ordinary drivers would rarely encounter Toyota anti-speed countermeasures. But as the old bumper sticker used to say, "Why Be Normal?"

With our huge and efficient Turbonetics turbo compressor providing massive amounts of air to the engine on demand (we

Calibrating the HKS adjustable cam sprockets to optimize valve timing for peak power.

saw 344 horses at the wheels after installation of Alamo Autosports' ported and polished exhaust manifold and a custom high-flow air-intake), our computer had finally noticed a disparity between reported maximum airflow as truncated by the FCD and its own manifold pressure sensor—and set a trouble code.

Planning to reprogram for performance and programmatically eliminate stock fuel cut parameters, Tadashi disconnected the FCD from our car, reset the computer by pulling a fuse in the engine compartment fuse box, and cleared the knock sensor and airflow meter trouble codes for good before beginning the process of converting the MR2 computer to user-programmability.

TOYOTA MR2 TURBO 3S-GTE ENGINE UPGRADES ANALYSIS
Stage 0. Stock 3S-GTE Engine: 163 rear-wheel horsepower (200 flywheel) at 7- to 9-psi boost

A stock 2.0-liter 16-valve MR2 Turbo makes 200 advertised flywheel horsepower at 6,000 rpm, and 200 ft-lbs flywheel torque at 3,250, which turns out to be 160–165 horsepower at the wheels on the Dynojet chassis dynamometer. This powerplant uses certain computer-based countermeasures designed to protect the engine from damage if the engine is driven hard under adverse (hot, maximum-boost) conditions, or if the wastegate malfunctions and cannot limit boost to normal (safe) levels. These countermeasures must be defeated to hot rod the engine to higher levels of power.

The 3S-GTE engine is normally limited by the Toyota EMS to roughly 9 psi maximum boost, or 7-psi boost in "hot" or fallback mode via a computer-controlled dual-diaphragm wastegate. In low-boost mode, the computer provides injection pulse width and ignition timing from entirely separate fallback fuel and timing tables that are extra-conservative to protect the engine from damage and Toyota from warranty claims.

The 3S-GTE engine management systems provides fuel and timing under boost conditions from internal tables at 2-, 4-, 6-, 8-, 10-, and 12-psi intervals. The wastegate opens at 7 or 9 psi to limit boost. If the wastegate fails to limit boost, at 12 psi (as measured by an auxiliary MAP sensor as well as the main velocity airflow meter), the computer activates a fuel cut that comes into play under boost conditions and stays active until the key is turned off. Although the computer also has internal tables to provide fueling and spark at 14- and 16-psi boost levels, these entries are clamped off (disabled) by an internal flag in computer PROM, so any turbo boost above 12 psi is still fueled with the 12-psi air/fuel and air-ignition tables.

Yet another countermeasure used by the Toyota engine management logic to help prevent engine damage is the computer fuel and timing calibration. Toyota provides extremely rich air/fuel calibration and retarded ignition timing at maximum-boost near 6,000 rpm. The factory knock-sensor provides aggressive ignition-timing retard if it detects detonation (for example, at maximum boost with 87 octane fuel). And if the computer detects a knock sensor fault (sensor voltages are out of range or static), antidamage countermeasures include disabling secondary intake runner opening (TVIS), *very* aggressive retard of ignition timing, and fuel enrichment—tactics that effectively prevent the stock turbocharger from making much boost at all. A really big turbo will force boost into the engine, but performance in this limp-home mode is vastly degraded.

The stock 435-cc/min fuel injectors are capable of fueling roughly 300 rear-wheel horsepower at maximum duty cycle and stock fuel pressure. This effectively matches the maximum stock fuel pump capacity at 55-psi fuel pressure.

The 3S-GTE is a race-bred powerplant that was used extensively in Toyota Atlantic and other factory racing. A variant of the cast-iron block, steel crank, and heads has been pushed to 980 horsepower on methanol at 80-psi boost in qualifying trim for Pikes Peak. Engine applications for the 3S-GTE include the 1991–95 MR2 Turbo (a completely different powerplant is used in the naturally aspirated 2.2-liter MR2) and in the 1989-plus Celica All-Trac.

The street 3S-GTE uses pressure-cast pistons, emissions turbo camshafts, and the four-to-eight-port TVIS (Toyota Variable Intake System) computer-controlled intake-runner system that blocks the secondary runners with throttle butterflies at low rpm for improved air velocity and better volumetric efficiency. The street stock turbocharger is a Toyota Twin-Entry unit in which phased exhaust pulses are delivered independently to the turbine to increase exhaust energy to the turbine (the limiting factor in producing boost at lower engine speeds) by preventing exhaust pulse collisions that subtract from total exhaust kinetic energy to improve low-end turbo response. Exhaust gases exit the turbine through a high-flow catalytic converter and single-entry, dual exit muffler.

In general, the street 3S-GTE uses all possible design measures to improve low-end torque. In higher-gear dyno roll-ons, measurable boost begins between 2,500 and 3,000 rpm, with full boost available by 3,250 rpm—maximum torque.

Stage 1. 200 rear-wheel horsepower (240@flywheel) at 13.5-psi boost with 93-octane fuel
Add:
- MAP-sensor conversion (HKS vane pressure converter)
- Adjustable-boost air bleed valve (BEGI)
- Optional two-position electronic boost control (BEGI)
- High-flow air cleaner (HKS)
- Fuel cut eliminator (HKS Fuel-Cut Defenser)

Analysis: In order to exceed 12-psi boost without immediately encountering a computer-mandated fuel cut, the adjustable-gain HKS FCD intercepts the airflow meter signal and reduces

readings by an adjustable percentage in order to tune the airflow meter's output to prevent the Toyota computer from seeing more electrical voltage than that equivalent to a 12-psi boost—and activating the fuel cut. Unfortunately, this strategy may result in suboptimal injection pulse width and air/fuel mixtures, but installation of a mixture-adjustable HKS vane pressure converter allows retuning for best air/fuel ratios. The VPC simultaneously eliminates the stock air meter's airflow bottleneck by converting to speed-density fueling. (The VPC's microprocessor provides a "virtual" airflow signal to the Toyota computer based on deducing airflow from a VPC manifold absolute pressure [MAP] sensor, engine speed, and a VPC intake air temperature sensor.) Adjustable VPC dials adjust mixture by plus or minus 12 percent in 2 percent increments for idle, full throttle, and transient enrichment (maximum adjustment is limited to maintain emissions legality). Interestingly, we made best power on the Dynojet (while observing air/fuel ratio with a Horiba fast air/fuel-ratio meter) by *removing* top-end fuel with the VPC!

The stock turbo is fully capable of supplying Stage 1's 13.5-psi boost, which we did initially by effectively lying to the stock wastegate by interfering with the manifold pressure reference line using a mechanically adjustable BEGI adjustable air-bleed device. Later, we controlled boost with the more sophisticated but more expensive cockpit-adjustable HKS electronic valve controller (EVC), which provides significantly more torque as manifold pressure approaches the wastegate spring threshold pressure at which the wastegate is beginning to open and limit boost.

With 4–7 psi additional turbo boost available, eliminating the stock air cleaner and muffler pays power benefits by eliminating these potential bottlenecks.

Stage 1 provides nearly 25 percent more rear-wheel horses at a cost in the $1,000–1,500 range.

Stage 2. 250 rear-wheel horsepower (305 flywheel) at 16- to 17-psi boost
Add:
- 3-inch cat-back exhaust (HKS single exit)
- High-power Twin Power ignition module/coil (HKS)
- High-flow, high-efficiency air-air intercooler (BEGI)
- Electronic boost controller (HKS EVC)
- Delete:
- BEGI manual boost controller

Analysis: The HKS cat-back exhaust was not dirt cheap, but we discovered that it immediately added 12 horsepower with no increase in boost pressure, which became *almost 40 horsepower* with VPC tuning on the Dynojet.

The car was equipped with a huge BEGI air-air intercooler to keep compressor-out air temperatures from going out of sight at higher boost pressures (increased heat with pressure is a law of thermodynamics), fighting high combustion temperatures and possible detonation, and making more power via increased air density. We measured air temperature on both sides of the intercooler during dyno runs. Interestingly, the intercooler functions partially as a big heat sink, storing cold while the turbo is not working when the intake air itself can helps cool the intercooler.

Above 13.5 psi, an air-bleed device can become problematic as a method of increasing boost. The HKS EVC allowed us to pump up the boost to 16–17 psi, at which time the stock ignition began to misfire. Installation of the HKS Twin Power high-power CD ignition at new plugs gapped to 0.28 really brought on the power!

Stage 2 adds 50 rear-wheel horsepower at a cost of $3,000–3,500.

MR2 TURBO

239

Stage 3. 300 rear-wheel horsepower (377@flywheel) at 19-psi boost

Add:
- High-flow Stage II turbocharger (HKS Sport)
- Optional Aerodyne VATN turbocharger
- Turbo plumbing kit/custom installation (BEGI)
- Fuel pump voltage/volume enhance. (K-B Boost-a-Pump)
- Dual-entry fuel rail (Alamo Autosports)
- High-flow throttle body (Alamo Autosports)

Analysis: The stock Toyota turbo compressor is "out of gas" by 16-plus-psi boost. The HKS Sport Turbo or the Aerodyne VATN turbo each have the compressor capacity to achieve another 3- to 5-psi boost *at high efficiency*—good for another 50 horsepower. VATN turbos are usually good for providing really excellent low-end boost at off-idle engine rpm, but in this case, the HKS Sport Turbo was as good or better, and both are good for 300 horsepower at full howl. We suspect we never fully optimized the Aerodyne's low-end response.

At this level of horsepower, the stock fuel pump is dangerously close to its capacity. (And remember, the maximum fuel flow of the fuel pump *decreases* as the fuel pressure regulator *increases* fuel pressure to maintain a constant pressure drop from the injectors into the intake manifold.) A device like the Kenne-Bell Boost-a-Pump "supercharges" and extends the stock high-pressure centrifugal electric pump's range by safely increasing the electrical voltage that drives the unit under boost conditions.

Stage 4: 335–350 rear-wheel horsepower (420–440@flywheel) at 20-psi boost (1.35 bar)—on unleaded race gasoline

Add:
- Clutch upgrade (TRD carbon disc and pressure plate)
- Limited slip differential (TRD)
- Stage III turbocharger (Turbonetics T04E-50 BB)
- Turbo system plumbing (BEGI)
- External wastegate turbine conversion (BEGI)
- External wastegate (Turbonetics Racegate)
- Auxiliary in-line fuel pump (NOS, 550 horsepower)
- VRG fuel pressure regulator/FMU (BEGI)
- Super-duty 4-inch air intake (Alamo)
- Port and polish exhaust manifold (Alamo)
- *Optional forged pistons (JE)
- *Optional steel head gasket (HKS)
- Delete:
- Stage II turbo
- Fuel pump voltage booster (K-B Boost-a-Pump)
- High-flow air cleaner (HKS)

Analysis: 330–350 rear-wheel horsepower is serious performance and is the absolute limit of performance from this engine without serious internal engine work, including installation of forged pistons. In fact, forged pistons are probably a good idea at anything over 300 horsepower, definitely anything over 335. This is also the limit of the stock fuel supply, and many changes are required to get from 335 to 350.

We found that the stock MR2 clutch began to go away at 300 horsepower, slipping noticeably at high boost on the dyno in third or fourth gears, but ours had never been changed since new, so possibly a very fresh stock clutch would do a little better. TRD's upgrade clutch kit stopped the slipping without adding any noticeable added pedal effort. Most MR2 turbos were sold in 1991 and 1992, but a limited slip differential was optional in the early years. Getting 350 horsepower to the ground is not going to work very well with one tire, so you could

call TRD's limited slip differential our anti-smoking campaign (tire smoke, that is).

To make this kind of horsepower, you need a serious turbocharger. Bell Engineering installed a state-of-the-art super-duty Turbonetics T04E-50/T3-turbine ball-bearing turbocharger. This unit really delivered the goods, eventually providing the airflow to add another 125 horsepower at the wheels in Stage 6. This ball-bearing turbo has the rugged construction to live under extreme acceleration, heat, and very high power limits near overspeed. BEGI built the custom turbo system plumbing and added a Turbonetics Racegate external wastegate to safely limit boost without creep problems in high-rpm/high-power conditions.

Very careful attention to turbo system air supply and exhaust gas flow is extremely important at 350-plus rear-wheel horsepower, including excellent exhaust gas flow through the wastegate, particularly when wide open. BEGI milled a hole in 0.48, 0.63, and 0.82 turbine housings and plumbed the Racegate flange to large-diameter black-steel pipe welded directly to the housings for optimal gas-flow geometry.

We tested three different A/R turbine housings from Turbonetics (lower number housings shoot exhaust gases through a narrower nozzle to provide improved turbine spool-up). The tradeoff is low-end response versus power-limiting high-end backpressure, but, later on, our Stage 5 stroker crank delivered so much exhaust energy that we made boost at 1,800 rpm with the medium-sized 0.63 housing and a virtually identical low-end torque curve with the 0.82! At Stage 4, however, on a stock-internals engine, the 0.48 housing definitely provides increased throttle response and low-end torque and boost in typical street driving.

At this stage, Alamo Autosports' exhaust manifold porting and big-mouth throttle body clearly helped the big turbo get air in the engine with high efficiency.

You need every trick in the book to deliver enough fuel at this stage without lean-mixture problems. We made 335 horsepower with an Alamo Autosports Twin-entry fuel rail kit, with the stock pump and a BEGI variable-rate-of-gain regulator, but the Boost-A-Pump or an in-line NOS auxiliary booster fuel pump is the way to go for anything over 300 horsepower. Any auxiliary fuel-supply device must be installed with fail-safe measures to cut fuel or boost if the fuel pressure somehow drops—as we learned the hard way when we smoked the project MR2's ring lands in a lean-mixture accident. By 330 horsepower, you are absolutely on the ragged edge of the stock fuel injector capacity even for short Dynojet pulls, and the best solution at this stage is installing larger late-model Turbo Supra 550 cc/min injectors and G-Force's ROMboard conversion for the stock computer.

The stock computer's air/fuel calibration is a limitation at 300–350 horsepower, and work-around countermeasures like the BEGI FMU and the VPC's limited tuning capability are problematic at these substantial levels of power. We found G-Force Engineering's ROM-board fully programmable conversion kit to be the elegant solution for serious MR2 hot rodding. G-Force replaces the stock Toyota PROM with a piggyback processor/dual-PROM daughterboard, and then optionally removes the stock fuel cut, stock redline, stock low-boost, fall-back fuel, and spark maps, enables the stock 14- and 16-psi internal boost tables, and recalibrates the high-boost fuel and spark maps for larger turbos and higher boost. G-Force will reprogram the stock computer and the VPC internal "virtual" airflow tables to correctly handle 550-cc/min injectors. A 1.5-bar boost will make 350 horsepower.

When you pull the motor (with its tight clearances) you notice this thing has a Yamaha 16-valve alloy head! Bigger cams can help with the top end, but it is difficult or impossible to make this thing breathe better with porting.

Stage 5. 375 rear-wheel horsepower at 20- to 25-psi boost (1.35–1.7 bar)

Add:
- Turbocharger air-foil (Alamo)
- Turbocharger internal polish and thermal coat turbine wheel (Norwood)
- Super-duty clutch (Clutchmasters)
- Port and polish head (Alamo)
- Forged pistons (JE)
- Thermal coat pistons (Performance Coatings, Norwood)
- Thermal coat combustion Chambers and Runners
- Thermal coat external exhaust manifold/turbine housing
- Upgrade stock connecting rods (shot-peen, and so on)
- Steel head gasket (HKS)
- Or copper head gasket (Norwood)
- And O-ring head (Norwood)
- High-pressure valve springs (Norwood)
- 550-cc/min fuel injectors ('95 Toyota Turbo Supra)
- Air-water intercooler system (Norwood Autocraft)
- Mild Stage I HKS 256 camshafts
- Program. Toyota computer upgrade (G-Force)
- Or Motec M4-8 performance EMS (many options, including mil-spec) (Norwood)
- MSD 6A Ignition/coil
- Delete:
- Air-air intercooler (BEGI)
- TRD clutch
- High-power Twin Power ignition module/coil (HKS)

Analysis: Stage 5 hot rodding tricks can be divided into two categories:

Tactics to improve durability on the dyno and in competition conditions: forged pistons, rod work, piston thermal coat-ings, improved head gasket, super-duty clutch, improved engine balancing, ultra-high-efficiency air/ice-water/tap-water/coolant intercooler, high installed-pressure valve springs.

Tactics to improve torque: displacement increase from boring and stroking, additional crank leverage from increased crank offset, turbo air foil, higher lift and duration camshaft, denser air from super-duty intercooler. Tactics to provide fuel and state of tune: 550-cc injectors, NOS in-line pump/Kenne-Bell Boost-a-Pump, G-Force computer recalibration.

Stage 6. 425–550 rear-wheel horsepower at 25- to 30-psi boost (1.7–2.0 bar)

Add:
- 83-lb/hr fuel injectors (Norwood Autocraft)
- Bosch-type fuel rail (Norwood)
- Super-duty ball-bearing turbo compressor upgrade (T04E-50 to -60)
- O-ring head (Norwood)
- Annealed-copper head gasket (Norwood/Alamo)
- Stroker crankshaft (Alamo)
- Adjustable cam sprockets (HKS)
- Motec M4-8 performance engine management system (many, *many* options, including mil-spec capability) (Norwood Autocraft)
- MSD 6A ignition/coil
- Delete:
- G-Force ROM-board
- Dual-entry fuel rail (Alamo)
- 550-cc/min fuel injectors ('95 Toyota Turbo Supra)

Analysis: The final efforts in this project were designed to contain combustion pressure (O-ring head), provide sufficient fuel supply and delivery (850-cc/min injectors in new rail, twin fuel pumps), provide maximum boost within sensor limits (upgrade turbo to T04E-60 specs), and provide optimized tuning. We knew the Stage 6 engine was very strong. We stopped here, but the next bottlenecks to more power would be the transaxle, the mild HKS cams, and, again, the turbocharger. Possibly the connecting rods.

Bigger displacement motors need more cam just to maintain rpm capability, and the built stroker engine was 10 to 20 percent larger, requiring more cam just to maintain VE. The current HKS cams on the larger displacement stroker engine are probably about equal to the stock cams on the stock engine with respect to VE at high rpm. Interestingly, peak power rpm now occurs at exactly the same speed as a stock MR2 turbo engine's peak power (6,000 rpm). Titanium valve retainers and cam followers would help enhance redline capability.

Some of the most radical Toyota 3S-GTE race cars run competition-quality Chevy small-block rod bearings and special TRD main bearings. With static compression of 8.5:1, the current effective compression ratio of the four-valve engine is extremely high at 30- to 32-psi boost. With the best Turbo Blue Extreme 115+ MON leaded fuel, the maximum possible knock-limited boost on gasoline with this engine is unknown, because we did not encounter knock. Suffice it to say, people who know were mighty impressed by a 32-psi boost. Beyond this, you're surely talking 1,600-cc/min Indy-type injectors and methanol injection.

REPROGRAMMING A VPC FOR BIG INJECTORS AND BIG BOOST

The HKS vane pressure converter (VPC) is an auxiliary micro-computer designed to allow elimination of a stock airflow meter by estimating engine airflow based on data from its own mani-

MR2 Project Dyno Results

Corrected Rear Wheel Power and Torque and the Dynojet Model
248C, best runs each configuration

Configuration	RWHP		Torque	
(Stock, 38K miles)	165.2	@5,400	179	@3,200
0.47 bar + BEGI equip	139.5	@5,750	137.8	@4,200
0.8 bar	152.8	@5,550	161.0	@4,200
0.8–0.9, HKS equip + new plugs, gap	192.5	@5,600	214.1	@3,400
1.1 + HKS Twin Power	201.8	@5,300	238.6	@3,500
1.1, + HKS Exhaust	239.1	@5,250	272.5	@4,200
1.1 + Sport Turbo	246.1	@5,200	274.0	@4,200
1.2 bar	284.0	@6,100	271.0	@5,000
1.30 bar, Aerodyne Turbo	305.0	@6,200	275.0	@4,700
1.3 bar, T04E-50 Turbo (bad clutch)	299.2	@5,750	300.3	@4,750
1.1-1.35 bar, Alamo T-body, tuning, etc.	335.9	@6,300	282.0	@5,900
1.1-1.35 bar, Alamo ported exhaust & custom air inlet, tuning, etc.	344.9	@6,250	294.6	@5,600
1.35 bar, G-Force computer mods	334.2	@6,500	281.1	@5,500
1.35 bar, above+ Supra injectors	335.9	@6,750	274.0	@6,250
1.5 bar, + above	345.6	@6,500	297.0	@4,750
2.0 bar, 2.19-liter stroker, Motec EMS	471	@6,000	438.5	@5,500

fold absolute pressure and inlet air temperature sensors. The VPC's microprocessor watches these sensors and continuously outputs a 0- to 5-volt signal to the main onboard computer corresponding to airflow and trim-able with idle, gain, and transition dials.

The MR2 computer's own internal air/fuel boost enrichment tables are designed to provide offsets to the base fuel calculations that continuously generate injector pulse width required to achieve a 14.7:1 or stoichiometric air/fuel ratio. The tables are defined in 16 engine loading ranges corresponding to airflow from roughly 30-inch manifold vacuum through 16-psi boost, although the 14- and 16-psi airflow tables are normally locked out in the factory calibration so all airflow above 12-psi boost is fueled via the 12-psi boost enrichment table.

Tadashi Nagata had activated the 3S-GTE computer's higher-load 14- and 16-psi air/fuel table entries via the Techtom ROM board and a laptop PC in order to provide correct fueling for higher-boost conditions. He had also turned off the low-boost air/fuel and spark tables that are activated as a fail-safe strategy so if the computer thinks the engine is too warm or overstressed (the fail-safe mode also reduces peak boost via a dual-ported wastegate that is no longer installed on our MR2).

In any case, if you can get to the computer's calibration, you can disable the fail-safe logic for consistent high performance. Tadashi had also added spark timing during transition into boost to improve naturally aspirated performance and low-boost performance on an engine with a large (and presumably slower-responding) turbocharger. He'd also modified certain high-end timing and fueling parameters, and performed other relatively minor recalibrations, given we were no longer using the fuel management unit to raise fuel pressure and planned even higher levels of boost. But from here out, we would be relying mostly on VPC reprogramming to compensate for the larger injectors at low rpm and to provide sufficient fuel enrichment for extremely high torque, high boost pressures in the 1.5- to 2.0-bar range (rather than simply fueling everything above 16 psi with the 16-psi airflow map).

Here is how we would do it:

The VPC map is a table of hex numbers used to convert various combinations of manifold pressure and inlet air temperatures into digital values that represent estimated airflow. These numbers are then translated by circuitry into a 0- to 5-volt electrical signal that is substituted for the stock airflow sensor signal to the Toyota computer. If we rescaled higher airflow ranges in the VPC table so increased manifold pressures would be required for the VPC to output maximum or near-maximum virtual airflow sensor voltages (near 5 volts)—and smoothed the transition to intermediate ranges in the table to represent a linear increase in airflow ranges, we could, in effect, map an increased dynamic range of engine airflow onto the 0- to 5-volt output signal of the VPC. Then, if we simultaneously scaled the lower range airflow conversion values downward so the Toyota computer was, in general, fed a lower (virtual) airflow voltage at idle or light cruise, the Toyota computer would simply assume airflow was down (as it would, for example, be at high elevation) and decrease injection pulse width, thereby compensating for the increased injector size. With this VPC strategy, we hoped to get by with only minimal changes to the MR2's main computer calibration, which we considered desirable.

INDIVIDUAL CYLINDER TIMING, KNOCK, AND FUEL CONTROLS: MOTEC M48 AND J&S SAFEGUARD

"Do you realize," says J&S owner John Pizutto, "that when you set the timing on your engine so it doesn't ping, you are really just optimizing it for the cylinder that pings the easiest? What do you think this does to the efficiency of the rest of the engine?"

The J&S SafeGuard is an individual-cylinder knock control ignition, designed to work on virtually any vehicle, even if it already has a factory knock sensor. Factory knock-control strategies always involve listening with a piezo-crystal knock sensor and retarding timing to all cylinders in aggressive increments (4 degrees at a time is typical of Chevy factory computers).

Aftermarket computers with knock control normally have a programmable knock retard increment and rate, but the cylinders virtually always retard as a group. On the other hand, racing and supercar programmable engine management systems like the Motec are normally tuned very frequently on a individual-cylinder basis. This type of tuning assumes a knowledgeable driver who will always provide gasoline of sufficient octane to prevent knock—although, naturally, the Motec can control virtually anything including knock sensors and anti-knock algorithmic countermeasures via user-definable auxiliary data channels. The Motec system provides individual-cylinder boost-retard via settings in the base timing tables.

The SafeGuard, which is good insurance for any engine—even a Motec-equipped one—is designed to make more power by permitting maximum raw timing advance and minimum combustion chamber fuel cooling. Octane, combustion chamber temperature, unequal fuel distribution, and other factors can vary the engine octane number requirement (ONR) such that it is not normally feasible to tune for maximum power on factory-type vehicles without individual-cylinder trim of spark timing and injection pulse width—without the installation of knock-sensing ignition retard devices. Manifold pressure-based ignition-retard alone will not produce best power and efficiency with factory-type engine management strategies. According to J&S, tests by Bosch have shown *a 30 percent change in relative humidity can change the knock threshold by 7 degrees.*

The Safeguard selectively detects detonation when it occurs, makes a note of which cylinder knocked, and retards *only that cylinder's spark* on the next ignition event. In the meantime, the other cylinders are able to continue operating at maximum efficiency. Compared to factory knock sensing and retard of all cylinders, the J&S unit can produce significant power gains on knock-limited engines, that is, all high-performance vehicles running on pump gas. The J&S SafeGuard allows programmed retard in 2-, 4-, or 6-degree increments, with recovery advance limited to 2 degrees per 10 revolutions (knock is much easier to prevent than eliminate once started due to the high rate of temperature increase from abnormal combustion). The SafeGuard includes a built-in ignition module (coil driver) with adjustable dwell to 7 milliseconds. A built-in digital rev-limiter drops every third cylinder, rotating the misfire throughout the firing order. J&S can provide an optional monitor that dynamically displays the amount of knock retard.

Replacing the factory knock control with a J&S is a prescription for more power and efficiency, especially on hot, humid days at maximum performance.

PROGRAMMING ONBOARD COMPUTERS FOR BIG INJECTORS

Really high-output engines—especially ones with only four cylinders—have a problem. It's called huge dynamic range, and it refers to the difference between the fuel used at maximum power and the fuel used at minimum or idle power. Large dynamic range can create serious problems for the fuel supply system, and ultimately, the onboard computer. To see why this is so, compare a 500-horsepower naturally aspirated Chevy 7.4-liter big block V-8 and a 500-horsepower turbocharged 2.0-liter MR2 Turbo. They both require the same theoretical amount of fuel at maximum power. Actually, the 454 might require more, due to frictional losses of all those rings and bearing surfaces and so forth, but, in any case, the MR2 has to supply 500 horses worth of fuel through just four injectors instead of eight like the Chevy. Fine, there are some *big* injectors out there. At the same time, make sure the fuel supply lines are large enough to flow the fuel, and make sure the fuel rail itself and the entry and pressure-regulator exit are optimized to prevent adverse resonation pulse effects. We don't want pressure waves surging back and forth through the rail, upsetting accurate fuel distribution.

Now let's try to get both 500-horse engines to idle right.

At idle, the 7.4 Chevy, which probably has a fairly healthy cam to make an honest 500 horsepower, will naturally *require* more fuel due to the increased friction and large displacement. The hot cam probably requires a richer than stoichiometric mixture due to increased valve overlap and inherent problems of dilution of the charge with exhaust gases, and unpredictable dissipation of some of the charge directly out the still-open exhaust valve. In any case, the 7.4 engine has eight relatively small injectors, which will require a relatively long injection pulse width to idle well—all for the good, because most aftermarket electronic control units (ECU) cannot produce a stable pulse below about 1.5 milliseconds. At such short injection periods, the injection event becomes very short in relation to other phenomena that can collectively be referred to as noise. Noise includes the effects of fuel pressure on injector open and close time, the effects of minor injector driver voltage fluctuations on open and close time, and minor variations in injector response due to manufacturing tolerances.

The MR2 has much larger injectors per liter than the Chevy because the specific power under high-boost turbocharged condi-

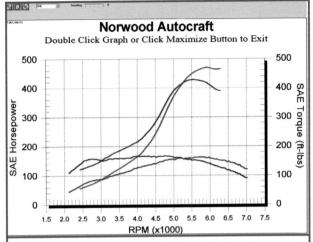

The bottom line: Peak power more than doubles from the stock 165 rear-wheel horsepower to 471 rear-wheel horsepower at 6,000 rpm.

tions is much higher. But at idle, there is no turbo boost and less displacement. The typical turbo cam profile is mild, with relatively little valve overlap (although probably featuring as much lift as possible), so a 14.7 air/fuel mixture is desirable. All these factors translate into a very short injection pulse.

Most injection fuel supply systems use certain countermeasures to prevent excessively short injection pulse width at idle (such as referencing the fuel pressure regulator to manifold pressure to pull pressure up and down with engine loading; lower idle rail pressure means a longer pulse is required, all other things being equal). Auxiliary variable-rate-of-gain fuel pressure regulator strategies that raise fuel pressure as a multiple of boost pressure limit the need for really big injectors to make peak power, because there is much more fuel injected at 85 psi than at 40. In some cases, there is little choice but to use multiple staged injectors per cylinder, with a primary set operating all the time and a secondary set coming into play at higher power levels. A variant of this is to add an additional injector controller, which controls one or more additional injectors to provide fuel enrichment at higher power levels.

The Motec M4, M48, and M8 solve the problem in a different way, by upgrading the ability of the ECU to provide very accurate fueling at very short pulse widths, enabling the use of very large injectors. Electronic fuel injectors use solenoid valves to open the injectors. If you design a powerful driver circuit that slams open the injector in a fast, predictable fashion, you are simultaneously developing a powerful electric field in the coils of the injector electromagnets. When driver circuitry drops voltage to close the injector, the magnetic field in the coils of the injector solenoid collapses, but not necessarily in an entirely predictable fashion. Motec uses patented circuitry that abruptly *switches* the flow of current to circuitry that is able to store the current for the next injection event. The bottom line is that the injectors open and close with great precision and big injectors idle well at required very short pulse widths.

CHAPTER 25
MR6: COMPOUND FORCED-INDUCTION

A small, light, good-looking, and *affordable* midengine sports car thrust into an intimate relationship with a V-6 powerplant 50 percent larger in displacement than the stock 2.0-liter turbo powerplant just seemed like a thing that should happen.

Not that there's anything wrong, mind you, with the stock 2.0-liter four-cylinder MR2 Turbo powerplant. The 3S-GTE is an iron-block, alloy-head phenomenon built as strong as an artillery piece that has been hot rodded to nearly 600 horsepower (471 rear-wheel horsepower) on gasoline in a project in this book (and more on methanol). But let's face it, a four-banger lacks the sex appeal of a V-6. Besides, this book needed another engine swap project.

But it came to pass that we glimpsed a rare opportunity. TRD markets a bolt-on blower-in-manifold roots-type supercharger kit for the Toyota-Lexus V-6 that leaves the throttle body in the stock position and plenty of room in the engine compartment around the engine. Plenty of room, that is, for the various components of a *turbo conversion* we might fabricate to be fully compatible with the TRD-supercharged 1MZ—or the stock all-motor 1MZ powerplant, sans blower. Or everything at once. Call it turbo-assisted supercharging, call it compound forced induction (compounded turbocharging, compound supercharging, what have you), call it twincharging, we're talking a big intercooled turbocharger breathing down the neck of a supercharged engine's throttle body, so to speak. Our twincharged system would give us not only the rare apples-to-apples opportunity to compare and contrast blower versus turbo power and torque, but the potential to investigate the mystical world of *compound forced induction*, a radical technology more akin to the world of super-high-output pulling tractors and piston-engine World War II fighter aircraft than the world of automotive performance.

The thinking was that if we did it right, compound forced induction on a 1MZ V-6 could create enough airflow to make a *lot* of power and plenty of torque—under pretty much all circumstances. Such a compound-supercharged powerplant could potentially be used to set a land speed record in the 3.0 class in a Gen II MR2. And the same engine would be completely streetable. In *rush-hour traffic*. The initial task would be to get a V-6 in an MR2, then get the blower kit on the V-6, and then get everything running nicely under programmable aftermarket engine management. Later on we'd build a turbo conversion. Easy-on, easy-off. This would be a test bed to try interesting *experiments*.

SHOTGUN WEDDING: THE V-6–MR2 SWAP

After considering the alternatives, we decided to work with a 1994 Lexus 1MZ-FE V-6, which, at 401 pounds, is similar in weight to the stock iron-block 3S-GTE MR2 Turbo powerplant and would not adversely affect handling. We contacted Greenleaf (Ford Recycling), a nationwide firm that pulls, tests, warehouses, and distributes used auto parts at a large network of wrecking yards around the United States, bringing economies of scale and an assembly-line mentality to recycling good used parts. They transferred an engine to their Fort Worth yard.

Our plan was to get the V-6 in the car, get it running right stock, do some dyno testing, and then design and fabricate the power-adder systems and get low-power dyno results. Because we knew we'd one day be cranking up the volume on our 1MZ-FE V-6 with high-boost power-adders from Toyota Racing Development and Majestic Turbo, we got the ball rolling by ordering in a set of stronger Wiseco forged pistons, and similarly made plans to upgrade the connecting rods, cylinder head gasket, and valve springs.

We also ordered some huge 630-cc/min Bosch injectors and super-duty Ignition Solutions coil and coil-driver components. And we obtained a sophisticated and powerful Motec M48 engine management computer from Norwood Autocraft to interface with the engine and vehicle sensors, actuators, and electrical system. We also acquired a Motec IEX direct-fire multiplexing unit that synchronized on a special ignition pulse from the M48 and thereafter sends each subsequent ignition pulse to the correct direct-fire coil.

We decided to mount the engine from the stock MR2 passenger-side attachment point via the V-6 engine's robust right-side torque-stay bracket. Norwood Performance used some Ferrari suspension parts to build a strong, lightweight mount that provided shock mounting but was otherwise extremely rigid. A mod-

This compound-supercharged 1MZ-FE V-6 runs a gigantic T76 turbocharger, intercooled, breathing down the neck of a TRD positive-displacement supercharger. The combo makes immediate low-end boost, but when the turbo starts to wail, the supercharger belt-drive actually puts power back into the crankshaft.

ified stock V-6 right-side torque stay mounted to the right MR2 shock tower to control engine twist. And, last, Norwood Performance modified a stock MR2 right-side axle bearing carrier to attach nicely to the Lexus V-6 to retain the right-side axle.

Indeed, as Toyota's Ron Reed had predicted, a 1MZ V-6 bolts up to the stock MR2 transaxle (with *nearly* all of the bell housing bolts). What's more, 3S-GTE four- and 1MZ-FE six-cylinder manual-trans Toyota flywheels are the same diameter and thickness, and a manual-trans MR2 flywheel even slides seamlessly onto the V-6 crank flange. *Unfortunately*, the V-6 crank flange uses smaller diameter flywheel bolts to mount the flywheel—clocked at identical positions but in a larger diameter circle.

We ordered up a multi-disc Kevlar aluminum MR2 clutch-flywheel kit from Clutchmasters manufactured with the correct custom bolt pattern for a 1MZ V-6. Clutchmasters' low-inertia setup was designed to provide light pedal pressure yet plenty of clamping force to handle serious horsepower without slipping. To handle even more power, Clutchmasters can provide a full-metal multi-disc clutch, but it's an animal for street driving in an MR2.

Meanwhile, we removed the stock Toyota engine management harness from the V-6 engine and painstakingly constructed a custom wiring harness for a Motec M48 engine management system operating a foreign V-6 in the MR2 environment.

STARTUP MAPPING AND FIRST DYNO RESULTS

Motec does not offer plug-and-play fuel-spark maps or any sort of predefined or automated calibrations. At this point Bob Norwood began constructing a custom Toyota V-6 startup map in the Motec as we prepared to fire up the engine.

Norwood selected a calibration from his own library with similar displacement per cylinder (a stock 2.0-liter MR2 turbo, in fact), and began reworking it for the MR6 application:

- Sensor Setup
  - Throttle position (linear or not, 0 percent voltage, 100 percent voltage)
  - MAP (300 kilopascals Delco)
  - Air temp (3,300 ohms Delco)
  - Engine temp (3,300 ohms Bosch)
  - Aux temperature (disabled)
  - Aux voltage (100 kilopascals Delco MAP sensor—for barometric pressure compensation)
  - Lambda sensor (Motec wideband)
  - Cam position sensor (mag input, mag sensor)
  - Crank position sensor (mag input, mag sensor, 36 teeth, 2 missing teeth, index pos. 520 deg)
- Number of cylinders (6)
- Firing order (1-2-3-4-5-6)
- Ignition (Special—Motec IEX direct-fire multiplexing)
- Coils (6)
- Dwell (2.5 milliseconds)
- Ignition delay (50 microsecs)
- RPM site setup (40 sites: 200 rpm intervals from 0 to 7,600, and 7,700)
- Stopped fuel (one x-percent normal injection pulse with engine stopped and throttle exceeds 90 percent)
- First injection (one x-percent added first injection)
- Ignition trim (1 percent)
- Injector scaling (10-msec pulse width corresponds to 100 in main fuel table [down from 13 used for the MR2's 550-cc/min Bosch injectors!])
- Injection battery compensation (0.5 ohm 6 amp injectors,

45-psi fuel pressure)
- Cold Enrichment (defaults)
- Acceleration enrichment (defaults)
- With the engine running, Norwood began building a custom Toyota V-6 engine calibration on the Dynojet chassis dynamometer for what was essentially a lightened stock 1MZ powerplant with a custom Alamo Autosports exhaust system. (For Norwood's tuning methods, see the chapter on tuning.)

The nearly stock naturally aspirated MR6 made 181 rear-wheel horsepower at 5,400 rpm, and 203.4 ft-lbs torque at 4,350 rpm, Norwood all the while working to optimize air/fuel ratios with conservative ignition timing and then adding back timing and pulling fuel to achieve maximum possible horsepower with 93-octane fuel. This power translates to 210–225 crankshaft horsepower and 240–256 ft-lbs torque compared with the stock MR2's 200 and 200 figures and a stock 1994 Lexus 1MZ's 188 horsepower and 209 ft-lbs torque. It's 9-wheel horsepower above that of a factory-stock Solara on the Dynojet and about what you might expect from premium-fuel tuning with a free-flow exhaust.

As this point, we made the car fully roadworthy, and I drove the car around Texas for a bit to shake out the new engine systems and look for bugs.

SUPERCHARGING THE MR6

The TRD 1MZ blower kit was designed and manufactured by supercharging specialists Magnuson Products and is available through Toyota dealers with a factory warranty for most Toyota vehicles equipped with the 1MZ V-6. This factory-type positive-displacement blower kit is normally a turnkey installation by Toyota dealers or aftermarket performance shops, but it can be installed by careful and experienced enthusiast-mechanics. On the 1MZ, the M62 roots-type blower is overdriven to produce a peak 5-psi boost (at the cost of 50–60 degrees of heating and perhaps 35 horsepower). If you had the right crank pulley, an M62 could easily boost the 1MZ to 10-psi boost at just over 15,000 blower rpm (which is well within the safe rpm range for a Gen 4 Eaton supercharger).

The TRD kit includes a replacement intake manifold/supercharger housing, which maintains the stock factory throttle body precisely in stock position, with the Eaton rotor assembly contained in a special replacement TRD intake manifold. Unfortunately, this negates the possibility of intercooling the blower discharge on its way into the engine (though it could potentially be cooled with water or methanol injection). The supercharger drive pulley protrudes from the right side of the TRD intake manifold and is precisely aligned with the plane of the stock crank pulley and the 1MZ's alternator and air conditioning compressor.

The supercharger is equipped with a blower-bypass that allows intake air to bypass the supercharger during idle or other high-vacuum conditions when boost is unnecessary or undesirable. This valve essentially eliminates all blower parasitic drag during cruise or idle conditions. The TRD kit includes a self-tensioning serpentine blower-drive system in which the crank pulley drives the stock alternator and other accessories *plus* the supercharger and two or three additional static idler pulleys plus the self-tensioning idler pulley. TRD's blower-drive system is sandwiched between two anodized aluminum plates, and this drive system is designed to bolt to the 1MZ engine in place of the stock right-side torque stay.

The TRD kit includes the replacement intake manifold/supercharger unit, a long, multi-ribbed serpentine blower belt, a blower drive system with various idler and tensioning pulleys, a bag full of hoses, hose barbs, cable ties, check

02121700/HARTMAN TOYOTA 1MZ-FE V6 SUPER/T
Fuel Main (% of IJPU)

Eff\RPM	0	200	400	600	800	900	1000	1200	1400	1600	1800	2200	2400	2600	2800	3000	3200	3400	3600	3800
200.0	38.5	39.5	41.0	42.5	44.0	45.5	46.5	47.5	48.5	49.0	49.5	50.0	50.5	51.5	52.5	53.5	54.5	55.5	56.5	57.5
180.0	38.5	40.0	41.5	42.5	44.0	45.0	46.5	47.5	48.5	49.5	50.0	50.5	51.0	51.5	52.0	52.5	53.5	54.5	56.0	57.5
160.0	39.0	40.0	41.5	43.0	44.5	46.5	47.5	48.5	49.5	50.0	51.0	51.5	52.5	53.0	53.5	54.5	56.5	59.5	60.5	59.0
140.0	38.5	39.5	40.5	42.0	43.5	45.0	46.5	47.5	48.0	49.5	50.5	51.5	53.5	54.5	55.0	55.5	57.0	57.5	57.5	58.0
120.0	37.0	38.0	40.0	41.5	43.0	44.0	45.0	47.5	49.5	51.5	53.5	54.5	54.0	54.0	53.0	51.5	52.0	57.0	58.0	58.0
100.0	38.0	40.0	41.5	42.5	43.5	44.5	45.5	46.5	47.0	47.0	47.5	48.0	51.0	48.5	49.5	50.0	51.0	57.0	58.0	58.0
80.0	40.0	41.5	42.5	43.0	43.5	44.0	44.0	45.0	45.0	46.0	46.5	47.0	47.5	48.0	48.5	48.5	49.5	50.0	50.5	50.5
60.0	40.0	41.5	42.5	43.0	43.0	43.5	44.0	44.5	45.5	45.5	46.5	47.0	47.5	48.0	48.5	48.5	49.0	49.5	50.0	50.5
40.0	41.5	42.5	43.0	43.5	43.5	44.0	44.5	44.5	45.0	45.5	46.0	46.5	46.0	46.5	47.0	47.5	47.5	47.0	47.0	47.0
20.0	41.5	42.0	42.5	42.5	42.5	42.5	42.5	38.0	37.5	44.5	43.0	43.0	43.5	44.0	44.0	44.0	44.5	44.5	44.5	44.5
0.0	38.0	38.0	38.0	38.0	38.0	39.5	39.5	39.5	41.5	41.5	42.0	42.5	42.5	42.5	42.5	44.5	44.5	44.5	44.5	44.5

Ignition Main (degrees BTDC)

Eff\RPM	0	200	400	600	800	900	1000	1200	1400	1600	1800	2200	2400	2600	2800	3000	3200	3400	3600	3800
200.0	23.0	23.0	23.0	23.0	23.0	23.0	23.0	23.0	23.0	23.0	23.0	23.0	23.0	23.0	23.0	23.0	23.0	23.0	23.0	23.0
180.0	24.0	24.0	24.0	24.0	24.0	24.0	24.0	24.0	24.0	24.0	24.0	24.0	24.0	24.0	24.0	24.0	24.0	24.0	24.0	24.0
160.0	25.0	25.0	25.0	25.0	25.0	25.0	25.0	25.0	25.0	25.0	25.0	25.0	25.0	25.0	25.0	25.0	25.0	25.0	25.0	25.0
140.0	26.0	26.0	26.0	26.0	26.0	26.0	26.0	26.0	26.0	26.0	26.0	26.0	26.0	26.0	26.0	26.0	26.0	26.0	26.0	26.0
120.0	27.0	27.0	27.0	27.0	27.0	27.0	27.0	27.0	27.0	27.0	27.0	27.0	27.0	27.0	27.0	27.0	27.0	27.0	27.0	27.0
100.0	25.5	26.5	27.0	27.5	28.0	28.0	28.0	28.0	28.0	28.0	28.0	28.0	28.0	28.0	28.0	28.0	28.0	28.0	28.0	28.0
80.0	20.0	21.5	23.0	24.0	25.0	26.5	27.5	29.0	30.0	31.0	31.0	31.0	31.0	31.0	31.0	31.0	31.0	31.0	31.0	31.0
60.0	14.5	16.0	18.0	19.5	21.0	22.0	23.5	25.0	26.5	29.0	30.5	32.0	33.0	33.0	33.0	33.0	33.0	33.0	33.0	33.0
40.0	11.5	13.0	14.5	16.0	17.5	19.0	20.5	22.5	24.0	25.5	26.5	28.5	30.0	31.5	33.0	34.5	35.0	35.5	36.0	36.0
20.0	17.0	19.0	20.0	20.5	21.5	23.0	24.5	23.5	20.5	29.0	30.5	33.0	35.5	37.5	38.5	39.5	39.5	39.5	39.5	39.5
0.0	16.0	17.5	19.5	21.0	23.0	25.0	26.5	28.5	30.0	32.5	34.0	36.5	38.0	39.0	40.0	40.0	40.0	40.0	40.0	40.0

valves, brass fittings and other hardware, and a small electronic module that triggers the opening of the supercharger bypass valve during shifts in an automatic transmission environment to limit stress on the trans clutch packs and other components (which we wouldn't be needing).

In prospect, TRD kit installation on the MR6 appeared simple and straightforward, but in actuality was complicated slightly by the location of the custom passenger-side mount we'd built to support the engine. Our custom engine mount bolted to the engine at the exact same location required by the blower-drive assembly. We eventually chose to raise the blower-drive assembly approximately half an inch and fasten both the engine mount and blower-drive to the engine with the same fasteners. Therefore, we were forced to remake the right-side mount to sandwich between the engine mount flange and the blower drive assembly. The whole business was straightforward, with the exception that Norwood Performance relocated a hole in the inboard blower drive plate through which a special long bolt was required to align with the alternator pivot boss and alternator. We welded the old hole closed and drilled and counter-sunk a hole at the new location.

Although engine cover clearance was not a factor in the MR2 with the naturally aspirated 1MZ engine, the supercharger pulley left scant clearance to spare. And the serpentine belt—on its way to and from the blower—interfered in a minor way with the engine-cover support-rod bracket on the MR2's right-side shock tower. We removed two small sections of the bracket.

DYNO-TUNING THE SUPERCHARGED MR6
TRD's 1MZ supercharger kit is designed to run at 5-psi boost and to deliver 62 horsepower on a stock 1MZ engine using the stock computer, stock calibration, stock injectors, stock fuel supply, stock ignition and spark timing, stock compression ratio, and just a mite more fuel pressure. At this level of boost, the Eaton M62 supercharger can be expected to raise intake temperature maybe 50–60 degrees, which does not *require* intercooling. Nor does it require larger injectors or additional fuel pressure. Well, unless you want to fully optimize *power*.

Tuning the Motec engine management on the Dynojet, Bob Norwood was able to build a calibration that delivered the advertised power, plus some. However, the 288-cc/min injectors were out of capacity before we'd fully explored the potential of the blower under Motec control, and at 100 percent duty cycle, the engine was still beginning to run a bit lean of the safest peak-power air/fuel ratios at the top end. This is fine, even desirable, for a low-boost application on an emissions-controlled automatic-trans Camry or similar models, but it will not make the maximum power. We installed a Cartech fuel management unit (variable-rate-of-gain fuel pressure regulator), which increases fuel pressure as a multiple of boost pressure to increase fuel enrichment under boost conditions to more optimal levels. In effect, for the moment, we treated the Motec like a nonprogrammable factory system that required completely external fuel enrichment.

With a *very* flat torque curve delivering more than 200 ft-lbs torque at the wheels from less than 2,000 rpm to more than 6,000 rpm (peaking at 221.8 ft-lbs at about 3,750 rpm) the supercharged engine delivered 250.1 rear-wheel horsepower at 6,750 rpm, 500 rpm over the stock 1MZ-FE redline of 6,250—and the power was still climbing straight toward 260 rear-wheel horsepower at 7,000. Bottom line, the power was up from 181 to 250 at the wheels, a gain of almost 70 rear-wheel horsepower, equivalent to a gain of between 80 and 90 horsepower at the crank-

02121700/HARTMAN TOYOTA 1MZ-FE V6 SUPER/T
Fuel Main (% of IJPU)

4000	4200	4400	4600	4800	5000	5200	5400	5600	5800	6000	6200	6400	6600	6800	7000	7200	7400	7600	7700	Eff\RPM
59.0	60.0	61.0	62.5	64.0	65.0	66.5	67.5	69.0	70.0	72.0	73.5	75.0	76.0	77.0	78.0	78.5	79.0	80.0	80.5	200.0
58.5	59.5	62.5	63.5	65.0	66.0	67.5	68.5	69.0	70.0	70.5	71.5	72.5	73.5	74.5	75.0	75.5	76.0	76.5	76.0	180.0
59.0	60.0	61.0	61.5	61.5	63.0	64.0	64.5	65.0	65.5	66.5	67.0	68.0	68.5	69.0	69.5	69.5	70.0	70.5	70.5	160.0
59.0	59.5	59.5	60.0	60.5	61.0	61.0	61.5	62.0	63.0	64.0	64.0	64.5	65.5	66.0	66.5	66.5	67.0	67.0	67.5	140.0
58.0	58.0	60.0	60.0	56.5	57.0	57.5	57.5	57.5	57.5	57.5	57.5	57.5	58.0	58.0	58.5	58.5	59.0	59.0	59.5	120.0
58.0	58.0	60.0	60.0	55.0	55.0	55.0	55.0	55.0	55.0	55.0	55.0	55.0	55.0	55.0	55.0	55.5	55.5	54.5	53.0	100.0
50.5	51.0	51.5	51.5	52.0	52.0	52.0	52.5	52.5	52.5	52.5	52.5	53.0	53.0	53.0	53.0	52.0	52.0	52.0	52.0	80.0
50.5	50.5	50.5	51.0	51.0	51.5	51.5	52.0	52.5	52.5	53.0	53.0	53.0	53.0	53.0	53.0	53.0	53.0	53.0	53.0	60.0
47.0	47.0	47.5	47.5	47.5	48.0	48.5	48.5	48.5	48.5	46.5	46.5	46.5	46.5	46.0	46.0	46.0	46.0	44.5	44.5	40.0
44.5	44.5	44.5	44.5	44.5	44.5	44.5	44.5	44.5	44.5	44.5	44.5	44.5	44.5	42.5	42.5	42.5	42.5	42.5	42.5	20.0
44.5	44.5	44.5	44.5	44.5	44.5	42.5	42.5	42.5	42.5	42.5	42.5	42.5	42.5	42.5	42.5	42.5	42.5	42.5	41.5	0.0
23.0	23.0	23.0	23.0	23.0	23.0	23.0	23.0	23.0	23.0	23.0	23.0	23.0	23.0	23.0	23.0	23.0	23.0	23.0	23.0	200.0
24.0	24.0	24.0	24.0	24.0	24.0	24.0	24.0	24.0	24.0	24.0	24.0	24.0	24.0	24.0	24.0	24.0	24.0	24.0	24.0	180.0
25.0	25.0	25.0	25.0	25.0	25.0	25.0	25.0	25.0	25.0	25.0	25.0	25.0	25.0	25.0	25.0	25.0	25.0	25.0	25.0	160.0
26.0	26.0	26.0	26.0	26.0	26.0	26.0	26.0	26.0	26.0	26.0	26.0	26.0	26.0	26.0	26.0	26.0	26.0	26.0	26.0	140.0
27.0	27.0	27.0	27.0	27.0	27.0	27.0	27.0	27.0	27.0	27.0	27.0	27.0	27.0	27.0	27.0	27.0	27.0	27.0	27.0	120.0
28.0	28.0	28.0	28.0	28.0	28.0	28.0	28.0	28.0	28.0	28.0	28.0	28.0	28.0	28.0	28.0	28.0	28.0	28.0	28.0	100.0
31.0	31.0	31.0	31.0	31.0	31.0	31.0	31.0	31.0	31.0	31.0	31.0	31.0	31.0	31.0	31.0	31.0	31.0	31.0	31.0	80.0
33.0	33.0	33.0	33.0	33.0	33.0	33.0	33.0	33.0	33.0	33.0	33.0	33.0	33.0	33.0	33.0	33.0	33.0	33.0	33.0	60.0
36.0	36.0	36.0	36.0	36.0	36.0	36.0	36.0	36.0	36.0	36.0	36.0	36.0	36.0	36.0	36.0	36.0	36.0	36.0	36.0	40.0
39.5	39.5	39.5	39.5	39.5	39.5	39.5	39.5	39.5	39.5	39.5	39.5	39.5	39.5	39.5	39.5	39.5	39.5	39.5	39.5	20.0
40.0	40.0	40.0	40.0	40.0	40.0	40.0	40.0	40.0	40.0	40.0	40.0	40.0	40.0	40.0	40.0	40.0	40.0	40.0	40.0	0.0

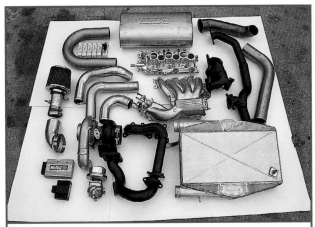

Here's the Alamo Autosports 1MZ intercooled turbo kit with Motec M48 EMS and big Bosch 630 cc/min peak-and-hold fuel injectors.

shaft, for an estimated total of 315 crankshaft horsepower at 6,750, maybe 325 at 7,000, a gain of 137 horsepower over the 1994 to 1996 stock 1MZ's 188 horsepower, that is, nearly a 75 percent power increase.

Driving the car on the street, boost was available instantly upon application of throttle at any rpm, with a slow increase in boost from 3-plus psi at 2,000 rpm to 5 psi by 5,000 rpm. With torque holding on very well across the rpm range, with a mild, gentle peak at 3,750, power increases in almost a straight line from 80 rear-wheel horsepower at 2,000 to 250 rear-wheel horsepower at 6,700. Acceleration rushes to a thrilling crescendo as you haul ass through 5,000 and sprint hard toward redline. We weren't sure what might fail—and when—above the stock 6,250 redline, so we ended the dynoruns at 6,750, 500 rpm on the wrong side of the yellow. Power was still increasing, and it would have been tempting to bump the redline a bit on the Motec to 7,000 or even 7,500, maybe overdrive the blower a bit more.

Instead, we installed a turbocharger to back up the blower.

DESIGNING A TURBO SYSTEM FOR COMPOUND SUPERCHARGING

Staged forced induction has historically been used for super-high-boost applications where the required pressure ratio for a single compressor would be too high for reasonable thermal efficiency, or where sufficient boost is impossible with one compressor: You step up the air pressure a reasonable, efficient amount (no more than 1–2 atmospheres) with the stage-1 compressor, intercool it, and then feed the stage-1 pressurized charge air into a second stage-2 turbo to pump up the boost to extreme levels with another reasonable pressure ratio (followed by a second stage of inter-cooling). Achieving, say, 50-psi boost with one turbocharger might require an enormous centrifugal compressor section and could result in excessive charge heating and significantly increased intercooler loading. And such a compressor could be in great danger of surging, depending on total engine air consumption. With twin staged compressors, an equal amount of air might be supplied to the same engine at the 50-psi level with greatly reduced heat of compression due to improved thermal efficiency, with the added benefit of improved spool time and responsiveness.

Our plan was not necessarily to run extreme levels of total boost (though we wanted that potential); it was to combine turbocharging with *supercharging* to get the best of both technologies. The great strength of a roots-type Eaton supercharger is that it provides instant boost when you open the throttle, avoiding turbo lag entirely. The displacement of the blower is selected to be greater than that of the engine. As soon as the throttle opens, the Eaton's equalization valve closes and the blower's rotor lobes close around a gulp of air and push it into the engine in greater quantities than the engine could achieve by itself just from atmospheric pressure pushing in air when the pistons go down.

A positive-displacement roots-type supercharger provides linear increases in pumping volume at the cost of maximum boost, achieving reasonable thermal efficiency in the 1- to 8-plus-psi range (though inferior to that of the centrifugal compressor in a turbocharger or centrifugal blower). For our application, a big turbocharger would take up where the roots blower left off, compounding 5–10 psi stage-1 supercharger boost to virtually *any* level of combined boost required— with reasonable thermal efficiency.

I find it useful to think of TRD's positive-displacement 1MZ blower kit as essentially transforming a 3.0-liter V-6 into a big-displacement, high-compression powerplant with 325 horsepower. Think of the TRD blower as transforming a 3.0 V-6 into a black box that acts from the outside like a 4.5-liter V-8. So when it came to turbocharging our black box, essentially we were looking for a compressor capable of spanking some serious power out of, say, a Ford Mustang DOHC 4.6-liter Cobra V-8.

And for this application we had a rigorous set of constraints that we hoped did not create an empty solution set. Specifically:
1. We needed to design around a turbo compressor capable of very high mass airflow (horsepower) in combination with the supercharged 1MZ.
2. Since we knew the nonintercooled TRD blower at its maximum pressure ratio of 0.33 was going to raise the temperature a minimum of about 45 degrees, we wanted an intercooled turbo system with extremely high thermal efficiency that could deliver compressed air to the supercharger at or near ambient temperature.
3. For test purposes, as far as possible, we wanted a turbocharger system that was also capable of high levels of power *without the supercharger installed*.
4. Under such conditions, it was extremely desirable that the turbocharger would be reasonably responsive in turbo-only testing.
5. It was essential that the turbo operate on the good side of the surge line under all circumstances, with or without the supercharger on the engine, meaning the turbo should not operate in the realm of producing high boost at very low airflows.
6. Therefore, the turbo needed to operate without surge over the range of foreseeable pressure ratios in turbo-only boost mode (5-psi boost, to, say, 20–30 psi), as well as *compounded,* from 5 psi turbo boost with 5-psi blower boost, all the way up to the 20- to 30-psi turbo boost. Without surge.

Perhaps our biggest compounding challenge was the following: How much boost could the 1MZ engine tolerate under Motec control without detonating on 93-octane premium pump

When it came time to make serious power, the stock 200-horse V-6 needed some super-duty parts, including Wiseco forged pistons, Crower forged rods, and a rebalanced 1MZ crankshaft.

gasoline—considering that the supercharger section of the compound super-turbo would be located *inside* the intake manifold, and could not, for all practical purposes, be intercooled? Well, OK, it could be intercooled to some degree with water or alcohol injection (or perhaps a very small air-water intercooler), but injection-cooling is a Band-Aid with complexity and disadvantages we preferred not to contend with.

Applying the compression heating formula and Eaton's published thermal efficiency of 60 percent, at 5-psi boost (pressure ratio of 0.33, that is, one-third of an atmosphere), the TRD Eaton blower was potentially going to raise intake charge temperature from 100 degrees Fahrenheit, to as much as 177 degrees Fahrenheit. So a maximum-performance MR6 required that the intercooled turbo subsystem of the compound forced-induction system compensate for blower thermal heating by keeping charge air *at or below* ambient temperature at 25-psi boost, with up to 500 wheel horsepower worth of air cfm and negligible pressure drop through the intercooler.

For this, we were going to need a *giant* custom air-water intercooler. Not only would an air-water cooler design allow us to optimize the layout of the air cooler separately from its water-cooling radiator (particularly valuable on a mid-engine vehicle), but we'd have the option to cool the air cooler with total-loss cold tap water on the dyno or with ice water during drag or land-speed-record competition. We decided fairly early on to locate the intercooler in the MR2 trunk behind the rear engine firewall. I siamesed two air-to-air cooler cores and boxed them with sheet metal to contain liquid coolant. The goal was zero pressure drop at 500 rear-wheel horsepower, with high efficiency.

Sizing turbos and blowers correctly is critical to extracting the maximum performance for your particular powerplant operating environment, and—I assumed—more critical when combining two compressors in series as a staged or compound system. So I read everything I could find on the Internet about the subject of staged or compounded compressors and studied every

We considered using Clark copper head gaskets and an O-ringed block essentially to avoid head gasket failure.

(usually minuscule) section of every book I could find on turbos or blowers that dealt with the subject in any way.

I consulted with forced-induction experts Bob Norwood and Corky Bell (author of technical books on turbocharging and supercharging *Maximum Boost* and *Supercharged!*). Bell ran some pressure- and density-ratio equations and suggested some criteria for the right compressor and turbine for turbo-assisted supercharging. Norwood discussed compounded engines he'd worked with in the past and made suggestions on compressor trim and turbine A/R. And I consulted with Texas turbocharger suppliers Majestic Turbo in Waco regarding selection of specific turbocharging equipment.

The TRD positive-displacement blower would knock out full boost the instant you opened the throttle, and this would immediately generate *enhanced exhaust energy and heat* to help spool a turbo, but proceeding along the line of thinking of our supercharged 3.0-liter V-6 as an enhanced-displacement high-compression V-8 *black box*, even the turbos on big V-8 engines still need to spool up before pumping efficiency becomes significant. So, OK, we could always change the turbo compressor if we guessed wrong the first time or if we actually used the 1MZ engine system at Bonneville.

But it was quickly clear that a defining question, of course, was how much maximum power did we truly need (which translates to mass airflow through the engine), and what pressure ratio would it take to achieve such an airflow? Would we *actually* run the engine at Bonneville?

In a compromise of our initial goal of a turbo capable of delivering 850 horsepower worth of air, we decided that our initial system should be capable of making at least *500* rear-wheel horsepower with or without the supercharger installed, which would provide more slack to meet requirements for turbo-only power and responsiveness. Five-hundred horsepower is a nice round number that is not in total nosebleed boost territory, but it is a good 30 rear-wheel horsepower above the 471 rear-wheel

horsepower we achieved in the project MR2 Turbo in this book with the stock 2.0-liter powerplant.

I felt it was politically correct that a compound-supercharged 3.0-liter V-6 MR2 should outperform a stock-block 2.0-liter turbo MR2. Of course, to achieve 500 horsepower at the wheels, you'd need to make more like 575–625 horsepower at the crankshaft, which at a ratio of 10 horsepower per pound of air implies roughly 60 pounds of air per minute from the turbo compressor. The smallest Garret-type compressor that will possibly achieve this airflow is a GT-61 trim compressor.

Centrifugal compressor performance is fairly predictable using compressor maps, but it can be difficult or impossible to predict *turbine* performance on a particular engine without a certain amount of testing. Unfortunately, it is largely *turbine* performance that determines the onset of measurable boost. The driving energy for a turbine is often discussed as if it were *free* power because exhaust energy is present in the form of both gas pressure and heat, and the exhaust heat is 100 percent wasted on a naturally aspirated or supercharged vehicle, whereas on a turbocharged powerplant some of the heat-energy is used to turn the turbine.

Turbine performance is actually a function of (to a lesser degree) exhaust backpressure blowing through the turbine *and* (to a much greater degree) energy imparted to the turbine as hot exhaust gases expand through the turbine nozzle and cool, the expanding molecules imparting terrific pressure independent of the exhaust pressure. To quantify this, we consulted turbo and supercharging expert Corky Bell, owner of Bell Experimental Group in San Antonio.

Bell points out that 70 to 80 percent of the energy required to drive a turbine comes from the heat in the exhaust rather than the pressure, and that the energy that can go down the shaft to drive the compressor is a function of the absolute turbine inlet temperature *to the fourth power* minus the absolute turbine outlet temperature to the fourth power. Since the temperature drop through the turbine nozzle can easily be 200 degrees, there is prodigious heat energy available to do work driving a centrifugal compressor.

Bell says you can calculate required compressor horsepower as pounds per square inch times cubic feet per minute divided by compressor efficiency (say 0.8). Meanwhile, the parasitic effect of exhaust back pressure, says Bell, is typically in the range of 1.5–2 tenths of a percent engine power robbed per psi exhaust backpressure. This means that 20-psi exhaust backpressure robs two percent engine power. So a centrifugal compressor that might require 20 to 40 horsepower through the shaft as a supercharger to make 10-psi boost on a 300 horsepower engine would only require 20 to 30 percent of this (maximum of 0.3 x 40 = 12 horsepower; min of 0.2 x 30 = 6 hp) as a turbocharger where parasitic losses from back pressure in the exhaust get transmitted into the crankshaft as power losses pumping against the backpressure.

So we see that the turbine energy available to do work is dependent in a *small* way on exhaust pressure and in a *big* way on exhaust heat. This is why insulating or coating exhaust plumbing on the way to a turbine can yield dividends in faster turbine response: It takes great energy to overcome the inertia of a turbo to accelerate it extremely rapidly to full working speed. What's more, because centrifugal compressor airflow increases *exponentially* with compressor speed, the compressor must be turning 30,000–40,000 rpm before it's capable of significant pumping at all.

Bottom line, on the way to driving the centrifugal compressor to make boost, the turbine needs to overcome the inertia of the turbine itself (plus that of the compressor wheel and turbo shaft) to accelerate the compressor to the relatively high rpm required to make measurable boost. Predicting required and actual turbine and compressor performance is made more difficult by the fact that a larger compressor wheel has more inertia that must be overcome to accelerate to boost-producing speeds, but, conversely, a bigger compressor wheel can sling more air at a lower rpm and therefore requires less acceleration before it arrives at the speed at which turbine power begins to be consumed to doing pumping work rather than just acceleration. A T-76 compressor, for example, can begin to make boost as low as 36,000 rpm, whereas a T04E-60, for example, requires as much as 46,000 rpm shaft speed to make any boost pressure.

We had debated employing two smaller low-inertia turbos located below the exhaust manifold on each bank of the V-6. But compounded with a TRD blower, a second turbo seemed unnecessary and needlessly complex. What's more, two turbos would've had to be located quite low on the engine, potentially raising insurmountable problems designing the required continuous downhill oil drain-back plumbing to return lubricant to the engine *above* the oil level in the sump. We decided early on to concentrate on making compound and turbo-only systems work with one—and only one—turbo. Fortunately, achieving really high power from one turbo has become easier in recent years: There are some Godzilla turbos around.

Majestic Turbo supplied a huge T-76 turbocharger to *make* boost and a Deltagate-type wastegate to *limit* total maximum boost delivered by both the blower and the turbo. The initial turbine housing was a Majestic P-trim 0.81 unit. Majestic's T-76 is a *large* turbo capable of delivering 90 pounds per minute of air at a pressure ratio between 2.6 and 3.2—good for supplying enough air for as about 900 horsepower. One would not necessarily expect such a big air pump to deliver much low-end boost, but the idea was to let the blower deliver the low-end boost. Bottom line, an intercooled T-76 turbo could deliver high midrange and massive top-end. Being large, the T-76 could also deliver a fair amount of air at a very low pressure ratio without surging—a requirement for streetability on our supercharged MR6.

Corky Bell calculated that an Eaton supercharger running 5-psi boost (0.33 pressure ratio) ideally provides a temperature increase of 8.5 percent; applied to an ambient temperature of 100 degrees Fahrenheit, the unit would heat the 100-degree air to 148 degrees. Correcting for the Eaton running as much as 62 percent efficiency, the blower will actually raise 100-degree air to 177 degrees. At 2,500 rpm, the supercharger provides a 16 percent gain in torque over the naturally aspirated 3.0L-liter 1MZ-FE V-6.

Bell noted that the TRD supercharger would significantly increase the low-end exhaust energy available from the 1MZ to drive a supplemental turbocharger, and that the roots-type Eaton TRD blower system provided some instantaneous additional torque in the 1,500–2,500 rpm range that would probably not be practical with other power-adders such as nitrous or turbocharging. Bell also pointed out that at 400 horsepower, the big Majestic T-76 turbo is barely getting into the maximum efficiency range. The surge line on the T-76 compressor map occurs at 33 pounds per minute of air. On the other hand, the T-76 has a tremendous upside if you need it.

FABRICATING THE TURBO SYSTEM

One fine spring day we drove the supercharged MR6 onto a lift at Alamo Autosports to begin construction of the turbo conversion—and immediately built some new muscles hefting the T-76 into various possible locations for a turbocharger.

Motec M48 and Motec M8 ECUs with "starter" wiring components.

We initially considered locating the turbo in the upper engine compartment above the transaxle and piping exhaust to it directly from both exhaust banks, but we eventually decided to locate it lower down behind the transaxle to the left of the engine near the rear-bank exhaust so that Alamo needed only merge cross-over exhaust from the front bank into a two-into-one collector to combine it with exhaust discharge from the rear bank header—and then extend the collector with a short tube terminating into a T04 turbine flange—to deliver exhaust gases to the T-76.

This arrangement would provide enough room behind the V-6 for a 3-inch turbine discharge pipe delivering exhaust gases exiting the turbo to a muffler or catalytic converter with passenger-side entry. The location placed the compressor conveniently close to the rear firewall, intercooler, and throttle body. Later we would reinforce critical areas of the header-to-turbo plumbing with gussets and TIG welding for maximum strength.

After mocking up and tack-welding the basic turbo plumbing, we pulled the 1MZ-FE V-6 engine for final turbo system welding and for machining operations on the block to install fittings to supply pressurized oil to the turbo center section and to drain oil back to the sump just above the maximum oil level. Pulling the engine also permitted good photography of the compound-supercharged system installed on the engine when you could really see it well.

In the final mock-up and installation of the turbo system, we welded a V-band discharge fitting to the T-76 turbine and carefully constructed a mandrel-bend exhaust system with Borla stainless tubing and super-high-flow muffler. With the exhaust system in place, we bolted the Majestic Deltagate to a flange welded onto the side of the collector-to-turbine inlet pipe so the wastegate had plenty of ground clearance and room for a discharge that—initially, at least—could bleed off exhaust directly into the atmosphere at as low as 5-psi total boost.

We installed braided-steel hoses for turbo oil-supply and drain (which required careful routing and shielding to maintain a continuous gentle downhill flow and to preclude damage from the hot exhaust system). To avoid damage from the super-hot turbine housing, we reworked the speedo-drive assembly to point rightward rather than upward over the transaxle, and then routed the cable itself below the rear exhaust manifold and forward;

Twincharged MR2 with 3.0-liter power and Motec controls.

we would eventually sacrifice *four* $60 Toyota speedo cables before we got an arrangement that could survive fierce soaking heat from the turbocharger/exhaust and hot Texas summer weather after shutdown or idling in traffic with insufficient airflow through the lower engine compartment.

HIGH-BOOST FUEL SUPPLY

While dyno testing the supercharged-only MR6, we had pushed the stock injectors to their fuel-delivery limit and beyond with the Bell Engineering fuel management unit (FMU). Now, with bigger power in mind, we obtained a set of twice-the-size 630-cc/min peak-and-hold injectors from Bosch. These injectors are low resistance 6-amp units with a form-factor similar enough to the stock 1MZ's 288s that—with O-ring changes—they are usable with the stock 1MZ fuel rail and lower-manifold injector bosses. At rated fuel pressure, these injectors can deliver sufficient fuel for as much as 750 horsepower. On the other hand, the stock MR2 Turbo fuel pump runs out of capacity at roughly 60 psi just above 400 flywheel horsepower, and requires boosted voltage (for example from a Kenne-Bell Boost-A-Pump) to deliver additional fuel (or installation of an auxiliary in-line fuel pump or substitution of a high-capacity in-tank unit).

COMPOUND-SUPERCHARGING TEST AND ANALYSIS

We ended up testing the twin-charged system before gaining any experience turbo-only. Compounded behavior was simple: With the wastegate referenced with a boost-pressure line routed directly from the compressor section of the turbo, the turbo began making additional boost and torque above and beyond that of the supercharger as low as 2,300 rpm—from which point torque and boost increased in a smooth, linear fashion until the wastegate opened at 4,000 rpm and stopped the boost increase at 5-psi turbo boost, 5- to 10-psi total boost (depending on wastegate setting). If we kept the wastegate closed, boost and torque headed directly for the top of the dyno chart. In one memorable early runaway dyno pull, the MR6 managed to hit 300 ft-lb rear-wheel torque at 4,000 rpm before Bob Norwood backed off the throttle.

It quickly became apparent that the forced-induction and fuel-delivery systems *vastly* exceeded the mechanical capacity of the 1MZ V-6 powerplant. And when a 15-psi dynorun apparently lifted the rear head just enough to produce a slight brief outbreath of coolant steam, we realized that further gains demanded better head fasteners and wire O-ringed cylinders with copper head gaskets to prevent catastrophic head gasket failure. We also knew that the stock pistons and connecting rods would not be long for this world at 15-plus-psi boost.

Thinking ahead, we already had a set of Wiseco custom super-duty forged pistons built with the stock 10.4:1 compression ratio but designed so the piston crowns could be machined to reduce compression while maintaining sufficient thickness. We began looking for a set of custom 1MZ rods. We also began investigating the feasibility of installing piston-cooling oil squirters or drilling the rod beams to spray oil from the crank bearings onto the underside of the pistons. The pistons would be coated with ceramic thermal coating for protection and to keep heat out of the water jacket and send it to the turbo where it would do some good.

With these super-duty parts on the engine, we were confident it would survive a lot of boost. But we also recognized that sooner or later the stock 1MZ valve springs would float and that valves and intake ports and cam profiles that flowed well on a 200-horsepower all-motor engine were going to be a major restriction on a 500-plus horsepower compound-supercharged engine. In fact, testing exactly how much air we could force through a super-duty 1MZ valvetrain with two power-adders was going to be . . . interesting. Just not yet.

In the meantime, we were eager to get some road-test results without flaming the engine, so we settled for a wastegate setting and a "compound" Motec calibration that reliably delivered 425 horsepower (350 to the wheels) and a flat torque curve representing about 270 ft-lb from 3,500 to 6,500 rpm.

The compound system had a character all its own, distinct from blower-only or turbo-only performance: bottom-end felt big-block massive, yet top-end developed with a turbo-rush. The compound super-turbo MR6 even had its own *sound*. According to Jerry Magnuson, who designed the 1MZ TRD blower kit, when a compound turbocharger is operating at full-howl—forcing tons of boost downstream into the TRD blower—turbo boost actually begins to *drive the blower's rotor assembly* something like exhaust gases driving a turbine, actually putting power *back* into *the supercharger drive belt*.

And, indeed, we observed that as the turbo component of boost spools up above 5-psi boost, the blower whine of the Eaton fades and what you hear is purely the whisper of turbo power and then the roar of a big-power, turbo-boosted engine as the wastegate begins dumping exhaust directly overboard into the atmosphere. A lingering and unresolved question was what the overdriving turbo does to the parasitic drag of the supercharger if turbo boost is pushing energy back into the blower belt.

TURBO-ONLY TESTING AND OPTIMIZATION

After we'd driven the compounded car around Texas for a while, we decided it was time to test the MR6 in turbo-only mode. Happily, removing the TRD supercharger system is easy on the MR6: Basically, you remove the intercooler-to-throttle body tube, de-tension the blower belt, and remove the intake manifold/supercharger (four bolts and two nuts). Now unbolt the blower drive-belt assembly and remove it, replacing two bolts that also help secure the passenger-side engine mount. Next, replace the OEM alternator tensioning bracket and reinstall the serpentine blower-alternator-air conditioning belt with the smaller OEM alternator and air conditioning belt, then retension the belt. Now remove the throttle body from the supercharger manifold and install it on the OEM intake manifold, and then bolt the upper manifold onto the engine.

As you might expect, in each of the possible permutations of power-adder modes (none, turbo-only, blower-only, compounded with both) there were unique areas of the Motec air/fuel map the engine would encounter that weren't reachable in certain of the other modes. For example, the turbo made lower boost at full-throttle and 2,000 rpm than you'd ever see with the supercharger at the same engine speed, yet the optimal turbo calibration in this range is not precisely the same as the NA engine. The loading required to drive the supercharger had an effect, as did the exhaust backpressure of a turbocharger.

The problem is, although a particular MR6 power-adder configuration might trace its own unique path through the main fuel map, nearby cells from another configuration will be averaged into the final fuel calculation if they happen to be adjacent to cells encountered in the current configuration. All of which meant that Bob Norwood had some dyno work to optimize turbo-only performance after we pulled the blower. It also meant that perhaps we'd have to change the main fuel map when we changed the power-adder configuration from NA to blower-only to turbo-only to compounded—or as we adjusted the maximum turbo boost.

The first turbo-only dynoruns were U-shaped: The T-76 with 0.81 A/R was sluggish in the bottom end, and turbo-only torque did not surpass blower-only torque until 3,500 rpm. At 4,500 rpm, the turbo-only torque temporarily surpassed the *compounded* torque, but by 5,700 rpm, turbo-only torque had fallen below that of blower-only torque, neither of which we'd expected to see.

With the blower off the MR6, we drove 1,000 miles from Dallas to Tucson in the turbo-only configuration, at which point we reinstalled the blower and continued on a road trip to Los Angeles, after which we again removed the supercharger for a drive to Silicon Valley. We eventually drove the 2,000 miles back to Austin uneventfully. Throughout the trip all systems were up to the best abuse that various magazine editors and I could extract from on-highway flogging. The Clutchmasters multi-disc clutch performed equally well stuck in LA freeway traffic or under hard, race-type flogging. Although the turbo was sluggish in the 2,000 to 3,000 range, the 1MZ motor is *not* sluggish, and people who drove the turbo-only car loved the low-end all-motor torque combined with the rush of additional power and torque as the turbo spooled up to make 10-psi boost in the midrange.

Various people advanced theories for the loss of turbo torque in the higher rpm ranges. One was that turbine energy was insufficient to drive the big T76 compressor wheel in this range. Another was that we had some kind of mismatch between the turbine and compressor in terms of available exhaust energy versus turbine speed. Yet another was that, for some reason, ignition energy was degrading in the upper speed-density ranges (coincidental to the removal of the blower). Perhaps the most likely cause was that exhaust backpressure was forcing open the wastegate.

We decided to focus first on improving turbo-only low-end spooling, and discover what this did to the disappearing torque above 5,000 rpm. We began looking for ways we might improve turbo-only spool-up in the 2,000–3,000 range without devastating compounded performance in the higher end. After considering a compressor change to a smaller wheel to reduce inertia and speed up the turbocharger, Majestic recommended swapping turbine housings to a 0.70 A/R and reducing spooling inertia with their new patent-pending low-inertia fastener in place of the standard nut on the turbine end of the turbo shaft to improve spooling response. In fact, the 0.70 housing is a twin-entry unit, and

We wrestled with spark gap misfire problems under boost conditions until we junked the stock Toyota coils, installed these Ignition Solutions plasma booster coils, and reset dwell in the Motec ECM.

we considered redesigning the exhaust feed to maintain separate exhaust pipes from the two banks all the way to the twin-entry turbine housing to keep exhaust pulses from conflicting with each other by separating temporally close exhaust pulses until they actually enter the turbine nozzle.

In the meantime, Ignition Solutions shipped us a set of six plasma-direct performance coils, which could be installed for coil-on-plug direct-fire or remotely with plug wires.

At this point, we did not yet have dyno results with the new ignition and turbo equipment. We still hoped to accomplish:

1. Further development of the turbo-only and compounded engine configurations.

2. Measuring power losses in the blower-only configuration.

3. Measuring the turbine power losses of the turbocharger due to backpressure or other factors in the turbo-only configuration.

4. Logging turbine speed.

5. Measuring temperature rise and intercooling efficiency associated with the turbo and the supercharger at various levels of boost.

We decided this was the right time to pull the engine for clutch examination and internal-engine super-duty upgrades. In general, we took it as an opportunity to examine the status of *everything* after 12,000 miles of hard driving and upgrade *any* questionable weak spots.

ANALYSIS AND TEARDOWN: TOYOTA'S 1MZ V-6

Alamo Autosports, builders of the MR6 turbo conversion, helped us remove and teardown the boosted 1MZ V-6 powerplant. Later on, we consulted with Bob Norwood, cylinder head expert Craig Gallant in Houston, and various performance equipment suppliers to design and implement the super-duty MR6 engine strategy.

We concluded that the following components were sufficiently strong to withstand a significant power boost to at least triple the stock 200 horsepower and 7,000 rpm:

• Forged crankshaft, six-bolt main caps, alloy block with structural upper oil pan and end plates, oil pump, and cam-drive system.

We decided the following parts required upgrading:

• Cast pistons, connecting rods, composition head gaskets, block deck surface.

The cylinder heads, camshafts, and valvetrain were a special case requiring further analysis at GTP. The goal was improved maximum rpm and improvements in airflow that maintained gas velocities commensurate with improvements to port-CFM flow and expected port volume increases.

IGNITION COILS

The stock Toyota coils did not like the lengthy dwell period we'd been feeding them to get the kind of ignition power needed to fire plugs through the vastly thicker mixtures of a compound-boosted powerplant. After a long Texas-to-California road trip, we replaced the stock Toyota coil-on-plug units with super-duty coil-on-plug plasma-direct coils from Ignition Solutions. Ignition Solutions uses the following chart for correlating dwell time (pro-grammable in the Motec M48 engine management system) to current and spark energy. Ignition Solutions recommended that we run a dwell of 2.5 milliseconds, which would provide a vast increase in spark energy compared to stock Toyota coils.

Coil Calculation

DWELL (msec)	Current (Ampere)	Energy (Milli Joule)
1.00	2.25	11.68
1.50	3.3	25.13
2.00	4.5	46.73
2.50	5.7	74.97
3.00	7.0	113.07

Ulf Arens of Ignition Solutions says, "The plasma-direct replacement ignition coil is built using Ignition Solutions' 'Boost-er' technology to amplify secondary spark current by 100 percent (two times as much as OEM) and release about four times as much energy at the initial moment of spark." The plasma-direct coils are designed to provide 100 percent secondary current (spark amperage) increase from 50 milliamperes to more than 100 milliamperes on most systems, with four times spark energy increase for the initial spark discharge ($E=1/2xLxixi$). An added benefit is each coil's external LED that indicates power firing of the coil.

ASSEMBLING THE SUPER-DUTY COMPOUND SUPER-TURBO 1MZ

Kim Barr Racing Engines in Garland near Dallas, handled our 1MZ pre-assembly machine work, which included honing the cylinders for standard overbore pistons and checking the top deck for flatness. Barr also machined the upper surface of the cast-iron cylinder liners for wire O-rings and polished the crank journals, which included removal of all traces of surface damage from the spun rod on the center left bank rod journal, as well as any funk-iness where balancing weights were attached to the crank rod jour-nals to balance the reciprocating assembly for lighter pistons and heavier rods and wrists pins. In fact, given the goal of making tremendous power in a daily-driver street vehicle, we balanced the crank to stock specs.

We began the build process by checking Wiseco's piston-to-bore clearance on all six bores, and then filing the compression ring end-gaps and checking each in the bore with a feeler gauge until it was perfect. At this point we installed the Wiseco pistons on the Crower rods using Wiseco's super-duty full-floating wrist pins and circlips. The Wiseco pistons are built from aluminum forgings, while the Crower rods are built from 4340 chromium-

This monster water-cooled intercooler sits in the trunk behind the engine, capable of greater than 100-percent thermal efficiency with total-loss tap water or ice bath.

molybdenum steel billet and are designed to survive at least 150 horsepower per rod.

We installed ARP main studs finger tight in the block with both sets of threads lubed with synthetic oil for smooth accurate torquing, installed a set of lubed stock Toyota main bearings, and then dropped in the crank for checking with Plastigauge. Toyota offers standard bearings in precision sizes from "Mark" 1 to 5 design to allow selection of bearings at initial engine build that provide 0.0015–0.0025 inches bearing oil clearance at the con-necting rods and 0.0010 to 0.0018 main bearing clearance.

A person rebuilding a stock 1MZ engine can simply replace bearings with the stock size "Mark" bearings, but with new rods, bearings, and polished crankshaft, we needed to check clearance. One good way to do that is with a piece of Plastigauge on each journal, which can be measured after it's crushed and flattened between a bearing and journal when you torque down the fas-teners. Stock-replacement Mark 2 rod and Mark 3 and 4 main bearings work nicely.

Since the stock undercut main bolts are designed to stretch for even loading when torqued in place, the MR6's new ARP fas-teners would not necessarily be optimally torqued using the stock technique of 16, 18, or 20 ft-lbs plus an additional 90 degrees. ARP recommends setting torque at 60 ft-lbs on the mains and 65 psi on the rods. Bob Norwood likes to watch the torque gauge and watch for the fastener turning without torque increasing, indicating the fasteners are now just stretching. Norwood jams the crankshaft forward with a screwdriver to load it against the thrust washers as it will be when the clutch engages under hard acceleration.

The pistons installed effortlessly on the rods, Norwood lub-ing the wrist pin with synthetic oil, locating the stamped num-bers on the rod outward away from the vee-bank, indentation dots on the piston crowns turned forward. With one circlip in place in each piston, we threaded the pin into the opposite side of the pis-ton, through the rod bushing, and home against the circlip, then inserted the second clips.

Norwood normally employs a ring funnel to compress the rings and pushes the piston-rod assemblies down into the bores after lubricating the bores with synthetic motor oil. We checked the con rod bearing oil clearance with the Plastigauge (just for fun). Clearance was perfect—right in the middle of Toyota's recommended range—after we opened up the clearance a hair using Toyota's Mark 1–5 bearings. Our goal was to home in on the slightly looser clearance required for a high-power application versus the tightest factory clearances you'd want on a stock Camry/Lexus OEM grocery-getter designed to last 200,000-plus miles.

With copper head gaskets, it was critical to provide sealant around the coolant passages through the gasket. Clark Copper Head Gaskets, Inc., had thoughtfully provided some secret sauce sealant exactly for this purpose, and we slathered a ring around both sides of the gasket and both the head and block deck surface, making sure the sealant did not plug the coolant passages.

We removed the camshafts from both heads in order to get a wrench on the head stud nuts, carefully lubricated all valvetrain components, and lowered the heads over the studs, tightening the nuts in factory order to 65 ft-lbs—in three stages.

SUPER-DUTY TESTING

With the engine rebuild and upgrades completed, we now had a *strong* Toyota 1MZ-FE 3.0-liter V-6 powerplant, complete with forged Crower rods, forged Wiseco pistons, super-strong ARP head and main studs, Clark copper head gaskets and block O-ringing, rebalanced crankshaft, GTP ported heads with super-duty springs and retainers, rebalanced crankshaft, and more—including new organic-metallic facing on the multi-disc Clutchmasters clutch system. Kim Barr Racing engines did the precision machining and Norwood Autocraft and Alamo Autosports handled the assembly and installation.

Break-in? One-hundred miles driving around Dallas. And then it was onto the Dynojet where we screwed a Motec wideband O$_2$ sensor into the V-6 front-bank O$_2$ sensor port, and Bob Norwood booted up his laptop, plugged in the communications cable to the Motec M48 ECU, launched the M48-v61 calibration software, and fired up the engine.

At this point we made a few dyno pulls and then I spent some time fighting electrical demons. It turns out Motec really means it when they recommend powering the coils with a relay triggered by the Motec fuel pump circuitry so that if the engine stops running the coils will *definitely* be turned off by the ECU along with the fuel pump. Depending on the random position at which the engine stops, if the key is still on, the Motec M48 *might* leave current running through a coil, and Ignition Solutions direct coils are not designed to survive being energized for a long time. What is a long time? Well, for example, more than two minutes. At which point a coil can melt down in a furious gust of electrical smoke (don't ask how I know). And if that happens, the meltdown can trigger secondary events like blown 40-amp fuses or destroyed fuseable links and all that good stuff.

This car was a test mule, and the engine had been in and out various times, we'd added or deleted new engine management appliances (like the Ignition Solutions plasma-boost direct-fire coils), and the wiring had been hacked and then the hacks hacked to the point that it was clearly time to refurbish the engine wiring, by this point a nontrivial task. What's more, the MR6 had spent several months gathering dust at Alamo, and the battery had been run down to Less Than Zero at least once, which is not good for battery longevity.

In order to provide accurate injection pulse width compensation for changes in battery voltage, Motec requires a tuner to specify injector current. We quickly got a good lesson titled, "Don't assume your default battery compensation table is right," in which running the car sans alternator with a powerful battery charger effectively illustrates how changing the charger from 2 amps to 20 amps, to 40 amps, to 125 amps affects available battery voltage and can affect the air/fuel ratio via defaults in the battery voltage compensation based on the specified injector current configuration, resistance, and fuel rail pressure. Make sure the battery compensation table is right for your fuel injectors. If it's not, if the voltage is normal when you calibrate the main fuel and timing tables and then the voltage changes, the air/fuels ratios can be off enough to cause trouble.

INJECTOR CURRENT, RESISTANCE, FUEL PRESSURE, AND BATTERY VOLTAGE COMPENSATION

All ECUs must compensate for varying battery voltage by adjusting the final injection pulse width calculation to increase injector open time when available battery/alternator power is lower and the injectors will open more slowly. Battery voltage compensation value is dependent on the injector type. As you can see from the table below, Motec included fuel pressure as a compensation factor in the newest compensation parameters.

Injector Number	(Ohms)	Fuel Pressure	Battery Comp
Motec Indy	(4.5)	300 kPa (45 psi)	6
Motec Indy	(4.5)	500 kPa (75 psi)	7
Bosch 036	(2.5)	300 kPa (45 psi)	8
Bosch 036	(2.5)	500 kPa (75 psi)	9
Bosch 363	(0.5)	300 kPa (45 psi)	10
Bosch 363	(0.5)	500 kPa (75 psi)	11
Bosch 803	(4.5)	300 kPa (45 psi)	12
Bosch 803	(4.5)	500 kPa (75 psi)	13
Bosch 911	(3.0)	300 kPa (45 psi)	14
Bosch 911	(3.0)	500 kPa (75 psi)	15

Older Battery Comp Settings (Note: If an engine has been tuned using any of the following comps, Motec does not recommend changing to the newer comps, (above), unless the engine will be re-tunretuned.)

Bosch 007	(2.5)	3
Bosch 035	(2.5)	3
Bosch 036	(2.5)	3
Bosch 217	(16.0)	4
Bosch 351	(0.5)	0
Bosch 363	(0.5)	0
Bosch 403	(2.0)	2
Bosch 706	(16.0)	4
Bosch 803	(4.5)	3
Bosch M/Sport 910	(2.4)	2
Bosch M/Sport 911	(3.0)	2
Bosch M/Sport 912	(4(4.5))	3
Rochester 988	(2.2)	0
Rochester 989A	(2.2)	0

EXTREME TEMPERATURES

The day we arrived at the Norwood Autocraft Dynojet chassis dynamometer to calibrate and dyno-test the new super-duty powerplant at higher power levels, the temperature in Dallas was over 100 degrees, and the relative humidity was above 50 percent. The MR6 made its way with reasonable drivability from Arlington to the dyno northeast of Dallas, but the calibration was

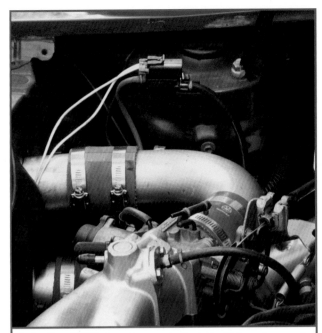

I wired in a GMC Typhoon boost controller to the Motec using one of the digital pulse width–modulated (PWM) outputs. The controller bleeds air at a specified pressure to the wastegate actuator diaphragm to alter maximum boost.

surely no longer optimal: the heads had been ported by GTP to open up the upper airflow (which can affect volumetric efficiency across the board) and the new engine was surely tighter than the old 100,000-plus factory engine. In any case, we had not checked the calibration since leaving on the California road trip months before.

The first thing we did was tune the engine with no power-adders to see what head-porting had done to the power and torque curve. Initial results indicated that the engine was, indeed, stronger in the upper-rpm range above peak power, although with the stock cams, peak unboosted power was only up about 4 rear-wheel horsepower at 5,500 rpm, with peak torque up by about 7 ft-lbs (conversely, power and torque were down by about 12 horsepower around 3,500).

On the other hand, previous testing had been in 80-degree weather, and now it was 20 degrees hotter. Although the Dynojet is equipped with a weather station that factors ambient temperature and humidity into the dyno calculation to correct for changes in air density and oxygen content, Dynojet's Winpep Performance Evaluation Program user interface explicitly warns you in this sort of case that "Comparisons of these runs may be misleading due to large intake air temperature differences." Bob Norwood, who handled all the Motec tuning on this project, points out that experience shows engines (and especially turbo engines) do not make as much power in really hot weather, and that the standard corrections never fully account for this.

When Norwood felt good about the NA calibration (the old calibration was—inexplicably—running rich in several broad ranges), we reinstalled the TRD blower system and moved back onto the dyno. After a bit of tuning, the blown super-duty engine results were an almost identical overlay with those of the old engine, with the exception that torque—and, therefore, power—was unexpectedly down in the 6,000–6,500 range by at least 15 rear-wheel horsepower.

At this point, during a long internal ECU datalog offload, the number five coil suddenly began smoking furiously, which signaled the end of dyno-tuning for that day. Fortunately, Ignition Solutions had another in stock—in San Diego. In the meantime, it was time to troubleshoot the electrical system.

ELECTRICAL TROUBLES

Motec tries very hard to sell a custom, factory-built wiring harness with every ECU they sell because robust engine management wiring is so critical to a reliable vehicle. It is also worth noting that there is a large difference between wiring that will have great longevity if you install it and leave it alone versus wiring that will survive repeated R&R for engine removal, blower or turbo system removal, and repeated modification as new features are added or subtracted from an R&D project.

Racing wiring requires super-duty components that can survive removal every week, and the MR6 wiring was a combination of stock and add-on wiring that certainly never met racing standards. In any case, the 1MZ V-6 had, by this time, been in or out of the MR2 at least 10–12 times during initial installation, engine mount changes, clutch changes, engine build up and so forth, and the TRD blower system and the Alamo-Majestic turbo system had each been installed and removed countless more times.

We had rewired the engine loom to add idle air control, to test various tach-driver experiments, to change the injector connectors when we replaced the stock 288-cc/min Toyota injectors with larger 630-cc/min Bosch injectors, to move the main engine compartment fuse box into the truck to prevent a proximity melt-down when Alamo built the turbo system, to move the intake air temp sensor when we built the turbo conversion, to replace the stock direct-fire coils with the three-wire-per Ignition Solutions coils, and much more.

It developed that when the coil began burning up, hasty fire-control countermeasures had caused a secondary short that overheated some wiring in the relocated engine fuse/relay box, now in the trunk near the Motec ECU. At this point I spent multiple hours troubleshooting, repairing, and upgrading the wiring harness, including the battery and the replacement coil. At least in part due to the extreme hot and humid Dallas weather working without air conditioning, I cannot say the work went especially quickly.

When the MR6 was running again, I immediately removed the modified TRD blower system (easy to do in less than an hour) and installed the Alamo turbo system (more difficult, requiring 2 to 4 hours). Since it is a lot easier to R&R the blower system than the turbo system, it always makes sense to R&R the supercharger rather than the turbo if there is a choice. Certainly there was more supercharger development possible (with modified 6,000-plus rpm AFRs, different pulley sizes, and so forth), but for now we decided to move on with turbo-only testing.

J&S SAFEGUARD

When it comes to high-boost testing and competition, we believe in Safeguards. J&S Safeguards, that is. The J&S Safeguard has the ability to detect detonation from individual cylinders and to retard timing to only the cylinder(s) knocking. Not only is this device excellent engine insurance, it is a good way to optimize the timing. Though the Motec M48 EMS has the ability to adjust individual cylinder timing, if you get a little too aggressive, a computer device like the Safeguard will detect knock and immediately initiate timing retard countermeasures to protect the engine from damage.

It can be difficult or impossible for a human being to hear knock and back off the throttle fast enough to keep detonation from hurting the engine when you are running close to the ragged edge of detonation and looking for every last horsepower. The J&S Safeguard can also make it possible to optimize power for ideal engine conditions, while relying on the Safeguard to protect the powerplant if you're running hotter than normal, or with lower-octane fuel, or reduced fuel pressure or if there are any other pro-knock factors in effect. For a six-coil application like the MR6, a compromise would be necessary in the form of wiring the coils in pairs as a wasted-spark application in which each coil fires on its own compression stroke as well as on its exhaust stroke with the compression stroke of the paired cylinder that is 360 degrees out of phase.

When it has detected detonation, the Safeguard lights up LEDs on a display, with a column dedicated each cylinder (or pair on a wasted-spark application). The Safeguard indicates the magnitude of the retard by the number of LEDs that light in each column. A tuner could potentially use this information to optimize individual-cylinder fueling or ignition timing on a sophisticated ECU like the Motec M48.

TURBOCHARGED!

So far, we had not yet done anything too fancy or impressive with the turbocharger, mainly because we had never previously considered the powerplant strong enough to handle much power. So far, we'd used no electronic boost controller or other means of dynamically altering maximum boost. We had ripped off a 330 rear-wheel horsepower *twin*-charged run with the old stock-internals engine, but at horsepower levels approaching double stock power, we'd been concerned about the looming possibility of catastrophic engine damage and had never ran more than about 11-psi total boost.

In the meantime, Majestic had supplied a special low-inertia T-76 turbocharger with a P-trim compressor and 0.7 A/R twin-entry turbine (and a super-lightweight shaft nut designed to cut inertia and improve spooling). The GT-76 has an extremely powerful compressor, capable of flowing as much as 90 pounds per minute of air (enough for a *lot* of power, probably at least 700 horsepower).

The Majestic wastegate has a Deltagate-type dual-ported actuator with the "bottom" port pushing open the wastegate poppet valve to provide traditional boost-opens-wastegate control. The wastegate's "top" port is optionally available to *offset* a proportion of the bottom-port pressure on the opposite side of the wastegate diaphragm to increase boost pressure using a pulse width–modulated (PWM) electronic boost controller or just a mini pressure regulator.

The default Majestic wastegate spring pressure used to limit intake manifold boost pressure (as well as to contain exhaust backpressure!) was 6 psi. Depending on where you referenced boost pressure for the wastegate (upstream or downstream of the supercharger), boost pressure would be either 6 psi or 11 psi (running with compound supercharging, with 6 psi added to the blower's 5-psi boost capacity with the current pulley).

With the Alamo turbo system installed on the MR6 and the blower removed, Bob Norwood made a series of turbo-only dyno runs to calibrate the Motec at 6 psi. The engine managed a best peak torque of 301.4 ft-lbs at 4,500 rpm, and a best rear-wheel horsepower of roughly 270 rear-wheel horsepower at 5,000 rpm. Compared to the last turbo-only dynoruns we'd made months earlier with the stock-internals engine and the old turbocharger (larger 0.86 A/R turbine nozzle on the old turbo, higher-inertia shaft nut), the new engine and turbocharger made an additional 5–20 horsepower across the board from 2,500 rpm to peak power at 5,000 (up by 15 rear-wheel horsepower to 270). What's more, the new setup hung on much better above peak power to make as much as 50 additional rear-wheel horsepower at 6,300 rpm.

The new setup made about 17 additional peak ft-lbs torque at 4,500, with torque up everywhere at least 10 to 50 ft-lbs.

Nonetheless, dyno testing coupled with Motec datalogs showed that torque above 4,500 rpm was dropping faster than we liked, and the datalog revealed that peak boost pressure rose to 6 psi in the midrange, but then fell back to 4 psi as dyno pulls raced toward redline. We wondered if exhaust backpressure was pushing open the wastegate in the higher rpm ranges.

Meanwhile, in order to test at higher levels of boost, we decided to install a GM Typhoon boost controller, a pulse width–modulated (PWM) air-valve used on 1992 and 1993 GMC Typhoon 3.8-liter turbocharged compact trucks that can be controlled by the Motec M48. Pulse width–modulation is what electronic fuel injectors do to put fuel in the engine, that is, open and close a certain percentage of time very quickly to change the stream of fuel entering the engine per time. By opening and closing a vacuum-boost reference line quickly a certain percentage of the time, a boost controller can regulate the amount of pressure the reference line delivers to a wastegate, from 100 percent of the vacuum/boost source pressure all the way down to 0-psi pressure. If the boost controller delivers 0 percent vacuum/boost to the secondary Deltagate wastegate port, then the wastegate will deliver maximum boost purely on the basis of the default spring pressure of the wastegate (6 psi, in our case); at 100 percent secondary-port pressure, the wastegate will stay closed until (and if) exhaust backpressure gets high enough to force it open.

Typhoon electronic boost-controller installation is quite simple: run a pressure reference from the compressor discharge (highest possible boost pressure in the turbo system) to the boost-controller inlet port, run a pressure reference from the controller outlet port to the wastegate delta port. Run a switched +12V signal to the positive terminal on the controller electrical connector, run a wire from the other terminal to a Motec M48 ECU PWM digital Aux output (#1 or #2).

We were also concerned that the stock MR2 fuel pump would soon run out of capacity as we raised the boost much above the default 6-psi level. Testing the other four-cylinder MR2 project in this book, we'd discovered that the stock pump was good to handle at least 335 rear-wheel horsepower, but we also hoped to leapfrog 335 in a hurry. Some people like to remove the stock MR2 in-tank fuel pump in favor of a stock Gen II Supra Turbo fuel pump, which is apparently good for over 500 horsepower—a good solution, but not one we could implement in time for dyno testing the same day.

Fortunately, MSD Fuel sells a 43-gallon/hour (at 40 psi) high-pressure in-line fuel pump that can be used to supplement the stock MR2 Turbo pump. MSD rates this pump as being good for 500 crankshaft horsepower. Rating the capacity of fuel pumps pumping in series is a tricky business, but combined with the stock pump's capacity of at least 400 crankshaft horsepower, we calculated that using the stock 5/16-inch fuel plumbing, working together, the two pumps could provide sufficient fuel for at least 500 rear-wheel horsepower.

BOOST CONTROLLER CONFIGURATION AND HIGH-BOOST DYNO-TESTING

Configuring a Motec digital output for boost control is not completely trivial. First you select boost controller as the aux output *function* (there are 17 different possible functions). Setting *gain* determines the magnitude of possible valve movement to eliminate differences between desired and actual pressure, and the method of configuration is to use as high a value as possible without hunting and then back off by at least 30 percent, keeping in

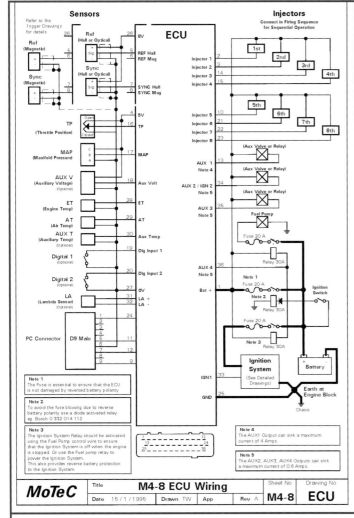

Motec M48 ECU wiring. We used every output to control something, and we also used most of the inputs. One PWM output got used to simulate a four-cylinder tach signal to the MR2 tachometer, with the other digital PWM output used to manage the boost controller.

able for most applications except where large changes in altitude will be encountered. The boost pressure or duty cycle is adjusted for various rpm and load points in the *boost limit main table.* You must also set the *frequency* of the boost-controller and the minimum and maximum useful *duty cycles.*

Unfortunately, a Motec M48 configured to run a six-cylinder powerplant is limited to driving two digital PWM output channels, both of which were already in use—the first to provide a four-cylinder tach pulse to drive the stock MR2 tachometer, the second to pulse a Ford idle air control (IAC) stepper motor used to provide idle speed stabilization control. For the time being, we decided that the MR6 could do for now without a tach and that later on we could make alternate arrangements with an aftermarket tach or adapting a 1MZ tach meter to the MR2 dash cluster.

Norwood used the more sophisticated proportional derivative boost control method and then set all the psi values in the boost limit-main table to the maximum target boost. Our initial target was 22-psi boost.

Following boost-controller installation, with the fuel tank nearly empty, we dumped five gallons of 114-octane unleaded race gasoline in the MR2 fuel tank and got ready to rock and roll.

Working on the Dynojet using judicious stabs of throttle to gradually produce varying amounts of engine loading and boost and keeping a close eye on the Motec wideband air/fuel ratio sensor via laptop, Bob Norwood gradually filled in various high-boost cells in the fuel-main table that we had not yet encountered. And then without a word, Norwood hit the record run button on the Dynojet and put the pedal to the metal at 2,600 rpm. Revolutions per minute climbed steadily to 3,500, and now you could hear the engine start to make serious boost. In the 1,500 rpm from 3,500 to 5,000, torque approximately doubled from 225 to more than 450 ft-lbs in a matter of a couple seconds.

Similarly, in the same interval, power approximately *tripled* from 150 rear-wheel horsepower to 430 and in the succeeding 500 rpm increases to more than 465! We changed to colder -9 plugs (versus the stock 11s) and tried closing the gap from 0.029 to 0.023 (and then returned to 0.029 when 0.023 made *less* power). In two successive tuning runs Norwood adjusted pulse width for optimal air/fuel ratios and increased the target boost pressure from 22 to 24 psi. Max power increased to almost 480 rear-wheel horsepower, with maximum torque over 475 ft-lbs. Assuming 20 percent drivetrain and tire rolling resistance losses, maximum power at the crankshaft translated to roughly 600 horsepower!

Unfortunately, the Motec datalog revealed that at maximum power the wastegate was receiving 100 percent PWM boost to the secondary Deltagate wastegate port, indicating that boost was not able to climb above 24 psi, and that the Motec never needed to reduce boost pressure to the secondary Deltagate wastegate port to limit boost. For now, we were stopped at 20 rear-wheel horsepower short of 500.

Analysis:

A number of interesting questions were posed by the latest dyno results:

• Why was NA torque (and power) down in the 2,500–4,500 rpm range by as much as 35 ft-lbs?

• Why did turbo-only torque fall off so quickly above 4,500 rpm, loosing 100 ft-lbs in 1,500 rpm—and why did boost fall off from 6 to 4 psi in this same range?

mind that too low a value may make it impossible to achieve the desired pressure.

Configuring a larger *derivation* improves the response to rapid pressure changes. Too large a value can cause erratic control due to pressure pulsations being interpreted as pressure changes. *Average position* should be set at the normal average duty cycle of the actuator, and differences between the average position and the actual duty cycle will cause a difference between the desired pressure and the actual pressure. *Boost control type* tells the Motec whether to use proportional derivative or duty cylinder control, whether closing or opening the valve reduces boost (PD control), or whether the valve is on or off at 100 percent of the boost table value (DC control).

Proportional derivative control allows absolute control of the boost pressure but is more difficult to tune as the PD gain and derivative factors must be adjusted to give stable control. These factors are very dependent on the particular type of installation. *Duty cycle control* directly sets the control valve duty cycle that indirectly controls the boost pressure. This method is easier to setup than the PD mode as there are no PD control parameters. This mode is suit-

• Why was Majestic's GT-76 turbo only capable of making 24-psi boost?

Given that a GT-76 turbo should clearly be capable of airflow for more than 600 horsepower, there are several possible problems. One would be that backpressure in the exhaust system was climbing high enough to hurt power and even push the wastegate open. Unfortunately, to this point we had no way of measuring exhaust backpressure, nor did we have any way of knowing if the wastegate was being forced open other than if we could hear the sound of exhaust gases exiting the unmuffled wastegate bleed. If this were the case, tightening down the wastegate adjustment spring would help. The boost controller can eliminate manifold pressure acting against the wastegate boost port as a factor in limiting boost (by neutralizing it with equal pressure on the Deltagate side of the wastegate diaphragm), but this will have zero effect on the tendency of exhaust pressure to force open the wastegate when pressure gets high enough.

Since exhaust backpressure is a function of both the total exhaust gas products of combustion and the amount instantaneously being diverted through the wastegate, exhaust backpressure is not perfectly predictable purely on the basis of intake boost pressure. If exhaust backpressure gets high enough—say in the upper rpm range—at even the default 6-psi boost pressure (with the wastegate spring adjusted to the minimal position), *backpressure* could start to force open the wastegate and cut boost from 6 psi to 4 psi in the upper rpm ranges when exhaust gas velocity is very high.

Evaluating the above, it was clear we needed to get back on the dyno with stronger spring pressure in the wastegate.

MR2/V-6 MISCEGENATION
Mating the Lexus-Toyota
V-6 Engine and MR2 Transaxle

You'll need to pull the stock MR2 drivetrain, handle prep work on the V-6 engine and MR2 manual transmission, mate them together, and stuff the whole works back in the car.

We quickly verified that the 1MZ V-6 ring-gear and 3S-GTE starter are fully compatible, that V-6 and I4 clutches are interchangeable, that four- and six-cylinder manual-trans Toyota flywheels are the same diameter and thickness, that the MR2 transmission input shaft and V-6 pilot bushing are compatible, and that a manual-trans MR2 flywheel slides seamlessly onto the V-6 crank flange. Unfortunately, we also verified that the V-6 crank flange uses smaller flywheel bolts to mount the flywheel—clocked at identical positions to the MR2's but in a larger diameter circle.

Options are:

Employ a relatively rare 1999 or later manual-trans V-6 flywheel (a new Solara flywheel from Toyota costs about $290).

Weld the MR2 flywheel bolt holes closed and mill-drill new flywheel bolt holes compatible with the crankshaft.

Build a custom flywheel-clutch assembly.

We decided the first two alternatives were suboptimal, in that the stock MR2 or Solara flywheel-clutch assembly has relatively high inertia (hurting revving response) and can run out of clamping force somewhere not far above 300 wheel horsepower. We called Clutchmasters and ordered up a multi-disc Kevlar aluminum MR2 clutch-flywheel kit manufactured with the 1MZ V-6 bolt pattern. This low-inertia setup delivers light pedal pressure while providing the clamping force needed to handle serious horsepower without slipping. To handle even more horsepower, a full-metal multi-disc clutch is available.

Meanwhile, with the turbo motor gone and the V-6 flywheel out of the picture, it was a simple matter to drop the V-6 into

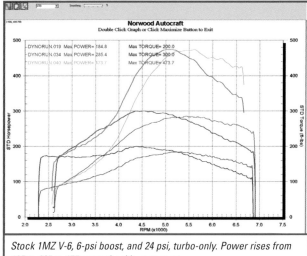

Stock 1MZ V-6, 6-psi boost, and 24 psi, turbo-only. Power rises from 185 to 285 to 475 rear-wheel horsepower.

place on the MR2 transaxle from *above* to experiment with space and connectivity issues. When an MR2 engine is fully intact, it is often easier to remove the complete engine and transaxle as a unit than just the engine. We verified that a 1MZ V-6 will indeed bolt to the stock MR2 transaxle with nearly all of the bell housing bolts and that there is a reasonable amount of space around a V-6 in an MR2. We also decided that a good custom exhaust system looked feasible.

MOUNTING THE ENGINE

With the clutch and starter in place on the V-6 engine/MR2 trans unit, you'll have to deal with hanging the engine and interfacing it to the various supporting fuel, electrical, and mechanical systems. You'll need to build a custom passenger-side motor mount or attach the 1MZ engine mounts to the MR2 chassis. You'll need to modify a V-6 torque-stay to attach to the passenger-side MR2 shock tower to keep the powerplant from twisting and shuddering under hard acceleration.

We decided it was simpler and more elegant to abandon the stock V-6 engine mounts and hang the 1MZ engine from the MR2 passenger-side engine attachment point using the V-6's torque-stay bracket (very robust). Norwood Performance used some old Ferrari suspension parts to fabricate a strong, lightweight rubber-cushioned engine mount that sandwiches between the torque-stay casting and the engine torque-stay bracket. We shortened a stock V-6 right-side torque stay rod to attach to the engine and passenger-side MR2 shock tower. Norwood Performance modified a stock MR2 passenger-side axle-bearing carrier to attach nicely to the Lexus V-6 and retain the right-side axle.

EXHAUST SYSTEM

Your MR6 will need a muffler and custom exhaust system. To be legal, you also need a catalytic converter with (third) heated O_2 sensor. The stock Lexus ES300 V-6 shorty headers *will* fit on the engine in an MR2, but we decided to modify the headers for optimal fit and access in the MR2 even though this might not, strictly speaking, meet the letter of the law in California.

Alamo amputated the two stock three-into-one stainless steel tube headers just below the collector and built twin stainless downpipes. The twin downpipes can be merged into a collector and from there into a single catalytic converter (like the stock Lexus/Camry), or you might consider constructing a dual-cat system with twin free-flow mufflers. We wanted a zero-restriction exhaust flow for high-output turbo- or supercharging, so, in the end, Alamo Autosports worked with a large Borla stainless muffler and Borla stainless mandrel-bent tubing to construct a large single-discharge exhaust system suitable for single-turbo turbocharging.

COOLING SYSTEM

Coolant suction and discharge hoses are at opposite ends of the 1MZ engine from those of the stock 3S-GTE four. We used a pipe cutter and an assortment of rubber hoses and metal tubing to route coolant to and from the V-6 powerplant and the stock MR2's 1.25-inch coolant pipes running up the center of the car to the radiator. Basically, we routed hot coolant exiting the 1MZ engine from a discharge located between the passenger-side vee-banks forward, left along the front firewall and vee-bank, and then downward next to a hose *returning* coolant from the radiator to a fitting between the vee-banks just above the transaxle. Make sure to form a lip on the end(s) of any metal coolant pipe in order to keep hose clamps from slipping and allowing rubber hoses to blow off when the engine gets hot and coolant pressure rises.

OTHER MISCELLANEOUS SWAP ISSUES

The two shifter cables must be reattached to the transmission, which is fairly straightforward. Use cable ties or other means of keeping them away from the front-bank exhaust manifold and downpipe.

Reroute the stock MR2 throttle cable to the 1MZ throttle body and secure with cable ties or other fasteners. If you care about cruise control, you'll need to run a longer throttle cable from the MR2 cruise control to the 1MZ throttle body or relocate the cruise actuator and lengthen the wiring. The MR2 has a dedicated cruise computer that can be made to work a V-6 even without the MR2 main computer on the car; you'll need to activate the cruise electronic controller with a voltage that mimics the voltage expected from the main ECU (more about this shortly).

The MR2 brake-booster hose (passenger-side near top shock tower) must be plumbed to the V-6 intake manifold.

The V-6 alternator must be wired into the stock MR2 electrical system.

If you want to be cool, you'll need to interface the 1MZ's R-134 air conditioning compressor to the stock MR2 R-12 air conditioning system. We connected stock 1MZ air conditioning hoses to the 1MZ compressor (the aluminum ends bolt in place and utilize O-ring sealing), then cut the aluminum pipes close to the fittings. Next we amputated the stock MR2 aluminum air conditioning fittings midpipe, marked the four ends of the aluminum pipes to get the orientation correct (making *certain* to get the direction of refrigerant flow correct on the suction and discharge lines), then pulled the hoses and compressor fitting off of the car and welded them together.

You'll need to submerge the rubber hoses under water or use other countermeasures to keep the extreme heat of welding from damaging the hoses where they are crimped onto the aluminum pipes/fittings. They absolutely *will* fail if the pipes are simply welded. An alternative is to weld the pipes together and then take the assemblies to an air conditioning specialist to get new rubber hoses fitted. We plumbed the compressor to the MR2 air conditioning system using the modified hoses, flushed the system, added R-134 lubricant, and charged the car with R-134 refrigerant. The system has worked fine ever since under the most extreme thermal loading.

The stock MR2 fuel pump is good to supply at least 335 wheel horsepower against 20-psi boost, and it is certainly more than adequate on a V-6 to similar power levels where less fuel pressure is required. We fabricated custom hoses to plumb fuel supply and return lines from the MR2's fuel tank to the 1MZ fuel rail and pressure regulator. We also welded a -4 connector to one of the 1MZ fuel-rail banjo fittings for easy attachment of a fuel pressure gauge to keep an eye on rail pressure at initial startup and during dyno tuning.

When using the stock 1MZ-V-6 plumb the mass airflow meter into the intake tract of the V-6 engine in the MR6 between the throttle body and the air cleaner. This should be done in a way that minimizes any spurious turbulence that might invalidate the MAF reading.

You'll need a good air cleaner; an aftermarket unit like our K&N filter will certainly do the job. If at all possible, devise a way to get *100 percent cold air* from a source *completely* isolated from the engine compartment. Remember: *Cold air makes power.*

We inspected and serviced various systems that has not been used on the basket-case car in a while, turning the brake rotors and installing new pads all around. On our car, the speedo cable and right-side emergency brake cable proved to be dead and required replacement.

ENGINE MANAGEMENT AND ELECTRICAL

The biggest challenge of this engine swap is making the V-6 engine management system and the MR2's electrical system and auxiliary computers work correctly together. The MR2 V-6 engine swap requires a computer and engine harness change, and this is definitely *not* plug and play. Unfortunately, the 1MZ engine management computer is designed to interface to a number of systems that *don't exist* on the MR2, and vice versa. In general, you'll need to know how to read a wiring diagram, how to perform proquality crimping and wiring, and probably how to use a volt meter. You'll need the 1MZ engine wiring diagrams and the MR2 engine and chassis wiring diagrams and documentation. These are available by calling Toyota Service Publications at (800) 622-2033.

There are a number of functions that will not translate seamlessly from the MR2 to the 1MZ computer, and this will be worse if you are trying to get a 1MZ powerplant and computer that had been mated to an auto transmission in a previous life to function properly in the five-speed MR2. A manual-trans V-6 harness is the way to go, but most U.S. 1MZ's lived their first life in a Camry, Lexus ES300, or Sienna van, all of which were automatic transmission only (more in a moment).

These are several special-purpose electronic controllers on the MR2 independent from the main engine management computer. These include a fan-control computer, an air conditioning computer-amplifier, a cruise control computer, and so forth. With minor rewiring, these will work even if you do a brain transplant on the main onboard computer.

Note: If you don't need to be street legal, a viable alternative to the stock 1MZ engine management is an aftermarket programmable engine management system. This would not be legal in terms of passing a visual emissions inspection, and even if it somehow did pass, many aftermarket engine management systems do not have the sophistication to run certain 1MZ emissions equipment probably required to make the carbon monoxide (CO) and hydrocarbon (HC) numbers needed to pass an emissions sniff test if your state and city requires one.

That said, an aftermarket EFI system is, by definition, flexible and configurable, and will easily accommodate performance modifications far beyond the scope of the stock computer. We experimented with the stock 1MZ engine management system on our MR6 and then—because we knew we'd eventually be running very high output turbocharging and supercharging on the MR6—converted to a Motec M48 aftermarket system that is extremely sophisticated and powerful and can be equipped with an expander module for direct-fire multi-coil ignition control you need for a distributor-less engine like the 1MZ. The M48 has the ability to manage virtually any external actuators and sensors, including emissions-control devices.

Assuming you've decided to work with the stock 1MZ computer:

• In order for the 1MZ computer to come alive at all, you'll need to supply switched +12V power from the MR2 ignition circuit at the main engine relay block to pin 14 on the ECU (computer) A-connector. You'll need to run constant +12V (battery) power to pin 14 on the D-connector of the 1MZ ECU. The 1MZ engine harness has various grounds that connect to the engine (pins A33, 34, 16, and 28), and—of course!—the engine must be very well grounded to the chassis with large gauge grounding straps.

You'll need to interface the 1MZ computer to the MR2 fuel pump relay—and keep in mind that the Turbo MR2 has both high-flow and low-flow fuel pump modes (via resister wiring) that are controlled by the 3S-GTE engine management computer. Unless you want to be clever, just interface the fuel pump wire from the 1MZ computer to drive the pump in high-flow mode only.

• You will definitely want the 1MZ's check engine light working so you'll have some idea what the computer thinks is going on in the external world of the V-6 engine and MR2 vehicle and electrical system. You may decide it is simpler wiring an add-on auxiliary check light for the 1MZ engine management computer than trying to make the 1MZ check signal work with the MR2's combo meter.

• You will discover there are certain electrical functions unrelated—strictly speaking—to engine management that are bundled into the MR2's engine harness—such as alternator wiring and engine gauge sensor wiring for coolant temperature, oil pressure, and so on.

• Note: The following pin assignments apply to the 1994 1MZ; later models may differ.

• Since the 3S-GTE engine and engine harness will not be present on a V-6 MR2 (you'll probably want to unplug it from the engine relay block and connectors in the trunk), you'll have to install and route wires to the 1MZ alternator, air conditioning compressor, and several engine gauge sending units. One of the alternator wires is the thick-gauge wire that delivers charging voltage from the alternator to the battery, while the other is a narrow-gauge wire the alternator needs to dynamically sense the level of charging that's required from moment to moment.

• You'll need to interface the vestigial MR2 starter wiring that remains after you've removed the 3S-GTE engine to the starter (either an MR2 or V-6 starter will bolt to the MR2 bell housing and mesh seamlessly to the V-6 ring gear). The 1MZ computer will want to know when the engine is cranking, so you'll need to run a wire from the starter solenoid to pin 13 on the A connector.

• The 1MZ computer also wants to know when you're braking, so run a wire from a rear stop light to pin 24 on the D connector of the computer.

Noncritical electrical issues you'll want to resolve include:

• The MR2's air conditioning system is controlled by a dedicated computer that requires a tach-signal input for evidence that the engine is running before it will permit the compressor to operate. You'll need to wire the black TACH0 signal originating at the V-6 igniter to MR2 connector EA3 pin 5, which conveys this signal forward to the tach, combo meter, and air conditioning amplifier. An alternate is to replace the 3S-GTE air conditioning amplifier with a 5S-FE amplifier, which does not require the tach signal.

• If you care about having cruise control working, you should interface the 1MZ computer's Pin 12 on the D connector to the MR2's cruise computer.

• The four-cylinder tach will *not* read correctly if driven by a V-6 igniter. There are good solutions but none are both easy *and* cheap. Assuming you're running the factory engine management system, your choices are:

1. Install an aftermarket tach.
2. Adapt a 1MZ V-6 tach to fit the MR2 dash module.
3. Have Palo Alto Speedometer or other specialists modify the four-cylinder tach to read correctly.
4. Find an electronic tach adapter or mechanically driven tach generator that will provide four—not six!—driver pulses to the MR2 tach per four-cycle.
5. Install a Motec M4-8 engine management computer to run the V-6 engine (which we eventually did when we decided to run compound forced induction on the car), and pulse modulate an auxiliary output on the Motec to provide four cylinder-type tach-driver data to the MR2 tach.
6. Use a Sharpie marker to draw new numbers on your MR2 tach (just kidding, though you might, indeed, be able to fabricate a replacement number panel for the MR2 tach).

The following are undefined wiring issues you'll probably want to resolve:

• On the MR2 side, the two-speed electric power steering pump will be expecting to be controlled by the 3S-GTE computer. Actually, power steering is not very important on an MR2, since the 42–58 percent front-rear weight bias means you absolutely don't need power steering. Your choice will be no P/S, jumpering a voltage to the electric P/S controller for constant low- or high-effort power steering, or (possibly) leaving the 3S-GTE computer onboard and connected to control the P/S (but probably *very* unhappy).

• On the 1MZ engine side of things, an auto-trans 1MZ ECU will be looking for (torque) converter lockup clutch slip sensor input, which has no meaning on the manual-trans MR2. Similarly, the main ECU will want reassurance that the auto-trans controller is happy about transmission fluid temperature. Best to use a manual-trans 1MZ. Alternative: contact Tadashi Nagata at TechnoSquare in Torrance, California, (310–787–0847) about a circuitry required to keep an auto trans 1MZ computer from freaking into limp-home mode when it doesn't see expected data from the 1MZ's auxiliary automatic transmission controller.

CHAPTER 26
OVERBOOSTED
VW GOLF 1.8T

If there were a 12-step program for people addicted to *power*, we could teach it—stand tall as a Bad Influence. Can't fool us. Once you've had that high—once you've felt the thrill of hard acceleration squishing your face backward, sloshing the blood out of your frontal lobes and straight back into the darkest crevasses of the reptilian brain—there's no going back to your mom's old Buick. You'll do *anything* for one more horsepower.

Alamo Autosports, developers of the 355-horse Golf 1.8T in this chapter, knows. Two Kraut KKK turbos gave their lives to make it so far. But like they say, you have to break some eggs to make an omelet. Only in the case of the Golf 1.8T R&D project, experience proves sometimes you have to trash some turbos to create a killer car capable of truly ugly deeds at the drags wars—all hotted up on 22-psi boost and cranking out 305 front-wheel horsepower on 110-octane unleaded Phillips race gas.

The engine's never been out of this car—never even been *open*—but numbers off the Alamo Autosports Dynojet chassis dyno indicate a measured 305 horses worth of power on 1.4-bar boost (which arrives by 3,800 rpm) at the front wheels—which translates to more than 350 at the flywheel, assuming 15 percent drivetrain and tire rolling losses! At the time of this writing, we did not know of a stronger streetable MK4 Golf 1.8T than this, and Alamo Autosports definitely explored strange new worlds on this project. So: We're talking power in rice-burner style, right?—a jillion horsepower in one huge whoop of turbo boost located in a single spike of power about 23 rpm below redline, right? You think? Read on.

This Golf 1.8T rolled off the lot in the late spring of 2000, a silver GLS with 20-valve turbo engine cranking out 150 advertised horsepower at the flywheel. A stock 1.8T is no slouch, deci-

sively outrunning the 174-horse 2.8-liter Golf GLX VR6 from 30–50 by *1.6 seconds*. The project 1.8T Golf actually made 142.4 before front-wheel horsepower on the Alamo Dynojet, which translates into—well, let's just say *more than 150* at the flywheel. Genuine flywheel horsepower disappears on the way to the wheels due in part to frictional losses in the drivetrain and rolling-resistance losses at the tires. In our experience, you can *count* on losing 13–21 percent power at the wheels, depending on the vehicle. The number 142.4 is awesome from a car with 150 *advertised*. But to a true power addict, nothing's ever fast enough until it truly *scares* you. And, really, not even then.

CHEAP TRICKS

The 1.8T project began in the usual way—with just a little PROM change, a software recalibration of the stock VW onboard computer via APR's K03 software (FedEx your onboard computer to APR overnight and they'll desolder and replace your stock PROM module with a proprietary daughter-board equipped with its own onboard microprocessor and multiple sockets for quick R&R of up to four PROM calibrations).

Today, stock naturally aspirated (NA) engines hardly ever benefit much from recalibrated upgrade PROMs, but turbo engines are a different story because the horsepower output of a turbo engine system is normally not limited by the breathing capacity of the engine system as it is on an NA engine, but by the ability of various highly stressed parts to survive under the harsh conditions of increasing specific power output at high boost—and automakers artificially de-rate power output by controlling the boost to very conservative levels that protect the life of the engine in the hands of careless owners to prevent warranty problems. Got 110-octane fuel for your turbo engine? Great: Turn up that boost. Willing to risk maybe breaking the transaxle or CV joints? Great: Pump up the volume a bit more. APR's software unleashed a massive 1.1 bar (16-psi worth) of midrange turbo boost in the Golf (with a consequent massive increase in torque) and even unleashed a few extra clicks of turbo boost at *peak power* engine speeds (full 1.1 boost all the way through peak-power rpm is unavailable because it is beyond the capability of the stock fuel system).

APR's 1.8T PROM recalibration—plus a few tricks to improve engine breathing on both the ingestion and exhalation sides of the turbo powerplant—allowed the project Golf to approach—then exceed—200 front-wheel horsepower on the Alamo dyno. Tricks included an Alamo custom 3-inch mandrel-bent

A 2000 Golf 20-valve 1.8T, good for 150 stock horsepower.

cat-back exhaust, a Thermal R&D muffler, a K&N high-flow air filter element, and so forth.

Alamo's Stage 1 VW 1.8T modifications alter maximum boost at peak torque and peak power, but still leave the factory VW computer in charge of boost control via the stock electronic wastegate controller. Limiting maximum boost on a street turbo car is virtually *always* required in order to prevent high effective compression ratios from rising to the point that the engine is destroyed by explosive detonation when the air/fuel mixture explodes rather than burning smoothly.

Wastegates limit boost by opening a valve to divert varying amounts of exhaust gases to bypass the turbocharger's turbine section. This limits the turbine's ability to drive the turbocharger's centrifugal compressor fan that force-feeds massive additional quantities of air to the engine by spooling up to speeds that can easily exceed 100,000 rpm and approach 200,000 rpm in the smallest turbos. As Alamo discovered on this project, to achieve very high levels of boost at maximum power, if you

A Golf 1.8T with extensive external modifications.

stick with the stock computer, the engine must be run with the factory mass airflow (MAF) unit disconnected in order to prevent the computer from freaking at the excessive airflow—and countering by decreasing the electronic throttle. The check-engine light will be on (if you ask really nicely, you may discover firms like APR can supply special Golf PROMs that will prevent a MAF disconnect from resulting in the Golf computer from turning on the check engine light!).

Hot rodding R&D on the 1.8T eventually killed the stock VW KKK-03 turbocharger, which simultaneously necessitated replacement and provided a pretty good excuse for getting a bigger one—especially since, at that time, it was tough to find KKK-03 rebuild parts and a new 2003 turbo cost something like $1,400! The same forces that consumed the original turbo also damaged the stock blow-off valve (a spring-loaded, pneumatically controlled dump-valve designed to safely bleed off the spikes of boost that can develop, for example, when the turbo is wailing along with a good head of steam at high boost and you suddenly close the throttle, say, to change gears).

Keep in mind, a Golf 1.8T's throttle is no longer connected to the accelerator by a cable; this one's a fly-by-wire design in which the accelerator pedal simply actuates a potentiometer that delivers a variable-resistance to the onboard computer, depending on accelerator position. A fast stepper motor under computer control actually moves the throttle. Hot rodders should beware of secret VW countermeasures against warranty-destroying levels of boost going on behind the scenes. You may only *think* you're at wide-open throttle when the pedal is down to the metal.

FULL-MONTY TRICKS

With the necessity of acquiring another turbocharger, Billy Tylas-

ka and Alamo Autosports went for broke (so to speak) in external air-flow mods:

1. Turbonetics T3-Super 60 turbocharger, good for over more than an estimated 356 crankshaft horsepower worth of air flowairflow at 22 psi boost-psi boost.

2. Greddy Type S blow-off valve, good for lots of blowing-off.

3. Alamo 2.5/3-inch turbine downpipe, good for moving a lot of exhaust at the drag strip if your cat somehow falls off on the way over.

4. Turbonetics high-flow cast-iron exhaust headers with T-3 bolt pattern at the turbine flange.

5. Tial competition-type wastegate, good—when fully open—for efficiently diverting massive quantities of exhaust around the turbine to control maximum boost without creeping as power goes wild.

6. Forced-performance high-flow, high-efficiency side-mount air-air intercooler, good for high CFMCFM air flowairflow at high boost with low pressure drop through the intercooler core.

7. Alamo-ported large-plenum modified-stock intake manifold, good for better breathing at high-rpm, high boost.

8. Alamo hand-ported throttle body assembly (everything helps).

9. Alamo Autosports modified large-plenum intake manifold with provision for four additional 370 370-cc/min secondary fuel injectors that are controlled by a Greddy Rebic IV additional injector controller unit—good to retain excellent stock drivability and emissions characteristics throughout startup, idle, low-speed and cruise driving conditions (off-boost), while simultaneously making provision for sufficient fuel to match the higher air-flow capacities of the new hot rod engine and cool combustion.

After the addition of lots of external bolt-on parts and some larger turbochargers, the car cranked out 335 rear-wheel horsepower on the Alamo Autosports Dynojet. With nitrous, the car later on managed more than 400 at the wheels.

such thing as too much power, right? Achieving a good launch from a standing start is both difficult and essential to achieving reasonable 0–60 and quarter-mile times. Any front-wheel-drive car with a ton of power is problematic to launch well because there is always both a torque and weight shift from the front to the rear wheels on an accelerating car. Drag racing FWD street-type vehicles is challenging (but fun), and a good launch depends on achieving a subtle balance between power and a thing called the *coefficient of friction*.

What does this all mean? For one thing, a good limited slip differential is a necessity on the Golf 1.8T. This one's got a Peloquin billet LSD. *Check.* Good, soft, drag radial tires are essential to a decent launch. *Check.* The project 1.8T has 225/50/16 Nitto drag radials up front on custom alloy wheels for daily driving, 205/50/15 BFG drag radials on stock steelies for drag racing. Alamo knew they'd need slicks for serious racing. A clutch that will slip a bit without fading away completely in the heat when you ride it a bit is vital. Alamo installed a Clutchmasters Stage 1 Clutch for a G60, which bolts right in the 1.8T tranny so long as you use a G60 flywheel, which also bolts right up to the 1.8T crank.

10. A one-way in-line mechanical air-bleed device (technically a precisely controlled air leak) installed upstream of the stock manifold absolute pressure (MAP) sensor manifold pressure reference line—good to effectively lie to the MAP sensor, understating manifold pressure by 4 psi to the onboard computer, thereby keeping the computer blissfully unaware of the wild party going on in the intake manifold at high boost and preventing the computer from initiating rash, anti-horsepower countermeasures (such as closing the throttle).

11. Greddy Profec B boost controller, good to relieve the VW onboard computer of all boost-control responsibilities, enabling us to adjust boost from the cockpit anywhere from 0.65 bar (9.5 psi) on up to blowup (located somewhere above the current 22 22-psi maximum dyno-tested and calibrated by Alamo with Phillips 110-octane unleaded).

GETTING FAST AND FURIOUS

How do you launch this thing? Typical high-output rice-burner turbo, right, all the power's up near redline, drive it like you're mad at it for any real performance, right? Well, no. Torque exceeds the stock peak 150 ft-lbs wheel torque across a broad swath of rpm from 3,000 through the stock 6,500 redline and beyond to 7,200 rpm. Torque exceeds 250 ft-lbs *at the wheels* from 4,000–6,000 rpm, reaching a peak nearly twice that of the stock Golf near 4,800 rpm. Torque is *not* a problem.

One could make a convincing argument that a 305-front-wheel-horsepower Golf definitely qualifies as overpowered. Yes, trying to run 305 front-wheel horsepower through two stock street radial tires can be a bit like sending an Olympic sprinter out to race on a waxed floor in slipper socks. Perhaps traction-challenged is a better description than overpowered, since, by definition, there's no

DRIVING THE ALAMO GOLF

I jack with the seat until I like the driving position, floor the clutch, turn the key. The five-valve-per-cylinder 1.8 comes to life and settles quickly into a crisp idle. The engine's still being controlled by stock Motronic engine management algorithms, even if the MAP sensor does fib a bit to the computer under boost conditions. True, the PROM-based air/fuel and timing maps directing the Motronic system how to fuel *this* particular vehicle are bootlegged across the rpm range at positive manifold pressure for airflow numbers inconceivable on a stock Golf 1.8T.

Blip the throttle; transient response is very nice. This engine likes the fuel, drinks down readily with the factory calibration. Notch the shifter into gear. The Clutchmasters Stage 1 G60 feels very light considering it's good for putting 400 horsepower to the road. I stab the throttle again for feel, and the engine responds smartly—just a tiny bit of that instant-on, instant-off feel you get when you twist the grip on a superbike. That's the Golf's 13-pound Alamo Autosports lightened and balanced flywheel, the lightweight Clutchmasters pressure plate and carbon disc setup, and the Unorthodox Racing lightweight underdrive pulleys.

I ease out the clutch and move rapidly through the darkened outskirts of Arlington, no traffic here, wider open now, OK to hammer it a bit, I figure. Did I mention this car has Alamo upgraded suspension (Koni coil-overs all around), tires, and wheels? OK, drop the clutch and push the Golf's five-speed slowly up through the gears. This is definitely not grandma's Buick; subtle bumps are super-conducted directly through the seats in a way that lends a new meaning to the words road feel. But road feel through the steering wheel is also excellent, and the 17-inch tires feel like they're nailed to the asphalt with railroad spikes.

I urge the turbo Golf a little harder through the turns, seeing what it can do, feeling the tires grip the road, searching for any handling problems. No way, this thing is a control freak, German suspension at its best with good aftermarket upgrades, alloy block with only four cylinders worth of weight set down nice and low and tipped back, car's balanced beautifully, suspension tuned nicely, set on tall wide wheels with soft meaty tires.

Fifth gear, 20 miles per hour, plant the throttle firmly at wide open. Boost builds quickly, but I'm not sure how many-psi boost to claim for 3,000 rpm because the revs climb so fast. Max boost is easy to know, given the Greddy 60 millimeter digital 2-bar boost gauge with peak hold memory and matching Greddy air/fuel gauge. The tach *streaks* through 4,000, hell-bent for redline, the boost gauge maxed at 2.5 atmospheres absolute, and at this point the 1.8T's torque curve simply goes *nuts*, and the car is lunging forward like a set of planetary gears has locked up and someone is dumping nitrous oxide straight down the throttle body. The tach plunges through redline and buries itself deep in forbidden territory like a TV evangelist with a 16-year-old virgin at Motel 6.

This car has already killed two turbos, a clutch, a stock differential, and a compressor-bypass valve on its way to glory. And it may kill more, but this serial killer may also kill a few unsuspecting Camaros or Mustangs at the drag wars or on the streets of Arlington, Texas.

What is there to say? Viva turbo power. And viva electronic engine management!

MANAGING YOUR ENGINE MANAGEMENT SYSTEM: HOW FAR CAN YOU GO

Go far enough in a quest for power, and any turbo-EFI engine hot rodding project is doomed to run up against several successive firewalls of resource limitations.

• Soft configurable parameters located within tables of data in the onboard computer's programmable read-only memory (PROM, or electrically erasable PROM) direct the logic of the onboard computer within a permissible operating range of possible boost, fuel delivery, and ignition timing options. These controllable parameters can be recalibrated if you know how and have the tools. The tools of PROM-hacking? Aftermarket replacement PROMs, EEPROM downloads via laptop computer or emulator, or complete replacement onboard computer boxes.

• There are virtually always hard (nonconfigurable) limits to what the stock computer's logic algorithms will permit or execute in the way of maximum boost pressure, ignition timing, electronic injection pulse width (injector open time) and so on. Tuners virtually always accept these harder logic-based limits as a given; modifying onboard computers' embedded real-time logic requires access to highly proprietary source code (or extremely sophisticated and prohibitively expensive reverse engineering that couldn't possibly make sense, given the alternative of installing sophisticated-albeit expensive-aftermarket replacement onboard computers, available as programmable open systems that are fully capable of delivering all functionality of the original engine management system as well as virtually any additional capabilities you could want or need). In lieu of aftermarket computers, hard lim-

its attributable to the logic of factory onboard computer algorithms can almost always be effectively defeated or modified by changing the actuators that actually manage the engine (that is, installing bigger fuel injectors), or by installing auxiliary interceptor microcomputers that intelligently modify the hard analog voltages or instructions to the actuators (that is, extend injection pulse width or introduce a delay in the ignition timing trigger voltage to a coil-driver or direct-fire coil). OBD-II computers tend to complicate the above because they may have sophisticated closed-loop logic coupled with secondary sensors that are capable, in many cases, of more directly measuring the actual results of engine management commands to actuators on the engine—in which case the onboard computer can execute secondary countermeasures if the initial command was (apparently) not effective (that is, did boost pressure—or the sensor voltage indicating boost pressure—appear to actually decline? If not, compare manifold pressure to mass airflow, and if boost and airflow appear to correlate correctly, assume the MAP sensor is correct and that boost is out of control. In this case, execute a fuel cut). In some cases, modern computers are configured with knowledge concerning the maximum acceptable rate of acceleration for a given engine in a certain car in a certain gear. The tools for defeating algorithms? Interceptors, auxiliary computers, fuel pressure changes, actuator changes—with or without computer recalibrations. Aftermarket engine management systems.

• Sooner or later turbo-EFI hot rodding projects run up against physical limitations in the engine's fuel- and air-supply systems. For example, at a given level of electrical voltage and pressure regulation, high-pressure electric fuel pumps have a ceiling to the volume of fuel they can deliver per unit of time. Similarly, spring-based fuel pressure regulators referenced to manifold pressure via vacuum hose have physical limits to the maximum or minimum fuel rail pressure they can accurately maintain, and the rate-of-gain with respect to manifold pressure is normally fixed and linear, like it or not. At a given pressure, fuel rails also have a maximum fuel flow they can accurately deliver evenly to all injectors before a pressure drop occurs at certain weak-link injectors. While modern turbochargers have a relatively broad range of pumping capacity at reasonably high levels of thermal efficiency, eventually you run into various absolute physical limitations of the turbocharger itself in the form of (1) Reduced thermal efficiencies (airflow above which the centrifugal compressor is having to thrash the air around excessively to meet the demands for boost pressure-overheating the inlet air-compared to a larger compressor that could deliver cooler, denser air at the required level of boost), followed by (2) Turbo overspeed, in which the turbo is having to spin so fast in order to pump enough air at high levels of boost and rpm that it is in danger of being damaged due to compressor surging or lubrication and thermal failure problems, followed by (3) Compressor choke, in which the compressor cannot pump any more air no matter how fast it turns. The tools of liberating an engine from physical resource limitations? Boosted fuel pump voltage, bigger fuel pumps, additional fuel pumps, bigger fuel lines and injector rails, larger injectors, higher fuel pressure, larger inducer and exducer size, or reworked compressor-wheel aerodynamics. Bigger turbochargers. More turbochargers. Improved engine breathing. Big-mouth throttle bodies.

CHAPTER 27
FRANK-M-STEIN: TURBO M3 ROADSTER

Combining variable valve timing, several hundred cubic centimeter's worth of added bore and stroke with increased compression, intake and exhaust breathing tricks, appropriate engine management, super-duty parts from top to bottom, and a magnificently tuned suspension, BMW succeeded with the M3 in upgrading the 3-series coupe into one of the most exciting cars on earth, a no-compromise machine that could lick the likes of 300ZXs and Toyota Supra sports cars in all areas of performance and still carry four people to the 7-Eleven in comfort.

By contrast, the 325ic is a *really nice car*. Good for laying down roughly 150 horsepower where the rubber meets the road on the Dynojet chassis dyno—about 190 peak flywheel horsepower as advertised by BMW. *Great* for top-down days in the spring or fall when Dallas isn't hideously hot or uncomfortably cold—at the cost of a little additional weight over the 325i coupe that's needed to stiffen a flat chassis lacking the reinforcing triangulation of a steel roof. A 2.5-liter in-line BMW six makes a maximum of just over 150 ft-lbs torque at 3,800 rpm (at the wheels). It's got the sex appeal of a genuine wind-in-your-face drop-top. It's a *really nice car*. It's just not an M3.

THE M3 SOLUTION

We began Project Frank-M-Stein in 1999 by testing the stock 325ic on Alamo Autosports' Dynojet chassis dyno. The 325ic proved to have 151.4 rear-wheel horses. Drivetrain and tire rolling resistance added up to a loss of roughly 21 percent from the advertised 189 factory flywheel horsepower.

Initially, Alamo techs installed a performance-calibration PROM and several enhancements designed to increase engine volumetric efficiency: The combination of B&B exhaust, RC Engineering big-bore throttle body, and Alamo custom high-flow cold-air intake was good for about 20 peak horsepower at the wheels and some torque increase from 4,000 to 6,500 rpm. Peak horsepower still occurred near 6,000 rpm, but the modifications produced 20 added horses available from 5,000–6,500 rpm. Essentially such modifications open up the top half of the car's performance envelope.

But horsepower is addictive, and the car needed more midrange torque. In the second stage of Project Frank-M-Stein, Alamo lowered compression with a thick steel head gasket and installed an Active Autowerke turbo kit on the stock. The results

"M3 Cab" in action.

were nothing short of dramatic: A power boost of roughly 100 rear-wheel horsepower, a 66 percent increase.

In the third major stage of the project, Alamo acquired everything required to turn the car into a 1998 BMW M3 and installed all applicable items on the 325ic 3.2-liter motor: transmission, driveshaft, limited slip differential, front and rear suspension components, and brake calipers. These items differed from equipment installed on the stock 325ic in the following way: Total displacement was up from 2,494 cubic centimeters to 3.2 liters, a gain of almost 30 percent. Thirty percent added displacement translated to massively increased low-end torque and improved power. It also translated to more high-energy exhaust gases to drive a turbocharger, which translated to increased responsiveness and reduced turbo lag.

While the original 325i engine made serious boost in the 3,500–3,700 range, the bigger M3 engine produced significant turbo spooling at 2,500 rpm. Like all other M3 drivetrain components, the transmission is strengthened to handle the increased torque and power of the M motor. The M3 driveshaft is beefier. The M3 limited slip differential is stronger and critical to getting to power to the ground without converting the tires directly from rubber to smoke. Tuned suspension components from the M3 include the lower front A-arms, which are built stiffer to handle the possibility of an M3 driver initiating an exhaustive search for the car's limits. M3 brakes improve stopping distances, and fade is reduced dramatically. Alamo substituted a set of drilled aftermarket brake rotors to further improve stopping performance.

Alamo also installed a super-duty clutch and lightweight flywheel. The accessory-drive system was converted to all under-drive pulleys in order to reduce accessory drive frictional losses. Alamo designed a custom 3-inch turbo downpipe to remove exhaust gases from the turbine section of the turbocharger to reduce backpressure and further improve power. When the car goes to the drag races, the stock cat is removed and the car runs on race gasoline. Around town, the car runs 11 to 12 pounds of turbo boost; on race gas the Mitsubishi turbo can produce even higher manifold pressure.

MODIFYING MOTRONICS

Hot rodders with Porsches, BMWs, Volvos, and other vehicles with Bosch Motronic EFI have several options when it comes to recalibrating fuel delivery and spark timing for turbo and blower conversions. The simplest is to use a rising-rate-of-gain fuel pressure regulator to increase fuel per injector squirt, without modifying the engine management system whatsoever. This strategy can be quite effective up to perhaps 7 psi—at which point you're probably running as much fuel pressure as is reasonable for any fuel delivery system. The variable-rate-of-gain regulator strategy effective limit is about 7 psi on the BMW, since BMW fuel pumps are normally equipped with bypass valves to prevent overpressurization at around 80 psi. Running really high pressures can also cause some injectors to malfunction—for example, refusing to close properly against the pressure, although many Bosch injectors have been run at 90 to 100 or more psi with no troubles. At 4- to 7-psi boost, the stock knock sensor can usually retard timing quickly enough at the onset of trace detonation to protect the engine, negating the requirement to programmatically provide boost timing retard.

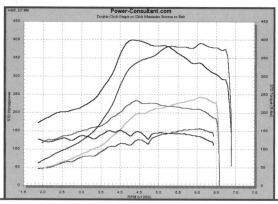

The bottom line: Stock 325-ci, turbo 325-ci, and turbo M3 dyno power. The combination of big-cube M3 torque and a big turbo quadrupled low-end and nearly tripled peak torque, pumping up peak horsepower from 160 to 385-plus at the wheels.

Engine compartment stuffed full of "M-Power"—and turbo boost!

At higher levels of boost, many tuners have used aftermarket engine management systems—Accel/DFI, Haltech, Electromotive, Motec, and others—to replace the stock engine management computer with one that is entirely programmable via laptop computer, building custom air/fuel and timing tables for any size injectors, no knock sensor required.

However, the increasing sophistication of late-model injection systems, and the integration of cruise control, transmission controls, or other functionality into the main onboard computer has made the aftermarket solution increasingly problematic for people who want to retain all stock functionality. Yes, there are aftermarket systems capable of managing many auxiliary tasks only indirectly related to engine fuel and spark, but they are expensive, and the programming effort may be expensive and difficult. The aftermarket systems may also be illegal for street use, but in any case, tuners and enthusiasts have increasingly embraced a concept that could be summarized as, "Why throw the baby out with the bathwater?" Why junk the whole engine management system if all you need is some minor reprogramming of timing and injection pulse width in certain wide-open-throttle situations?

Tuners with Motronic engine management controls have several options when it comes to recalibration. Autothority and others have offered custom chips or PROM recalibrations for certain specific classes of engine modifications on frequently hot rodded engines, such as the Porsche 911 and some BMWs. The problem is, you're here, with your car, and the custom chip programmer is not here, so you may not get an optimal calibration for your specific engine—at least not without several iterations of custom chips followed by diagnostic testing and evaluation.

Since the stock M3 has totally different wiring from the 325i system, Alamo Autosports retained the stock 325ic computer on Project Frank-M-Stein to control the 3.2-liter M3 engine, but worked to recalibrate the ECU using downloaded files from Active Autowerke and local PROM-blowing equipment. The fuel injectors were the bottleneck, with the 440-cc/min injectors pushing 100 percent duty cycle. Alamo's own Dynojet chassis was essential to achieving the optimal custom computer calibration. The trick was, the EMS still thought it was running a 325 motor.

TURBO M3

With the M3 donor equipment in place, Alamo reinstalled the turbo system that had recently been force-feeding just 2.5 liters of BMW in-line-six. The turbo was a Mitsubishi TD06 with a 12-centimeter exhaust housing and was mated to a 3-inch custom downpipe with external wastegate. A complete package of external equipment increases or handles the added power: Alamo Autosports custom cold-air intake with big mass airflow (MAF) meter and RC Engineering high-flow throttle body eliminated the intake system as an airflow bottleneck. An Active Autowerke lightweight aluminum flywheel and high-performance racing clutch set helped to improve engine responsiveness on sudden application of wide-open throttle and as an active countermeasure against clutch slip in a situation where you've got double the horsepower of the stock M3. Alamo added a stainless steel exhaust system to improve turbo spool-up response and improve peak power, Eiback Pro-kit springs, front and rear anti-roll bars, and front stress bar to increase chassis stiffness.

All forced-induction systems with at least 4-psi boost should have intercooling to return charge air temperature to within (ideally) 20 degrees of stock in order to control combustion temperatures and knock. This is particularly important on high-compression engines like the M3 with its 10.5:1 squeeze, though Alamo had installed a thicker steel head gasket to lower compression to 8.8:1 on the M3 powerplant. The Alamo M3 system uses a custom intercooler mounted in front of the radiator and air conditioning condenser, requiring removal of the front pusher radiator fan—and for some climates, possibly an upgrade to the rear sucker fan—although, so far, the stock rear fan has been entirely sufficient in 95-plus-degree summer weather in Dallas. The intercooler is designed to handle the thermal loading of boost pressures as high as 15–20 psi, 1.0–1.36 bar.

PERFORMANCE RESULTS

The M3 Turbo-Cab is so rock stable at three-digit speeds that swashbuckling through twisting turns feels easy, and mediocre drivers will feel like great drivers. With Project Frank-M-Stein, when you want to go fast, you gently ask the car and it turns into a monster violently and immediately. Boost begins in the mid 2s and reaches the full 10 psi available on 93-octane Texas pump gas by 4,000 rpm. As the engine climbs to redline, you've got almost 400 rear wheel horses that make going very fast look very easy. That's nearly 500 crankshaft horsepower. In race trim, the powerplant produces as much as 430 rear-wheel horsepower at boost level of 1.0-plus bar. Project Frank-M-Stein was a testament to the truism that you can't be too rich, and you can't be too thin, and you can't have too much horsepower.

EMS/EFI SUPPLIERS

034 Supplementary ECU
(Piggyback EMS)
www.034efi.com

Accel/DFI (standalone engine
management systems)
(888) MR GASKET, ext 9999
www.mrgasket.com/dfimain.html

AEM (standalone engine
management systems)
2205 126th Street, Unit A
Hawthorne, CA 90250
(310) 484-2322
(310) 484-0152
www.aempower.com

Aeroquip (performance engine
system plumbing)
Eaton Corporation
14615 Lone Oak Road
Eden Prairie, MN 55344
www.aeroquip.com

Airflow Performance, Inc. (racing
multi-point fuel injection)
111 Airflow Drive
Spartanburg, SC 29306
(864) 576-4512
(864) 576-0201 (fax)
www.airflowperformance.com

Alamo Autosports (forced induction
and EMS installation and dyno
tuning)
1218 Colorado Lane, Suite C3
Arlington, TX 76015
(817) 860-4300
www.alamoautosports.com

Anderson Ford Motorsports
PO Box 638
Route 10 West
Clinton, IL 61727
(888) 715-6487
www.andersonfordmotorsport.com

APEXi Integration (piggyback
systems and programmable air
sensors)
330 W. Taft
Orange, CA 92865
(714) 685-5700
(714) 685-5701
www.apexi-usa.com

APR (direct-port programming and
performance-upgrades for VW,
Porsche, and Audi)
1027-B Opeliaka Road
Auburn, AL 36830-4025
(800) 680-7921
apr@goapr.com
www.goapr.com

Arizona Speed and Marine (GM
EFI recalibration, performance
equipment, and upgrades)
6313 W. Commonwealth Ave.nue
Chandler, AZ 85226
(480) 753-0208
www.azspeed-marine.com

AutoTap (OBD-II diagnostic
scanner)
B&B Electronics Mfg. Co.
707 Dayton Road, P.O. Box 1040
Ottawa, IL 61350
(815) 433-5100
(815) 433-5105 (fax)
www.autotap.com

Autothority (ECU upgrades, Mass-
flow kits, performance intakes, VAF-
MAF conversion chip)
3769-B Pickett Road
Fairfax, VA 22031
(703) 323-6000 (showroom)
(703) 323-0919 (orders)
www.autothority.com

Auto-Ware (racing, engine analyzer,
and diagnostic software)
7624 Verona NW
Albuquerque, NM 87120-4500
(800) 647-2392 x201
www.auto-ware.com

Autronic SMC/SM2 (standalone
engine management systems and
components)
(61) 7 4051 6672 (Australia: Ray
Hall Turbs)
(909) 245-9511 (California: Strader
Performance Engineering)
www.autronic.com

AVL Powertrain Engineering, Inc.
(DI-gasoline, powertrain engineering,
simulation, instrumentation and test
systems)
47519 Halyard Drive
Plymouth, Michigan 48170-2438
(734) 414-9618
(734) 414-9690 (fax)
www.avl.com/pei

BG Fuel Systems (4-barrel venturi-
EFI)
1450 McDonald Road
Dahlonega, Georgia 30533
(706) 864-8544
(706) 864-2206 (fax)
www.bgfuel.com

BBK Performance Inc. (throttle
bodies, fuel pumps, forced-induction)
1871 Delilah Street,
Corona, CA 92879-1800
(909) 735-2400
(909) 735-8892 (fax)
www.bbkperformance.com

BBR-GTi (Piggyback Interceptor
2000)
Unit One, Oxford Road,
Brackley, Northants,
NN13 7DY, England
(44) 1280 700800
(44) 1280 705339 (fax)
www.bbrgti.demon.co.uk

Bell Experimental Group, Inc.
(VRG fuel pressure regulator, forced
induction)
203 Kestrel Drive.
Spring Branch, TX 78070
(830) 438-2890
(830) 438-8361 (fax)
www.bellengineering.net

BG Fuel Systems (four-barrel
venturi-EFI)
1450 McDonald Road
Dahlonega, Georgia 30533
(706) 864-8544
(706) 864-2206 (fax)
www.bgfuel.com

Bill Craddock Engineering (BCE)
(intake and head porting, intake
manifold design)
10808 Foothill BlvdBoulevard.
Ranch Cucamonga, CA 91730
(800) 853-0035 or
(909) 944-8097
Fax (909) 944-8933 (fax)
raceum@earthlink.net
www.bceracing.com/

Blower Drive Service (mechanical FI
and EFI kits and components for
blown and NA applications)
12140 E. Washington Blvd.Boulevard
Whittier, CA 90606
(562) 693-4302
(562) 696-7091 (fax)
www.blowerdriveservice.com

Bonneville Motor Werks (Bosch
Motronic recalibration)
lndshrk@bonnevillemotorwerks.com
www.bonnevillemotorwerks.com

Borla Performance Industries
(performance mufflers and exhaust)
5901 Edison Drive
Oxnard, CA 93033
(877) GO BORLA
(877) 462-6752
www.borla.com

Bosch USA (EFI components)
2800 S. 25th Street.
Broadview, IL 60153
www.boschusa.com

Camden Superchargers
(superchargers and air-filters)
16715 Meridian East, Bldg.Building
K-A
Puyallup, WA 98375
(253) 848-7776
www.camdensuperchargers.com

**Carroll Supercharging
Superpumper** (superchargers,
programmable electronic VRG
regulators)
1803 Union Valley Road
West Milford, NJ 07480
(973) 728-9505
(973) 728-9506 (fax)
www.carrollsupercharging.com

ChipTorque (Xede Interceptor
piggyback, performance chips,
turbos, and performance parts)
1/50 Lawrence Drive
Nerang, Queensland 4211 Australia
(61) 7 5596 4204
(61) 7 5596 4228 (fax)
www.chiptorque.com.au

Creative Performance Racing
(wiring, fuel pumps, injectors)
106 E. Ten Mile Rd.oad
Hazel Park, MI 48030
(248) 398-2397
(248) 398-4464 (fax)
www.cprracing.com

Dastek (Unichip piggyback and AIC,
dynamometers)
The Racers Group (U.S. Distributor)
29181 Arnold Drive
Sonoma, CA 95476
(707) 935-3999
(707) 935-5889 (fax)
www.theracersgroup.com
www.dastek.co.za

Diablosport (ECU programming
tools, scan tools, Cummins
performance module)
(877) 866-5726
nick@diablosport.com
www.diablosport.com

Rinda Technologies, Inc. (Diacom
diagnostic equipment, software, and
instrumentation)
4563 N. Elston Ave.
Chicago, IL 60630
(773) 736-6633
(773) 736-2950 (fax)
custserv@rinda.com
http://216.230.203.183/company.htm

Fluke Corporation (digital
multimeters, scan tools, etc.)
6920 Seaway Blvd.
Everett, WA USA 98203
PO Box 9090, Everett, WA 98206
(425) 347-6100 or (800) 44-FLUKE
(425) 446-5116 (fax)
www.fluke.com

diy-efi (Do-It-Yourself EFI) (tech-
savvy email list and references)
www.diy-efi.org

**DTA Competition Engine
Management Systems** (engine and
vehicle management systems)
Florence Street
Bradford BD3 8EX UK
(44) 1274 667960
(44) 1612 785959
(44) 1274 663627 (fax)
www.dtafast.co.uk

267

Earl's Performance Products
(performance engine and fuel
plumbing components)
189 West Victoria Street
Long Beach, CA 90805
(310) 609-1602
www.earlsperformance.com

ECUTek (recalibration hardware and
software for Subaru WRX)
8 Union Buildings
Wallingford Road
Uxbridge UB8 2FR UK
(44) 1895 811200
(44) 1895 811611 (fax)
www.ecutek.com

Edelbrock Corporation
(performance EFI kits and
components, turbo systems, intakes
and headers)
Media Relations
2700 California Street
Torrance, CA 90503
(310) 781-2222
(310) 533-0844 (fax)
www.edelbrock.com

EFI Technology (standalone engine
management systems and
components)
3838 Del Amo Blvd.Boulevard, #201
Torrance, CA 90503
(310) 793-2505
(310) 793-2514 (fax)
www.efitechnology.com

EFILive (performance data
acquisition software and tools)
info@efilive.com
www.efilive.com

Electromotive, Inc. (standalone
engine management systems and
components)
9131 Centreville Road
Manassas, VA 20110
(703) 331-0100
(703) 331-0161
www.getfuelinjected.com
www.electromotive-inc.com

Enderle Fuel Injection (fuel
metering systems for racing
applications)
1830 N. Voyager Avenue.
Simi Valley, CA 93063
(805) 526-3838
(805) 526-2885 (fax)
www.enderlefuel.com

Extrude Hone (pressure-media
intake and head porting)
1 Industry Boulevardlvd.
P.O. Box 1000
Irwin, PA 15642
(800) 367-1109)
(800) 613-0983 (service)
(724) 863-5900 (Pennsylvania)
(724) 863-8759 (fax)
exhone@extrudehone.com
www.extrudehone.com

FAST Electronics (standalone engine
management systems and
components)
151 Industrial Drive
Ashland, MS 38603
(662) 224-3495
(662) 224-8255
www.fuelairspark.com

Fastchip.com (Custom GM
programming and chips)
1114 West 41st Street.
Tulsa, OK 74107
(918) 446-3019
(918) 446-1233 (fax)
ewright@gorilla.net
www.fastchip.com

Feuling Engineering (VE,
combustion system, and EFI R&D,
consulting, products and patents)
21561 Pioneer Way
El Cajon, Ca 92020
(877) 338-5464 (order)
(619) 579-5700 (tech)
(619) 579-5701 (fax)
www.feuling.com

FJO Racing (auto electronics:
wideband controller, datalogging
software)
839 E. New Haven Avenue.
Melbourne, FL 32901-5458
www.fjoracing.com/index2.htm

Flowmaster (performance mufflers)
100 Stony Point Road, Suite 125
Santa Rosa, CA 95401
(800) 544-4761
www.flowmastermufflers.com

Fluke Corporation (digital
multimeters, scan tools, etc.)
6920 Seaway Boulevard
Everett, WA USA 98203
PO Box 9090, Everett, WA 98206
(425) 347-6100
(800) 44-FLUKE
(425) 446-5116 (fax)
www.fluke.com

Force Fuel Injection (EFI for Chevy
V-8s, carb intake EFI conversions,
stack manifolds)
4285 SW 37th Avenue
Miami, FL 33133
(305) 665-8300
(305) 665-4049 (fax)
www.force-efi.com

Ford Racing (Ford EMS/EFI and
performance components)
www.fordracing.com/parts/
(586) 468-1356 (tech)

Fuel Injection Specialties
(performance GM EMS Tuning and
parts)
11051 Wye Drive
San Antonio, TX 78217
(210) 654-0774
(210) 654-1444 (fax)
www.fuelinjection.com

Fuel System Enterprises (FSE)
(performance fuel pumps, regulators,
filters, wiring, throttle bodies)
Glencoe House, Drake Avenue,
Gresham Road
Staines Middex, TW18 2AW,
England
(44) 1784 493555
(44) 1784 493222 (fax)
www.fuelsystem.co.uk

GM Performance Parts (EMS and
performance upgrade equipment)
www.gmgoodwrench.com/PartsAccess
ories/GMPerformanceParts.html

Greddy (eManage piggyback and
performance upgrades for Japanese
engines)
9 Vanderbilt
Irvine, CA 92618
(949) 588-8300
(949) 588-6318 (fax)
www.greddy.com

G-Tech (Tesla Electronics) (inertial
onboard power-measurement
equipment)
1749 14th Street
Santa Monica, CA 90404
(310) 452-0030
www.gtechpro.com

Haltech (standalone engine
management systems and
components)
10 Bay Rd Taren Point
NSW Australia, 2229
(61) 2 9525 2400
(61) 2 9525 2991 (fax)
www.haltech.com.au

Hilborn Fuel Injection (mechanical
and EFI manifolds, throttle bodies,
and components)
22892 Glenwood Drive
Aliso Viejo, CA 92656
(949) 360-0909
(949) 360-0991 (fax)
www.hilborninjection.com

HKS USA (standalone and
piggyback EMS programmable
sensors, AIC and performance parts)
2355 Miramar Avenue
Long Beach, CA 90815
(310) 763-9600
(310) 763-9775 (fax)
www.hksusa.com

Holley Fuel Systems (multi-port and
throttle body EMS kits and
components)
(800) 2-HOLLEY (sales)
(270) 781-9741 (tech)

Home Dyno (onboard power-
measurement hardware and software)
mchaney@charm.net
www.charm.net/~mchaney/homedyn
o/dynokit.htm

Hondata (Honda ECU upgrade,
recalibration, and datalogging)
2341 West 205th Streett,
Suite 106
Torrance, CA 90501
(310) 782-8278

Horiba Instruments (wideband AFR
meter)
1080 E. Duane Ave. nue, Suite A
Sunnyvale, CA 94086
(408) 730-4772
(408) 730-8975 (fax)
www.emd.horiba.com/engmeas/mexa
700lambda

Howell Engine Developments, Inc.
(Chevy V-8 multi-port EFI kits,
wiring, and components)
6201 Industrial Way
Marine City, MI 48039
(810) 765-5100
(810) 765-1503 (fax)
www.howell-efi.com/

HPC (high-performance thermal
coatings)
14788 S. Heritagecrest Way
Bluffdale, UT 84065
(801) 501-8303
(800) 456-4721 (tech)
(801) 501-8315 (fax)
www.hpcoatings.com

Hypertech, Inc. (performance chips
and ECU recalibrations)
3215 Appling Roadd.
Bartlett, TN 38133
(901) 382-8888
www.hypertech.com

IMCO (performance mufflers)
P.O. Box 397
1121 North Main
Schulenburg, Texas 78956
(979) 743-4521
(979) 743-4444 (fax)
(800) 231-2390 (sales)
(979) 743-4520 (sales fax)
www.imcoweb.com/index2.html

Injec Racing Developments
(Standalone EMS and EFI kits)
137 Parer Road, Airport West,
Victoria 3042, Australia.
(61) 3 9338 4133
(61) 3 9338 4955 (fax)
www.injec.com

Injection Logic (standalone engine
management systems and
components)
Injection Logic, LLC
San Luis Obispo, CA
(805) 781-9022
sslow@injectionlogic.com
www.injectionlogic.com

Innovative Turbo Systems (performance turbos and components)
845 Easy Street, Unit 102
Simi Valley, CA 93065
(805) 526-5400
(800) 526-9240
www.innovativeturbo.com/cgi-bin/start.cgi/turbo/index.html

JET Performance Products (performance chips and ECU upgrades)
17491 Apex Circle
Huntington Beach, CA 92647
(714) 848-5515
(714) 847-6290 (fax)
www.jetchip.com

Jethot (high-performance thermal coatings)
(800) 432-3379 (order)
(610) 277-5646 (tech)
www.jet-hot.com

Jims Performance (TPI parts, sensors, PROMs, wiring, etc.)
Old Frederick Roadd.
Ellicott City, Maryland 21043
(410) 465-9569
(877) 465-9569
www.jimsperformance.com

JTR (GM EFI V-8 conversion kits and components for Jaguars)
P.O. Box 66,
Livermore, California 94551
(800) 858-3333
(909) 826-4000
JTR@JagsThatRun.com
www.jagsthatrun.com

K&N Engineering, Inc.
(performance air filters, intakes, and piggybacks for bikes and automotive)
P.O. Box 1329
1455 Citrus Ave.
Riverside, California 92502
(800) 858-3333
(909) 826-4000
www.knfilters.com

Kaufmann Products (Ford EFI engine swap kits and performance components, forced induction)
7591 Acacia Avenue
Garden Grove, CA 92841
(714) 903-9717
www.kaufmannproducts.com

Kinsler Fuel Injection (Mechanical FI and EFI components)
1834 Thunderbird
Troy, MI 48084
(248) 362-1145
www.kinsler.com

Lambdaboy (low-cost wideband AFR meters)
1442A Walnut Street,. #109
Berkeley, CA, 94709
(650) 575-0400
www.lambdaboy.com

Link ElectroSystems USA (standalone and piggyback engine management systems and components, wideband controller)
1643 Monrovia Ave.nue
Costa Mesa, CA 92627
(949) 646-7461
(949) 646-7471 (fax)
www.link-electro-usa.com

LS1 Edit (GM PCM programming software)
info@carputing.com
http://carputing.tripod.com/index.htm

Lucas Aftermarket Operations (injectors and EFI components)
P.O. Box 7079
5600 Crooks Road
Troy, MI 48007
(248) 288-2000
(248) 288-8280

Magnaflow (performance mufflers and exhaust)
www.magnaflow.com/01home.htm

Magnusen Products (Eaton superchargers and conversion kits for EFI applications)
3172 Bunsen Ave., nue, Unit K
Ventura, CA 93003
(805) 642-8833
(805) 642-1936
www.magnusonproducts.com

Majestic Turbo (performance turbos and system components)
815 Jefferson
Waco, TX 76701
(800) 231-5566
(254) 757-1008 (fax)
www.majesticturbo.com

Mallory Ignition (performance ignitions and components)
www.mrgasket.com/mallorymain.html

Marren Fuel Injection (fuel management systems and components, injector services)
49 Burtville Ave., nue, Unit 3A
Derby, CT 06418
(203) 732-4565
www.injector.com/

Microtech (standalone engine management systems and components)
www.mazkwik.com.au/products/microtech/

Motec Systems USA (standalone engine management systems and components)
5355 Industrial Drive
Huntington Beach, CA 92649
(714) 897-6804
(714) 897-8782 (fax)
www.motec.com/home.htm

MSD Fuel Management (fuel and ignition management components)
12120 Esther Lama, Suite 114
El Paso, Texas 79936
(915) 855.-7123
(915) 857-3344 (fax)
www.msdignition.com/fuel_intro.htm

New Age Automotive Electronics/Air Automotive (EFI and automotive diagnostic components)
P.O. Box 212
Ingleburn, NSW 1890 Australia
(61) 2 9829 1666
(61) 2 9618 3519 (fax)
www.airautomotive.com/

Nitrous Oxide Systems (Nitrous nitrous and fuel systems and components for EFI and carbureted engines)
www.holley.com

Norwood Autocraft (EMS design, installation, and dyno calibration, engine systems design)
(214) 683-9966
www.bobnorwood.com

Painless Performance Products (EFI harnesses and accessories)
9505 Santa Paula Drive
Fort Worth, TX 76116
(817) 244-6212
(817) 244-4024 (fax)
(800) 423-9696 (tech)
www.painlessperformance.com

Paladin Tools/Connectool Inc. (Crimping/wiring tools)
10446 Lakeridge Parkway
Ashland, VA 23005
(800) 272-8665
(800) 272-5257
www.paladin-tools.com

Pectel Control Systems Ltd. (racing ECUs)
Brookfield Motorsports Centre
Twentypence Road
Cottenham, Cambridge
United Kingdom CB4 8PS
(44) 1954 253610
(44) 1954 253601 (fax)
enquiries@pectel.co.uk.
www.pectel.co.uk/

Perfect Power (standalone and piggyback engine management systems and components)
600 Texas Road
Morganville, NJ 07751
(732) 591-1245
www.perfectpower.com

PMS (Ford EEC Piggyback EMS)
Anderson Ford Motorsports
P.O. Box 638, Route 10 West
Clinton, IL 61727
(217) 935-2384
(217) 935-4611
www.andersonfordmotorsport.com

Power Train Electronics Co. (EMS calibration, AFR meters, engine systems design)
6600 Toro Creek Road
Atascadero, CA 93422
(805) 466-5252
(805) 466-5292 (fax)
www.powertrain.net

ProECM Powerchip PC-3 (harness-connected piggyback computer chips)
info@proecm.com
www.proecm.com

Professional Flow Tech (performance and programmable MAF sensors)
(248) 541-4780
(248) 541-0438 (fax)
sales@pro-flow.com
www.pro-mracing.com/index.html

Python Injection (performance EMS systems and components)
8625 Central Ave.nue
Stanton, CA 90680
(714) 828-1406
(800) 959-2865
info@python-injection.com
www.python-injection.com

Race Logic (2-stroke performance)
8706 Round Lake Hwy.Highway
Addison, MI 49220
(517) 467-8112 or
(866) RACELOGIC (722-3564)
www.racelogic.com

RC Engineering (performance injectors, throttle bodies, EFI components, and diagnostics, injector servicing and upgrades)
1728 Border Avenue
Torrance, CA 90501
(310) 320-2277
(310) 782-1346 (fax)
www.rceng.com

Redline Weber (fuel calibration displays, throttle bodies, programmable SEFI)
19630 Pacific Gateway Drive.
Torrance, CA 90502
(800) 733-2277 x 7345
www.redlineweber.com

Rinda Technologies, Inc. (Diacom diagnostic equipment, software, and instrumentation)
4563 N. Elston Avenue
Chicago, IL 60630
(773) 736-6633
(773) 736-2950 (fax)
custserv@rinda.com
http://216.230.203.183/company.htm

Russell Performance Products Inc. (fuel management and engine systems and plumbing components)
2301 Dominguez Way,
Torrance, CA 90501
Phone:(310) 781-2222; Technical Support: (800) 416-8628 (tech)
(310) 320-1187 (fax)
www.russellperformance.com

Siemens Automotive (automotive electronics and electronic fuel system components)
www.siemensauto.com

Simple Digital Systems (Racetech Inc.)
(standalone engine management systems and components)
1007-55 Ave. nue NE
Calgary, Alberta, Canada T2E 6W1
(403) 274-0154 (Canada)
(403) 274-0556 (fax)
www.sdsefi.com

Skunk2 Racing (ECU upgrades and conversion harnesses)
2050 5th Street
Norco, CA 92860
(909) 808-9888
www.skunk2.com

Speartech Fuel Injection Systems
(EFI retrofit kits, harnesses, calibration, custom EFI)
Anderson, IN
speartech@insightbb.com
www.speartech.com

Split Second (standalone and piggyback engine management systems and components, AIC, programmable sensors and clamps)
1949 Deere Ave.nue
Santa Ana, CA 92705
(949) 863-1359
(949) 863-1363 (fax)
www.splitsec.com

Street and Performance Electronics
(Chevy V-8 EFI intakes, harnesses, ECU recalibration, and EFI engine swap components)
304 Smokey Lane
North Little Rock, Arkansas 72117
(501) 945-0354
(501) 945-0370
http://streetandperformanceelectronic
s.com

Superchips Inc. (performance PROMs and ECU upgrades)
2609 S. Sanford Avenue
Sanford, FL 32773
(407) 547-0130
(407) 260-9106 (fax)
www.superchips.com

Techlusion TFI (TFI piggyback)
668 Middlegate Road
Los Vegas, NV 89105
(702) 558-5142 or
(877) 764-3337
(702) 558-5396

Technomotive (Eclipse ECU modifications and datalogger)
341 Bollay Drive
Goleta, CA 93117
(805) 961-9444
(805) 685-5007 (fax)
www.tmo.com

Technosquare (OEM ECU modification, recalibration and upgrade)
22521 Normandie Ave.nue
Torrance, CA 90501
(310) 787-0847
(310) 787-0948 (fax)
www.technosquareinc.com/

Thermo-Tec Automotive Products
(exhaust and heat protection)
P.O. Box 96
Greenwich, Ohio 44837
(800) 274-8437
(419) 962-4556 (outside the U.S.)
(419) 962-4013 (fax)
www.thermotec.com

Thunder Racing (late-model performance V-8 upgrades)
6960 N. Merchant Court
Baton Rouge, LA 70809
(225) 754-RACE
(225) 754-7220 (fax)
www.thunderracing.com

Top End Performance (EMS and EFI components and systems)
7452 Varna Ave. nue
North Hollywood, CA 91605
(818) 764-1901
(818) 764-0155 (fax)
topend@racetep.com
www.racetep.com

TPI Specialties, Inc. (performance GM fuel system and EFI components)
4255 Creek Road (County Rd 110)
Chaska, MN 55318
(952) 448-6021
(952) 448-7230
www.tpis.com

TTS Power Systems (chips, OBD-II ECU reprogrammer, diagnostic software, headers and exhaust)
1280 Kona Drive
Compton, CA 90220
(310) 669-8101
www.ttspowersystems.com

Tunercat (GM ECM calibration software)
C.A.T.S.
14327 Dogwood Lane
Belle Haven, VA 23306
www.tunercat.com

Turbonetics Inc. (performance turbos and system components)
2255 Agate Courtt.
Simi Valley, CA 93065
(805) 581-0333
(805) 584-1913 (fax)
www.turboneticsinc.com

Turbosmart (electronic and pneumatic Fuel Cut Defenders, turbo boost controllers)
Mailing: P.O. Box 264
Croydon, NSW 2132 Australia
Home Office: 32 Milton Street North
Ashfield, NSW, Australia
(61) 2 9798 2866
(61) 2 9798 2826 (fax)
www.turbosmart.com.au

TurboXS (electronic boost controllers, Subaru WRX User-Tunable Engine Computer (UTEC) and staged power kits)
8041 Queenair Drive, Unit #2
Gaithersburg, MD 20879
(877) 887-2679
(301) 977-4727
(301) 977-6507 (fax)
www.turboxs.com

TWM Induction (competition EFI conversion systems and accessories, Weber-type throttle bodies)
325D Rutherford Street
Goleta, CA 93117
(805) 967 9478
(805) 683-6640 (fax)
www.twminduction.com

Vortech Engineering (supercharger and fuel system conversions for EFI engines)
1650 Pacific Avenue
Channel Islands, CA 93033-9901
(805) 247-0226
www.vortechsuperchargers.com

Walbro (standalone EMS and fuel systems and components)
6242 Garfield Street
Cass City, MI 48726-1325
(989) 872-2131
(989) 872-4957 (fax)
custserv@walbro.com
www.walbro.com

Waterman Racing Components
(performance racing fuel injection components)
37250 Church Street
Gualala, CA 95445-0148
(707) 884-4181
(707) 884-4183 (fax)
www.watermanracing.com

Weapon R Competition Products
(ECU OBD-I jumper harness, headers, etcand more.)
www.weapon-r.com/main02.html

Westers Garage (performance chips, custom programming, ECU reprogramming)
218 Centre Street, Box 159
Tilley, Alberta, Canada T0J 3K0
(888) WESTER-1
http://westers_garage.eidnet.org/

White Racing and Marine
(programmable ECMs and fuel system components)
12820 Nine Mile Road
Warren, Michigan 48089
(586) 756-3026
(586) 756-8350 (fax)
www.whiteracing.com/
http://members.aol.com/bigturbo1/pl
ugplay.html

Winbin (Windows editor for GM EFI controllers)
www.passtimeracing.com/eric/Cars/E
FI/

Wolf Engine Management Systems
(standalone engine management systems)
enquiries@wolfems.com.au
www.wolfems.com.au/home

ZDyne ECU Conversions (Honda ECU upgrade, modifications and recalibration)
6931 Topanga Canyon Blvd
Boulevard, Suite 8
Canoga Park, CA 91303
(818) 888-5350
www.zdyne.com

Z-Industries, Inc. (85-03 GM ECM recalibrations)
31200 Santiago Road
Temecula, Ca.92592
(909) 303-6857
(909) 303-6257 (fax)
www.z-industries.com

Zytek (standalone engine management systems and components)
www.zytek.co.uk/

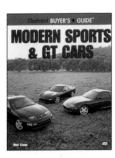